CLASSI~ ~~~~

UNI
WO

D1389878

WITHDRAWN

WP 1032048 2

GRAPH THEORY

CLASSIC TEXT

GRAPH
THEORY

CLASSIC TEXT

WATARU MAYEDA

University of Illinois

WILEY-INTERSCIENCE, a Division of John Wiley & Sons, Inc.
New York · London · Sydney · Toronto

MAIN LIBRARY   POLYTECHNIC
WOLVERHAMPTON WV1 1LY

511·5 MAY

83517

Copyright © 1972, by John Wiley & Sons, Inc.

All rights reserved. Published simultaneously in Canada.

No part of this book may be reproduced by any means, nor transmitted, nor translated into a machine language without the written permission of the publisher.

Library of Congress Catalog Card Number: 70–37366

ISBN 0–471–57950–5

Printed in the United States of America.

10 9 8 7 6 5 4 3 2 1

PREFACE

Many of the technical papers that have appeared in recent years contain words related to linear graph theory such as "topological," "graph theoretical," "cut-sets," and "trees" in their titles. This is no accident. The theory of linear graphs, itself currently in the process of mathematical development, is being applied in a variety of apparently unrelated fields such as engineering system science, social science and human relations, business administration and scientific management, political science, chemistry, and psychology.

The purpose of this book is not only to provide an introduction to the fascinating study of linear graph theory but to bring the reader far enough along the way to enable him to embark on a research problem of his own, whether it be in the theory of linear graphs itself or in one of its manifold applications.

It would be impossible to discuss all of the applications of linear graphs in one book; instead we will concentrate on electrical network theory, switching theory, communication net and transportation theory, and system diagnosis.

I would like to thank Dr. N. Wax for invaluable suggestions and reading and correcting the final manuscript. Thanks are also due Dr. M. E. Van Valkenburg and all past and present members of the systems group in the Coordinated Science Laboratory at the University of Illinois for their direct and indirect support for the accomplishment of this book.

WATARU MAYEDA

Urbana, Illinois
October 1971

v

MAIN LIBRARY   POLYTECHNIC
WOLVERHAMPTON WV1 1LY

CONTENTS

GRAPH THEORY

INTRODUCTION

There are many physical systems whose performance depends not only on the characteristics of the components but also on the relative locations of the elements. An obvious example is an electrical network. If we change a resistor to a capacitor, generally some of the properties (such as an input impedance of the network) also change. This indicates that the performance of a system depends on the characteristics of the components. If, on the other hand, we change the location of one resistor, the input impedance again may change, which shows that the topology of the system is influencing the system's performance. There are systems constructed of only one kind of component so that the system's performance depends only on its topology. An example of such a system is a single-contact switching circuit. Similar situations can be seen in nonphysical systems such as structures of administration. Hence it is important to represent a system so that its topology can be visualized clearly.

One simple way of displaying a structure of a system is to draw a diagram consisting of points called "vertices" and line segments called "edges" which connect these vertices so that the vertices and edges indicate components and relationships between these components. Such a diagram is a linear graph. A linear graph often is known by another name, depending on the kind of physical system we are dealing with; it may be called a network, a net, a graph, a circuit, a diagram, a structure, and so on.

Instead of indicating the physical structure of a system, we frequently indicate its mathematical model or its abstract model by a linear graph. Under such a circumstance, a linear graph is referred to as a flow graph, a signal flow graph, a flow chart, a state diagram, a simplical complex, a sociogram, an organization diagram, and so forth.

The earliest known paper on linear graph theory (1736) is due to Euler, who gave a solution to the Könisberger bridge problem by introducing the concept of linear graphs. In 1847, Kirchhoff employed linear graph theory for an analysis of electrical networks, known today as the topological formulas for driving point impedances and transfer admittances. This probably is the

1

first paper that applies the theory of linear graphs to engineering problems. However, it is not Kirchhoff's paper but Möbins' conjecture (about 1840) concerning the four-color problem that seems to attract many scholars to devote themselves to linear graph theory.

The four-color problem is to prove or disprove that four colors are sufficient to color any planar map such that no two adjacent regions have the same color. Place one vertex inside of each region of a planar map, and then connect two vertices by an edge if and only if the regions containing these vertices are adjacent to each other. These operations yield what is known as a planar graph. In other words, for a given planar map, there is a planar graph such that a region of the map corresponds to a vertex in the linear graph and the boundary between two regions corresponds to the edge connected between two vertices which represent these two regions. We may restate the four-color problem as follows: Prove or disprove that four colors are sufficient to color vertices of any planar graph such that no two vertices connected by an edge have the same color.

If you try to attack the four-color problem, you will immediately face the difficulty of distinguishing planar graphs from nonplanar graphs, and you will start to study the properties of planar graphs. In spite of the work done by Kuratowski and Whitney, who discovered fundamental properties of planar graphs, the four-color problem is still unsolved and attracting many scholars to devote themselves to search for more properties of linear graphs. Of course, some properties have been found because of the necessity of specific applications.

The first part of this book is the study of properties of linear graphs for beginners. This does not mean that we are studying only elementary and simple properties. In fact, it covers the most advanced materials such as the following:

1. The property of collection $\{P_{ij}\}$ of paths, suitable for generating all possible paths, and properties among collections of paths.

2. How to generate cut-sets; especially, how to generate all possible cut-sets separating two specific vertices. These cut-sets are very important in communication nets and traffic systems.

3. Proofs of realizability conditions of cut-set matrices including Tutte's condition.

4. A proof of Kuratowski's conditions for nonplanar graphs and a proof of Whitney's condition for planar graphs (duality).

5. An introduction to pseudo-cuts, which become the dual of paths when a linear graph is planar.

6. An algorithm for testing the existence of directed circuits in oriented linear graphs.

7. The development of two types of generation of all possible trees without duplications.

When we discuss applications of linear graphs, we often use weighted linear graphs where edges and (or) vertices have specifications known as weights. For example, we can represent a passive linear bilateral lumped electrical network by a linear graph where each edge has three weights (e.g., voltage, current, and a proportionality factor) as discussed in Chapter 7. To study maximum flow in a communication net or a traffic system, the corresponding linear graph often needs only one weight on each edge indicating the maximum capability of handling traffic by the edge (in Chapter 12). We can see that suitable weighted linear graphs can represent many other systems such as switching circuits, logic circuits, air traffic networks, and computer systems.

There are cases when such a weighted linear graph may be used only for representation of a system. However, in this book, we study how to employ weighted linear graphs in order to analyze systems, particularly one such method known as the topological method of analysis. There are two distinct types of topological method; one is to use rules known as topological formulas so that the property of the system that is in question can be studied directly from a weighted linear graph, and the other is to employ so-called equivalent transformations successively so that a weighted linear graph of a system will be simplified to consist of only one edge whose weight indicates the property. Examples of the first type include the following: (1) calculate electrical network functions by finding all possible special subgraphs of the weighted linear graph corresponding to a given linear lumped electrical network (given in Chapters 7 and 8); (2) find the maximum flow by locating a so-called minimum cut in a communication net and a traffic system (shown in Chapter 12); and (3) obtain a switching function by finding all possible paths between specified terminals in a switching circuits (discussed in Chapter 11). Some examples of the second type are (1) the node elimination technique to obtain a simpler signal flow graph (indicated in Chapter 10) and (2) equivalent transformations for a linear electrical network (given in Chapter 8).

The topological analysis of the first type gives a clear relationship between a property of a system and the locations of edges (components). In several cases, this relationship is enough to design or improve a system which satisfies a given specification, and this is called a topological synthesis of systems. The synthesis of switching functions and that of communication nets (in Chapters 11, 12, and 13) are good examples. The system diagnosis discussed in Chapter 14 again indicates that linear graph theory is an essential tool in the system theory area.

CHAPTER
1
NONORIENTED LINEAR GRAPH

1-1 INTRODUCTION

In this chapter, some properties of paths and circuits of a nonoriented linear graph are discussed. The paths and circuits are rather important subgraphs of linear graphs. For example, we will see later that paths determine the properties of switching networks, and circuits are related to Kirchhoff's voltage law in electrical network theory.

For defining linear graphs, it would be easier if we consider the familiar tetrahedron shown in Fig. 1-1-1. There are four vertices 1, 2, 3, and 4 and six

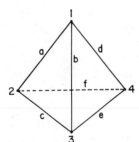

Fig. 1-1-1. A tetrahedron.

edges, a, b, c, d, e, and f. Each edge is located between two vertices; edge a is between vertices 1 and 2, edge b is between 1 and 3, edge c is between 2 and 3, and so on. In combinatorial topology, a collection of vertices and edges is known as a simplical one-dimensional (*linear*) complex, which we call a linear graph. However, the definitions of vertices and edges are more general than those of polyhedrons.

To expand the concept of vertices and edges geometrically, consider an n-dimensional Euclidean space. First, a vertex is a point in the space. With a given set Ω of vertices, an edge e is a curve between two vertices v and v' in Ω which passes no other vertices in Ω. The vertices v and v' are called the endpoints of edge e. When v and v' are the same, then edge e as shown in Fig. 1-1-2a is called a "self-loop." If we give a direction to a curve as shown in Fig. 1-1-2b, then an edge represented by the curve is called an oriented edge. Otherwise, it is a nonoriented edge.

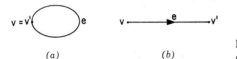

(a)   (b)

Fig. 1-1-2. Edges. (a) Self-loops; (b) oriented edge.

We may now use the preceding geometrical concept to derive an abstract definition of edges, vertices, and linear graphs. Instead of chosen points (in a space) being vertices, we take a set such that members in the set are given vertices. Geometrically, an edge is a curve between two vertices. Since there are no other vertices in the curve, we can consider an edge to correspond to a pair of vertices. On the other hand, we would like to allow several edges having the same endpoints. Hence the definition of vertices edges and a linear graph become abstractly as follows.

Definition 1-1-1. Let $\mathscr{E}$ and Ω be sets. If every $e \in \mathscr{E}*$ corresponds to exactly one pair (v, v') where $v, v' \in \Omega$, then every member in $\mathscr{E}$ is an *edge*, every member in Ω is a *vertex*, and $\mathscr{E} \cup \Omega^\dagger$ is a *linear graph*.

With this definition, the endpoints, oriented edges, and nonoriented edges are defined abstractly as follows.

Definition 1-1-2. Let e be an edge corresponding to a pair (v, v') of vertices. Then the two vertices v and v' are called the endpoints of edge e. If the pair (v, v') is ordered, then e is said to be *oriented* or called an *oriented edge*. Otherwise, e is said to be *nonoriented* or is called a *nonoriented edge*.

Furthermore, we define oriented and nonoriented (linear) graphs as follows.

Definition 1-1-3. If all edges in a linear graph are oriented, then the linear graph is said to be oriented or called an oriented (linear) graph. If all edges are nonoriented, a linear graph is said to be nonoriented or called a nonoriented (linear) graph.

---

* The symbol $\in$ means "belong to" or "in". $e \in \mathscr{E}$ means e in $\mathscr{E}$.

† $\mathscr{E} \cup \Omega$ is a set of all members in either $\mathscr{E}$ or Ω or both.

Example 1-1-1. We use the symbol $\alpha \to \beta$ for indicating that α corresponds to β. Consider two sets $\mathcal{E} = (a, b, c, d, e, f, g)$ and $\Omega = (1, 2, 3, 4, 5)$ where

$$a \to (1, 2)$$
$$b \to (1, 3)$$
$$c \to (2, 3)$$
$$d \to (1, 4)$$
$$e \to (1, 4)$$
$$f \to (4, 2)$$

and

$$g \to (2, 4)$$

Since each member in $\mathcal{E}$ corresponds to exactly one pair of vertices in Ω, a, b, c, d, e, f, and g are edges, 1, 2, 3, 4, and 5 are vertices, and $\mathcal{E} \cup \Omega$ is a linear graph by Definition 1-1-1. Vertices 1 and 2 are the endpoints of edge a, vertices 1 and 3 are the endpoints of edge b, and so on (Definition 1-1-2).

Instead of using the symbol $\alpha \to \beta$ as in the foregoing example, we use a drawing to indicate edges and the corresponding pairs of vertices. For this, we make the following agreements.

1. A vertex will be indicated by a small circle. When the name of a vertex must be indicated, it will be written either at a place near a circle or inside the circle as shown in Fig. 1-1-3a and b.

(a)

(b)

(c)

(d)

Fig. 1-1-3. Representation of vertices and edges. (a) Representation of a vertex 1; (b) representation of a vertex 1; (c) nonoriented edge e (v, v'); (d) oriented edge e (v, v').

2. When an edge is nonoriented, the edge is represented by a line between two vertices which are the endpoints of the edge. The name of an edge will be given at a place near the line if needed. As an example, a nonoriented edge $e \to (v, v')$ will be represented by a line shown in Fig. 1-1-3c.

3. When an edge is oriented, the edge will be represented by a line with an

arrow to indicate its orientation. As an example, the representation of an oriented edge $e \rightarrow (v, v')$ is shown in Fig. 1-1-3*d*.

Note that when v and v' of $e \rightarrow (v, v')$ are the same, then edge e is a self-loop and the line representing edge e is a loop starting from vertex v (a small circle representing vertex v) and terminated at the same vertex.

Since we need know only a pair of vertices for each edge, a shape of line representing an edge is immaterial. For example, an edge in Fig. 1-1-3*c* and those in Fig. 1-1-4*a* and *b* represent the same edge.

(*a*)   (*b*)

Fig. 1-1-4. Representation of edge e (v, v'). (*a*) Edge e; (*b*) edge e.

In a drawing, crossing points of edges other than those represented by small circles are also immaterial. For example, even though edges a and b in Fig. 1-1-5*a* are crossing each other, this drawing indicates only that non-

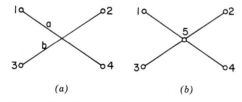

(*a*)   (*b*)

Fig. 1-1-5. Crossing of edges. (*a*) Edges a and b; (*b*) four edges.

oriented edge a corresponds to pair (1, 4) and nonoriented edge b corresponds to pair (2, 3). Note the difference between those in Fig. 1-1-5*a* and *b*.

With these agreements, we can specify a linear graph by a drawing. As an example, the linear graph $\mathscr{E} \cup \Omega$ in Example 1-1-1, where $\mathscr{E} = (a, b, c, d, e, f, g)$ and $\Omega = (1, 2, 3, 4, 5)$ with each pair of vertices being considered as a nonordered pair, can be represented by Fig. 1-1-6. Note that vertex 5 is not in any pair in this example. Hence there will be no lines connected to vertex 5.

Since a drawing gives a clear and simple representation of a linear graph, we use such drawings for specifying linear graphs in this book.

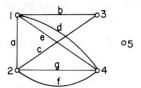

Fig. 1-1-6. A linear graph.

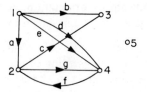

Fig. 1-1-7. An oriented graph.

With Definitions 1-1-2 and 1-1-3, we can see that a linear graph in Fig. 1-1-6 is nonoriented and that in Fig. 1-1-7 is an oriented (linear) graph. Note that Fig. 1-1-7 is the linear graph when all pairs of vertices in Example 1-1-1 are ordered pairs.

Until the later chapters, we will consider only nonoriented linear graphs. In other words, in the next few chapters, a *linear graph means a nonoriented linear graph.*

1-2 PATHS AND CIRCUITS

An edge is said to be *incident* or *connected* at a vertex if the vertex is one of the two endpoints of the edge. For example, edges *a*, *b*, and *c* in the linear graph Fig. 1-2-1 are incident (or connected) at vertex *A*. Edges *a*, *d*, *e*, and *f* are incident at vertex *B*.

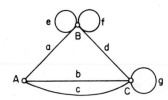

Fig. 1-2-1. A linear graph with self-loops.

The number of edges incident at each vertex is very important to characterize linear graphs such as paths, circuits, and Euler graphs. So we define the degree of a vertex as follows.

Definition 1-2-1. *The degree of a vertex v*, symbolized by $d(v)$, is defined as

$$d(v) = 2n_s + n_n \tag{1-2-1}$$

where n_s is the number of self-loops incident at vertex v and n_n is the number of edges other than self-loops incident at vertex v.

For example, the degree of vertex A of a linear graph in Fig. 1-2-1 is $d(A) = 3$ because $n_s = 0$ and $n_n = 3$ at this vertex. The degree of vertex B is $d(B) = 2(2) + 2 = 6$ where $n_s = 2$ and $n_n = 2$ at vertex B. The degree of vertex C is $d(C) = 5$.

Suppose edge e is connected between vertices p and q (i.e., the two endpoints of edge e are p and q). Then we count edge e as 1 for both $d(p)$ and $d(q)$ for $p \neq q$. When $p = q$, edge e (which is a self-loop) will be counted as 2 for $d(p)$. This is true for every edge in a linear graph. Hence the summation of the degrees of all vertices is equal to twice of the number of edges in a linear graph. That is,

$$\sum_{v \in G} d(v) = 2n_e \qquad (1\text{-}2\text{-}2)$$

where $\sum\limits_{v \in G}$ means the summation for all vertices in linear graph G and n_e is the number of edges in G. For example, in Fig. 1-2-2, $d(A) = 3$, $d(B) = 2$, $d(C) = 3$, $d(D) = 2$, and $d(E) = 4$. Thus $\sum\limits_{i \in G} d(i) = 14$. The number of edges in the linear graph is 7.

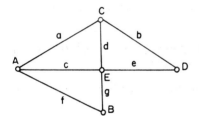

Fig. 1-2-2.   A linear graph.

Consider the linear graph in Fig. 1-2-2 as a map in which vertices indicate cities and edges indicate highways. We can see that there are several highways going from one city to another. Suppose we are planning to travel from a city A to a city D. If we list highways according to the order by which we are going to travel from city A to city D, we will have a sequence of edges which specifies a particular route from A to D. The vertex corresponding to the origin is called the initial vertex and the vertex corresponding to the destination is called the final vertex. As an example, with initial vertex A and final vertex D, some sequences of edges of a linear graph in Fig. 1-2-2 are (c, e), (a, d, c, f, g, e), and (c, d, b). It must be noted that each edge in the sequence discussed has one vertex in common with the preceding edge and the other vertex in common with the succeeding edge. For example, in sequence (c, d, b), vertex E of edge d is an endpoint of preceding edge c and vertex C

of edge *d* is an endpoint of succeeding edge *b*. When each edge in such a sequence appears only once, this sequence is called an edge train.

Definition 1-2-2.   An *edge train* is a sequence of edges with the following properties:

1. For any edge *e* other than the first and the last edges in the sequence, one endpoint of *e* is an endpoint of the preceding edge and the other endpoint of *e* is an endpoint of the succeeding edge.

2. One endpoint of the first edge is an endpoint of the succeeding edge and the other endpoint of the first edge is the initial vertex.

3. One endpoint of the last edge is an endpoint of the preceding edge and the other endpoint is the final vertex.

4. *Every edge appears exactly once.*

Definition 1-2-3.   An edge train is called an *open edge train* if the initial vertex and the final vertex are distinct. Otherwise it is a *closed edge train.*

For example, in Fig. 1-2-2, (*c d b e g*) is an open edge train whose initial and final vertices are *A* and *B*, sequence (*a d e b d g*) is not an edge train because edge *d* appears twice, and (*a d c*) is a closed edge train. Note that sequence (*c f g d b e*) is a closed edge train but (*a c d b e d*) is not a closed edge train.

Definition 1-2-4.   The linear graph corresponding to an edge train is the linear graph which consists of all edges in the edge train.

For example, the linear graph corresponding to edge train (*a d g f c e*) is shown in Fig. 1-2-3.

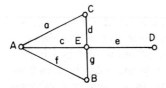

Fig. 1-2-3.   Linear graph corresponding to (*a d g f c e*).

Definition 1-2-5.   The degree of a vertex in an edge train means that the degree of the vertex in the linear graph corresponding to the edge train.

For example, the degree of vertex *C* of edge train (*a d g f c e*) is 2 and that of vertex *E* is 4.

When an open edge train has the properties that the degree of every vertex except the initial and the final vertices is 2 and the degrees of the initial and the final vertices are 1, then the set of all edges in the edge train is called a path between the initial and the final vertices.

Definition 1-2-6.   *A path* between vertices p and q is a set of all edges in an open edge train which satisfies the following conditions: (1) the initial and the final vertices are p and q and (2) every vertex other than p and q is of degree 2 and vertices p and q are of degree 1.

For example, a set of all edges in edge train $(a\,d\,e)$ in Fig. 1-2-2 is a path between vertices A and D, whereas a set of all edges in edge train $(a\,d\,g\,f\,c\,e)$ is not a path.

Let an edge train be $(e_1 e_2 \cdots e_n)$ as shown in Fig. 1-2-4. Suppose a set of

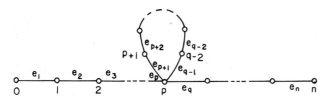

Fig. 1-2-4.   Linear graph corresponding to edge train $(e_1 e_2 \cdots e_n)$.

all edges in this edge train is not a path between vertices 0 and n. Then, by the definition of a path, there must be at least one vertex whose degree is more than 2. Let vertex p be one of those whose degree is more than 2. Suppose the first edge in the edge train whose endpoint is vertex p is e_p and the last edge whose endpoint is p is e_q as shown in Fig. 1-2-4. Then it can be seen that sequence $(e_1 e_2 \cdots e_p e_q \cdots e_n)$ is an edge train in which the degree of vertex p is 2. For convenience, we say that edge train T is a proper subedge train of edge train U if every edge in T is U and T and U are different. With this terminology, $(e_1 e_2 \cdots e_p e_q e_{q+1} \cdots e_n)$ is a proper subedge train of $(e_1 e_2 \cdots e_n)$. Furthermore, edges that have been deleted to obtain this subedge train form a closed edge train, $(e_{p+1} e_{p+2} \cdots e_{q-2}, e_{q-1})$. Hence an open edge train whose edges will not form a path can be decomposed into two, one a closed edge train and the other an open edge train whose initial and final vertices are the same as those of the original edge train. Furthermore, every edge in the original edge train is in one but not both of these new edge trains.

If there exists a vertex whose degree is more than 2 in edge train $(e_1 e_2 \cdots e_p e_q \cdots e_n)$, we can obtain a proper subedge train by the same process so that the degree of the vertex will be reduced to 2, and so on. Finally, we can obtain an open edge train whose initial and final vertices are the same as those of the original edge train such that the set of all edges in this edge train is a path. Thus we have Theorem 1-2-1.

Theorem 1-2-1. If a set of all edges in an open edge train whose initial and final vertices are p and q is not a path between vertices p and q, then the edge train can be decomposed into one open edge train and closed edge trains such that (1) every edge in the original edge train is in exactly one of these edge trains and (2) a set of all edges in the resultant open edge train is a path between vertices p and q.

For example, open edge train ($a\,d\,g\,f\,c\,e$) in Fig. 1-2-2 can be decomposed into one open edge train ($a\,d\,e$) and a closed edge train ($g\,f\,c$). Note that a set of all edges in the open edge train ($a\,d\,e$) is a path between A and D.

Definition 1-2-7. A linear graph G' is called a *subgraph* of a linear graph G if G' consists only of edges and vertices of G. A subgraph G' is called a proper subgraph of G if G' is not identical to G.

For example, consider linear graphs G, G', and G'' in Fig. 1-2-2 and Fig. 1-2-5a and b. Linear graph G' is a proper subgraph of linear graph G but linear graph G'' is not a subgraph of G because edge h in G'' is not in G.

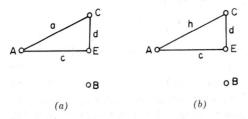

Fig. 1-2-5.   (a) Linear graph G'; (b) linear graph G''.

Definition 1-2-8.   *An isolated vertex* is a vertex of degree 0.

Many cases, considering a set of edges as a linear graph is convenient. Conversely, there are cases when treating a linear graph as a set of edges is more convenient. Hence we have the following definition.

Definition 1-2-9.   A linear graph corresponding to a set of edges is a linear graph consisting of all edges and all vertices which are the endpoints of edges in the set. A set of edges corresponding to a linear graph is a set of all edges in the linear graph.

When we treat a set of edges as a linear graph, we are considering a linear graph corresponding to the sets of edges. For example, a path is a set of edges by definition. However, we often treat a path as a linear graph. Conversely, when we treat a linear graph as a set of edges, we are considering a set of edges corresponding to the linear graph. Note that if a linear graph contains isolated vertices, a set of edges corresponding to the linear graph cannot indicate existence of isolated vertices. Hence usually only those graphs that do not contain isolated vertices will be treated as sets of edges. For example, treating the linear graph in Fig. 1-2-2 as a set (a, b, c, d, e, f, g) will not provide any information on vertices in the linear graph.

The linear graph in Fig. 1-2-6 consists of two subgraphs G_1 and G_2. Note

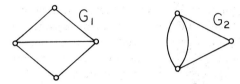

Fig. 1-2-6. Linear graph consisting of two subgraphs G_1 and G_2.

that there are no edges connected between a vertex in G_1 and a vertex in G_2. Hence there are no paths between a vertex in G_1 and a vertex in G_2. Such a linear graph is said to be separated.

Definition 1-2-10. A linear graph is said to be separated or is called a separated graph if there exist two vertices such that there are no paths between them. A linear graph is said to be connected or is called a connected graph if it is not separated.

The linear graph in Fig. 1-2-2 is connected because there are paths between any two vertices. A particular connected subgraph called a circuit is an important subgraph. We will study properties of circuits later but it is convenient to define a circuit now.

Definition 1-2-11. *A circuit* is a connected linear graph in which every vertex is of degree 2.

From this definition, we can see that a set of all edges in a closed edge train will be a circuit if the degree of every vertex is 2. For example, the set of all edges in closed edge train $(a\ c\ d)$ in Fig. 1-2-2 is a circuit but the set of all edges in closed edge train $(f\ g\ e\ b\ d\ c)$ is not a circuit. (Here we are treating a circuit as a set of edges. See Definition 1-2-9.) Let T be a closed edge train such that a set of all edges in T is not a circuit. Then by the same method employed previously for decomposing open edge trains, we can easily show that T can be decomposed into a number of closed edge trains T_1, $T_2, \ldots$, such that (1) every edge in T is in exactly one of $T_1, T_2, \ldots$, and (2) a set of all edges in each of these closed edge trains $T_1, T_2, \ldots$, is a circuit. We define an edge disjoint union of circuits as follows.

Definition 1-2-12.   An edge disjoint union of circuits is a collection of circuits such that no two of these circuits have edges in common.

Now we can state that a set of all edges in a closed edge train is either a circuit or an edge disjoint union of circuits. For example, a set of all edges in edge train $(f\, g\, e\, b\, d\, c)$ in Fig. 1-2-2 is not a circuit but an edge disjoint union of circuits (f, g, c) and (b, d, e). There may be a set of edges such that decomposition of the set into circuits is not unique. However, if there is at least one way of decomposing the set into circuits no two of which have edges in common, then the set is an edge disjoint union of circuits. For example, set (a, b, c, d) of edges in Fig. 1-2-7 is an edge disjoint union of circuits, but set (a, b, c, d, e) in Fig. 1-2-8 is not an edge disjoint union of circuits.

   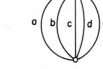

Fig. 1-2-7.   Edge disjoint union of circuits.   **Fig. 1-2-8.**   Linear graph.

Consider subgraphs G_1 and G_2 of a linear graph, which have no isolated vertices. By considering G_1 and G_2 as sets, we can use set operations such as $\cup$ (union), $\cap$ (intersection), $\oplus$ (ring sum), and $-$ (minus). That is,

$G_1 \cup G_2$:   Linear graph which consists of all edges in either G_1 or G_2.

$G_1 \cap G_2$:   Linear graph which consists of all edges in both G_1 and G_2 (those edges are usually called the common edges of G_1 and G_2).

$G_1 - G_2$:   A subgraph of G_1 obtained by deleting all edges which are also in G_2.

$$G_1 \oplus G_2 = (G_1 \cup G_2) - (G_1 \cap G_2) = (G_1 - G_2) \cup (G_2 - G_1)$$

$$(1\text{-}2\text{-}3)$$

where $=$ means identical.

Example 1-2-1.   In Fig. 1-2-9, two linear graphs G_1 and G_2 are shown in (1). The union of G_1 and G_2, represented by $G_1 \cup G_2$, is (2). Note that we are

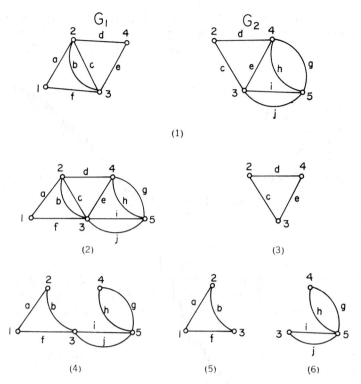

Fig. 1-2-9. Linear graphs and operations. (1) G_1 and G_2; (2) $G_1 \cup G_2$; (3) $G_1 \cap G_2$; (4) $G_1 \oplus G_2$; (5) $G_1 - G_2$; (6) $G_2 - G_1$.

treating set $G_1 \cup G_2$ as a linear graph in order to draw it. The intersection of G_1 and G_2, indicated by $G_1 \cap G_2$, is shown in (3); the ring sum of G_1 and G_2, written as $G_1 \oplus G_2$, is (4); the subtraction of G_2 from G_1, indicated by $G_1 - G_2$, is (5); and $G_2 - G_1$ is (6).

1-3 EULER GRAPH

Suppose each edge in a linear graph G represents a highway. If there is a closed edge train such that every edge in G is in the edge train, then this edge train indicates a continuous route which traverses each highway exactly once and returns to the starting point. What kind of linear graphs have such a closed edge train?

In the preceding section, we found that a closed edge train is either a circuit or an edge disjoint union of circuits. Hence if a connected linear graph is either a circuit or an edge disjoint union of circuits, there is a closed edge train which contains all edges in the linear graph. How can we test whether a connected linear graph is either a circuit or an edge disjoint union of circuits?

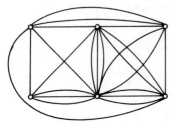

Fig. 1-3-1. A linear graph.

For example, is the linear graph in Fig. 1-3-1 either a circuit or an edge disjoint union of circuits? Of course, one way to answer this question is by actually trying to obtain a closed edge train which contains all edges in a given linear graph. Here we will study another way, which is by checking only the degree of vertices to answer the foregoing question. For this, we define an Euler graph as follows.

Definition 1-3-1.   A linear graph is called an *Euler graph* if every vertex is of even degree.

Since a circuit consists of vertices of degree 2, a circuit is an Euler graph. Consider an edge disjoint union of circuits. Suppose a vertex v is in at least one of these circuits. Since there are no edges in more than one of these circuits, the degree of vertex v is equal to twice the number of circuits that contain vertex v. Hence the degree of vertex v is even. This is true for any vertex in an edge disjoint union of circuits. Hence an edge disjoint union of circuits is an Euler graph.

Theorem 1-3-1.   A set of all edges in a closed edge train is an Euler graph.

Let us show the converse of this theorem. That is, we would like to find whether there exists a closed edge train consisting of all edges of an Euler graph. It is obvious that there are no such closed edge trains if an Euler graph is separated. Thus we assume that a given Euler graph E is connected. Because every vertex in Euler graph E is of even degree, there exist at least two edges connected to any vertex. Thus starting from any vertex, we can find a closed edge train. Let the set of all edges in this closed edge train be E_1. Note that E_1 is an Euler graph. It is clear that $E - E_1$ is an Euler graph. Thus unless $E - E_1$ is empty, we can find another closed edge train in $E - E_1$.

Let the set of all edges in this edge train be E_2. Then $E - E_1 - E_2$ is again an Euler graph, and so on. We can obtain a set of Euler graphs $E_1, E_2, \ldots, E_n$ which have the property that every edge in E is in exactly one of $E_1, E_2, \ldots, E_n$. (See Example 1-3-1.)

Now we combine these closed edge trains of $E_1, E_2, \ldots, E_n$ as follows: Let us take any one of $E_1, E_2, \ldots, E_n$, say E_1. Since E is connected, there exists at least one E_p in $E_2, E_3, \ldots, E_n$ such that $E_1 \cup E_p$ is connected. This means that there exists at least one vertex which is in both E_1 and E_p. Let v be a vertex in both E_1 and E_p. Note that if $T = (e_1 e_2 \cdots e_p \cdots e_r)$ is a closed edge train, then $(e_p e_{p+1} \cdots e_r e_1 e_2 \cdots e_{p-1})$ is also a closed edge train for any e_p. Hence any vertex in T can be the initial vertex of the closed edge train. Since v is a vertex in E_1 and E_p by assumption, we can make vertex v the initial vertex of both closed edge trains of E_1 and E_p. Let $(e_1 e_2 \cdots e_m)$ and $(f_1 f_2 \cdots f_n)$ be closed edge trains of E_1 and E_p, both of which have v as their initial vertex. Then $(e_1 e_2 \cdots e_m f_1 f_2 \cdots f_n)$ is a closed edge train consisting of all edges in $E_1 \cup E_p$. Again, there must be at least one Euler graph E_q in $E_2, E_3, \ldots, E_{p-1}, E_{p+1}, \ldots$, such that $E_1 \cup E_p \cup E_q$ is connected, and so on. Finally, we can obtain a closed edge train consisting of all edges in the given Euler graph E. This proves that for a connected Euler graph, there exists a closed edge train containing all edges in the Euler graph.

Theorem 1-3-2. A linear graph is a connected Euler graph if and only if there exists a closed edge train containing all edges in the linear graph.

Example 1-3-1. Suppose Fig. 1-3-2a is a given Euler graph E. Starting with vertex 1, suppose we obtain a closed edge train $(a\,b\,g\,e)$. Then $E - E_1$ is as shown in Fig. 1-3-2b where $E_1 = (a, b, g, e)$. Suppose we form a closed edge train $(c\,d\,f)$ from $E - E_1$. Then $E - E_1 - E_2$ will be as shown in Fig. 1-3-2c where $E_2 = (c, d, f)$. The last closed edge train will be $(h\,i)$. Now we

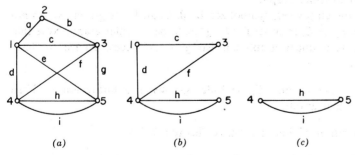

(a)   (b)   (c)

Fig. 1-3-2. Euler graph E and subgraphs. (a) Euler graph E; (b) $E - E_1$; (c) $E - E_1 - E_2$.

are going to obtain the desired closed edge train which consists of all edges in
E. $E_1 \cup E_2$ is connected; hence there exists at least one vertex in both E_1 and
E_2. Let us take vertex 3. Then the closed edge train of E_1 can be rotated such
that vertex 3 becomes the initial vertex as $(g\,e\,a\,b)$. Note that the initial vertex
of $(c\,df)$ is vertex 3. Thus we can combine them as $(g\,e\,a\,b\,c\,df)$ to obtain a
closed edge train which consists of all edges in $E_1 \cup E_2$. Similarly, by taking
vertex 5 as a common vertex of $E_1 \cup E_2$ and $E_3 = (h, i)$ and rotating $(g\,e\,a\,b\,c$
$df)$ as $(e\,a\,b\,c\,df\,g)$ so that the initial vertex of the closed edge train becomes
vertex 5, we can combine $(e\,a\,b\,c\,df\,g)$ and $(h\,i)$ as $(h\,i\,e\,a\,b\,c\,df\,g)$, which
is the desired closed edge train.

In general, there are several Euler graphs which are subgraphs of a linear
graph. For example, the linear graph in Fig. 1-3-2 has not only $E_1 =$
(a, b, e, g), $E_2 = (c, d, f)$, and $E_3 = (h, i)$ as subgraphs but also several other
Euler graphs such as $E_4 = (a, b, c)$, $E_5 = (c, e, g, h, i)$ and $E_6 = (c, e, g)$ as
its subgraphs. An important property of Euler graphs is that they form an
Abelian group so that all Euler graphs in a linear graph can be calculated by
the combination of a few Euler graphs called generators. To show this, we
will investigate the ring sum of two Euler graphs.

Let Euler graph E_1 and E_2 be subgraphs of a linear graph. Consider a
linear graph $E_1 \oplus E_2$ where $E_1 \neq E_2$. To find the degree of every vertex in
$E_1 \oplus E_2$, let $e_{11}, e_{12}, \ldots, e_{1r}$ be the edges in E_1 which are incident at vertex p.
Also let $e_{21}, e_{22}, \ldots, e_{2s}$ be the edges incident at vertex p in E_2. Without the
loss of generality, let $e_{11} = e_{21}, e_{12} = e_{22}, \ldots, e_{1t} = e_{2t}$, but $e_{1t+1}, e_{1t+2}, \ldots,$
e_{1r} and $e_{2t+1}, e_{2t+2}, \ldots, e_{2s}$ are all different. Hence in $E_1 \oplus E_2$, only edges
$e_{1t+1}, e_{1t+2}, \ldots, e_{1r}, e_{2t+1}, e_{2t+2}, \ldots,$ and e_{2s} are incident at vertex p. Thus
the degree of vertex p in $E_1 \oplus E_2$ is $r + s - 2t$. Since E_1 and E_2 are Euler
graphs, r and s are even numbers. Hence $r + s - 2t$ is even. This is true for
the degree of every vertex in $E_1 \oplus E_2$. Hence $E_1 \oplus E_2$ is an Euler graph. For
example, G_1 and G_2 in Fig. 1-2-9(1) are Euler graphs and $G_1 \oplus G_2$ in Fig.
1-2-9(4) is also an Euler graph.

By defining the empty set, symbolized by $\emptyset$, as an Euler graph, we do not
need to specify $E_1 \neq E_2$ in order for $E_1 \oplus E_2$ to be an Euler graph. Note that
the empty set is a linear graph consisting of no edges. Hence we have
Theorem 1-3-3.

Theorem 1-3-3. Let E_1 and E_2 be Euler graphs in a linear graph. Then
$E_1 \oplus E_2$ is an Euler graph.

The generalization of this theorem is Theorem 1-3-4.

Theorem 1-3-4. Let E_p for $p = 1, 2, \ldots, n$ be Euler graphs in a linear graph.
Then $E_1 \oplus E_2 \oplus \cdots \oplus E_n$ is an Euler graph.

For convenience, we define the symbol $\{E\}$ as follows.

Definition 1-3-2.   The symbol $\{E\}$ indicates a collection of all possible Euler graphs which are subgraphs of a linear graph.

With Theorems 1-3-3 and 1-3-4, we can see that the collection $\{E\}$ satisfies the following conditions:

1. If E_i and E_j are in $\{E\}$, then $E_i \oplus E_j = E_j \oplus E_i$ is in $\{E\}$.
2. There exists $E_0 (= \emptyset)$ such that for any E_i in $\{E\}$,

$$E_0 \oplus E_i = E_i$$

3. For any E_i in $\{E\}$,

$$E_i \oplus E_i = E_0$$

4. For any E_i, E_j, and E_k in $\{E\}$,

$$E_i \oplus (E_j \oplus E_k) = (E_i \oplus E_j) \oplus E_k$$

This leads to Theorem 1-3-5.

Theorem 1-3-5.   The collection $\{E\}$ of a linear graph is an Abelian group under the ring sum.

This theorem indicates that there are generators in $\{E\}$ by which all Euler graphs in $\{E\}$ can be obtained by the ring sum of these generators.

Example 1-3-2.   Consider a linear graph as shown in Fig. 1-3-3. From a set

Fig. 1-3-3.   A linear graph.

of generators which consists of E_1, E_2, and E_3 as shown in Fig. 1-3-4, we can obtain all remaining nonempty Euler graphs by

$$E_1 \oplus E_2 = E_7$$
$$E_1 \oplus E_3 = E_4$$
$$E_2 \oplus E_3 = E_6$$

and

$$E_1 \oplus E_2 \oplus E_3 = E_5$$

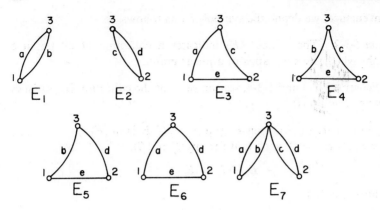

Fig. 1-3-4. {E} of the linear graph in Fig. 1-3-3.

The properties of {E} and how to obtain a set of generators in {E} will be studied in later chapters. Here we need only note that there is a simple algorithm of generating all possible Euler graphs in a linear graph.

In order to discuss an Euler graph which is separated, we define a "maximal connected subgraph" as follows.

Definition 1-3-3. A set of *maximal connected subgraphs* of a linear graph G is a set of connected subgraphs $g_1, g_2, \ldots, g_\rho$ such that every edge and every

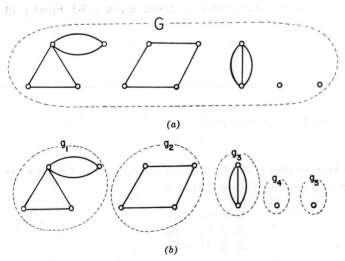

(a)

(b)

Fig. 1-3-5. Maximal connected subgraphs. (a) Linear graph G; (b) maximal connected subgraphs.

vertex in G is in exactly one of these subgraphs where an isolated vertex is a connected subgraph by definition. The *number* ρ is called the number of maximal connected subgraphs of linear graph G.

Note that for any two distinct maximal connected subgraphs g_r and g_s, $g_r \cup g_s$ is separated. For example, in a linear graph shown in Fig. 1-3-5a, there are five maximal connected subgraphs as shown in Fig. 1-3-5b. Thus $\rho = 5$. It can be seen that if a linear graph G is connected, we will have only one maximal connected subgraph. Hence we can say that a linear graph is connected if and only if $\rho = 1$.

We have learned that the set of all edges in a closed edge train is either a circuit or an edge disjoint union of circuits. Moreover, we know that for any Euler graph that is connected, there exists a closed edge train which consists of all edges in the Euler graph. In addition, it is clear that every maximal connected subgraph of an Euler graph is an Euler graph.

Theorem 1-3-6. Any Euler graph is either a circuit or an edge disjoint union of circuits, and collection $\{E\}$ is a collection of all possible circuits, edge disjoint unions of circuits of a linear graph, and the empty set.

We next consider some properties of paths between a pair of vertices. Recall that every vertex in a path except the initial and the final vertices is of degree 2 and the initial and the final vertices are of degree 1. An important property of paths is given by the following theorem.

Theorem 1-3-7. If P_1 and P_2 are paths between the same pair of vertices, then $P_1 \oplus P_2$ is an Euler graph.

Proof. Let P_1 and P_2 be paths between vertices i and j in a linear graph G. We form new linear graph G' from G by inserting an edge y whose endpoints are vertices i and j. Note that P_1 and P_2 are paths in G'. Consider $P_1 \cup (y)$ and $P_2 \cup (y)$, where $P_1 \cup (y)$ is a set of edges in P_1 and edge y and $P_2 \cup (y)$ consists of all edges in P_2 and edge y. Clearly these sets are Euler graphs. The ring sum of these two sets

$$[P_1 \cup (y)] \oplus [P_2 \cup (y)] = P_1 \oplus P_2 \qquad (1\text{-}3\text{-}1)$$

is an Euler graph by Theorem 1-3-3. Since all edges in $P_1 \oplus P_2$ are in G, $P_1 \oplus P_2$ is an Euler graph in G. QED

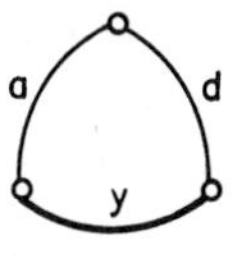

(a)

(b)

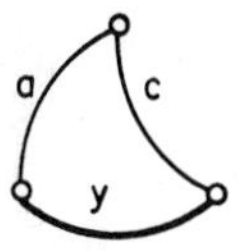

Fig. 1-3-6. Linear graphs (a) $P_1 \cup (y)$ and (b) $P_2 \cup (y)$.

For example, let $P_1 = (a, d)$ and $P_2 = (a, c)$ in linear graph G in Fig. 1-3-3. Then $P_1 \cup (y)$ and $P_2 \cup (y)$ are shown in Fig. 1-3-6a and b. Note that both are Euler graphs. Also note that edge y is not in G and $P_1 \oplus P_2 = (c, d)$ is an Euler graph in G.

By generalizing the preceding theorem, we find Theorem 1-3-8.

Theorem 1-3-8. If P_r for $r = 1, 2, \ldots, 2k$ are paths between the same pair of vertices, then $P_1 \oplus P_2 \oplus \cdots \oplus P_{2k}$ is an Euler graph.

Proof. We have

$$P_1 \oplus P_2 \oplus \cdots \oplus P_{2k} = (P_1 \oplus P_2) \oplus (P_3 \oplus P_4) \oplus \cdots \oplus (P_{2k-1} \oplus P_{2k})$$

Since $P_{i-1} \oplus P_i$ is an Euler graph by Theorem 1-3-7 and the ring sum of Euler graphs is an Euler graph by Theorem 1-3-3, the right-hand side of the equation is an Euler graph.   QED

Example 1-3-3. All possible paths between vertices 1 and 2 of the linear graph in Fig. 1-3-3 are shown in Fig. 1-3-7. We have

$$P_1 \oplus P_2 = E_6$$
$$P_2 \oplus P_3 = E_7$$
$$P_3 \oplus P_4 = E_1$$
$$P_1 \oplus P_2 \oplus P_3 \oplus P_4 = E_5$$

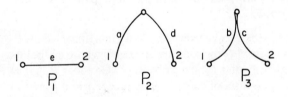

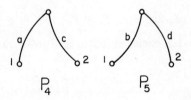

Fig. 1-3-7. Paths between 1 and 2.

where E's are shown in Fig. 1-3-4. However, $P_1 \oplus P_2 \oplus P_3$ is not an Euler graph but is identical to a given linear graph.

1-4 *M*-GRAPH

We have found that the ring sum of two paths between the two specified vertices is an Euler graph. However, the ring sum of a path and an Euler graph may not become a path. Hence the collection of all possible Euler graphs and all possible paths between two specified vertices may not form a group under the ring sum. This means that it is more difficult to obtain all possible paths between two specified vertices than it is to generate all possible Euler graphs in $\{E\}$ which we have seen in the previous section. Here we introduce an *M*-graph so that the collection of all possible Euler graphs and *M*-graphs between two specified vertices form an Abelian group under the ring sum. Furthermore, we will see that the collection of all possible paths between two specified vertices is in the collection of all possible *M*-graphs between the same specified vertices.

Let M, P, and E be subgraphs of a linear graph where P is a path between vertices i and j, E is an Euler graph, and M is equal to $P \oplus E$. We are going to show that every vertex in M except vertices i and j is of even degree and vertices i and j are of odd degree.

Consider a new linear graph G', which is obtained from a given linear graph by inserting one edge y whose endpoints are i and j. Since y is a new edge, Euler graph E in G does not contain y. Hence edge y is in $[P \cup (y)] \oplus E$. It is also seen easily that $P \cup (y)$ is an Euler graph in G'. By Theorem 1-3-3, $[P \cup (y)] \oplus E$ is an Euler graph. Hence every vertex in $[P \cup (y)] \oplus E$ is of even degree. By deleting edge y from $[P \cup (y)] \oplus E$, we have $P \oplus E = M$. Deletion of edge y reduces the degrees $d(i)$ and $d(j)$ of vertices i and j by one but will not influence the degree of any other vertex. Thus we can conclude that every vertex except two vertices in M is of even degree. Such a linear graph is called an *M*-graph.

Definition 1-4-1. An *M-graph* is a linear graph in which there exist exactly two vertices of odd degree. These vertices (of odd degrees) are called the *terminal vertices* of an *M*-graph.

Let i and j be the two vertices in an *M*-graph which are of odd degrees. Then by inserting a new edge whose endpoints are i and j to the *M*-graph, we will have an Euler graph because every vertex becomes of even degree. Thus an *M*-graph can be considered a linear graph obtained from an Euler graph by deleting one edge. One of the properties of *M*-graphs is as follows.

Theorem 1-4-1. There exists an open edge train which contains all edges of a connected *M*-graph.

Proof. Let M be a connected *M*-graph and i and j be the terminal vertices. Let y be an edge whose endpoints are i and j which is not in M. Then $M \cup (y)$

is a connected Euler graph. Since there exists a closed edge train which consists of all edges in a connected Euler graph, let $(e_1 e_2 \cdots e_n)$ be such a closed edge train of $M \cup (y)$. Thus y is in the edge train. Without the loss of generality, let y be located between edges e_p and e_{p+1} in the edge train as $(e_1 e_2 \cdots e_p y e_{p+1} \cdots e_n)$. By rotating the edge train, we can obtain the closed edge train with y being the first edge as $(y e_{p+1} \cdots e_n e_1 e_2 \cdots e_p)$. Deleting edge y, the resultant sequence $(e_{p+1} \cdots e_n e_1 e_2 \cdots e_p)$ is an open edge train which contains all edges in M.   QED

Example 1-4-1. Consider the M-graph in Fig. 1-4-1 in which the terminal vertices are clearly 1 and 2. By inserting edge y, whose endpoints are vertices 1 and 2, we have the resultant graph shown in Fig. 1-4-2, which is obviously an

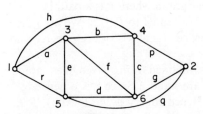

Fig. 1-4-1. M-graph M.

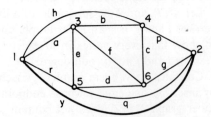

Fig. 1-4-2. Euler graph $M \cup (y)$.

Euler graph. Thus we can obtain a closed edge train $(a\,b\,c\,d\,e\,f\,g\,y\,h\,p\,q\,r)$, which consists of all edges in $M \cup (y)$. By rotating the edge train, we have $(y\,h\,p\,q\,r\,a\,b\,c\,d\,e\,f\,g)$. Without y, the sequence $(h\,p\,q\,r\,a\,b\,c\,d\,e\,f\,g)$ is an open edge train which consists of all edges in M.

Definition 1-4-2. An *edge disjoint union of a path and circuits* is a set of edges which can be decomposed into a path and an edge disjoint union of circuits such that the path and the edge disjoint union of circuits have no edges in common.

For example, set $(a, b, c, d, e, f, g, h, p, q, r)$ of edges, which is a set of all edges in M-graph M in Fig. 1-4-1, can be decomposed into path (a, b, p) and an edge disjoint union of circuits (c, d, e, f, g, h, q, r). Thus the set is an edge disjoint union of a path and circuits.

Theorem 1-4-2. An M-graph is either a path or an edge disjoint union of a path and circuits.

Proof. We know that there exists an open edge train which contains all edges in connected M-graph by Theorem 1-4-1. We also know that if a set of all edges in an open edge train is not a path, then we can decompose the edge train into an open edge train and a closed edge train such that the set of all edges in this open edge train is a path. Furthermore, the set of all edges in a closed edge train is an Euler graph which is either a circuit or an edge disjoint union of circuits. Thus a connected *M*-graph is either a path or edge disjoint union of a path and circuits.

Suppose a given *M*-graph *M* is separated. Let $g_1, g_2, \ldots, g_\rho$ be the set of maximal connected subgraphs of *M*. Since one-half of the sum of all degrees of vertex is the number of edges in a linear graph, if a vertex of odd degree is in g_r ($1 \le r \le \rho$), then the other vertex whose degree is odd must also be in g_r. Thus one of $g_1, g_2, \ldots, g_\rho$, say g_1, is an *M*-graph and all others must be Euler graphs. Hence we can conclude that g_1 is either a path or an edge disjoint union of a path and circuits and each of $g_2, \ldots, g_\rho$ is either a circuit or an edge disjoint union of circuits. Thus a given *M*-graph *M* is an edge disjoint union of a path and circuits.   QED

Consider a familiar puzzle, which is to draw a given picture without lifting your pencil. If a picture can be drawn without lifting your pencil, then there must be an edge train which contains all edges in the picture when we consider a given picture as a linear graph. Thus if it is an open edge train, the picture must be a connected *M*-graph. On the other hand, if we have a closed edge train, the picture must be a connected Euler graph. Furthermore, if you cannot draw a given picture without lifting your pencil, then it is impossible to obtain an edge train. Thus a picture is neither a connected *M*-graph nor a connected Euler graph. This means that a picture can be drawn without lifting your pencil if and only if the picture is either a connected *M*-graph or a connected Euler graph. For example, the linear graph in Fig. 1-4-3a is a connected *M*-graph. Hence we can draw it without lifting a pencil. However, the linear graph in Fig. 1-4-3b is neither an *M*-graph nor an Euler graph. Thus we cannot draw it without lifting a pencil.

(*a*)

(*b*)   **Fig. 1-4-3.** Puzzles.

We have seen that the ring sum of two paths between the same vertices is an Euler graph. Since an *M*-graph is an edge disjoint union of a path and circuits, the ring sum of two *M*-graphs having the same terminal vertices will also be an Euler graph.

Theorem 1-4-3. Let M_1 and M_2 be M-graphs which are subgraphs of a linear graph and which have the same terminal vertices. Then $M_1 \oplus M_2$ is an Euler graph.

Proof. Let y be an edge whose endpoints are the terminal vertices of M_1 (and M_2) and that is in neither M_1 nor M_2. Consider $M_1 \cup (y)$ and $M_2 \cup (y)$, both of which are obviously Euler graphs. Thus

$$[M_1 \cup (y)] \oplus [M_2 \cup (y)] = M_1 \oplus M_2 \qquad (1\text{-}4\text{-}1)$$

is an Euler graph.   QED

We have learned that the ring sum of a path and a circuit may not be a path. However, the ring sum of an M-graph and an Euler graph will always be an M-graph, which is shown by the next theorem.

Theorem 1-4-4. Let M be an M-graph and E be an Euler graph of a linear graph. Then $M \oplus E$ is an M-graph.

Proof. By Theorem 1-4-2, M can be written as $M = P \oplus E'$ where P is a path and E' is an Euler graph. Thus

$$M \oplus E = (P \oplus E') \oplus E = P \oplus E'' = M' \qquad (1\text{-}4\text{-}2)$$

where $E' \oplus E = E''$ is an Euler graph (by Theorem 1-3-3).   QED

From Theorems 1-3-3, 1-4-3, and 1-4-4, we can form the ring sum table for an M-graph M and an Euler graph E as

$\oplus$	E	M
E	E	M
M	M	E

Note the symmetry between this table and the addition table for binary numbers given below:

$+$	0	1
0	0	1
1	1	0

For convenience, we define the symbols $\{P_{ij}\}$ and $\{M_{ij}\}$ as follows.

Definition 1-4-3. The symbol $\{P_{ij}\}$ is a collection of all possible paths between two vertices i and j in a linear graph G. The symbol $\{M_{ij}\}$ indicates a collection of all possible subgraphs in G which are M-graphs with i and j being terminal vertices.

Note that we have defined the symbol $\{E\}$ as a collection of all possible subgraphs which are Euler graphs in a linear graph. By Theorem 1-4-2, we have

$$\{M_{ij}\} = \{P \oplus E; \quad P \in \{P\}, E \in \{E\}\} \qquad (1\text{-}4\text{-}3)$$

Also because of Theorems 1-3-3, 1-4-3, and 1-4-4, we have Theorem 1-4-5.

Theorem 1-4-5. Collection $\{E, M_{ij}\}$ of all subgraphs in $\{M_{ij}\}$ and $\{E\}$ is an Abelian group under the ring sum.

Example 1-4-2. We already know $\{E\}$ and $\{P\}$ of the linear graph in Fig. 1-3-3 which are shown in Figs. 1-3-4 and 1-3-7. Those are

$$\{E\} = \{0, (a, b), (c, d), (a, c, e), (b, c, e), (b, d, e), (a, d, e), (a, b, c, d)\}$$

and

$$\{P_{12}\} = \{(e), (a, d), (b, c), (a, c), (b, d)\}$$

Hence all *M*-graphs with terminal vertices being 1 and 2 are

$$\begin{aligned}\{M_{12}\} &= \{P \oplus E; \quad P \in \{P\}, E \in \{E\}\} \\ &= \{(e), (a, d), (b, c), (a, c), (b, d), (a, b, e), (c, d, e), (a, b, c, d, e)\}\end{aligned}$$

where the first five *M*-graphs are paths and the rest are edge disjoint unions of a path and circuits as shown in Fig. 1-4-4. Let $P_1, P_2, P_3, P_4,$ and P_5 be

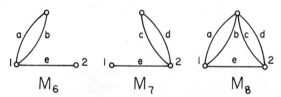

$$M_6 \qquad\qquad M_7 \qquad\qquad M_8$$

Fig. 1-4-4. *M*-graphs.

those defined in Fig. 1-3-7 and $M_6 = (a,b,e)$, $M_7 = (c,d,e)$, and $M_8 = (a,b, c,d,e)$. Then a set of generators E_1, E_2, E_3, and P_1 will produce all possible members in collection $\{E, M_{12}\}$ by

$$\begin{aligned} E_1 \oplus E_2 &= E_7 \\ E_1 \oplus E_3 &= E_4 \\ E_2 \oplus E_3 &= E_6 \\ E_1 \oplus E_2 \oplus E_3 &= E_5 \\ E_1 \oplus P_1 &= M_6 \end{aligned}$$

$$E_2 \oplus P_1 = M_7$$
$$E_3 \oplus P_1 = P_4$$
$$E_1 \oplus E_2 \oplus P_1 = M_8$$
$$E_1 \oplus E_3 \oplus P_1 = P_3$$
$$E_2 \oplus E_3 \oplus P_1 = P_2$$
$$E_1 \oplus E_2 \oplus E_3 \oplus P_1 = P_5$$

and

$$E_1 \oplus E_1 = E_2 \oplus E_2 = \cdots = P_1 \oplus P_1 = \emptyset$$

1-5 NONSEPARABLE GRAPH

There are several sets of generators such that the ring sum of these generators give all members of $\{E, M_{ij}\}$. It is not difficult to pick several paths between two specified vertices in a linear graph. But once these paths have been found, we still must determine whether they can form a set of generators for $\{E, M_{ij}\}$. In Example 1-4-2, we can choose paths P_1, P_2, P_3, and P_4 as generators for $\{E, M_{12}\}$ because

$$P_3 \oplus P_4 = E_1$$
$$P_2 \oplus P_4 = E_2$$

and

$$P_1 \oplus P_4 = E_3$$

On the other hand, if a given linear graph is M_6 in Fig. 1-4-4, then there is only one path between vertices 1 and 2, which is (e). Obviously, path (e) alone cannot be a set of generators for

$$\{E, M_{12}\} = \{[\emptyset, (a, b), (a, b, e), (e)]\}$$

These examples indicate that whether or not there is a set of paths which becomes a set of generators for $\{E, M_{ij}\}$ depends on the structure of a given linear graph. One such structure where paths alone can form a set of generators is called a nonseparable graph. In order to define a nonseparable graph, we will define a notion of cutting vertices as follows.

Definition 1-5-1. *Cutting vertex p into two new vertices p_1 and p_2 means* changing the endpoints of all edges that have p as their endpoints from p to either p_1 or p_2.

For example, several ways of cutting vertex p in a linear graph in Fig. 1-5-1a are shown in Fig. 1-5-1b, c, d, and e.

Consider the connected linear graph in Fig. 1-5-2a. If we cut vertex p into two vertices p_1 and p_2 as shown in Fig. 1-5-2b, the linear graph becomes separated. However, if a connected linear graph is one shown in

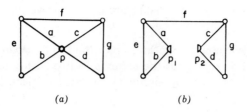

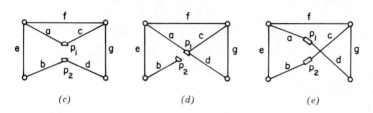

Fig. 1-5-1. Different ways of cutting vertex p. (a) Linear graph G; (b) G_1; (c) G_2; (d) G_3; (e) G_4.

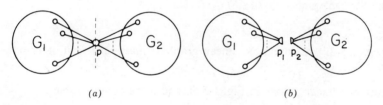

Fig. 1-5-2. Separable graph. (a) Cut vertex p; (b) separated graph.

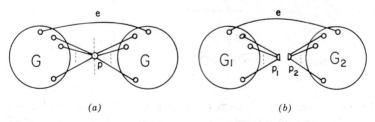

Fig. 1-5-3. Nonseparable graph. (a) A linear graph; (b) cutting vertex p.

Fig. 1-5-3*a*, then it cannot be separated regardless of how we cut vertex *p* into two.

Definition 1-5-2. A connected linear graph is said to be *separable* if there exists a vertex *p* and a way to cut the vertex into two such that the linear graph becomes separated without producing isolated vertices. Vertex *p* is called a *cut vertex*.

For example, linear graphs in Fig. 1-5-4*a*, *b*, and *c* are separable. Note that vertices *p* and *p'* are cut vertices in these linear graphs.

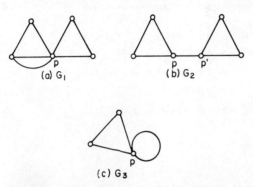

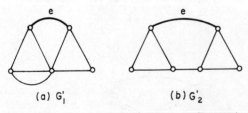

Fig. 1-5-4. Separable graphs. (*a*) G_1; (*b*) G_2; (*c*) G_3.

Definition 1-5-3. A connected linear graph is *nonseparable* if it is not separable.

For example, linear graphs in Fig. 1-5-5*a* and *b* are nonseparable.

Fig. 1-5-5. Nonseparable graphs. (*a*) G'_1; (*b*) G'_2.

To find some properties of nonseparable graphs, consider a linear graph *G* consisting of two maximal connected subgraphs g_1 and g_2, each of which contains at least one edge. It is easily seen that the linear graph G becomes connected by inserting one edge e between a vertex in g_1 and a vertex in g_2 as shown in Fig. 1-5-6. However, the resulting graph is separable. Conversely, if G' is a nonseparable graph, then deleting one edge will not produce a separated graph.

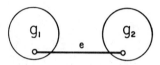

Fig. 1-5-6.   A separable graph.

Let e be an edge connected between vertices i and j in a nonseparable graph G'. Since G' will not be separated when we delete edge e, there exists a path P_{ij} between vertices i and j in G'_e which is obtained from G' by deleting edge e as shown in Fig. 1-5-7. This path P_{ij} and edge e form a circuit in G'.

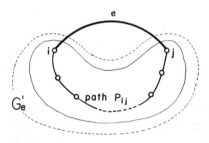

Fig. 1-5-7.   A nonseparable graph G'.

Thus we have Theorem 1-5-1.

Theorem 1-5-1.   Any edge in a nonseparable graph must be in at least one circuit.

Note that some separable and separated graphs will have the property given by Theorem 1-5-1 (i.e., every edge is in at least one circuit). For example, the separable and separated graphs in Fig. 1-5-8a and b have this property. Thus this property is not a property possessed only by nonseparable graphs.

There are two ways of removing an edge from a linear graph. One is *deleting an edge* (or *opening an edge*), and the other is *shorting an edge*. The term "shorting edge e" means to delete edge e and coincide the two vertices, which are the endpoints of edge e, to make one vertex. For example,

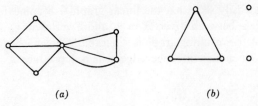

(a)   *(b)*

Fig. 1-5-8. Linear graphs. *(a)* Separable graph; *(b)* separated graph.

by deleting (opening) edge *e* in the linear graph in Fig. 1-5-9*a*, we have the resultant graph shown in Fig. 1-5-9*b*. However, shorting edge *e* in the same linear graph will give the linear graph shown in Fig. 1-5-9*c*.

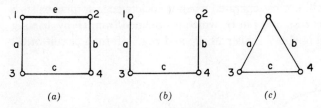

(a)   *(b)*   *(c)*

Fig. 1-5-9. Deleting and shorting edge *e*. *(a)* A linear graph; *(b)* deleting edge *e*; *(c)* shorting edge *e*.

We next show that any edge can be either deleted or shorted to maintain a linear graph nonseparable. Suppose deleting edge *e* from a nonseparable graph G' makes it separable. Let the resultant graph be G'_e (Fig. 1-5-10) where vertex *p* is a cut vertex. Also let the two endpoints of edge *e* be *i* and *j*. Then it can be seen that one of *i* and *j* must be in G'_{1e} and the other must be in G'_{2e}. Furthermore, neither *i* nor *j* can be vertex *p* because the original linear graph G' is nonseparable. Thus if instead of deleting edge *e* we short it, the resultant graph will be nonseparable as shown in Fig. 1-5-11.

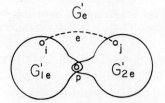

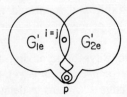

Fig. 1-5-10. Separable graph G'_e.   **Fig. 1-5-11.** Shorting edge *e*.

Suppose that by shorting edge e, a nonseparable graph G'' becomes separable. Let the resultant separable graph be G''_e shown in Fig. 1-5-12. Then the original nonseparable graph must consist of edges e and two subgraphs G_1 and G_2 joined by two vertices p_1 and p_2 as shown in Fig. 1-5-13.

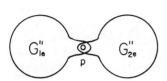 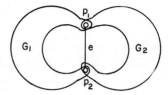

Fig. 1-5-12. A separable graph G' result-ing from shorting an edge. **Fig. 1-5-13.** A nonseparable graph.

Note that vertices p_1 and p_2 become cut vertex p when edge e is shorted. Thus if instead of shorting edge e we delete the edge, the resultant graph will be nonseparable. This is true as long as a nonseparable graph consists of at least three vertices.

Theorem 1-5-2. If G is a nonseparable graph containing at least three vertices, any edge can be either deleted or shorted to make the resultant graph nonseparable.

Now we are ready to discuss one of the properties which only nonseparable graphs can have.

Theorem 1-5-3. For any edge e and any pair of vertices i and j, there exists a path between i and j which contains edge e if and only if a linear graph is nonseparable.

Proof. If a nonseparable graph G has exactly three vertices, the theorem is obvious. Thus suppose the theorem is true for all nonseparable graphs having k vertices. Consider a nonseparable graph G having $k + 1$ vertices. By Theorem 1-5-2, we can pick any edge and either short or delete it to keep the resultant graph nonseparable. Pick an edge e_1, which is not edge e and is not connected to i in G, and short it. If the resultant graph is separable, we delete edge e_1 instead of shorting it. Then we pick another edge e_2, which is not edge e and is not connected to i, and short it. If the resultant graph is separable, we delete e_2 rather than short it, and so on. Soon we will have an edge e_r, which is not edge e and is not connected to i by shorting; the resultant graph is nonseparable. Let this resultant graph be G'. It is clear that G' is of order k. Hence the theorem is true for G', that is, there exists a path between

i and j which contains edge e. Then the same path, except that edge e_r may be included, will be a path between i and j in G and contain edge e.   QED

From this theorem, we can obtain the following theorem.

Theorem 1-5-4. For any pair of vertices i and j and any circuit C in a non-separable graph, there exist two paths P_1 and P_2 between i and j such that

$$C = P_1 \oplus P_2 \qquad (1\text{-}5\text{-}1)$$

This theorem guarantees that there is a set of paths which can be a set of generators for $\{E, M_{ij}\}$. Note that the linear graph in Fig. 1-3-3 is non-separable and a set of generators can consist of paths P_1, P_2, P_3, and P_4 shown in Fig. 1-3-7.

1-6 COLLECTION OF PATHS

As we have seen, a collection $\{M_{ij}\}$ of M-graphs whose terminal vertices are i and j contains all possible paths between i and j. We also know that $\{E, M_{ij}\}$ is an Abelian group under the ring sum. Furthermore, by Theorems 1-4-3 and 1-4-4, if we choose generators properly, we can obtain $\{M_{ij}\}$ easily. For example, by choosing E_1, E_2, E_3, and P_1 (Figs. 1-3-4 and 1-3-7) as generators for a linear graph in Fig. 1-3-3, we know that all M-graphs in $\{M_{12}\}$ are P_1, $P_1 \oplus E_1$, $P_1 \oplus E_2$, $P_1 \oplus E_3$, $P_1 \oplus E_1 \oplus E_2$, $P_1 \oplus E_1 \oplus E_3$, $P_1 \oplus E_2 \oplus E_3$, and $P_1 \oplus E_1 \oplus E_2 \oplus E_3$. Hence one way to obtain all possible paths between vertices i and j will be to generate collection $\{M_{ij}\}$ by generators in collection $\{E, M_{ij}\}$. Then we extract only paths somehow from $\{M_{ij}\}$. One operation, called a "minimum operation," will be introduced here which will extract all paths from $\{M_{ij}\}$. The same operation can extract all possible circuits from collection $\{E\}$ of Euler graphs.

This operation is also useful to obtain important properties among collections of paths. Furthermore, these properties will help us develop another simple algorithm of generating all possible paths between two specified vertices which we will study in this section. First, we define a "minimum operation" on a collection as follows.

Definition 1-6-1. Let A be a collection of sets. Then min A is a sub-collection of A such that any set α in A is in min A if every set β in A satisfies $\beta \not\subset \alpha$ as long as β is not the empty set. ($\beta \not\subset \alpha$ means that b is not a subset of α, that is, there exists at least one member in set β which is not in set α.)

For example, suppose $A = \{(a,b,c), (a,d), (a), (b,c), \emptyset\}$. Then min A consists of sets (a), (b,c), and the empty set $\emptyset$. It can be seen that the following process will give the min A of a collection A.

Test α_p ($\neq \phi$) for $p = 1, 2, 3, \ldots, n$ whether there exists a nonempty set α_r ($\neq \alpha_p$) such that $\alpha_r \subset \alpha_p$. If there exists such a set α_r, delete α_p. Otherwise leave α_p in the collection. ($\alpha_r \subset \alpha_p$ means that α_r is a subset of α_p, that is, every member of α_r is in α_p.) For example, the minimum of $\{E\}$ in Example 1-3-2 is

$$\min \{E\} = \{\emptyset, (a, b), (c, d), (a, c, e), (b, c, e), (a, d, e), (b, d, e)\}$$

obtained by deleting (a, b, c, d). This is because only (a, b, c, d) has another nonempty set (a, b) satisfying $(a, b) \subset (a, b, c, d)$.

Recall that $\{E\}$ is the collection of all distinct circuits, edge disjoint unions of circuits, and the empty set. Let $E_1 \in \{E\}$ be an edge disjoint union of circuits $C_1, C_2, \ldots, C_k$. Then since C_1 is a circuit, C_1 is in $\{E\}$. Moreover, $C_1 \subset E_1$. Thus E_1 will be deleted to obtain $\min \{E\}$. Since any circuit C in $\{E\}$ is a connected subgraph in which every vertex is of degree 2, no proper subset of C is a circuit.

Theorem 1-6-1.

$$\min \{E\} = \{C\} \qquad (1\text{-}6\text{-}1)$$

where $\{C\}$ is a collection of all possible circuits and the empty set.

Collection $\{M_{ij}\}$ is defined as a collection of all possible M-graphs whose terminal vertices are i and j. We have found that a M-graph M_{ij} is either a path P_{ij} between vertices i and j or an edge disjoint union of a path P_{ij} and circuits. Let M_1 in $\{M_{ij}\}$ be an edge disjoint union of path P_{ij} and circuits. Since P_{ij} is in $\{M_{ij}\}$, M_1 will not be in $\min \{M_{ij}\}$.

Suppose P'' is a proper subset of a path P'_{ij} in $\{M_{ij}\}$. Then by Theorem 1-3-7, $P'' \oplus P'_{ij}$ is either a circuit or an edge disjoint union of circuits. However, P'' is a proper subset of P'_{ij}. Hence $P'' \oplus P'_{ij}$ is a subgraph of P'_{ij} which cannot contain circuits. Thus such P'' will not be in $\{M_{ij}\}$.

Theorem 1-6-2.

$$\min \{M_{ij}\} = \{P_{ij}\} \qquad (1\text{-}6\text{-}2)$$

where $\{P_{ij}\}$ is the collection of all possible paths between vertices i and j.

We use an operation called the ring product $\otimes$ to multiply two collections as follows.

Definition 1-6-2. The *ring product* of collections A and B of sets, denoted by $A \otimes B$, is defined as

$$A \otimes B = \min \{\alpha_p \oplus \beta_q; \quad \alpha_p \in A, \beta_q \in B\} \qquad (1\text{-}6\text{-}3)$$

For example, suppose $A = \{(a), (b, c), (d, e)\}$ and $B = \{(a, b), (b, c, d), (e)\}$. Then

$$A \otimes B = \min \{(b), (a, b, c, d), (a, e), (a, c), (d), (b, c, e), (a, b, d, e)\}$$
$$= \{(b), (a, e), (a, c), (d)\}$$

One of the applications of the ring produce is as follows. By Theorem 1-5-4, any circuit can be expressed as the ring sum of two paths between vertices i and j in a nonseparable graph. Thus if we collect the ring sum of all possible combinations of paths between i and j two at a time, we will have all possible circuits in a nonseparable graph.

Theorem 1-6-3.   For a nonseparable graph

$$\{P_{ij}\} \otimes \{P_{ij}\} = \{C\} \qquad (1\text{-}6\text{-}4)$$

Note that

$$\{P_{ij}\} \otimes \{P_{ij}\} = \min \{P_r \oplus P_s; \quad P_r, P_s \in \{P_{ij}\}\} \qquad (1\text{-}6\text{-}5)$$

Since the empty set $\emptyset$ is in $\{C\}$ and the ring sum of any set c and the empty set $\emptyset$ is the set c, we can see that

$$\{C\} \subset \{C\} \otimes \{C\} \qquad (1\text{-}6\text{-}6)$$

Furthermore, for C_r and C_s in $\{C\}$, $C_r \oplus C_s$ is either a circuit or an edge disjoint union of circuits. We have

$$\{C_r \oplus C_s; \quad C_r, C_s \in \{C\}\} \subset \{E\} \qquad (1\text{-}6\text{-}7)$$

Note that $\{E\}$ is the set of all distinct circuits, edge disjoint unions of circuits, and the empty set. Thus

$$\{C\} \otimes \{C\} = \min \{C_r \oplus C_s; \quad C_r, C_s \in \{C\}\} \subset \min \{E\} \qquad (1\text{-}6\text{-}8)$$

However, $\min \{E\}$ is $\{C\}$ from Eq. 1-6-1. Thus with Eq. 1-6-6, we have

$$\{C\} \subset \{C\} \otimes \{C\} \subset \{C\} \qquad (1\text{-}6\text{-}9)$$

Hence we have Theorem 1-6-4.

Theorem 1-6-4.

$$\{C\} \otimes \{C\} = \{C\} \qquad (1\text{-}6\text{-}10)$$

Since the empty set is in $\{C\}$, we can see that

$$\{P_{ij}\} \subset \{P_{ij}\} \otimes \{C\} \qquad (1\text{-}6\text{-}11)$$

On the other hand,

$$\{p \oplus C; \quad p \in \{P_{ij}\}, C \in \{C\}\} \subset \{M_{ij}\} \qquad (1\text{-}6\text{-}12)$$

where $\{M_{ij}\}$ is the collection of all distinct subgraphs which are M-graphs with terminal vertices being i and j. Since min $\{M_{ij}\}$ is $\{P_{ij}\}$ by Eq. 1-6-2, we have

$$\{P_{ij}\} \otimes \{C\} = \min \{p \oplus C; \quad p \in \{P_{ij}\}, C \in \{C\}\} \subset \min \{M_{ij}\} = \{P_{ij}\}$$
$$(1\text{-}6\text{-}13)$$

This, with Eq. 6-1-11 gives Theorem 1-6-5.

Theorem 1-6-5.

$$\{P_{ij}\} \otimes \{C\} = \{P_{ij}\} \qquad (1\text{-}6\text{-}14)$$

This theorem can be further modified.

Theorem 1-6-6.

$$\{P_{ij}\} \otimes \{P_{ij}\} \otimes \{P_{ij}\} = \{P_{ij}\} \qquad (1\text{-}6\text{-}15)$$

Proof. Since the empty set $\emptyset$ is in $\{P_{ij}\} \otimes \{P_{ij}\}$,

$$\{P \oplus Q; \quad P \in \{P_{ij}\}, Q \in \{P_{ij}\} \otimes \{P_{ij}\}\} \supset \{P_{ij}\} \qquad (1\text{-}6\text{-}16)$$

Since Q is an Euler graph, $P \oplus Q$ is an M-graph. QED

Example 1-6-1. In Example 1-3-3, $\{P_{12}\}$ is given as

$$\{P_{12}\} = \{(e), (a,d), (b,c), (a,c), (b,d)\}$$

By calculating $\{P_{12}\} \otimes \{P_{12}\}$, we should obtain all possible circuits in the nonseparable graph in Fig. 1-3-3:

$$\begin{aligned}
\{P_{12}\} \otimes \{P_{12}\} &= \min \{(e) \oplus (e), (e) \oplus (a, d), (e) \oplus (b, c), (e) \oplus (a, c), (e) \oplus \\
&\qquad (b, d), (a, d) \oplus (b, c), (a, d) \oplus (a, c), (a, d) \oplus (b, d), \\
&\qquad (b, c) \oplus (a, c), (b, c) \oplus (b, d), (a, c) \oplus (b, d)\} \\
&= \min \{\emptyset, (a, d, e), (b, c, e), (a, c, e), (b, d, e), (a, b, c, d), (c, d), \\
&\qquad (a, b)\} \\
&= \{\emptyset, (a, d, e), (b, c, e), (a, c, e), (b, d, e), (c, d), (a, b)\} \\
&= \{C\}
\end{aligned}$$

Note that the subscripts r and s of $\{P_{rs}\}$ indicate the two vertices between which every path in $\{P_{rs}\}$ exists. That is, the symbol $\{P_{rs}\}$ is a collection of all possible paths between vertices r and s. Equation 1-6-4 shows that the ring product of $\{P_{rs}\}$ and $\{P_{rs}\}$ gives $\{C\}$. However, it should be clear that the ring product of $\{P_{rs}\}$ and $\{P_{tu}\}$ for $r \neq t, u$ will not give $\{C\}$.

Theorem 1-6-7. Let G be a connected graph. Also let $i, j,$ and k be the distinct vertices in G. Then

$$\{P_{ij}\} \otimes \{P_{jk}\} = \{P_{ik}\} \qquad (1\text{-}6\text{-}17)$$

where $\{P_{ij}\}$ is a collection of all possible paths between i and j, $\{P_{jk}\}$ is a collection of all possible paths between j and k, and $\{P_{ik}\}$ is a collection of all possible paths between i and k in G.

Proof. For any path P_{ij} in $\{P_{ij}\}$ and any path P_{jk} in $\{P_{jk}\}$, $P_{ij} \oplus P_{jk}$ is an M-graph M_{ik}. Hence all we need to show is that for any path P_{ik} in $\{P_{ik}\}$, there exist paths P_{ij} and P_{jk} in $\{P_{ij}\}$ and $\{P_{jk}\}$, respectively, such that

$$P_{ij} \oplus P_{jk} = P_{ik} \tag{1-6-18}$$

Since G is connected, there exists a path from j to any vertex in path P_{ik}. Let P' be a path from j to a vertex r in path P_{ik} in G which contains no other vertex in P_{ik}. Let P_{ir} be the path between i and r in subgraph P_{ik} and P_{rk} be the path between r and k in subgraph P_{ik} as shown in Fig. 1-6-1. Then $P_{ij} = P_{ir} \cup P'$ and $P_{jk} = P_{rj} \cup P'$ are the desired paths.   QED

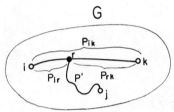

Fig. 1-6-1. Paths P_{ik}, P_{ir}, P_{rk}, and P'.

This theorem can be extended as follows.

Theorem 1-6-8. For a connected linear graph G,

$$\{P_{ij_1}\} \otimes \{P_{j_1 j_2}\} \otimes \{P_{j_2 j_3}\} \otimes \cdots \otimes \{P_{j_m k}\} = \{P_{ik}\} \tag{1-6-19}$$

where $i, j_1, \ldots, j_m$ and k are vertices in G.
With Eq. 1-6-4 we can modify the above theorem.

Theorem 1-6-9. For a nonseparable graph,

$$\{P_{ij_1}\} \otimes \{P_{j_1 j_2}\} \otimes \cdots \otimes \{P_{j_m k}\} \otimes \{P_{ki}\} = \{C\} \tag{1-6-20}$$

Example 1-6-2. In a connected graph in Fig. 1-3-3, $\{P_{23}\}$ is

$$\{P_{23}\} = \{(c), (d), (a, e), (b, e)\}$$

With $\{P_{12}\}$ in Example 1-6-1, we have

$$\begin{aligned}
\{P_{13}\} = \{P_{12}\} \otimes \{P_{23}\} &= \{(e), (a, d), (b, c), (a, c), (b, d)\} \otimes \{(c), (d), (a, e), (b, e)\} \\
&= \min \{(c, e), (d, e), (a), (b), (a, c, d), (a, b, d, e), (b, c, d), (a, b, c, e)\} \\
&= \{(c, e), (d, e), (a), (b)\}
\end{aligned}$$

Consider a nonseparable graph G'. Let e be an edge in G' and the endpoints of edge e be vertices i and j as shown in Fig. 1-5-7. Let G'_e be the linear graph obtained from G' by deleting edge e. Since G'_e is connected, there exists at least one path between vertices i and j. Let $\{P_{ij}\}$ be a collection of all possible paths between i and j in G'_e. Note that (e) itself is a path between i and j in G'. Hence $(e) \oplus P$, where P is in $\{P_{ij}\}$, is a circuit. In other words, for any P in $\{P_{ij}\}$, there exists a circuit C in G' such that

$$C \oplus (e) = P \tag{1-6-21}$$

Let $\{C\}$ be a collection consisting of the empty set and all possible circuits in G'. Since $(e) \oplus \emptyset = (e)$, the collection $\min \{(e) \oplus C;\ C \in \{C\}\}$ is the collection of all possible paths between vertices i and j.

Definition 1-6-3. The *symbol* $\{P_{[e]}\}$ is a collection of all possible paths between the vertices which are the endpoints of edge e.

Theorem 1-6-10. For any edge e in a linear graph,

$$\{P_{[e]}\} = \min \{(e) \oplus C;\quad C \in \{C\}\} \tag{1-6-22}$$

Proof. Let $P_1\ (\neq \emptyset)$ be a path between vertices i and j. Then $P_1 \cup (e)$ is a circuit in $\{C\}$. The empty set $\emptyset$ is in $\{C\}$. Thus

$$\{(e) \oplus C;\quad C \in \{C\}\} \supset \{P_{[e]}\} \tag{1-6-23}$$

$(e) \oplus C$ is an M-graph with terminal vertices being the endpoints of edge e. Thus

$$\min \{(e) \oplus C;\quad C \in \{C\}\} = \{P_{[e]}\} \tag{1-6-24}$$

QED

Example 1-6-3. The set $\{C\}$ of a nonseparable graph in Example 1-6-1 is

$$\{C\} = \{\emptyset, (a, b), (c, d), (a, c, e), (b, c, e), (b, d, e), (a, d, e)\}$$

Then

$$\begin{aligned}
\{P_{[a]}\} &= \min \{(a) \oplus C;\quad C \in \{C\}\} \\
&= \min \{(a), (b), (a, c, d), (c, e), (a, b, c, e), (a, b, d, e), (d, e)\} \\
&= \{(a), (b), (c, e), (d, e)\}
\end{aligned}$$

which is equal to $\{P_{13}\}$ in Example 1-6-2. It can be seen that from Example 1-6-2

$$\{P_{12}\} \otimes \{P_{23}\} \otimes \{P_{13}\} = \{P_{13}\} \otimes \{P_{13}\} = \{C\}$$

Theorem 1-6-11. For any edge e in a linear graph,

$$\{P_{[e]}\} = \{(e) \oplus C;\quad \text{every circuit } C \text{ which is either the} \tag{1-6-25}$$
$$\text{empty set or containing } e\}$$

Proof. Let i and j be the endpoints of edge e. From Theorem 1-6-10, $(e) \oplus C$ is an M-graph M_{ij}. In order that M_{ij} be a path, C must contain edge e. That is, if edge e is in C, $(e) \oplus C$ cannot be an edge disjoint union of a path P_{ij} and circuits. Moreover, for any path P_{ij}, $(e) \oplus P_{ij}$ is a circuit which contains edge e.   QED

In Example 1-6-3, circuits (a, b), (a, c, e), and (a, d, e) contain edge a. Hence $\{P_{[a]}\}$ consists of (a), (b), (c, e), and (d, e) by Theorem 1-6-11.

Suppose edges $e_1, e_2, \ldots, e_k$ form a path between vertices i and j. Then, by Theorem 1-6-8, collection $\{P_{ij}\}$ of all possible paths between i and j is

$$\{P_{ij}\} = \{P_{[e_1]}\} \otimes \{P_{[e_2]}\} \otimes \cdots \otimes \{P_{[e_k]}\} \qquad (1\text{-}6\text{-}6)$$

Consider a path P in $\{P_{ij}\}$. We can see that

$$\min \{P \oplus C; \quad C \in \{C\}\}$$

may not be equal to $\{P_{ij}\}$ except when P consists of one edge only. For example, consider linear graph G in Fig. 1-6-2. Suppose we take (a, b) as a path

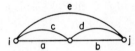

Fig. 1-6-2.   A linear graph.

P. Then there is no circuit C in G such that $P \oplus C = (c, d)$. On the other hand, consider Theorem 1-6-12.

Theorem 1-6-12.   Let $\{E\}$ be a collection of all Euler graphs and the empty set of a linear graph G. Suppose P is a path in $\{P_{ij}\}$. Then

$$\{P_{ij}\} = \min \{P \oplus E; \quad E \in \{E\}\} \qquad (1\text{-}6\text{-}7)$$

Proof.   Let P' be any path in $\{P_{ij}\}$. Then $P \oplus P'$ is an Euler graph. Hence $P \oplus P'$ is in $\{E\}$.   QED

One application of sets of paths is in the field of switching theory. Consider a linear graph as a switching network in which each edge indicates a switching variable. It is true that a switching function F_{ij} from i to j can be expressed as

$$F_{ij} = \sum_{(q)} \text{path } P_{q_{ij}} \text{ product} \qquad (1\text{-}6\text{-}8)$$

where $\sum_{(q)}$ means the sum for all possible paths from i to j and a "path $P_{q_{ij}}$ product" is the product of all switching variables associated with edges in path $P_{q_{ij}}$. Thus, if we know all possible paths between vertices i and j, we can obtain switching function F_{ij}.

Example 1-6-4.   Consider the linear graph in Fig. 1-6-3 as a switching network where a, b, c, d, e, f, and g are switching variables. Suppose we need to

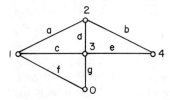

Fig. 1-6-3.   A switching network.

find switching functions F_{12} and F_{14}. We can obtain all possible paths between
vertices 1 and 2 and all possible paths between vertices 1 and 4 directly from
the network. However, we can use Eq. 1-6-2 and Theorems 1-6-7 and 1-6-12
to obtain those paths as follows. We obtain $\{E\}$ first as

$$\{E\} = \{\emptyset, (a, c, d), (c, f, g), (b, d, e), (a, d, f, g), (a, b, c, e), (b, c, d, e, f, g),$$
$$(a, b, e, f, g)\}$$

Then

$$\{C\} = \min \{E\} = \{\emptyset, (a, c, d), (c, f, g), (b, d, e), (a, d, f, g), (a, b, c, e),$$
$$(a, b, e, f, g)\}$$

Since $\{P_{12}\} = \{P_{[a]}\}$, we can obtain

$$\{P_{12}\} = \{P_{[a]}\} = \{(a), (c, d), (d, f, g), (b, c, e), (b, e, f, g)\}$$

Thus

$$F_{12} = a + cd + dfg + bce + befg$$

Since $\{P_{24}\} = \{P_{[b]}\}$,

$$\{P_{24}\} = \min \{(b) \oplus c; \quad c \in \{C\}\} = \{(b), (d, e), (a, c, e), (a, e, f, g)\}$$

By Theorem 1-6-7,

$$\{P_{14}\} = \{P_{12}\} \otimes \{P_{24}\} = \{(a), (c, d), (d, f, g), (b, c, e), (b, e, f, g)\}$$
$$\otimes \{(b), (d, e), (a, c, e), (a, e, f, g)\}$$
$$= \{(a, b), (a, d, e), (c, e), (e, f, g), (b, c, d), (b, d, f, g)\}$$

or

$$F_{14} = ab + ade + ce + efg + bcd + bdfg$$

Since (a, b) is a path between vertices 1 and 4, by using Theorem 1-6-12, we
can obtain $\{P_{14}\}$ as

$$\{P_{14}\} = \{P_{[ab]}\} = \min \{(a, b) \oplus E; \quad E \in \{E\}\}$$
$$= \{(a, b), (b, c, d), (a, d, e), (b, d, f, g), (c, e), (e, f, g)\}$$

which gives the same answer.

1-7 τ-GRAPH

We can classify the ring sum of a pair of paths into three types. One is the ring sum $P_1 \oplus P_2$ of paths where both P_1 and P_2 are in the same collection $\{P_{ij}\}$. In this case we have found that $P_1 \oplus P_2$ is either a circuit or an edge disjoint union of circuits. Another type is the ring sum of paths P_{ij} and P_{jk} where P_{ij} is a path between vertices i and j and P_{jk} is a path between vertices j and k. In the previous section, we learned that $P_{ij} \oplus P_{jk}$ is either a path P_{ik} or an edge disjoint union of a path P_{ik} and circuits, where P_{ik} is a path between vertices i and k. The other type we are going to study now is the ring sum of path P_{rs} and P_{tu} where P_{rs} is a path between r and s, P_{tu} is a path between t and u, and vertices r, s, t, and u are all different. For example, in Fig. 1-6-3, (d, e) is a path between vertices 2 and 4 and (f, g) is a path between vertices 1 and 3. The ring sum of these two paths is $(d, e) \oplus (f, g) = (d, e, f, g)$ as shown in Fig. 1-7-1, which is clearly not a path.

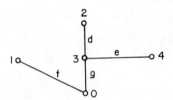

Fig. 1-7-1. $(d, e) \oplus (f, g)$.

It is clear that the degrees of vertices r, s, t, and u of the resultant linear graph $P_{rs} \oplus P_{tu}$ are all odd. Thus $P_{rs} \oplus P_{tu}$ is not a path. Let us now define a τ-graph.

Definition 1-7-1. A linear graph is called a τ-graph if there exist no circuits and no isolated vertices. The symbol $\tau_{(i_1 i_2 \cdots i_m)}$ indicates a τ-graph in which vertices of odd degree are $i_1, i_2, \ldots, i_m$.

For example, the linear graph in Fig. 1-7-1 is a τ-graph in which the vertices of odd degree are 1, 2, 3, and 4. Thus we can use the symbol $\tau_{(1234)}$ to indicate the linear graph.

Theorem 1-7-1. Let P_{rs} be a path between vertices r and s and P_{tu} be a path between vertices t and u in a linear graph where r, s, t, and u are all distinct. Then $P_{rs} \oplus P_{tu}$ is either a $\tau_{(rstu)}$ or an edge disjoint union of a $\tau_{(rstu)}$ and circuits.

We can now generalize this theorem.

Theorem 1-7-2. Suppose $(i_1, i_2, \ldots, i_m) \oplus (j_1, j_2, \ldots, j_n)$ is not empty. Also suppose τ-graph $\tau_{(i_1 i_2 \cdots i_m)}$ and $\tau_{(j_1 j_2 \cdots j_n)}$ are subgraphs of a linear graph. Then $\tau_{(i_1 i_2 \cdots i_m)} \oplus \tau_{(j_1 j_2 \cdots j_n)}$ is either a τ-graph $\tau_{((i_1 i_2 \cdots i_m) \oplus (j_1 j_2 \cdots j_n))}$ or an edge disjoint union of τ-graph $\tau_{((i_1 i_2 \cdots i_m) \oplus (j_1 j_2 \cdots j_n))}$ and circuits.

The proof can be accomplished by considering the degree of every vertex. Note that we will omit commas between members in a set as $(i_1 i_2 \cdots i_m)$ if the set appears as a subscript, for simplicity.

Example 1-7-1. The ring sum of $\tau_{(1240)}$ in Fig. 1-7-2a and $\tau_{(1234)}$ in Fig. 1-7-1 is

$$\tau_{(1240)} \oplus \tau_{(1234)} = (b, f) \oplus (d, e, f, g) = (b, d, e, g)$$

which is an edge disjoint union of $\tau_{(30)}$ and circuit (b, d, e) as shown in Fig.

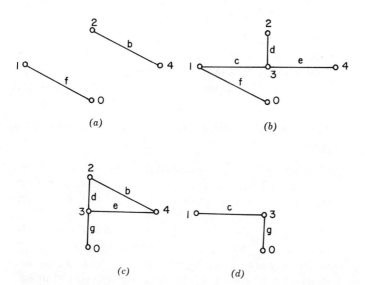

(a)   (b)

(c)   (d)

Fig. 1-7-2. τ-Graphs and edge disjoint union of a τ-graph and a circuit. (a) $\tau_{(1240)}$; (b) $\tau_{(2340)}$; (c) edge disjoint union of $\tau_{(30)}$ and a circuit; (d) $\tau_{(10)}$.

1-7-2c. On the other hand, the ring sum of a $\tau_{(1234)}$ in Fig. 1-7-1 and a $\tau_{(2340)}$ in Fig. 1-7-2b is

$$\tau_{(1234)} \oplus \tau_{(2340)} = (d, e, f, g) \oplus (c, d, e, f) = (c, g)$$

which is $\tau_{(10)}$ as shown in Fig. 1-7-2d.

Another way of generalizing Theorem 1-7-1 is as follows.

Theorem 1-7-3. Let $P_{i_1j_1}, P_{i_2j_2}, \ldots, P_{i_kj_k}$ be paths between vertices i_p and j_p for $p = 1, 2, \ldots, k$ in a linear graph. Also let $(i_1, j_1) \oplus (i_2, j_2) \oplus \cdots \oplus (i_k, j_k)$ $\neq \emptyset$. Then $P_{i_1j_1} \oplus P_{i_2j_2} \oplus \cdots \oplus P_{i_kj_k}$ is either a $\tau_{((i_1j_1)\oplus(i_2j_2)\oplus\cdots\oplus(i_kj_k))}$ or an edge disjoint union of a $\tau_{((i_1j_1)\oplus(i_2j_2)\oplus\cdots\oplus(i_kj_k))}$ and circuits.

The proof of this theorem can be accomplished by considering the degree of every vertex in the resultant graph.

Now we are ready to discuss the collection of sets $\{P_{i_rj_r}\}$ (for $r = 1, 2, \ldots, k$). One of the important properties of a collection of τ-graphs is given by the following theorem.

Theorem 1-7-4. Let $\{P_{rs}\}$ be a collection of all possible paths between vertices r and s and $\{P_{tu}\}$ be a collection of all possible paths between vertices t and u in a connected linear graph G where r, s, t, and u are all different. Then

$$\{P_{rs}\} \otimes \{P_{tu}\} = \{\tau_{(rstu)}\} \tag{1-7-1}$$

where $\{\tau_{(rstu)}\}$ is a collection of all τ-graphs τ_{rstu} in G.

The proof of this theorem can be accomplished by showing that for any $\tau_{(rstu)}$, there exists a pair of paths P_{rs} and P_{tu} such that $P_{rs} \oplus P_{tu}$ is the τ-graph. Before proving the existence of such paths, we will study some important properties of a τ-graph.

Lemma 1-7-1. Let τ be a τ-graph which is connected and consists of n_v vertices. Then τ contains exactly $n_v - 1$ edges.

Proof. We prove this lemma by the mathematical induction. Let $\tau^{(2)}$ be a connected τ-graph of 2 vertices. Since there is no circuit, $\tau^{(2)}$ must have exactly one edge. Hence the lemma is true for this case. Suppose the lemma is true for all τ-graphs which are connected and each of which has k vertices. Then we need to prove that a $\tau^{(k+1)}$, which is connected and has $k + 1$ vertices, has exactly k edges. Since $\tau^{(k+1)}$ has no circuit, shorting an edge e in $\tau^{(k+1)}$ (which means removing edge e and making two vertices which are the endpoints of edge e coincide) will produce a connected τ-graph consisting of k vertices. By assumption, this resultant τ-graph contains exactly $k - 1$ edges. Hence $\tau^{(k+1)}$ has exactly k edges which proves the lemma.   QED

Lemma 1-7-2. For distinct vertices i and j, any τ-graph $\tau_{(ij)}$ is a path between i and j.

Proof. We can prove this lemma by inserting one edge as follows. By definition of $\tau_{(ij)}$, all vertices other than i and j are of even degree. Hence by inserting an edge y whose endpoints are vertices i and j, $\tau_{(ij)} \cup (y)$ will be an Euler graph. If $\tau_{(ij)} \cup (y)$ is not a circuit but an edge disjoint union of circuits, then $\tau_{(ij)}$ must contain at least one circuit, which contradicts the assumption that $\tau_{(ij)}$ is a τ-graph. Thus $\tau_{(ij)}$ is a path. This also indicates that $\tau_{(ij)}$ is connected. QED

Proof of Theorem 1-7-4. Since $P_{rs} \oplus P_{tu}$ is either a τ-graph $\tau_{(rstu)}$ or an edge disjoint union of a τ-graph $\tau_{(rstu)}$ and circuits by Theorem 1-7-1, we only need to prove that for any $\tau_{(rstu)}$ in $\{\tau_{(rstu)}\}$, there exist a path P_{rs} in $\{P_{rs}\}$ and a path P_{tu} in $\{P_{tu}\}$ such that $P_{rs} \oplus P_{tu}$ is $\tau_{(rstu)}$.

Suppose $\tau_{(rstu)}$ consists of k maximal connected subgraphs. Note that there are four vertices of odd degree in $\tau_{(rstu)}$. Also recall that a connected graph must contain even number of vertices of odd degree. If $k > 2$, there will be a maximal connected subgraph consisting of vertices of even degree. This subgraph is obviously an Euler graph. Hence this subgraph is either a circuit or an edge disjoint union of circuits. However, $\tau_{(rstu)}$ cannot have circuits by definition. Thus k must be no more than 2.

CASE 1. Suppose $k = 2$. Let g_1 and g_2 be the two maximal connected subgraphs of $\tau_{(rstu)}$. Then exactly two of vertices r, s, t, and u must be in g_1 and the other two vertices must be in g_2. Since there exist no circuits in both g_1 and g_2, g_1 and g_2 are τ-graphs. Thus by Lemma 1-7-2, g_1 and g_2 are paths. Depending on which vertices are in g_1, we have the following two cases:

1. When vertices r and s are in g_1, we have $g_1 = P_{rs}$ and $g_2 = P_{tu}$ by Lemma 1-7-2. It is clear that these paths are the desired paths.
2. When vertices r and t are in g_1, we have $g_1 = P_{rt}$ and $g_2 = P_{su}$ as shown in Fig. 1-7-3a. Since we assume that there are paths P_{rs} and P_{tu}, there exists a

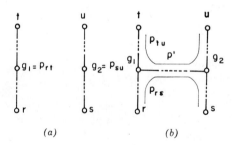

Fig. 1-7-3. $\tau_{(rstu)}$ and paths P_{rs} and P_{tu}. (a) $\tau_{(rstu)}$ of Case 1(2); (b) existence of path P'.

path between a vertex in g_1 and a vertex in g_2 in a linear graph from which P_{rs} and P_{tu} are obtained. Let this path be P' as shown in Fig. 1-7-3b. Thus clearly we have two desired paths P_{rs} and P_{tu} such that $P_{rs} \oplus P_{tu} = \tau_{(rstu)}$.

CASE 2. Suppose $k = 1$. Then $\tau_{(rstu)}$ is connected. Let us obtain an open edge train starting with vertex r. This edge train will obviously terminate at one of vertices s, t, and u. Let g be the linear graph consisting of all edges in this open edge train. Then g is a path because $\tau_{(rstu)}$ contains no circuits. Furthermore, linear graph $[\tau_{(rstu)} - g]$ contains no circuits and has exactly two vertices of odd degree. Hence $[\tau_{(rstu)} - g]$ must be a τ-graph and is a path by Lemma 1-7-2. Let these two paths be P_1 and P_2. Note that P_1 and P_2 have no edges in common. If P_1 and P_2 have at least two vertices in common, then $P_1 \cup P_2 = \tau_{(rstu)}$ will have at least one circuit, which contradicts the assumption that $\tau_{(rstu)}$ is a τ-graph. If P_1 and P_2 have no vertices in common, then $P_1 \cup P_2 = \tau_{(rstu)}$ is a separated graph, which is again a contradiction. Hence P_1 and P_2 must have exactly one vertex in common, as shown in Fig. 1-7-4a and b.

If $\tau_{(rstu)}$ is one vertex in Fig. 1-7-4a, then it is clear that there exist P_{rs} and P_{tu}

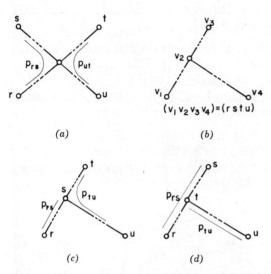

Fig. 1-7-4. Structure of $\tau_{(rstu)}$. (*a*) A structure of $P_1 \cup P_2$; (*b*) a structure of $P_1 \cup P_2$; (*c*) location of vertices r, s, t, and u; (*d*) location of vertices r, s, t, and u.

such that $P_{rs} \oplus P_{tu} = \tau_{(rstu)}$. Hence the theorem is true. Suppose $\tau_{(rstu)}$ is one vertex in Fig. 1-7-4b. Without the loss of generality, let $v_1 = r$. Then there are only two different types, one of which is when $v_2 = s$ and the other when $v_2 = t$. When $v_2 = s$, we can obtain desired paths as shown in Fig. 1-7-4c, which proves the theorem. When $v_2 = t$, we can also choose paths as shown in Fig. 1-7-4d such that the theorem will hold.   QED

From the proof of Theorem 1-7-4, we can see that the following holds.

Lemma 1-7-3. For a connected graph,

$$\{P_{rs}\} \otimes \{P_{tu}\} = \{P_{rt}\} \otimes \{P_{su}\} \qquad (1\text{-}7\text{-}2)$$

For example, $\tau_{(1240)}$ in Fig. 1-7-2a obviously can be obtained from $P_{10} \oplus P_{24}$ where $P_{10} = (f)$ and $P_{24} = (b)$. However, it can also be obtained from $P_{12} \oplus P_{40}$ where $P_{12} = (d, f, g)$ and $P_{40} = (b, d, g)$ of the linear graph in Fig. 1-6-3.

Another interesting property of a τ-graph is that a τ-graph can always be decomposed into an edge disjoint path which is given by the following theorem.

Theorem 1-7-5. A $\tau_{(i_1 i_2 \cdots i_{2k})}$ is an edge disjoint union of k paths $P_{j_1 j_2}$, $P_{j_3 j_4}, \ldots, P_{j_{2k-1} j_{2k}}$ where set $(i_1, i_2, \ldots, i_{2k})$ is equal to set $(j_1, j_2, \ldots, j_{2k})$.

Proof. We prove this theorem by mathematical induction. When $K = 1$, $\tau_{(i_1 i_2)}$ is a path $P_{i_1 i_2}$ by Lemma 1-7-2. Suppose the theorem is true for $k = m$. Then for $k = m + 1$, we form an open edge train of $\tau_{(i_1 i_2 \cdots i_{2m+2})}$ starting from vertex i_1 which will obviously terminate at one of vertices $i_2, i_3, \ldots, i_{2m+2}$. Suppose it is terminated at vertex i_r where $2 \le r \le 2m + 2$. Let g be a linear graph consisting of all edges in this open edge train. It is clear that g is a path $p_{i_1 i_r}$. It is further clear that the linear graph $(\tau_{(i_1 i_2 \cdots i_{2m+2})} - g)$ is a $\tau_{(i_2 \cdots i_{r-1} i_{r+1} \cdots i_{2m+2})}$ consisting of $2m$ vertices of odd degree. Hence, by assumption, $(\tau_{(i_1 i_2 \cdots i_{2m+2})} - g)$ is an edge disjoint union of m paths $p_{j_1 j_2}, \ldots, P_{j_{2m-1} j_{2m}}$. Thus with a path $P_{j_{2m+1} j_{2m+2}} = g$ where $j_{2m+1} = i_1$ and $j_{2m+2} = i_r$, we can say that $\tau_{(i_1 i_2 \cdots i_{2m+2})}$ is an edge disjoint union of $m + 1$ paths $P_{j_1 j_2}, \ldots, P_{j_{2m-1} j_{2m}}$, $P_{j_{2m+1} j_{2m+2}}$ where $(i_1, i_2, \ldots, i_{2m+2}) = (j_1, j_2, \ldots, j_{2m+2})$. QED

Example 1-7-2. From Example 1-6-4, $\{P_{12}\}$ of the linear graph in Fig. 1-6-3 is

$$\{P_{12}\} = \{(a), (c, d), (d, f, g), (b, c, e), (b, e, f, g)\}$$

We can obtain $\{P_{34}\}$ as

$$\{P_{34}\} = \{P(e)\} = \{(e), (b, d), (a, b, c), (a, b, f, g)\}$$

Thus

$$\{\tau_{(1234)}\} = \{P_{12}\} \otimes \{P_{34}\} = \min \{P_r \oplus P_s; \quad P_e \in \{P_{12}\}, P_s \in \{P_{34}\}\}$$
$$= \{(a, e), (a, b, d), (b, c), (b, f, g), (c, d, e), (d, e, f, g)\}$$

Each of these τ-graphs is shown in Fig. 1-7-5.

Theorem 1-7-4 also may be generalized as follows.

Theorem 1-7-6. For distinct vertices $i_1, i_2, \ldots, i_k$ in a connected linear graph,

$$\{P_{i_1 i_2}\} \otimes \{P_{i_3 i_4}\} \otimes \cdots \otimes \{P_{i_{k-1} i_k}\} = \{\tau_{(i_1 i_2 \cdots i_k)}\} \tag{1-7-3}$$

The proof can be accomplished by using Theorem 1-7-5 and Lemma 1-7-3. We now can generalize further.

Theorem 1-7-7. Let $(i_1, i_2) \oplus (i_3, i_4) \oplus \cdots \oplus (i_{k-1}, i_k)$ be a nonempty set where $i_1, i_2, \ldots, i_k$ are the vertices in a connected linear graph. Then

$$\{P_{i_1 i_2}\} \otimes \{P_{i_3 i_4}\} \otimes \cdots \otimes \{P_{i_{k-1} i_k}\} = \{\tau_{((i_1 i_2) \oplus (i_3 i_4) \oplus \cdots \oplus (i_{k-1} i_k))}\} \tag{1-7-4}$$

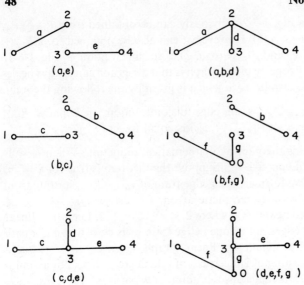

Fig. 1-7-5. $\{\tau_{(1234)}\}$.

Proof. Let $(i_1, i_2) \oplus (i_3, i_4) \oplus \cdots \oplus (i_{k-1}, i_k) = (j_1, j_2, \ldots, j_r)$ where $r \leq k$. By Theorem 1-6-7, $\{P_{i_t i_u}\} \otimes \{P_{i_m i_n}\} = \{P_{i_t i_n}\}$ if $i_u = i_m$. Hence we can express

$$\{P_{j_1 j_2}\} \otimes \{P_{j_3 j_4}\} \otimes \cdots \otimes \{P_{j_{r-1} j_r}\} = \{P_{i_1 i_2}\} \otimes \{P_{i_3 i_4}\} \otimes \cdots \otimes \{P_{i_{k-1} i_k}\}$$

By Theorem 1-7-4, the left-hand side of the preceding equation is equal to $\{\tau_{(j_1 j_2 \cdots j_r)}\}$, which is $\{\tau_{((i_1 i_2) \oplus \cdots \oplus (i_{k-1} i_k))}\}$. QED

Our final generalization of Theorem 1-7-4 is as follows.

Theorem 1-7-8. Let G be a connected linear graph. Suppose the ring sum of sets of vertices $(i_1, i_2, \ldots, i_m) \oplus (j_1, j_2, \ldots, j_n)$ is not empty. Then

$$\{\tau_{(i_1 i_2 \cdots i_m)}\} \otimes \{\tau_{(j_1 j_2 \cdots j_n)}\} = \{\tau_{((i_1 i_2 \cdots i_m) \oplus (j_1 j_2 \cdots j_n))}\} \qquad (1\text{-}7\text{-}5)$$

The proof can be accomplished by using Theorems 1-6-7 and 1-7-6.

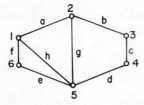

Fig. 1-7-6.   A connected linear graph.

Example 1-7-3. Consider a linear graph in Fig. 1-7-6. Now $\{\tau_{(1245)}\}$ and $\{\tau_{(1346)}\}$ are, respectively,

$$\{\tau_{(1245)}\} = \{(a, d), (a, b, c, g), (b, c, h), (b, c, e, f), (d, g, h), (d, e, f, g)\}$$

and

$$\{\tau_{(1346)}\} = \{(a, b, d, e), (b, d, f, g), (a, b, d, f, h), (a, c, e, g), (c, e, h), (c, f),$$
$$(b, d, e, g, h)\}$$

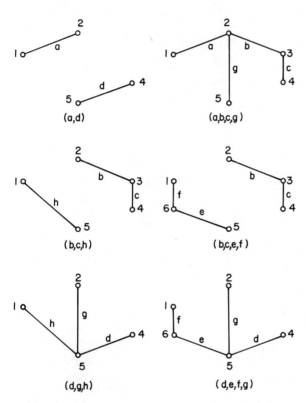

Fig. 1-7-7. $\{\tau_{(1245)}\}$.

Each τ-graph in $\{\tau_{(1245)}\}$ is shown in Fig. 1-7-7 and each τ-graph in $\{\tau_{(1346)}\}$ is shown in Fig. 1-7-8. Then the ring product of $\{\tau_{(1245)}\}$ and $\{\tau_{(1346)}\}$ is

$$\{\tau_{(1245)}\} \otimes \{\tau_{(1346)}\} = \{\tau_{(2356)}\}$$
$$= \{(b, e), (a, b, f, g), (b, f, h), (c, d, e, g), (a, c, d, e, h), (a, c, d, f),$$
$$(c, d, f, g, h)\}$$

Suppose $(1, 2, 3, \ldots, n)$ is the set of all vertices in a connected linear graph.

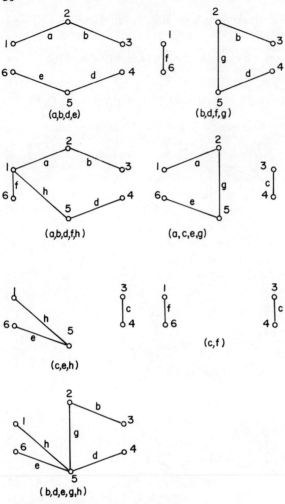

Fig. 1-7-8. $\{\tau_{(1346)}\}$.

Let the symbol $\{\{\tau\}, \{C\}\}$ be a set which consists of collection $\{C\}$ and all possible collection $\{\tau_{(i_1 i_2 \cdots i_{2r})}\}$ as members where $1 \leq i_1 < i_2 < \cdots < i_{2r} \leq n$ for $r = 1, 2, \ldots, [n/2]$ and $[n/2]$ is the integer closest to but not larger than $n/2$. Then we can see that the set $\{\{\tau\}, \{C\}\}$ satisfies the following conditions:

1. For any collections $\mathscr{U}_1$ and $\mathscr{U}_2$ in $\{\{\tau\}, \{C\}\}$,

$$\mathscr{U}_1 \otimes \mathscr{U}_2 = \mathscr{U}_2 \otimes \mathscr{U}_1 \quad \text{is in } \{\{\tau\}, \{C\}\}$$

2. There exists a collection $\mathcal{U}_0 = \{C\}$ in $\{\{\tau\}, \{C\}\}$ such that for any collection $\mathcal{U}$ in $\{\{\tau\}, \{C\}\}$,

$$\mathcal{U} \otimes \mathcal{U}_0 = \mathcal{U}$$

3. For any collection $\mathcal{U}$ in $\{\{\tau\}, \{C\}\}$,

$$\mathcal{U} \otimes \mathcal{U} = \mathcal{U}_0 \qquad \text{(self-inverse)}$$

4. For any collection $\mathcal{U}_1$, $\mathcal{U}_2$, and $\mathcal{U}_3$ in $\{\{\tau\}, \{C\}\}$,

$$(\mathcal{U}_1 \otimes \mathcal{U}_2) \otimes \mathcal{U}_3 = \mathcal{U}_1 \otimes (\mathcal{U}_2 \otimes \mathcal{U}_3)$$

Hence $\{\{\tau\}, \{C\}\}$ is an Abelian group under the ring product.

PROBLEMS

1. Draw linear graphs (if possible) in which the degree of vertices are (*a*) 2, 2, 2, 2, 2, (*b*) 2, 3, 4, 5, 6, and (*c*) 3, 3, 4, 5, 5, 5, 6.

2. Can you find an open edge train which contains all edges in each of the following linear graphs?

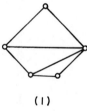

(1)   (2)   (3)

Fig. P-1-2.

3. Prove that a connected linear graph having n vertices contains at least $n - 1$ edges.

4. Prove that a connected linear graph having n vertices which has more than $n - 1$ edges has at least one circuit.

5. Suppose a connected linear graph has $n\ (> 1)$ vertices. Prove that if G consists of $n - 1$ edges, there exists at least one vertex of odd degree.

6. Let P be a path. Prove that there exists exactly one path between any two vertices in P.

7. Let G be a connected linear graph. Let i and j be the only vertices in G which are of odd degree. Prove that for any path P between i and j, $G - P$ is an Euler graph.

8. Let G be a linear graph containing no self-loops. Prove that a vertex p is a cut vertex if and only if there exists a subgraph g of G such that every path between a vertex in g and a vertex in $G - g$ passes vertex p.

9. Let P be a path between vertices i and j in a connected linear graph G. Also let $\{C\}$ be a collection of all possible circuits in G and the empty set. Prove that, in general, collection $\{P \oplus C; \quad C \in \{C\}\}$ will not contain all possible paths between i and j.

10. Let $\{P_{ij}\}$ be a collection of all possible paths between vertices i and j in a *separable* linear graph G. Find conditions that min $\{P_r \oplus P_s; \quad P_r, P_s \in \{P_{ij}\}\}$ is a collection of all possible circuits of G.

11. A *complete graph* is a connected linear graph in which there exists exactly one edge between any two vertices. Let G be a complete graph of order n. Then prove or disprove that
(a) There are

$$\sum_{k=3}^{n} \frac{(k-1)!}{2} \binom{n}{k}$$

circuits,
(b) The number of circuits which contain an edge e is

$$\sum_{k=1}^{n-2} K! \binom{n-2}{K}$$

and
(c) The number of paths between any two vertices is

$$\sum_{k=1}^{n-2} \frac{(n-2)!}{(n-k-2)!} + 1$$

12. Prove Theorem 1-7-1.

13. Prove Theorem 1-7-2.

14. Prove Theorem 1-7-3.

15. Obtain collection $\{E\}$ of all possible Euler graphs in the linear graph in Fig. P-1-15.

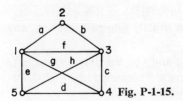

Fig. P-1-15.

16. Obtain collection $\{C\}$ of all possible circuits and the empty set in a linear graph in Fig. P-1-15.

17. Obtain collection $\{P_{12}\}$ of all possible paths between vertices 1 and 2 in a linear graph in Fig. P-1-15.

18. Obtain collections $\{P_{12}\}$, $\{P_{23}\}$, and $\{P_{13}\}$ in a linear graph in Fig. P-1-15. Then show that

$$\{P_{12}\} \otimes \{P_{23}\} = \{P_{13}\}$$

19. Obtain $\{\tau_{(1234)}\}$ by (a) getting $\{P_{13}\}$ and $\{P_{24}\}$ and calculating $\{P_{13}\} \otimes \{P_{24}\}$, and by (b) getting $\{C\}$ then calculating $\{P_{[a]}\}$ and $\{P_{[c]}\}$ from which you can obtain $\{P_{[a]}\} \otimes \{P_{[c]}\} = \{\tau_{(1234)}\}$.

2

INCIDENCE SET AND CUT-SET

2-1 INCIDENCE SET

We have considered two collections of subgraphs that form groups under the ring sum. These collections are $\{E\}$ and $\{E, M_{ij}\}$. There are other subgraphs that form groups under the ring sum, called cut-sets and pseudo-cuts. The properties of cut-sets are very similar to those of circuits and the properties of pseudo-cuts are almost identical to those of paths. We are concerned here with the properties of cut-sets, but first we will study a special type of a cut-set called incidence sets. Because neither cut-sets nor incidence sets contain self-loops, we assume, for convenience, that a given linear graph contains no self-loops unless specified.

ASSUMPTION. *A linear graph is assumed to have no self-loops unless they are specified.*

Definition 2-1-1. A set of all edges which are incident at a vertex is called an *incidence set* corresponding to the vertex. The symbol $S(v)$ indicates the incidence set corresponding to vertex v.

From this definition, if there are no edges incident at a vertex v, the incidence set corresponding to the vertex is the empty set. Moreover, we can see that there is one incidence set for every vertex in a linear graph. For example, the linear graph in Fig. 2-1-1 consists of five vertices. Hence there

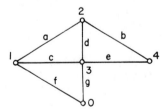

Fig. 2-1-1. A linear graph.

are five incidence sets. The incidence set $S(0)$ corresponding to vertex 0 is (f, g). The incidence sets $S(1)$, $S(2)$, $S(3)$, and $S(4)$ corresponding to vertices 1, 2, 3, and 4, respectively, are (a, c, f), (a, b, d), (c, d, e, g), and (b, e).

We know that the ring sum of two Euler graphs is an Euler graph. Hence the ring sum of two circuits is either a circuit or an edge disjoint union of circuits. On the other hand, the ring sum of two incidence sets is, in general, neither an incidence set nor an edge disjoint union of incidence sets. To see this, we consider incidence sets $S(v_1)$ and $S(v_2)$ of a linear graph G. In order that the ring sum of these incidence sets, $S(v_1) \oplus S(v_2)$, be an incidence set, every edge in $S(v_1) \oplus S(v_2)$ must be incident at a vertex v'. Furthermore, these edges in $S(v_1) \oplus S(v_2)$ must be only the edges incident at vertex v' in linear graph G. In other words, suppose

$$S(v_1) = (e_1, e_2, \ldots, e_p, a_1, a_2, \ldots, a_m) \text{ and } S(v_2) = (e_1, e_2, \ldots, e_p, b_1, b_2, \ldots, b_n)$$

Note that edges $e_1, e_2, \ldots, e_p$ are in both $S(v_1)$ and $S(v_2)$. Then $S(v_1) \oplus S(v_2) = (a_1, a_2, \ldots, a_m, b_1, b_2, \ldots, b_m)$. Since edges $a_1, a_2, \ldots, a_m$ are not incident at vertex v_2 [because they are not in $S(v_2)$], $S(v_1) \oplus S(v_2)$ is not the incidence set corresponding to vertex v_2. Similarly, $S(v_1) \oplus S(v_2)$ is not the incidence set corresponding to vertex v_1. Only when $S(v_1) \oplus S(v_2)$ is an incidence set is every edge in the set incident at a vertex other than v_1 and v_2 and is no other edge incident at the vertex in a given linear graph. Hence the given graph must have the linear graph shown in Fig. 2-1-2 as a maximal connected subgraph.

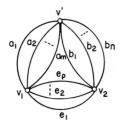

Fig. 2-1-2. Linear graph which has $S(v_1) \oplus S(v_2)$ as an incidence set.

Thus, in general, the ring sum of two incidence sets is not an incidence set. Suppose $S(v_1) \oplus S(v_2)$ is not an incidence set. Then, is $S(v_1) \oplus S(v_2)$ an edge disjoint union of incidence sets? If $S(v_1)$ and $S(v_2)$ have no edges in common, then it is obvious that $S(v_1) \oplus S(v_2)$ is an edge disjoint union of incidence sets. However, such a case is a special case. So we assume that $S(v_1)$ and $S(v_2)$ have edges in common, as we did in the previous discussion. In order that $S(v_1) \oplus S(v_2)$ be an edge disjoint union of incidence sets, $S(v_1) \oplus S(v_2)$ must be able to decompose into k sets $S(v_1')$, $S(v_2')$, \ldots, $S(v_k')$ for $k \geq 2$ such that every edge in $S(v_1) \oplus S(v_2)$ is in exactly one of these sets. Furthermore, set $S(v'_r)$ must be the incidence set corresponding to vertex v'_r in a given linear graph for all $r = 1, 2, \ldots, k$. Thus the given linear graph must contain the subgraph shown in Fig. 2-1-3 as a maximal connected subgraph. Hence we can conclude that, in general, the ring sum of two incidence sets is neither an incidence set nor an edge disjoint union of incidence sets.

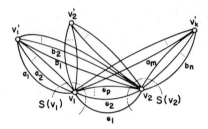

Fig. 2-1-3. Linear graph which has $S(v_1) \oplus S(v_2)$ as an edge disjoint union of incidence sets.

2-2   CUT-SET

In order to define cut-sets, we need a new terminology, called a rank of a linear graph.

Definition 2-2-1. The *rank of a linear graph* is $n_v - \rho$ where n_v is the number of vertices and ρ is the number of maximal connected subgraphs in the linear graph. For example, the rank of the linear graph in Fig. 2-1-1 is 4, that of the linear graph in Fig. 1-2-6 is 5, and that of the linear graph in Fig. 1-3-5a is 7.

Suppose we delete all edges in an incidence set $S(v_1)$ from a linear graph G. Then the resultant graph G' contains all vertices of G except vertex v_1. Thus if the number of maximal connected subgraphs in the resultant graph G' is the same as that of the original graph G, the rank of G' is one less than that of G.

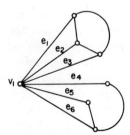

Fig. 2-2-1. A separable graph G.

However, there are linear graphs such that the ranks of the resultant graphs after deleting all edges in an incidence set will not be one less than those of the original graphs. For example, incidence set $S(v_1)$ of a linear graph G in Fig. 2-2-1 consists of edges e_1, e_2, e_3, e_4, e_5, and e_6. By deleting these edges, the resultant graph G' is that shown in Fig. 2-2-2. The rank of G is 6. The rank

Fig. 2-2-2. The graph obtained from G by deleting all edges in $S(v_1)$.

of G', on the other hand, is $n_v - \rho = 6 - 2 = 4$ which is two (not one) less than that of the original graph G.

A set of edges called a "cut-set" has the property that the deletion of all edges in the cut-set reduces the rank by exactly one. One of the well-known definitions of cut-sets is as follows.

Definition 2-2-2. A set S of edges of a linear graph G is a *cut-set* if S has two properties: (1) deletion of all edges in S from G reduces the rank by exactly one; and (2) any proper subset* of S does not have property (1).

Example 2-2-1. Consider set (b, c, d, f) of a linear graph in Fig. 2-1-1. We can see that (1) deleting all edges in (b, c, d, f) produces the linear graph as shown in Fig. 2-2-3a whose rank is $5 - 2 = 3$ (note that the rank of the original linear graph is 4) and (2) the deletion of some but not all edges in (b, c, d, f) keeps the linear graph connected. For example, deleting edges b, c, and d from the original graphs gives the graph shown in Fig. 2-2-3b, whose rank is 4. Hence the rank is not changed. Thus, by definition, the set (b, c, d, f) is a cut-set.

Consider a set (b, e, g) of the same linear graph. The deletion of all edges in (b, e, g) produces the graph shown in Fig. 2-2-3c, whose rank is $4 - 1 = 3$. Hence the deletion of all edges in this set reduces the rank by exactly one. However, there exists a proper subset (b, e) of (b, e, g) such that the deletion of all edges in the subset (b, e) also reduces the rank by exactly one, which

---

* A proper subset of a set S is a subset of S which is not identical with S.

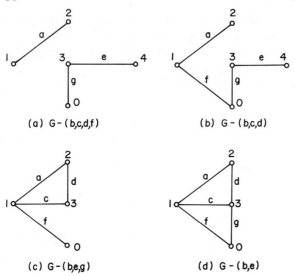

Fig. 2-2-3. Subgraphs of linear graph G. (a) $G - (b, c, d, f)$; (b) $G - (b, c, d)$; (c) $G - (b, e, g)$; (d) $G - (b, e)$.

can be seen from the resultant graph in Fig. 2-2-3d. Hence, by definition, (b, e, g) is not a cut-set.

Another example will be incidence set $S(v_1)$ of the linear graph in Fig. 2-2-1. Recall that the rank of the linear graph is 6, and the deletion of all edges in $S(v_1)$ given the resultant graph as shown in Fig. 2-2-2, whose rank is 4. Hence, by definition, $S(v_1)$ is not a cut-set.

Definition 2-2-3. A *complete graph* is a linear graph in which there exists exactly one edge between any pair of vertices in the linear graph.

For example, the linear graph in Fig. 2-2-4 is a complete graph consisting of five vertices. It is clear that a complete graph of more than two vertices is nonseparable.

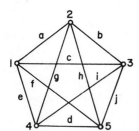

Fig. 2-2-4. A complete graph.

Let G_c be a complete graph. Let $\Omega_c = (v_1, v_2, \ldots, v_n)$ be the set of all vertices of G_c. The rank of G_c is therefore $n - 1$. Also let Ω_1 be a proper subset of Ω_c, and $\overline{\Omega}_1$ be the set of all vertices in Ω_c other than those in Ω_1. Hence

$$\Omega_1 \cup \overline{\Omega}_1 = \Omega_c \tag{2-2-1}$$

and

$$\Omega_1 \cap \overline{\Omega}_1 = \emptyset \tag{2-2-2}$$

For example, if $\Omega_1 = (v_1, v_2, \ldots, v_k)$ for $k < n$, then $\overline{\Omega}_1 = (v_{k+1}, v_{k+2}, \ldots, v_n)$.

Definition 2-2-4. The symbol $\mathscr{E}(\Omega_p \times \Omega_q)$ indicates a set of all edges each of which is connected from a vertex in Ω_p to a vertex in Ω_q.

For a nonoriented graph, $\mathscr{E}(\Omega_p \times \Omega_q)$ is a set of all edges each of which is connected between a vertex in Ω_p and a vertex in Ω_q. For example, Ω_c of a complete graph in Fig. 2-2-4 is $(1, 2, 3, 4, 5)$. If we choose $\Omega_1 = (1, 2, 3)$, then $\overline{\Omega}_1 = (4, 5)$. With these sets of vertices, we have

$$\mathscr{E}(\Omega_1 \times \Omega_1) = (a, b, c)$$
$$\mathscr{E}(\overline{\Omega}_1 \times \overline{\Omega}_1) = (d)$$
$$\mathscr{E}(\Omega_1 \times \overline{\Omega}_1) = (e, f, g, h, i, j)$$

and obviously

$$\mathscr{E}(\Omega_c \times \Omega_c) = (a, b, c, d, e, f, g, h, i, j)$$

which is all the edges in the given complete graph. The subgraph $\mathscr{E}(\Omega_1 \times \overline{\Omega}_1)$ is shown in Fig. 2-2-5.

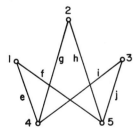

Fig. 2-2-5. Subgraph $\mathscr{E}(\Omega_1 \times \overline{\Omega}_1)$.

Now consider a subgraph $\mathscr{E}(\Omega_1 \times \Omega_1)$ of G_c. Since G_c is a complete graph, there exists an edge connected between any two vertices in Ω_1 and this edge will be in subgraph $\mathscr{E}(\Omega_1 \times \Omega_1)$. Hence there exists an edge between any two vertices in subgraph $\mathscr{E}(\Omega_1 \times \Omega_1)$, which means that $\mathscr{E}(\Omega_1 \times \Omega_1)$ is a complete graph. For example, suppose we choose $\Omega_1 = (1, 3, 4, 5)$ of the complete

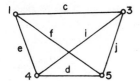

Fig. 2-2-6.   A subgraph.

graph in Fig. 2-2-4. Then $\mathscr{E}(\Omega_1 \times \Omega_1)$, as shown in Fig. 2-2-6, is obviously a complete graph consisting of four vertices.

Let Ω_1 and Ω_2 be proper subsets of Ω_c such that

$$\Omega_1 \cap \Omega_2 = \emptyset \qquad\qquad (2\text{-}2\text{-}3)$$

Let us consider subgraph $\mathscr{E}(\Omega_1 \times \Omega_2)$. Since subgraph $\mathscr{E}(\Omega_1 \times \Omega_2)$ consists only of edges each of which is connected between a vertex in Ω_1 and a vertex in Ω_2 in G_c, this subgraph has no edges between any two vertices in Ω_1. Thus subgraph $\mathscr{E}(\Omega_1 \times \Omega_2)$ is not a complete graph. The next question is whether subgraph $\mathscr{E}(\Omega_1 \times \Omega_2)$ is connected or not. Let v_1 be a vertex in Ω_1 and v_2 be a vertex in Ω_2. There exists an edge between v_1 and v_2 because G_c has an edge between v_1 and v_2 and $\mathscr{E}(\Omega_1 \times \Omega_2)$ is a set of all edges connected between a vertex in Ω_1 and a vertex in Ω_2. Hence there exists a path (consisting of one edge) between vertices v_1 and v_2. Thus we need to examine only whether $\mathscr{E}(\Omega_1 \times \Omega_2)$ has a path between any two vertices in Ω_1 and a path between any two vertices in Ω_2 in order to answer the question. Let v_{1a} and v_{1b} be two vertices in Ω_1. Also let v_2 be a vertex in Ω_2. Since, in subgraph $\mathscr{E}(\Omega_1 \times \Omega_2)$, there exist an edge between v_{1a} and v_2 and an edge between v_2 and v_{1b}, we can say that there exists at least one path between v_{1a} and v_{1b}. Hence subgraph $\mathscr{E}(\Omega_1 \times \Omega_2)$ has a path between any pair of vertices in Ω_1. Similarly, subgraph $\mathscr{E}(\Omega_1 \times \Omega_2)$ has a path between any pair of vertices in Ω_2. Thus we can conclude that subgraph $\mathscr{E}(\Omega_1 \times \Omega_2)$ is connected. For example, suppose we choose $\Omega_1 = (1, 2)$ and $\Omega_2 = (3, 4)$ of the complete graph in Fig. 2-2-4. Then subgraph $\mathscr{E}(\Omega_1 \times \Omega_2)$ is one shown in Fig. 2-2-7, which is clearly connected.

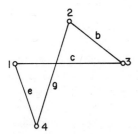

Fig. 2-2-7.   Subgraph $\mathscr{E}(\Omega_1 \times \Omega_2)$.

Let Ω_1 be a proper subset of Ω_c and $\bar{\Omega}_1 = \Omega_c - \Omega_1$. We have studied that subgraph $\mathscr{E}(\Omega_1 \times \Omega_1)$ is a complete graph. Similarly, subgraph $\mathscr{E}(\bar{\Omega}_1 \times \bar{\Omega}_1)$ is a complete graph. Also, because $\Omega_1 \cap \bar{\Omega}_1 = \emptyset$, complete graphs $\mathscr{E}(\Omega_1 \times \Omega_1)$ and $\mathscr{E}(\bar{\Omega}_1 \times \bar{\Omega}_1)$ have no edges and no vertices in common. For example, suppose $\Omega_1 = (1, 2)$ where the given complete graph is as shown in Fig. 2-2-4. Then $\bar{\Omega}_1 = (3, 4, 5)$ and the subgraph $\mathscr{E}(\Omega_1 \times \Omega_1) \cup \mathscr{E}(\bar{\Omega}_1 \times \bar{\Omega}_1)$ is shown in Fig. 2-2-8. Note that $\mathscr{E}(\Omega_1 \times \Omega_1)$ and $\mathscr{E}(\bar{\Omega}_1 \times \bar{\Omega}_1)$ have no edges and no vertices in common.

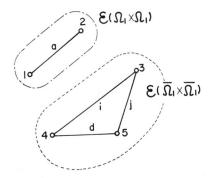

Fig. 2-2-8. A subgraph.

Consider three subgraphs $\mathscr{E}(\Omega_1 \times \Omega_1)$, $\mathscr{E}(\Omega_1 \times \bar{\Omega}_1)$, and $\mathscr{E}(\bar{\Omega}_1 \times \bar{\Omega}_1)$ of a complete graph G_c. Let e be any edge in G_c whose endpoints are p and q. In order to see which of these three subgraphs contains edge e, we need to consider the following three cases:

1. Suppose both vertices p and q are in Ω_1. Then edge e is in subgraph $\mathscr{E}(\Omega_1 \times \Omega_1)$ but neither $\mathscr{E}(\Omega_1 \times \bar{\Omega}_1)$ nor $\mathscr{E}(\bar{\Omega}_1 \times \bar{\Omega}_1)$.

2. Suppose both p and q are in $\bar{\Omega}_1$. Then it is clear that edge e is in $\mathscr{E}(\bar{\Omega}_1 \times \bar{\Omega}_1)$ but not in the other two subgraphs.

3. Suppose vertex p is in Ω_1 and vertex q is in $\bar{\Omega}_1$. Then edge e must be in $\mathscr{E}(\Omega_1 \times \bar{\Omega}_1)$ but is neither in $\mathscr{E}(\Omega_1 \times \Omega_1)$ nor in $\mathscr{E}(\bar{\Omega}_1 \times \bar{\Omega}_1)$.

Thus we can conclude that every edge in G_c is in exactly one of subgraphs $\mathscr{E}(\Omega_1 \times \Omega_1)$, $\mathscr{E}(\Omega_1 \times \bar{\Omega}_1)$, and $\mathscr{E}(\bar{\Omega}_1 \times \bar{\Omega}_1)$. That is,

$$\mathscr{E}(\Omega_1 \times \Omega_1) \cup \mathscr{E}(\Omega_1 \times \bar{\Omega}_1) \cup \mathscr{E}(\bar{\Omega}_1 \times \bar{\Omega}_1)$$
$$= \mathscr{E}(\Omega_1 \times \Omega_1) \oplus \mathscr{E}(\Omega_1 \times \bar{\Omega}_1) \oplus \mathscr{E}(\bar{\Omega}_1 \times \bar{\Omega}_1) = \mathscr{E}(\Omega_c \times \Omega_c) \quad (2\text{-}2\text{-}4)$$

Recall that $\mathscr{E}(\Omega_c \times \Omega_c)$ is a given complete graph G_c by definition. Hence, if we delete all edges in $\mathscr{E}(\Omega_1 \times \bar{\Omega}_1)$ from G_c, we will have $\mathscr{E}(\Omega_1 \times \Omega_1) \cup \mathscr{E}(\bar{\Omega}_1 \times \bar{\Omega}_1)$. That is,

$$\mathscr{E}(\Omega_c \times \Omega_c) - \mathscr{E}(\Omega_1 \times \bar{\Omega}_1) = \mathscr{E}(\Omega_1 \times \Omega_1) \cup \mathscr{E}(\bar{\Omega}_1 \times \bar{\Omega}_1) \quad (2\text{-}2\text{-}5)$$

Now we are ready to show that the set $\mathscr{E}(\Omega_1 \times \bar{\Omega}_1)$ is a cut-set. By the definition of cut-sets, we need to show that the deletion of all edges in $\mathscr{E}(\Omega_1 \times \bar{\Omega}_1)$ reduces the rank by exactly one. Furthermore, we must show that the deletion of some but not all edges in $\mathscr{E}(\Omega_1 \times \bar{\Omega}_1)$ from G_c does not reduce the rank.

Suppose Ω_1 consists of at least two vertices and $n \geq 4$. From Eq. 2-2-5, we can see that the deletion of all edges in the set $\mathscr{E}(\Omega_1 \times \bar{\Omega}_1)$ from G_c produces the subgraph $\mathscr{E}(\Omega_1 \times \Omega_1) \cup \mathscr{E}(\bar{\Omega}_1 \times \bar{\Omega}_1)$. Hence the resultant graph contains all vertices of G_c. We have seen that the subgraphs $\mathscr{E}(\Omega_1 \times \Omega_1)$ and $\mathscr{E}(\bar{\Omega}_1 \times \bar{\Omega}_1)$ have no vertices in common. Hence the subgraph $\mathscr{E}(\Omega_1 \times \Omega_1) \cup \mathscr{E}(\bar{\Omega}_1 \times \bar{\Omega}_1)$ consists of two maximal connected subgraphs. Thus the rank of the resultant graph $\mathscr{E}(\Omega_1 \times \Omega_1) \cup \mathscr{E}(\bar{\Omega}_1 \times \bar{\Omega}_1)$ is $n - 2$. Note that the rank of G_c is $n - 1$. This shows that the deletion of all edges in the set $\mathscr{E}(\Omega_1 \times \bar{\Omega}_1)$ reduces the rank by exactly one.

Since any edge in $\mathscr{E}(\Omega_1 \times \bar{\Omega}_1)$ is connected between a vertex in Ω_1 and a vertex in $\bar{\Omega}_1$, if we add any edge in the set $\mathscr{E}(\Omega_1 \times \bar{\Omega}_1)$ to the resultant graph, these two maximal connected subgraphs $\mathscr{E}(\Omega_1 \times \Omega_1)$ and $\mathscr{E}(\bar{\Omega}_1 \times \bar{\Omega}_1)$ will be joined together to become a connected graph whose rank is clearly $n - 1$. Thus the deletion of some but not all edges in $\mathscr{E}(\Omega_1 \times \bar{\Omega}_1)$ from G_c will not reduce its rank. Thus, by definition, the set $\mathscr{E}(\Omega_1 \times \bar{\Omega}_1)$ is a cut-set.

We have assumed that Ω_1 and $\bar{\Omega}_1$ consist of at least two vertices. Now we will consider the case when Ω_1 consists of one vertex, say v_1, that is $\Omega_1 = (v_1)$. Then the set $\mathscr{E}[(v_1) \times (\bar{v}_1)]$ is the set of all edges which are incident at vertex v_1, which means that this set is the incidence set corresponding to vertex v_1. Because $\mathscr{E}[(v_1) \times (v_1)] = \emptyset$, we can see from Eq. 2-2-5 that the deletion of all edges in this incidence set from G_c produces a connected graph $\mathscr{E}[(\bar{v}_1) \times (\bar{v}_1)]$, which consists of $n - 1$ vertices. Thus the rank of the resultant graph is one less than that of G_c. Furthermore, by inserting any edge in this incidence set $\mathscr{E}[(v_1) \times (\bar{v}_1)]$ to the resultant graph $\mathscr{E}[(\bar{v}_1) \times (\bar{v}_1)]$, the result is a connected linear graph containing vertex v_1. Hence the rank of this linear graph is $n - 1$. Thus, by the definition of cut-sets, incidence set $\mathscr{E}[(v_1) \times (\bar{v}_1)]$ is a cut-set. In other words, any incidence set of a complete graph is a cut-set.

Theorem 2-2-1. Let Ω_1 be a proper subset of Ω_c of a complete graph. Also set $\bar{\Omega}_1 = \Omega_c - \Omega_1$. Then $\mathscr{E}(\Omega_1 \times \bar{\Omega}_1)$ is a cut-set.

Example 2-2-2. Consider the complete graph in Fig. 2-2-4. Note that $\Omega_c = (1, 2, 3, 4, 5)$. Let $\Omega_1 = (1, 4)$. Then $\bar{\Omega}_1 = (2, 3, 5)$ and $\mathscr{E}(\Omega_1 \times \bar{\Omega}_1) = (a, c, d, f, g, i)$. The deletion of all edges in $\mathscr{E}(\Omega_1 \times \bar{\Omega}_1)$ gives the linear graph shown in Fig. 2-2-9, from which we can see clearly that $\mathscr{E}(\Omega_1 \times \bar{\Omega}_1)$ is a cut-set.

Let $\Omega_2 = (1)$. Then $\bar{\Omega}_2 = (2, 3, 4, 5)$ and $\mathscr{E}(\Omega_2 \times \bar{\Omega}_2) = (a, c, e, f)$, which is an incidence set as well as a cut-set.

Let $\Omega_3 = (1, 3)$. Then $\bar{\Omega}_3 = (2, 4, 5)$ and $\mathscr{E}(\Omega_3 \times \bar{\Omega}_3) = (a, b, e, f, i, j)$.

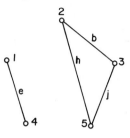

Fig. 2-2-9. $G_c - \mathscr{E}(\Omega_1 \times \bar{\Omega}_1)$.

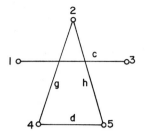

Fig. 2-2-10. A linear graph obtained from a complete graph by deleting a cut-set.

The resultant graph obtained by deleting all edges in $\mathscr{E}(\Omega_3 \times \bar{\Omega}_3)$ is shown in Fig. 2-2-10.

2-3 RING SUM OF CUT-SETS

In Chapter 1 we learned that the collection $\{E\}$ of circuits, edge disjoint unions of circuits, and the empty set of a linear graph is an Abelian group under the ring sum. Here we will see that the collection of cut-sets, edge disjoint unions of cut-sets, and the empty set is also an Abelian group under the ring sum. But first let us discuss a decomposition of a set of edges, which is necessary before we study the important property of the ring sum of cut-sets.

Consider proper nonempty disjoint subsets Ω_a, Ω_b, Ω_c, and Ω_d of set Ω of vertices in a linear graph. That is,

$$\Omega_p \subset \Omega \qquad (2\text{-}3\text{-}1)$$

$$\Omega_p \neq \emptyset \qquad (2\text{-}3\text{-}2)$$

for $p \in (a, b, c, d)$, and

$$\Omega_p \cap \Omega_q = \emptyset \qquad (2\text{-}3\text{-}3)$$

for $p, q \in (a, b, c, d)$ and $p \neq q$.

Let

$$\Omega_A = \Omega_a \cup \Omega_b \qquad (2\text{-}3\text{-}4)$$

and

$$\Omega_B = \Omega_c \cup \Omega_d \qquad (2\text{-}3\text{-}5)$$

Then

$$\mathscr{E}(\Omega_A \times \Omega_B) = \mathscr{E}(\Omega_a \cup \Omega_b \times \Omega_c \cup \Omega_d) \qquad (2\text{-}3\text{-}6)$$

We can decompose $\mathscr{E}(\Omega_a \cup \Omega_b \times \Omega_c \cup \Omega_d)$ into the following four sets:

1. The set $\mathscr{E}(\Omega_a \times \Omega_c)$ of all edges connected between a vertex in Ω_a and a vertex in Ω_c.

2. Set $\mathscr{E}(\Omega_a \times \Omega_d)$, consisting of all edges connected between a vertex in Ω_a and a vertex in Ω_d.

3. $\mathscr{E}(\Omega_b \times \Omega_c)$ is a set of all edges connected between a vertex in Ω_b and a vertex in Ω_c.

4. Set $\mathscr{E}(\Omega_b \times \Omega_d)$ consists of all edges connected between a vertex in Ω_b and a vertex in Ω_d.

In other words, $\mathscr{E}(\Omega_a \cup \Omega_b \times \Omega_c \cup \Omega_d)$ can be expressed as

$$
\begin{aligned}
\mathscr{E}(\Omega_a \cup \Omega_b &\times \Omega_c \cup \Omega_d) \\
&= \mathscr{E}(\Omega_a \times \Omega_c) \cup \mathscr{E}(\Omega_a \cup \Omega_d) \cup \mathscr{E}(\Omega_b \times \Omega_c) \cup \mathscr{E}(\Omega_b \times \Omega_d) \quad (2\text{-}3\text{-}7)
\end{aligned}
$$

It can be seen that any two of sets $\mathscr{E}(\Omega_a \times \Omega_c)$, $\mathscr{E}(\Omega_a \times \Omega_d)$, $\mathscr{E}(\Omega_b \times \Omega_c)$, and $\mathscr{E}(\Omega_b \times \Omega_d)$ have no edges in common. Hence Eq. 2-3-7 can be written as

$$
\begin{aligned}
\mathscr{E}(\Omega_a \cup \Omega_b &\times \Omega_c \cup \Omega_d) \\
&= \mathscr{E}(\Omega_a \times \Omega_c) \oplus \mathscr{E}(\Omega_a \times \Omega_d) \oplus \mathscr{E}(\Omega_b \times \Omega_c) \oplus \mathscr{E}(\Omega_b \times \Omega_d) \quad (2\text{-}3\text{-}8)
\end{aligned}
$$

Similarly, $\mathscr{E}(\Omega_A \times \Omega_A)$ can be expanded:

$$
\begin{aligned}
\mathscr{E}(\Omega_A \times \Omega_A) &= \mathscr{E}(\Omega_a \cup \Omega_b \times \Omega_a \cup \Omega_b) \\
&= \mathscr{E}(\Omega_a \times \Omega_a) \oplus \mathscr{E}(\Omega_a \times \Omega_b) \oplus \mathscr{E}(\Omega_b \times \Omega_b) \quad (2\text{-}3\text{-}9)
\end{aligned}
$$

Note that

$$
\mathscr{E}(\Omega_a \times \Omega_b) = \mathscr{E}(\Omega_b \times \Omega_a) \qquad (2\text{-}3\text{-}10)
$$

because the linear graph considered here is nonoriented.

Now we are ready to study the ring sum of two cut-sets in a complete graph. We divide Ω_c into four disjoint sets Ω_{11}, Ω_{12}, Ω_{21}, and Ω_{22} such that any vertex in Ω_c will be in exactly one of these four sets. That is,

$$
\Omega_{11} \cup \Omega_{12} \cup \Omega_{21} \cup \Omega_{22} = \Omega_c \qquad (2\text{-}3\text{-}11)
$$

and

$$
\Omega_{pq} \cap \Omega_{rs} = \emptyset \qquad (2\text{-}3\text{-}12)
$$

for $p, q, r, s = 1, 2$ except $(p, q) = (r, s)$.

Consider set $\Omega_1 = \Omega_{11} \cup \Omega_{12}$ and set $\Omega_2 = \Omega_{11} \cup \Omega_{21}$. Hence $\bar{\Omega}_1 = \Omega_{21} \cup \Omega_{22}$ and $\bar{\Omega}_2 = \Omega_{12} \cup \Omega_{22}$ as shown in Fig. 2-3-1.

The two cut-sets S_1 and S_2 are $S_1 = \mathscr{E}(\Omega_1 \times \bar{\Omega}_1)$ and $S_2 = \mathscr{E}(\Omega_2 \times \bar{\Omega}_2)$. From Eq. 2-2-4, these cut-sets can be expressed as

$$S_1 = \mathscr{E}(\Omega_1 \times \bar{\Omega}_1) = \mathscr{E}(\Omega_c \times \Omega_c) \oplus \mathscr{E}(\Omega_1 \times \Omega_1) \oplus \mathscr{E}(\bar{\Omega}_1 \times \bar{\Omega}_1) \quad (2\text{-}3\text{-}13)$$

and

$$S_2 = \mathscr{E}(\Omega_2 \times \bar{\Omega}_2) = \mathscr{E}(\Omega_c \times \Omega_c) \oplus \mathscr{E}(\Omega_2 \times \Omega_2) \oplus \mathscr{E}(\bar{\Omega}_2 \times \bar{\Omega}_2) \quad (2\text{-}3\text{-}14)$$

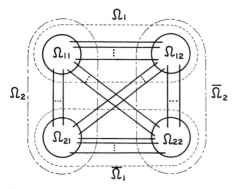

Fig. 2-3-1. Sets of vertices.

We assume that these two cut-sets are distinct because if they are the same, the ring sum of these two gives the empty set. From Eqs. 2-3-13 and 2-3-14, the ring sum of these two cut-sets $S_1 \oplus S_2$ is

$$S_1 \oplus S_2 = \mathscr{E}(\Omega_1 \times \Omega_1) \oplus \mathscr{E}(\bar{\Omega}_1 \times \bar{\Omega}_1) \oplus \mathscr{E}(\Omega_2 \times \Omega_2) \oplus \mathscr{E}(\bar{\Omega}_2 \times \bar{\Omega}_2)$$

$$(2\text{-}3\text{-}15)$$

By substituting $\Omega_1 = \Omega_{11} \cup \Omega_{12}$, $\bar{\Omega}_1 = \Omega_{21} \cup \Omega_{22}$, $\Omega_2 = \Omega_{11} \cup \Omega_{21}$, and $\bar{\Omega}_2 = \Omega_{12} \cup \Omega_{22}$ in Eq. 2-3-15 and using the relationship given by Eqs. 2-3-8 and 2-3-9, we have

$$\begin{aligned} S_1 \oplus S_2 &= \mathscr{E}(\Omega_{11} \cup \Omega_{12} \times \Omega_{11} \cup \Omega_{12}) \oplus \mathscr{E}(\Omega_{21} \cup \Omega_{22} \times \Omega_{21} \cup \Omega_{22}) \\ &\quad \oplus \mathscr{E}(\Omega_{11} \cup \Omega_{21} \times \Omega_{11} \cup \Omega_{21}) \oplus \mathscr{E}(\Omega_{12} \cup \Omega_{22} \times \Omega_{12} \cup \Omega_{22}) \\ &= \mathscr{E}(\Omega_{11} \times \Omega_{12}) \oplus \mathscr{E}(\Omega_{21} \times \Omega_{22}) \oplus \mathscr{E}(\Omega_{11} \times \Omega_{21}) \oplus \mathscr{E}(\Omega_{12} \times \Omega_{22}) \end{aligned}$$

$$(2\text{-}3\text{-}16)$$

It can be seen that any two of sets $\mathscr{E}(\Omega_{11} \times \Omega_{12}), \mathscr{E}(\Omega_{21} \times \Omega_{22}), \mathscr{E}(\Omega_{11} \times \Omega_{21})$, and $\mathscr{E}(\Omega_{12} \times \Omega_{22})$ have no edges in common. Hence we can use Eq. 2-3-8 to change Eq. 2-3-16 to

$$S_1 \oplus S_2 = \mathscr{E}(\Omega_{11} \cup \Omega_{22} \times \Omega_{21} \cup \Omega_{12}) = \mathscr{E}(\Omega_3 \times \bar{\Omega}_3) \quad (2\text{-}3\text{-}17)$$

where $\Omega_3 = \Omega_{11} \cup \Omega_{22}$. Hence, by Theorem 2-2-1, the ring sum of two cut-sets S_1 and S_2 is a cut-set.

Theorem 2-3-1.   The ring sum of two cut-sets of a complete graph is a cut-set.

Example 2-2-1.   Consider the complete graph in Fig. 2-2-4. Suppose we choose $\Omega_1 = (1, 2, 3)$ and $\Omega_2 = (1, 2, 4)$. Then $\bar{\Omega}_1 = (4, 5)$ and $\bar{\Omega}_2 = (3, 5)$. Also, we can see that $\Omega_{11} = (1, 2)$, $\Omega_{12} = (3)$, $\Omega_{21} = (4)$ and $\Omega_{22} = (5)$ in order that $\Omega_1 = \Omega_{11} \cup \Omega_{12}$ and $\Omega_2 = \Omega_{11} \cup \Omega_{21}$. Then, from the given graph,

$$S_1 = \mathscr{E}(\Omega_1 \times \bar{\Omega}_1) = (e, f, g, h, i, j)$$

and

$$S_2 = \mathscr{E}(\Omega_2 \times \bar{\Omega}_2) = (b, c, d, f, h, i)$$

Hence

$$S_1 \oplus S_2 = (b, c, d, e, g, j)$$

This must be equal to the cut-set given by Eq. 2-3-17 where $\Omega_3 = \Omega_{11} \cup \Omega_{22} = (1, 2, 5)$. We have

$$\mathscr{E}(\Omega_3 \times \bar{\Omega}_3) = (b, c, d, e, g, j)$$

obtained directly from the linear graph in Fig. 2-2-4, which is exactly the same as $S_1 \oplus S_2$ shown above.

Definition 2-3-1.   An *edge disjoint union of cut-sets* is a union of cut-sets no two of which have edges in common.

For example, two cut-sets (a, c, f) and (b, e) of the linear graph in Fig. 2-1-1 have no edges in common. Hence set (a, b, c, e, f) is an edge disjoint union of cut-sets.

Now we are going to study a property of a set obtained by the ring sum of two cut-sets of a linear graph when the linear graph is not a complete graph. Let Ω be a set of all vertices in a linear graph G. Let Ω_1 be a proper subset of Ω and Ω_2 be another proper subset of Ω which is not equal to Ω_1. Then two cut-sets S_1 and S_2 with these subsets are

$$S_1 = \mathscr{E}(\Omega_1 \times \bar{\Omega}_1) \tag{2-3-18}$$

and

$$S_2 = \mathscr{E}(\Omega_2 \times \bar{\Omega}_2) \tag{2-3-19}$$

Let $\Omega_1 = \Omega_{11} \cup \Omega_{12}$, $\bar{\Omega}_1 = \Omega_{21} \cup \Omega_{22}$, $\Omega_2 = \Omega_{11} \cup \Omega_{21}$, and $\bar{\Omega}_2 = \Omega_{12} \cup \Omega_{22}$ where Ω_{11}, Ω_{12}, Ω_{21}, and Ω_{22} have no vertices in common, as shown in Fig. 2-3-1. Note that the linear graph in Fig. 2-3-1 is assumed to be a non-complete graph at this time. Since the representations of S_1 and S_2 will hold for a noncomplete graph as well as for a complete graph, the result of the ring sum of these two cut-sets $S_1 \oplus S_2$ will be the same as that given by Eq. 2-3-17. Hence we have

$$S_1 \oplus S_2 = \mathscr{E}(\Omega_{11} \cup \Omega_{22} \times \Omega_{21} \cup \Omega_{12})$$
$$= \mathscr{E}(\Omega \times \Omega) - [\mathscr{E}(\Omega_{11} \cup \Omega_{22} \times \Omega_{11} \cup \Omega_{22})$$
$$\cup \mathscr{E}(\Omega_{21} \cup \Omega_{12} \times \Omega_{21} \cup \Omega_{12})] \qquad (2\text{-}3\text{-}20)$$

or

$$\mathscr{E}(\Omega \times \Omega) - S_1 \oplus S_2$$
$$= \mathscr{E}(\Omega_{11} \cup \Omega_{22} \times \Omega_{11} \cup \Omega_{22}) \cup \mathscr{E}(\Omega_{21} \cup \Omega_{12} \times \Omega_{21} \cup \Omega_{12}) \quad (2\text{-}3\text{-}21)$$

This shows that the deletion of all edges in $S_1 \oplus S_2$ from a linear graph G gives two subgraphs $\mathscr{E}(\Omega_{11} \cup \Omega_{22} \times \Omega_{11} \cup \Omega_{22})$ and $\mathscr{E}(\Omega_{21} \cup \Omega_{12} \times \Omega_{21} \cup \Omega_{12})$. When we are dealing with a complete graph, these subgraphs are themselves complete graphs. However, because G is not a complete graph by assumption, there is no guarantee that $\mathscr{E}(\Omega_{11} \cup \Omega_{22} \times \Omega_{11} \cup \Omega_{22})$ and $\mathscr{E}(\Omega_{21} \cup \Omega_{12} \times \Omega_{21} \cup \Omega_{12})$ are complete graphs. As a matter of fact, these subgraphs may be separated even if a given linear graph G is nonseparable, as shown in the next example.

Example 2-3-2. Consider the linear graph G in Fig. 2-3-2 in which each

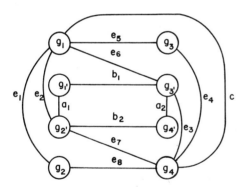

Fig. 2-3-2. A nonseparable graph.

circle represents a connected subgraph. Consider two cut-sets S_1 and S_2 where

$$S_1 = (e_1, e_2, e_3, e_4, a_1, a_2, c)$$

and

$$S_2 = (e_5, e_6, e_7, e_8, b_1, b_2, c)$$

The ring sum of these cut-sets is

$$S_1 \oplus S_2 = (e_1, e_2, e_3, e_4, e_5, e_6, e_7, e_8, a_1, a_2, b_1, b_2)$$

By defining Ω_{11} = (all vertices in g_1 and $g_{1'}$), Ω_{12} = (all vertices in g_3 and $g_{3'}$), Ω_{21} = (all vertices in g_2 and $g_{2'}$), and Ω_{22} = (all vertices in g_4 and $g_{4'}$), we can express these cut-sets as

$$S_1 = \mathscr{E}(\Omega_{11} \cup \Omega_{12} \times \Omega_{21} \cup \Omega_{22})$$

and

$$S_2 = \mathscr{E}(\Omega_{11} \cup \Omega_{21} \times \Omega_{12} \cup \Omega_{22})$$

Thus, by Eq. 2-3-17, the ring sum of S_1 and S_2 becomes

$$S_1 \oplus S_2 = \mathscr{E}(\Omega_{11} \cup \Omega_{22} \times \Omega_{21} \cup \Omega_{12})$$

or, by Eq. 2-3-21,

$$\mathscr{E}(\Omega \times \Omega) - S_1 \oplus S_2$$
$$= \mathscr{E}(\Omega_{11} \cup \Omega_{22} \times \Omega_{11} \cup \Omega_{22}) \cup \mathscr{E}(\Omega_{21} \cup \Omega_{12} \times \Omega_{21} \cup \Omega_{12})$$

where Ω is the set of all vertices in G. The linear graph shown in Fig. 2-3-3 is

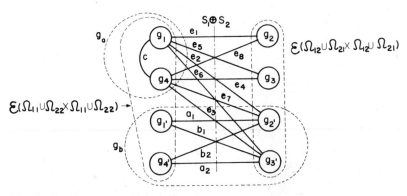

Fig. 2-3-3. A nonseparable graph.

obtained from the given linear graph by relocating subgraphs so that subgraph $\mathscr{E}(\Omega_{11} \cup \Omega_{22} \times \Omega_{11} \cup \Omega_{22})$ will be on one side and subgraph $\mathscr{E}(\Omega_{21} \cup \Omega_{12} \times \Omega_{21} \cup \Omega_{12})$ will be on the other side of the drawing.

When we delete all edges in $S_1 \oplus S_2$, we will have the linear graph of Fig. 2-3-4. Note that subgraph $\mathscr{E}(\Omega_{11} \cup \Omega_{22} \times \Omega_{11} \cup \Omega_{22})$ consists of three maximal connected subgraphs g_a, $g_{1'}$, and $g_{4'}$, and subgraph $\mathscr{E}(\Omega_{21} \cup \Omega_{12} \times$

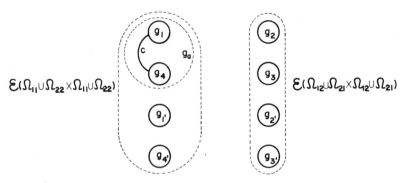

$\mathcal{E}(\Omega_{11}\cup\Omega_{22}\times\Omega_{11}\cup\Omega_{22})$

$\mathcal{E}(\Omega_{12}\cup\Omega_{21}\times\Omega_{12}\cup\Omega_{21})$

Fig. 2-3-4. After deleting $S_1 \oplus S_2$.

$\Omega_{21} \cup \Omega_{12})$ consists of four maximal connected subgraphs g_2, g_3, $g_{2'}$, and $g_{3'}$. Hence it is clear that $S_1 \oplus S_2$ is not a cut-set.

Now consider a subset S_a of $S_1 \oplus S_2$ which consists of all edges connected to vertices in maximal connected subgraph g_a, that is,

$$S_a = (e_1, e_2, e_3, e_4, e_5, e_6, e_7, e_8)$$

This subset S_a is not a cut-set because the deletion of all edges in S_a produces a linear graph which consists of four maximal connected subgraphs g_a, g_b, g_2, and g_3, as shown in Fig. 2-3-5.

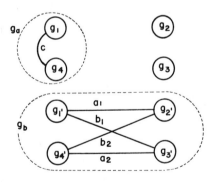

Fig. 2-3-5. After deleting S_a.

Consider a subset S_{a1} of S_a which consists of all edges connected to vertices in g_2, that is,

$$S_{a1} = (e_1, e_8)$$

The deletion of all edges in S_{a1} produces a linear graph consisting of two

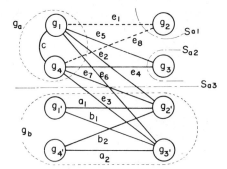

Fig. 2-3-6. After deleting S_{a1}.

maximal connected subgraphs as shown in Fig. 2-3-6. Hence we can easily show that S_{a1} is a cut-set. Similarly, a subset $S_{a2} = (e_4, e_5)$ of S_a consisting of all edges which are connected to g_3 is a cut-set, and a subset $S_{a3} = (e_2, e_3, e_6, e_7)$ of S_a which consists of all edges incident at vertices in g_b is a cut-set. Thus S_a is an edge disjoint union of cut-sets S_{a1}, S_{a2}, and S_{a3}. We can also see that a subset S_b of $S_1 \oplus S_2$ consisting of all edges that are connected to $g_{1'}$ is a cut-set and subset $S_c = (a_2, b_2)$ of $S_1 \oplus S_2$ consisting of all edges which are incident at vertices in $g_{4'}$ is also a cut-set. Thus $S_1 \oplus S_2$ is an edge disjoint union of cut-sets S_{a1}, S_{a2}, S_{a3}, S_b, and S_c.

To go back to a general case, from Eq. 2-3-21, it is clear that if subgraph $\mathscr{E}(\Omega_{11} \cup \Omega_{22} \times \Omega_{11} \cup \Omega_{22})$ and subgraph $\mathscr{E}(\Omega_{21} \cup \Omega_{12} \times \Omega_{21} \cup \Omega_{12})$ each is connected, then $S_1 \oplus S_2$ is a cut-set. Suppose subgraph $\mathscr{E}(\Omega_{11} \cup \Omega_{22} \times \Omega_{11} \cup \Omega_{22})$ is a separated graph consisting of k maximal connected subgraphs g_{a1}, $g_{a2}, \cdots, g_{ak}$. Let Ω_{ap} be the set of all vertices in g_{ap} for $p = 1, 2, \ldots, k$ as shown in Fig. 2-3-7. Remember that $\mathscr{E}(\Omega_{21} \cup \Omega_{12} \times \Omega_{21} \cup \Omega_{12})$ may also be

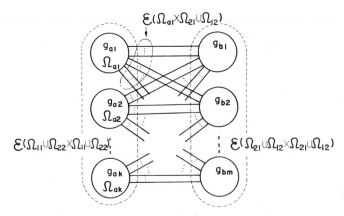

Fig. 2-3-7. Subgraphs $\mathscr{E}(\Omega_{11} \cup \Omega_{22} \times \Omega_{11} \cup \Omega_{22})$ and $\mathscr{E}(\Omega_{21} \cup \Omega_{12} \times \Omega_{21} \cup \Omega_{12})$.

separated. Note that every edge in $S_1 \oplus S_2 = \mathscr{E}(\Omega_{11} \cup \Omega_{22} \times \Omega_{12} \cup \Omega_{21})$ is connected to a vertex in one of $g_{a1}, g_{a2}, \ldots, g_{ak}$. We divide set $\mathscr{E}(\Omega_{11} \cup \Omega_{22} \times \Omega_{12} \cup \Omega_{21})$ into k subsets $S_{a1}, S_{a2}, \ldots, S_{ak}$ such that all edges in set S_{ap} ($1 \leq p \leq k$) are connected to vertices in g_{ap}. It is easily found that S_{ap} is equal to

$$S_{ap} = \mathscr{E}(\Omega_{ap} \times \Omega_{21} \cup \Omega_{12}) \tag{2-3-22}$$

Furthermore, we can see that every edge in set $\mathscr{E}(\Omega_{11} \cup \Omega_{22} \times \Omega_{12} \cup \Omega_{21})$ will be in exactly one of sets $S_{a1}, S_{a2}, \ldots, S_{ak}$. Hence set $\mathscr{E}(\Omega_{11} \cup \Omega_{22} \times \Omega_{12} \cup \Omega_{21})$ can be considered as an edge disjoint union of sets $S_{a1}, S_{a2}, \ldots, S_{ak}$. Since there are no edges connected between a vertex in Ω_{ap} and a vertex in Ω_{aq} for $1 \leq p \leq k$, set $\mathscr{E}(\Omega_{ap} \times \Omega_{12} \cup \Omega_{21})$ can be expressed as

$$
\begin{aligned}
&\mathscr{E}(\Omega_{ap} \times \Omega_{12} \cup \Omega_{21}) \\
&= \mathscr{E}(\Omega_{ap} \times \Omega_{12} \cup \Omega_{21} \cup \Omega_{a1} \cup \Omega_{a2} \cup \cdots \cup \Omega_{ap-1} \cup \Omega_{ap+1} \cup \cdots \cup \Omega_{ak})
\end{aligned}
\tag{2-3-23}
$$

However, by definition of $\bar{\Omega}_{ap}$, we can see that

$$\Omega_{12} \cup \Omega_{21} \cup \Omega_{a1} \cup \Omega_{a2} \cup \cdots \cup \Omega_{ap-1} \cup \Omega_{ap+1} \cup \cdots \cup \Omega_{ak} = \bar{\Omega}_{ap} \tag{2-3-24}$$

Hence

$$\mathscr{E}(\Omega_{ap} \times \Omega_{12} \cup \Omega_{21}) = \mathscr{E}(\Omega_{ap} \times \bar{\Omega}_{ap}) \tag{2-3-25}$$

for all $p = 1, 2, \ldots, k$. Thus $\mathscr{E}(\Omega_{11} \cup \Omega_{22} \times \Omega_{12} \cup \Omega_{21})$ can be expressed as

$$
\begin{aligned}
&\mathscr{E}(\Omega_{11} \cup \Omega_{22} \times \Omega_{12} \cup \Omega_{21}) \\
&= \mathscr{E}(\Omega_{a1} \times \bar{\Omega}_{a1}) \cup \mathscr{E}(\Omega_{a2} = \bar{\Omega}_{a2}) \cup \cdots \cup \mathscr{E}(\Omega_{ak} \times \bar{\Omega}_{ak}) \\
&= \mathscr{E}(\Omega_{a1} \times \bar{\Omega}_{a1}) \oplus \mathscr{E}(\Omega_{a2} \times \bar{\Omega}_{a2}) \oplus \cdots \oplus \mathscr{E}(\Omega_{ak} \times \bar{\Omega}_{ak})
\end{aligned}
\tag{2-3-26}
$$

By Eq. 2-2-4, set $\mathscr{E}(\Omega_{ap} \times \bar{\Omega}_{ap})$ can be expressed as

$$\mathscr{E}(\Omega_{ap} \times \bar{\Omega}_{ap}) = \mathscr{E}(\Omega \times \Omega) \oplus \mathscr{E}(\Omega_{ap} \times \Omega_{ap}) \oplus \mathscr{E}(\bar{\Omega}_{ap} \times \bar{\Omega}_{ap}) \tag{2-3-27}$$

or by Eq. 2-2-5 we have

$$\mathscr{E}(\Omega \times \Omega) - \mathscr{E}(\Omega_{ap} \times \bar{\Omega}_{ap}) = \mathscr{E}(\Omega_{ap} \times \Omega_{ap}) \cup \mathscr{E}(\bar{\Omega}_{ap} \times \bar{\Omega}_{ap}) \tag{2-3-28}$$

From this equation, we can see that if the subgraph $\mathscr{E}(\bar{\Omega}_{ap} \times \bar{\Omega}_{ap})$ is connected, then $\mathscr{E}(\Omega_{ap} \times \bar{\Omega}_{ap})$ is a cut-set. Note that subgraph $\mathscr{E}(\Omega_{ap} \times \Omega_{ap})$ is connected by assumption. Without the loss of generality, consider a set $\mathscr{E}(\Omega_{a1} \times \bar{\Omega}_{a1})$ as shown in Fig. 2-3-8. From Eq. 2-3-28, the deletion of all edges in $\mathscr{E}(\Omega_{a1} \times \bar{\Omega}_{a1})$ produces two subgraphs $\mathscr{E}(\Omega_{a1} \times \Omega_{a1})$ and $\mathscr{E}(\bar{\Omega}_{a1} \times \bar{\Omega}_{a1})$. We know that if $\mathscr{E}(\bar{\Omega}_{a1} \times \bar{\Omega}_{a1})$ is connected, $\mathscr{E}(\Omega_{a1} \times \bar{\Omega}_{a1})$ is a cut-set. Suppose $\mathscr{E}(\bar{\Omega}_{a1} \times \bar{\Omega}_{a1})$ is separated. Then the subgraph $\mathscr{E}(\Omega_{a1} \times \Omega_{a1})$ must be connected by edges in $\mathscr{E}(\Omega_{a1} \times \bar{\Omega}_{a1})$ as shown in Fig. 2-3-9 where $\mathscr{E}(\bar{\Omega}_{a1} \times \bar{\Omega}_{a1})$ is assumed to have n maximal connected

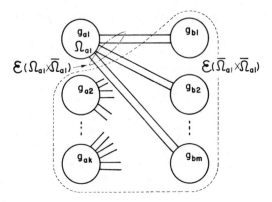

Fig. 2-3-8.   Subset $\mathscr{E}(\Omega_{a1} \times \bar{\Omega}_{a1})$.

subgraphs $g_{c1}, g_{c2}, \ldots, g_{cn}$. Let Ω_{cr} be the set of all vertices in maximal connected subgraphs g_{cr} for $r = 1, 2, \ldots, n$. Then $\mathscr{E}(\Omega_{a1} \times \bar{\Omega}_{a1})$ can be considered an edge disjoint union of sets $\mathscr{E}(\Omega_{a1} \times \Omega_{c1})$, $\mathscr{E}(\Omega_{a1} \times \Omega_{c2})$, \ldots, $\mathscr{E}(\Omega_{a1} \times \Omega_{cn})$, or

$$\begin{aligned}
\mathscr{E}(\Omega_{a1} \times \bar{\Omega}_{a1}) &= \mathscr{E}(\Omega_{a1} \times \Omega_{c1}) \cup \mathscr{E}(\Omega_{a1} \times \Omega_{c2}) \cup \cdots \cup \mathscr{E}(\Omega_{a1} \times \Omega_{cn}) \\
&= \mathscr{E}(\Omega_{a1} \times \Omega_{c1}) \oplus \mathscr{E}(\Omega_{a1} \times \Omega_{c2}) \oplus \cdots \oplus \mathscr{E}(\Omega_{a1} \times \Omega_{cn})
\end{aligned}$$
$$(2\text{-}3\text{-}29)$$

Since there exist no edges connected between a vertex in g_{cp} and a vertex in g_{cq} for $1 \le p < q \le n$, $\mathscr{E}(\Omega_{a1} \times \Omega_{cp})$ can be expressed as

$$\begin{aligned}
\mathscr{E}(\Omega_{a1} \times \Omega_{cp}) &= \mathscr{E}(\Omega_{a1} \cup \Omega_{c1} \cup \cdots \cup \Omega_{cp-1} \cup \Omega_{cp+1} \cup \cdots \cup \Omega_n \times \Omega_{cp}) \\
&= \mathscr{E}(\bar{\Omega}_{cp} \times \Omega_{cp})
\end{aligned}$$
$$(2\text{-}3\text{-}30)$$

where

$$\bar{\Omega}_{cp} = \Omega_{a1} \cup \Omega_{c1} \cup \cdots \cup \Omega_{cp-1} \cup \Omega_{cp+1} \cup \cdots \cup \Omega_{cn} \qquad (2\text{-}3\text{-}31)$$

By Eq. 2-2-5, we have

$$\mathscr{E}(\Omega \times \Omega) - \mathscr{E}(\bar{\Omega}_{cp} \times \Omega_{cp}) = \mathscr{E}(\Omega_{cp} \times \Omega_{cp}) \cup \mathscr{E}(\bar{\Omega}_{cp} \times \bar{\Omega}_{cp}) \quad (2\text{-}3\text{-}32)$$

From this equation, it is clear that the deletion of all edges in $\mathscr{E}(\bar{\Omega}_{cp} \times \Omega_{cp})$ from the original graph G gives two subgraphs $\mathscr{E}(\Omega_{cp} \times \Omega_{cp})$ and $\mathscr{E}(\bar{\Omega}_{cp} \times \bar{\Omega}_{cp})$. By assumption, $\mathscr{E}(\Omega_{cp} \times \Omega_{cp})$ is connected. From the linear graph in Fig. 2-3-9, we can see that $\mathscr{E}(\bar{\Omega}_{cp} \times \bar{\Omega}_{cp})$ is connected. Otherwise the original linear graph G is not connected. Thus $\mathscr{E}(\bar{\Omega}_{cp} \times \Omega_{cp})$ is a cut-set. Hence from Eq. 2-3-29, set $\mathscr{E}(\Omega_{a1} \times \bar{\Omega}_{a1})$ is an edge disjoint union of cut-sets. Therefore we can say that for every $p = 1, 2, \ldots, k$, $\mathscr{E}(\Omega_{ap} \times \bar{\Omega}_{ap})$ is

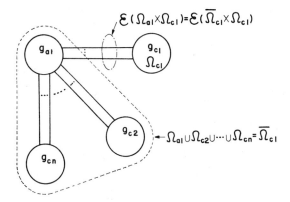

Fig. 2-3-9. Separable graph $\mathscr{E}(\overline{\Omega}_{a1} \times \overline{\Omega}_{a1})$.

either a cut-set or an edge disjoint union of cut-sets. Thus by Eq. 2-3-26, set $\mathscr{E}(\Omega_{11} \cup \Omega_{22} \times \Omega_{12} \cup \Omega_{21})$ is an edge disjoint union of cut-sets.

If we observe the requirements of two sets $\mathscr{E}(\Omega_1 \times \overline{\Omega}_1)$ and $\mathscr{E}(\Omega_2 \times \overline{\Omega}_2)$ to obtain the previous results, we can see that the sets $\mathscr{E}(\Omega_1 \times \overline{\Omega}_1)$ and $\mathscr{E}(\Omega_2 \times \overline{\Omega}_2)$ need not be cut-sets. In fact, the only requirements are that Ω_{11}, Ω_{12}, Ω_{21}, and Ω_{22} are distinct sets of vertices such that

1. $$\Omega_{11} \cup \Omega_{12} \cup \Omega_{21} \cup \Omega_{22} = \Omega \qquad (2\text{-}3\text{-}33)$$

2. Any two sets of Ω_{11}, Ω_{12}, Ω_{21}, and Ω_{22} have no edges in common.

3. $$\Omega_1 = \Omega_{11} \cup \Omega_{12} \quad \text{and} \quad \Omega_2 = \Omega_{11} \cup \Omega_{21} \qquad (2\text{-}3\text{-}34)$$

Thus either S_1 or S_2 or both S_1 and S_2 can be edge disjoint unions of cut-sets. Hence we can say that for a connected graph the ring sum of a cut-set and an edge disjoint union of cut-sets is either a cut-set or an edge disjoint union of cut-sets. Furthermore, the ring sum of two edge disjoint unions of cut-sets is either a cut-set or an edge disjoint union of cut-sets.

Suppose linear graph G is separated. Let S_1 be either a cut-set or an edge disjoint union of cut-sets of one maximal connected subgraph of G and S_2 be either a cut-set or an edge disjoint union of cut-sets of another maximal connected subgraph of G. Then it is clear that $S_1 \oplus S_2$ is an edge disjoint union of cut-sets. Thus we can state the following theorems without restricting a linear graph to be connected.

Theorem 2-3-2. Let S_1 be either a cut-set or an edge disjoint union of cut-set. Let S_2 be either a cut-set or an edge disjoint union of cut-sets different from S_1. Then $S_1 \oplus S_2$ is either a cut-set or an edge disjoint union of cut-sets.

Theorem 2-3-3. Let $\{S\}$ be the collection of all cut-sets, edge disjoint unions of cut-sets, and the empty set. Then $\{S\}$ is an Abelian group under the ring sum.

Example 2-3-3. Consider the linear graph in Fig. 2-1-1. By taking cut-sets $S_1 = (a, c, f)$, $S_2 = (a, d, e)$, $S_3 = (b, e)$, and $S_4 = (f, g)$ as generators, we can obtain all members in the collection $\{S\}$ of all cut-sets, edge disjoint unions of cut-sets, and the empty set as follows:

$$S_1 \oplus S_2 = (c, d, e, f)$$
$$S_1 \oplus S_3 = (a, b, c, e, f)$$
$$S_1 \oplus S_4 = (a, c, g)$$
$$S_2 \oplus S_3 = (a, b, d)$$
$$S_2 \oplus S_4 = (a, d, e, f, g)$$
$$S_3 \oplus S_4 = (b, e, f, g)$$
$$S_1 \oplus S_2 \oplus S_3 = (b, c, d, f)$$
$$S_1 \oplus S_2 \oplus S_4 = (c, d, e, g)$$
$$S_1 \oplus S_3 \oplus S_4 = (a, b, c, e, g)$$
$$S_2 \oplus S_3 \oplus S_4 = (a, b, d, f, g)$$
$$S_1 \oplus S_2 \oplus S_3 \oplus S_4 = (b, c, d, g)$$

and

$$S_1 \oplus S_1 = S_2 \oplus S_2 = S_3 \oplus S_3 = S_4 \oplus S_4 = \emptyset$$

Note that $S_1 \oplus S_3$, $S_2 \oplus S_4$, $S_1 \oplus S_3 \oplus S_4$, and $S_2 \oplus S_3 \oplus S_4$ are edge disjoint unions of cut-sets.

2-4   LINEARLY INDEPENDENT CUT-SETS

We have learned that collection $\{S\}$ is a group under the ring sum. If a linear graph is finite, that is, if the number of edges in the linear graph is finite, then $\{S\}$ will consist of a finite number of sets. Here we will show a way of obtaining the minimum number of generators called linearly independent cut-sets by which all sets in $\{S\}$ can be generated. It is true that collections $\{E\}$ and $\{E, M_{ij}\}$ also have generators. However, it will be easier to show the number of generators in $\{E\}$ and in $\{E, M_{ij}\}$ after studying the matrix representation of linear graphs in the next chapter.

One of the important properties of cut-sets is given by the following theorem.

Theorem 2-4-1. Any cut-set can be expressed as the ring sum of incidence sets.

Proof. Suppose a linear graph G consists of vertices $v_1, v_2, \ldots, v_n$. Hence $\Omega = (v_1, v_2, \ldots, v_n)$. Let $\Omega_{11} = \emptyset$, $\Omega_{12} = (v_1, v_2, \ldots, v_{k-1})$, $\Omega_{21} = (v_k)$, and

$\Omega_{22} = \Omega - \Omega_{12} \cup \Omega_{21} = (v_{k+1}, \ldots, v_n)$. Also let $\Omega_1 = \Omega_{11} \cup \Omega_{12}$ and $\Omega_2 = \Omega_{11} \cup \Omega_{21}$. Then by Eq. 2-3-17, the ring sum of two sets $\mathcal{E}(\Omega_1 \times \bar{\Omega}_1)$ and $\mathcal{E}(\Omega_2 \times \bar{\Omega}_2)$ is

$$\mathcal{E}(\Omega_1 \times \bar{\Omega}_1) \oplus \mathcal{E}(\Omega_2 \times \bar{\Omega}_2)$$
$$= \mathcal{E}(\Omega_{11} \cup \Omega_{12} \times \overline{\Omega_{11} \cup \Omega_{12}}) \oplus \mathcal{E}(\Omega_{11} \cup \Omega_{21} \times \overline{\Omega_{11} \cup \Omega_{21}})$$
$$= \mathcal{E}[(v_1, v_2, \ldots, v_{k-1}) \times \overline{(v_1, v_2, \ldots, v_{k-1})}] \oplus \mathcal{E}[(v_k) \times \overline{(v_k)}]$$
$$= \mathcal{E}[(v_{k+1}, \ldots, v_k) \times \overline{(v_1, v_2, \ldots, v_{k-1}, v_k)}]$$
$$= \mathcal{E}[(v_1, v_2, \ldots, v_{k-1}, v_k) \times \overline{(v_1, v_2, \ldots, v_{k-1}, v_k)}] \qquad (2\text{-}4\text{-}1)$$

From this equation, we can obtain the following equations. For $k = 2$,

$$\mathcal{E}[(v_1) \times \overline{(v_1)}] \oplus \mathcal{E}[(v_2) \times \overline{(v_2)}] = \mathcal{E}[(v_1, v_2) \times \overline{(v_1, v_2)}] \qquad (2\text{-}4\text{-}2)$$

For $k = 3$,

$$\mathcal{E}[(v_1, v_2) \times \overline{(v_1, v_2)}] \oplus \mathcal{E}[(v_3) \times \overline{(v_3)}]$$
$$= \mathcal{E}[(v_1) \times \overline{(v_1)}] \oplus \mathcal{E}[(v_2) \times \overline{(v_2)}] \oplus \mathcal{E}[(v_3) \times \overline{(v_3)}]$$
$$= \mathcal{E}[(v_1, v_2, v_3) \times \overline{(v_1, v_2, v_3)}] \qquad (2\text{-}4\text{-}3)$$

In general, we have

$$\mathcal{E}[(v_1) \times \overline{(v_1)}] \oplus \mathcal{E}[(v_2) \times \overline{(v_2)}] \oplus \cdots \oplus \mathcal{E}[(v_k) \times \overline{(v_k)}]$$
$$= \mathcal{E}[(v_1, v_2, \ldots, v_k) \times \overline{(v_1, v_2, \ldots, v_k)}] \qquad (2\text{-}4\text{-}4)$$

On the other hand, any cut-set is of the form $\mathcal{E}[(v_1, v_2, \ldots, v_k) \times \overline{(v_1, v_2, \ldots, v_k)}].]$ Thus any cut-set can be obtained by the ring sum of incidence sets.   QED

From the preceding theorem, we can obtain the following theorem.

Theorem 2-4-2. Let G be a connected graph. Then any incidence set is equal to the ring sum of all remaining incidence sets in G.

Proof. Without the loss of generality, let $(v_1, v_2, \ldots, v_n)$ be the set of all vertices in G. Also let v_1 be the chosen vertex. Then, by Eq. 2-4-4,

$$\mathcal{E}[(v_2, v_3, \ldots, v_n) \times \overline{(v_2, v_3, \ldots, v_n)}]$$
$$= \mathcal{E}[(v_2) \times \overline{(v_2)}] \oplus \mathcal{E}[(v_2) \times \overline{(v_3)}] \oplus \cdots \oplus \mathcal{E}[(v_n) \times \overline{(v_n)}] \qquad (2\text{-}4\text{-}5)$$

On the other hand, we know that

$$\overline{(v_2, v_3, \ldots, v_n)} = (v_1) \qquad (2\text{-}4\text{-}6)$$

Hence

$$\mathcal{E}[(v_2, v_3, \ldots, v_n) \times \overline{(v_2, v_3, \ldots, v_n)}] = \mathcal{E}[\overline{(v_1)} \times (v_1)] \qquad (2\text{-}4\text{-}7)$$

which is the incidence set corresponding to vertex v_1. Thus the right-hand side of Eq. 2-4-5 is equal to the incidence set corresponding to vertex v_1.   QED

By Theorem 2-4-1, we can easily see that if we ring sum all incidence sets of a connected graph, we will have the empty set. Thus even if a linear graph G is not connected, Theorem 2-4-2 will hold.

Theorem 2-4-3. An incidence set S is equal to the ring sum of all incidence sets except S in a linear graph.

For convenience, we define linearly dependent sets and linearly independent sets as follows.

Definition 2-4-1. Let $D_1, D_2, \ldots, D_m$ be sets. Then $D_1, D_2, \ldots, D_m$ are said to be *linearly dependent* or are called *linearly dependent sets* if

$$D_1 \oplus D_2 \oplus \cdots \oplus D_m = \emptyset \qquad (2\text{-}4\text{-}8)$$

If there are no linearly dependent sets among $D_1, D_2, \ldots, D_n$, then they are *linearly independent*.

For example, if $D_1 = \emptyset$, then $D_1, D_2, \ldots, D_n$ are clearly not linearly independent.

Example 2-4-1. Cut-sets (a, c, f), (a, b, d), (b, e), and (b, c, d, f) of the linear graph in Fig. 2-1-1 are not linearly independent because

$$(a, c, f) \oplus (a, b, d) \oplus (b, c, d, f) = \emptyset$$

On the other hand, cut-sets (a, c, f), (a, b, d), (b, e), and (c, d, e, g) are linearly independent.

By Theorem 2-4-1, any cut-set can be obtained by the ring sum of incidence sets. Furthermore, by Theorem 2-4-3, one incidence set is equal to the ring sum of all remaining incidence sets.

Theorem 2-4-4. Let G be a connected linear graph consisting of n_v vertices Then there exist at most $n_v - 1$ linearly independent incidence sets.

Then by Theorems 2-4-1 and 2-4-4, we have Theorem 2-4-5.

Theorem 2-4-5. There exist at most $n_v - 1$ linearly independent cut-sets in a connected linear graph which contains n_v vertices.

Suppose $n_v - 1$ of incidence sets of a connected linear graph G, which consists of n_v vertices, are not linearly independent. Let $S(v_1), S(v_2), \ldots, S(v_{k+1})$ be the incidence sets among those $n_v - 1$ incidence sets such that

$$S(v_1) \oplus S(v_2) \oplus \cdots \oplus S(v_k) \oplus S(v_{k+1}) = \emptyset \qquad (2\text{-}4\text{-}9)$$

This equation can be rewritten as

$$S(v_{k+1}) = S(v_1) \oplus S(v_2) \oplus \cdots \oplus S(v_k) \qquad (2\text{-}4\text{-}10)$$

However, from Eq. 2-4-4,

$$S(v_1) \oplus S(v_2) \oplus \cdots \oplus S(v_k) = \mathcal{E}[(v_1, v_2, \ldots, v_k) \times \overline{(v_1, v_2, \ldots, v_k)}] \quad (2\text{-}4\text{-}11)$$

Furthermore, there exists n_v vertices in G. Hence $\overline{(v_1, v_2, \ldots, v_k)}$ contains at least two vertices. Note that $k + 1 \le n_v - 1$. This means that $S(v_1) \oplus S(v_2) \oplus \cdots \oplus S(v_k)$ is not an incidence set, which contradicts the assumption that $S(v_{k+1})$ in Eq. 2-4-10 is an incidence set. Thus these $n_v - 1$ incidence sets must be independent.

Theorem 2-4-6. Let G be a connected linear graph consisting of n_v vertices. Then there exist exactly $n_v - 1$ linearly independent incidence sets.

Suppose a linear graph G consists of ρ maximal connected subgraphs g_1, $g_2, \ldots, g_\rho$. Also suppose g_r consists of n_r vertices for $r = 1, 2, \ldots, \rho$, and $\sum_{r=1}^{\rho} n_r = n_v$, where n_v is the number of vertices in G. From Theorem 2-4-6, there exist exactly $n_r - 1$ linearly independent incidence sets in g_r. Moreover, it is clear that linearly independent incidence sets of g_p and linearly independent incidence sets of g_q for $1 \le p < q \le \rho$ form linearly independent incidence sets.

Theorem 2-4-7. Let G be a separated linear graph consisting of n_v vertices and ρ maximal connected subgraphs. Then there exist exactly $n_v - \rho$ linearly independent incidence sets.

Because of Theorem 2-4-1, these $n_v - \rho$ linearly independent incidence sets are enough to obtain all cut sets of a given linear graph by the ring sum operation. In other words, the set $\{S\}$ of all cut-sets, edge disjoint unions of cut-sets, and the empty set of a linear graph can be expressed as

$$\{S\} = \{S(v_{i_1}) \oplus S(v_{i_2}) \oplus \cdots \oplus S(v_{i_r});$$
$$S(v_{i_1}), S(v_{i_2}), \ldots, S(v_{i_r}) \in \{S(v)\} \text{ for } r = 1, 2, \ldots, n_v - \rho\} \quad (2\text{-}4\text{-}13)$$

where $\{S(v)\}$ is a collection of $n_v - \rho$ linearly independent incidence sets of the linear graph (consisting of n_v vertices and ρ maximal connected subgraphs).

Example 2-4-2. Consider the linear graph in Fig. 2-1-1. Since $n_v - \rho = 4$, there are four linearly independent incidence sets. Suppose we choose $S(1) = (a, c, f)$, $S(2) = (a, b, d)$, $S(3) = (c, d, e, g)$, and $S(4) = (b, e)$ as a set of four linearly independent incidence sets. Then

$$S(1) \oplus S(2) = (b, c, d, f)$$
$$S(1) \oplus S(3) = (a, d, e, f, g)$$
$$S(1) \oplus S(4) = (a, b, c, e, f)$$
$$S(2) \oplus S(3) = (a, b, c, e, g)$$
$$S(2) \oplus S(4) = (a, d, e)$$
$$S(3) \oplus S(4) = (b, c, d, g)$$

$$S(1) \oplus S(2) \oplus S(3) = (b, e, f, g)$$
$$S(1) \oplus S(2) \oplus S(4) = (c, d, e, f)$$
$$S(1) \oplus S(3) \oplus S(4) = (a, b, d, f, g)$$
$$S(2) \oplus S(3) \oplus S(4) = (a, c, g)$$
$$S(1) \oplus S(2) \oplus S(3) \oplus S(4) = (f, g)$$

plus $S(1)$, $S(2)$, $S(3)$, $S(4)$, and the empty set will be the entire sets in $\{S\}$. The reader can compare these with the sets in Example 2-3-3.

Since we cannot say that every incidence set is a cut-set, we cannot say that there exist exactly $n_v - \rho$ linearly independent cut-sets just by Theorem 2-4-7. It would be much easier to show the number of linearly independent cut-sets if we know the properties of a particular subgraph called a "tree," which will be studied later. Hence we will not study this problem at this time. However, it is not difficult to prove the exact number of linearly independent cut-sets without using the knowledge of trees.

PROBLEMS

1. Prove that any cut-set of a connected linear graph can be expressed as $\mathscr{E}(\Omega_1 \times \bar{\Omega}_1)$.

2. Let S be a cut-set and C be a circuit. Prove that $S \cap C$ contains even number of edges.

3. Prove that the intersection of any three different incidence sets is the empty set.

4. Let e be an edge in a complete graph. Then (a) calculate the number of cut-sets which contain edge e and (b) calculate the number of cut-sets that do not contain edge e.

5. Prove that any edge can be in a cut-set.

6. Prove that any pair of edges in a nonseparable graph can be in a cut-set.

7. Prove that for any path P, there exists set S which is either a cut-set or an edge disjoint union of cut-sets such that $P \subset S$.

8. Let S be a set of edges in a linear graph G. Suppose deletion of all edges in S from G reduces its rank by $k \geq 1$. Prove or disprove that if and only if the deletion of all edges in any proper subset of S reduces the rank by at most $k - 1$, S is either a cut-set or an edge disjoint union of cut-sets.

9. Suppose there exists at most one edge between any pair of vertices in a linear graph G. Let Ω be the collection of all vertices in G. Prove that if and only if $\mathscr{E}(\Omega_1 \times \bar{\Omega}_1)$ is a cut-set for every proper subset Ω_1 of Ω, G is a complete graph.

10. Let G be connected linear graph consisting of n_v vertices. Then $\{S\}$ consists of 2^{n_v-1} sets. Note that $\{S\}$ is the collection of all cut-sets, edge disjoint unions of cut-sets, and the empty set.

11. Let e_1 and e_2 be two edges in a complete graph consisting of n_v vertices. Show that the number of cut-sets which contain both edges e_1 and e_2 will be 2^{n_v-3}.

12. Let G be a linear graph and S be a cut-set which contains edge e. Let G_e be the linear graph obtained from G by deleting edge e. Show that if S contains more than one edge, $S - (e)$ is a cut-set of G_e.

13. Let G be a linear graph and S be a cut-set. Suppose edge e is in G but not in S. Let $G_{\bar{e}}$ be a linear graph obtained from G by shorting edge e (i.e., coinciding vertices which are the endpoints of edge e). Show that S is a cut-set of $G_{\bar{e}}$.

14. Prove that if an incidence set $S(v)$ is not a cut-set, then the vertex v is a cut vertex.

3

MATRIX REPRESENTATION OF LINEAR GRAPH AND TREES

3-1 INCIDENCE MATRIX

Instead of using sets, we can use matrices to represent a linear graph. Suppose we use row p and column q of a matrix to represent subgraph g_p and edge e_q, respectively. Then we can show which edges belong to subgraph g_p by giving 1 at the (p, q) entry if edge e_q is in subgraph g_p and 0 at the (p, q) entry if edge e_q is not in subgraph g_p.

If all rows of such a matrix represent incidence sets of a linear graph, then the matrix is called an incidence matrix. Similarly, a cut-set matrix is one whose rows indicate cut-sets and edge disjoint unions of cut-sets. All rows of a circuit matrix, moreover, represent circuits and edge disjoint unions of circuits. Here we discuss some properties of cut-set and circuit matrices. We also develop a procedure to test whether a given matrix is a cut-set (circuit) matrix. First we consider incidence matrices.

Definition 3-1-1. An *exhaustive incidence matrix*, symbolized by A_e, is defined by

$$a_{pq} = \begin{cases} 1 & \text{if edge } q \text{ is incident at vertex } p \\ 0 & \text{otherwise} \end{cases} \tag{3-1-1}$$

where a_{pq} is the (p, q) entry of A_e.

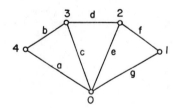

Fig. 3-1-1.   Linear graph G.

For example, exhaustive incidence matrix A_e of the linear graph in Fig. 3-1-1 is

$$
A_e = \begin{array}{c} \\ 1 \\ 2 \\ 3 \\ 4 \\ 0 \end{array}
\begin{array}{c} \begin{array}{ccccccc} a & b & c & d & e & f & g \end{array} \\
\left[\begin{array}{ccccccc}
0 & 0 & 0 & 0 & 0 & 1 & 1 \\
0 & 0 & 0 & 1 & 1 & 1 & 0 \\
0 & 1 & 1 & 1 & 0 & 0 & 0 \\
1 & 1 & 0 & 0 & 0 & 0 & 0 \\
1 & 0 & 1 & 0 & 1 & 0 & 1
\end{array}\right]
\end{array}
$$

Note that the incidence set $S(1)$ indicated by row 1 is (f, g). The incidence set $S(2)$ corresponding to vertex 2 is by row 2, and so on. We can see that each row of A_e indicates a vertex as well as the incidence set corresponding to the vertex.

Since an edge has two endpoints, there exist exactly two 1's in each column of an exhaustive incidence matrix. In order to obtain the rank of an incidence matrix, we first investigate an operation on an incidence matrix of a linear graph G so that the resultant matrix becomes an incidence matrix of a new linear graph obtained from G by coinciding vertices. Since we are considering only linear graphs that contain no self-loops and since it is possible to produce self-loops when vertices are made to coincide, we must make a further restriction on linear graphs as follows.

AGREEMENT 3-1-1. Whenever self-loops are produced by some operations on a linear graph, we will delete them so that the resultant graph will have no self-loops.

Suppose the vertices 4 and 0 of the linear graph in Fig. 3-1-1 are made to coincide; then edge a becomes a self-loop in the resultant graph. By the agreement, edge a will be deleted, resulting in the graph of Fig. 3-1-2.

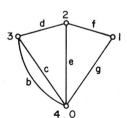

Fig. 3-1-2. A linear graph obtained by coinciding vertices 4 and 0 of G in Fig. 3-1-1.

The exhaustive incidence matrix of the resultant graph is

$$A_{e2} = \begin{array}{c} \\ 1 \\ 2 \\ 3 \\ 4 \text{ and } 0 \end{array} \overset{\displaystyle b \;\; c \;\; d \;\; e \;\; f \;\; g}{\begin{bmatrix} 0 & 0 & 0 & 0 & 1 & 1 \\ 0 & 0 & 1 & 1 & 1 & 0 \\ 1 & 1 & 1 & 0 & 0 & 0 \\ 1 & 1 & 0 & 1 & 0 & 1 \end{bmatrix}}$$

Let A'_e be an exhaustive incidence matrix of a linear graph G' obtained by coinciding vertices i and j of a linear graph G, and let A_e be an exhaustive incidence matrix of G. Let row i be the row representing vertex i and row j be the row representing vertex j of $A_e = [a_{pq}]$. If we add row i to the row j of A_e, then each entry of the resulting row j is $a_{ir} + a_{jr}$, which is one of the following:

CASE 1. $a_{ir} + a_{jr} = 1$ (either a_{ir} or a_{jr} is 1)   (3-1-2)

CASE 2. $a_{ir} + a_{jr} = 0$ (both a_{ir} and a_{jr} are 0)   (3-1-3)

CASE 3. $a_{ir} + a_{jr} = 2$ (both a_{ir} and a_{jr} are 1)   (3-1-4)

Case 1 occurs if edge e_r corresponding to column r has either i or j as one of the two endpoints but not both vertices i and j as its endpoints. If neither vertex i nor vertex j is one of the endpoints of edge r, Case 2 results. Hence row j of A'_e is equal to the addition of row i and row j of A_e for these cases.

Case 3 $(a_{ir} + a_{jr} = 2)$ means that edge e_r is connected between vertices i and j in G. Thus when vertices i and j coincide, edge e_r should disappear by Agreement 3-1-1, which means that $a_{ir} + a_{jr}$ must be equal to 0 rather than 2. Hence we can obtain exhaustive incidence matrix A'_e by adding row i to row j of A_e with the rule that $1 + 1 = 0$, then deleting row i from A_e. Such an algebra is the residue class modulo 2, and it is employed to operate matrices in this chapter. The multiplication and the addition tables of modulo 2 algebra are as follows:

Addition Table		
	1	0
1	0	1
0	1	0

Multiplication Table		
	1	0
1	1	0
0	0	0

Recall that each row of an exhaustive incidence matrix represents a vertex as well as the incidence set corresponding to the vertex. Let $S(i)$ be the incidence set corresponding to vertex i and $S(j)$ be the incidence set corresponding to vertex j in G. Then $S(i) \oplus S(j)$ is the set of edges which are incident at either vertex i or vertex j but not both i and j in G. This means that a row of a matrix representing $S(i) \oplus S(j)$ must have 1 at the column corresponding to edge e if and only if edge e is incident at either vertex i or vertex j but not both i and j. Since we are using modulo 2, the addition of row i and row j will give a row corresponding to $S(i) \oplus S(j)$. In other words, addition of rows in an exhaustive incidence matrix A_e is exactly the same as the ring sum of incidence sets. Hence the rank* of A_e is equal to $n_v - \rho$ by Theorem 2-4-7 where n_v is the number of vertices and ρ is the number of maximal connected subgraphs of G.

We can show that the rank of A_e is $n_v - \rho$ directly from A_e as follows. Since there are exactly two nonzeros in each column of A_e (because each edge has two endpoints), by adding all rows (except the last row) to the last row (modulo 2), the last row becomes a row of zeros. Hence the rank of A_e must be less than the number of rows in A_e. However, the number of rows in A_e is the number n_v of vertices in G. Therefore the rank $R(A_e)$ of A_e is

$$R(A_e) \leq n_v - 1 \tag{3-1-5}$$

Suppose G is a connected linear graph. Let the first column of A_e represent edge e, whose two endpoints are vertices i and j. By rearranging rows, we can make row i to be the first row:

$$
A_e = \begin{array}{c} \\ i \\ \\ \\ \\ \\ j \\ \\ \\ \\ \end{array}
\left[\begin{array}{c|c}
\begin{matrix} 1 \\ 0 \\ \vdots \\ 0 \\ 1 \\ 0 \\ \vdots \\ 0 \end{matrix} & A_{e1} \\
\end{array} \right]
\overset{e}{}
\tag{3-1-6}
$$

Note that in column e there are 1's at row i and row j, and all other entries in column e are 0. By adding row i (which is the first row) to row j, we have

---

* The rank of a matrix is the order of the largest nonsingular submatrix of the matrix. A nonsingular submatrix is a square submatrix whose determinant is nonzero.

$$A'_e = \begin{array}{c} i \\ \\ \\ \\ \\ j \\ \\ \\ \end{array} \left[\begin{array}{c|c} 1 & \\ \hline 0 & \\ \vdots & \\ 0 & A'_{e1} \\ 0 & \\ 0 & \\ \vdots & \\ 0 & \end{array} \right] \tag{3-1-7}$$

Since we have already found that matrix A'_{e1} is an exhaustive incidence matrix of a linear graph G' obtained from G by making vertices i and j coincide. G' is connected because shorting an edge in a connected graph produces a connected graph. Thus A'_{e1} has the same property that A_e has. Hence by rearranging rows and adding one row to another, we can make the first column of A'_{e1} a unit column in which there is only one 1 at the first row:

$$A''_e = \left[\begin{array}{c|c|c} 1 & & \\ \hline 0 & 1 & \\ \hline 0 & 0 & \\ \vdots & \vdots & A'_{e2} \\ 0 & 0 & \end{array} \right] \tag{3-1-8}$$

Furthermore A'_{e2} is an exhaustive matrix of a connected graph. Thus A'_{e2} has the same property A_e has, and so on.

In general, we can change A_e to the form

$$A_e^{(k)} = \left[\begin{array}{ccccc} 1 & & \cdots & & \\ 0 & 1 & \cdots & & \\ 0 & 0 & & & \\ & & \ddots & & \\ 0 & 0 & & 1 & \\ & 0 & & & A'_{ek} \end{array} \right] \tag{3-1-9}$$

where A'_{ek} is an exhaustive incidence matrix of a connected graph. We can keep modifying A until the submatrix A'_{ek} consists of one row. When A'_{ek} consists of one row, the corresponding linear graph contains no edges because we are excluding self-loops. Thus A'_{ek} must be one row of zeros. This means that A_e of a connected linear graph can be changed to

$$A_e^{(n_v-1)} = \begin{bmatrix} 1 & & \cdots & & \cdots & \\ 0 & 1 & \cdots & & \cdots & \\ 0 & 0 & & & & \\ \vdots & \vdots & \ddots & & & \\ 0 & 0 & . & 0 & 1 & \cdots \\ 0 & 0 & \cdots & 0 & 0 & 0 & \cdots & 0 \end{bmatrix} \tag{3-1-10}$$

It is clear that this operation does not change the rank of matrices. Hence we can see from the result that there exists at least one submatrix of order $n_v - 1$ whose determinant is nonzero.

Theorem 3-1-1. The rank $R(A_e)$ of an exhaustive incidence matrix A_e of a connected linear graph consisting of n_v vertices is

$$R(A_e) = n_v - 1 \tag{3-1-11}$$

Example 3-1-1. Consider the following matrix which is an exhaustive incidence matrix of a linear graph in Fig. 3-1-1.

$$A_e = \begin{array}{c} \\ 1 \\ 2 \\ 3 \\ 4 \\ 0 \end{array} \begin{array}{c} a\ \ b\ \ c\ \ d\ \ e\ \ f\ \ g \\ \begin{bmatrix} 0 & 0 & 0 & 0 & 0 & 1 & 1 \\ 0 & 0 & 0 & 1 & 1 & 1 & 0 \\ 0 & 1 & 1 & 1 & 0 & 0 & 0 \\ 1 & 1 & 0 & 0 & 0 & 0 & 0 \\ 1 & 0 & 1 & 0 & 1 & 0 & 1 \end{bmatrix} \end{array}$$

By interchanging columns g and a and adding row 1 to row 0, we have

$$\begin{array}{c} \\ 1 \\ 2 \\ 3 \\ 4 \\ 0 \end{array} \begin{array}{c} g\ \ b\ \ c\ \ d\ \ e\ \ f\ \ a \\ \begin{bmatrix} 1 & 0 & 0 & 0 & 0 & 1 & 0 \\ 0 & 0 & 0 & 1 & 1 & 1 & 0 \\ 0 & 1 & 1 & 1 & 0 & 0 & 0 \\ 0 & 1 & 0 & 0 & 0 & 0 & 1 \\ 0 & 0 & 1 & 0 & 1 & 1 & 1 \end{bmatrix} \end{array}$$

By interchanging columns d and b and adding row 2 to row 3, we have

$$
\begin{array}{c}
 \\
1 \\
2 \\
3 \\
4 \\
0
\end{array}
\begin{array}{ccccccc}
g & d & c & b & e & f & a \\
\end{array}
\begin{bmatrix}
1 & 0 & 0 & 0 & 0 & 1 & 0 \\
0 & 1 & 0 & 0 & 1 & 1 & 0 \\
0 & 0 & 1 & 1 & 1 & 1 & 0 \\
0 & 0 & 0 & 1 & 0 & 0 & 1 \\
0 & 0 & 1 & 0 & 1 & 1 & 1
\end{bmatrix}
$$

By adding row 3 to row 0, we have

$$
\begin{array}{c}
 \\
1 \\
2 \\
3 \\
4 \\
0
\end{array}
\begin{array}{ccccccc}
g & d & c & b & e & f & a \\
\end{array}
\begin{bmatrix}
1 & 0 & 0 & 0 & 0 & 1 & 0 \\
0 & 1 & 0 & 0 & 1 & 1 & 0 \\
0 & 0 & 1 & 1 & 1 & 1 & 0 \\
0 & 0 & 0 & 1 & 0 & 0 & 1 \\
0 & 0 & 0 & 1 & 0 & 0 & 1
\end{bmatrix}
$$

Finally, adding row 4 to row 0, we have

$$
\begin{array}{c}
 \\
1 \\
2 \\
3 \\
4 \\
0
\end{array}
\begin{array}{ccccccc}
g & d & c & b & e & f & a \\
\end{array}
\begin{bmatrix}
1 & 0 & 0 & 0 & 0 & 1 & 0 \\
0 & 1 & 0 & 0 & 1 & 1 & 0 \\
0 & 0 & 1 & 1 & 1 & 1 & 0 \\
0 & 0 & 0 & 1 & 0 & 0 & 1 \\
0 & 0 & 0 & 0 & 0 & 0 & 0
\end{bmatrix}
$$

Suppose G is a separated linear graph which consists of ρ maximal connected subgraphs $g_1, g_2, \ldots, g_\rho$. Let the number of edges and the number of vertices of g_r be n_{er} and n_{vr}, respectively, for $r = 1, 2, \ldots, \rho$. By rearranging rows and columns, A_e of G will have the following properties: (1) the first n_{v1} rows represent vertices in g_1, the next n_{v2} rows represent vertices in $g_2, \ldots$, the last $n_{v\rho}$ rows represent vertices in g_ρ; (2) the first n_{e1} columns represent edges in g_1, the next n_{e2} columns represent edges in $g_2, \ldots$, the last $n_{e\rho}$ columns represent edges in g_ρ. This exhaustive incidence matrix A_e should be able to partition as

$$
A_e =
\begin{array}{r}
\text{Vertices in } g_1 \\
\text{Vertices in } g_2 \\
\vdots \\
\text{Vertices in } g_\rho
\end{array}
\begin{array}{c}
\begin{array}{ccc}
\text{Edges} & \text{Edges} & \text{Edges} \\
\text{in } g_1 & \text{in } g_2 & \cdots \quad \text{in } g_\rho
\end{array} \\
\begin{bmatrix}
A_{e1} & & \\
& A_{e2} & \\
& & \ddots \\
0 & & A_{e\rho}
\end{bmatrix}
\end{array}
\qquad (3\text{-}1\text{-}12)
$$

where A_{er} is an exhaustive incidence matrix of g_r for $r = 1, 2, \ldots, \rho$. Since A_{er} is an exhaustive incidence matrix of a connected graph g_r, the rank of A_{er} is $n_{vr} - 1$. Thus the rank $R(A_e)$ of G is $n_v - \rho$.

Theorem 3-1-2. The rank of an exhaustive incidence matrix is equal to the rank of a linear graph.

By using Theorem 3-1-2, we can define an incidence matrix A of a linear graph as follows.

Definition 3-1-2. An incidence matrix A of a linear graph consists of $n_v - \rho$ rows, which are obtained from an exhaustive incidence matrix by deleting one row from each A_{er} in Eq. 3-1-12 so that the rank of A is equal to the rank of the linear graph.

Note that the incidence matrix A of a connected linear graph is obtained from an exhaustive incidence matrix A_e by deleting any one row. Since each row of an incidence matrix represents a vertex, when a linear graph is connected, the vertex, which corresponds to the row in an exhaustive incidence matrix and is deleted to obtain an incidence matrix A of the linear graph, is called the reference vertex.

Definition 3-1-3. For an incidence matrix, the *reference vertex* in a connected graph is the vertex that is not represented by any row in the incidence matrix of the linear graph.

For example, suppose an incidence matrix A of the connected graph in Fig. 3-1-1 is

$$
A = \begin{array}{c} \\ 1 \\ 2 \\ 3 \\ 4 \end{array}
\begin{array}{c} \begin{array}{ccccccc} a & b & c & d & e & f & g \end{array} \\
\left[\begin{array}{ccccccc}
0 & 0 & 0 & 0 & 0 & 1 & 1 \\
0 & 0 & 0 & 1 & 1 & 1 & 0 \\
0 & 1 & 1 & 1 & 0 & 0 & 0 \\
1 & 1 & 0 & 0 & 0 & 0 & 0
\end{array}\right] \end{array}
\qquad (3\text{-}1\text{-}13)
$$

Then vertex 0 is the reference vertex.

3-2 TREE

There is a subgraph known as a tree, which is very important not only theoretically but also in several applications of linear graphs such as analysis of electrical networks. Here we introduce trees and consider a relationship between an incidence matrix and trees. But first we define a major submatrix of a matrix as follows.

Definition 3-2-1. For a matrix of order $p \times q$, if the order of a square submatrix is equal to the minimum of p and q, the submatrix is called a *major submatrix*. The determinant of a major submatrix is called a *major determinant*.

For example, a major submatrix of incidence, matrix A in Eq. 3-1-13, consists of four rows because the number of rows is smaller than the number of columns in A. One of the major submatrices is

$$
\begin{array}{c}
 \\ 1 \\ 2 \\ 3 \\ 4
\end{array}
\begin{array}{cccc}
a & b & c & d \\
\end{array}
\left[
\begin{array}{cccc}
0 & 0 & 0 & 0 \\
0 & 0 & 0 & 1 \\
0 & 1 & 1 & 1 \\
1 & 1 & 0 & 0
\end{array}
\right]
$$

To show a relationship between trees and major submatrices of an incidence matrix, consider a connected linear graph G having n_v vertices. Let A be an incidence matrix of G. Since the rank of A is equal to the number of rows of A by Definition 3-1-2, there is at least one major submatrix whose determinant is nonzero. Consider a subgraph g_m of G whose edges correspond to the columns of a nonsingular major submatrix A_m. Since A_m is nonsingular, A_m must be an incidence matrix of g_m. Since the number of rows of A_m is $n_v - 1$ and each row corresponds to a vertex except the reference vertex, g_m has n_v vertices. Since A_m is a square matrix and each column corresponds to an edge, g_m has $n_v - 1$ edges. Furthermore, the rank of g_m is $n_v - 1$. Hence g_m must be connected. Such a subgraph is called a tree.

Definition 3-2-2. A *tree* of a connected linear graph G of n_v vertices is a connected subgraph having n_v vertices and $n_v - 1$ edges.

It should now be clear that if a subgraph g_m of a linear graph G is a tree, then a major submatrix A_m whose columns correspond to all edges of g_m is nonsingular. On the other hand, if a subgraph g_m consisting of n_v vertices and $n_v - 1$ edges is separated, the rank of g_m is $n_v - \rho$ where $\rho > 1$. Furthermore, because every major submatrix consists of $n_v - 1$ rows, the rank of a major submatrix A_m whose columns correspond to all edges of g_m is less than the number of rows of A_m. Hence A_m is singular.*

Theorem 3-2-1. Let A_m be a major submatrix of an incidence matrix of a connected graph. Then A_m is nonsingular if and only if the subgraph consisting of all edges corresponding to the columns of A_m is a tree.

* A major submatrix is singular if the determinant of the submatrix is zero.

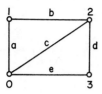

Fig. 3-2-1. A linear graph.

Example 3-2-1. Incidence matrix A of the linear graph in Fig. 3-2-1 with vertex 0 being the reference vertex is

$$A = \begin{array}{c} \\ 1 \\ 2 \\ 3 \end{array} \begin{array}{ccccc} a & b & c & d & e \\ \begin{bmatrix} 1 & 1 & 0 & 0 & 0 \\ 0 & 1 & 1 & 1 & 0 \\ 0 & 0 & 0 & 1 & 1 \end{bmatrix} \end{array}$$

Nonsingular submatrices of A and the corresponding subgraphs which are trees are shown in Table 3-2-1. Note that, for example, a square submatrix

$$\begin{array}{ccc} a & b & c \\ \begin{bmatrix} 1 & 1 & 0 \\ 0 & 1 & 1 \\ 0 & 0 & 0 \end{bmatrix} \end{array}$$

is singular because the subgraph corresponding to the submatrix consists of edges a, b, and c and vertices 0, 1, 2, and 3, which is separated.

It is clear from the definition of trees that there are no trees in a separated graph. However, there is a subgraph corresponding to a nonsingular major submatrix of an incidence matrix of a separated graph. Such a subgraph is called a forest.

Definition 3-2-3. Consider a separated graph G consisting of ρ maximal connected subgraphs $G_1, G_2, \ldots, G_\rho$. A subgraph τ consisting of ρ maximal connected subgraphs $\tau_1, \tau_2, \ldots,$ and τ_ρ is called a *forest* if τ_i is a tree of G_i for all $i = 1, 2, \ldots, \rho$.

For example, consider the linear graph G in Fig. 3-2-2, which consists of

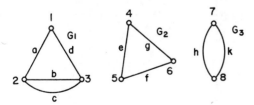

Fig. 3-2-2. A separated linear graph G.

Table 3-2-1 Nonsingular Submatrices and Trees

Nonsingular submatrix	Tree

t_1

$$\begin{array}{cccc} & a & b & d \\ 1 & \begin{bmatrix} 1 & 1 & 0 \\ 0 & 1 & 1 \\ 0 & 0 & 1 \end{bmatrix} \\ 2 & \\ 3 & \end{array}$$

t_2

$$\begin{array}{cccc} & a & b & e \\ 1 & \begin{bmatrix} 1 & 1 & 0 \\ 0 & 1 & 0 \\ 0 & 0 & 1 \end{bmatrix} \\ 2 & \\ 3 & \end{array}$$

t_3

$$\begin{array}{cccc} & a & c & d \\ 1 & \begin{bmatrix} 1 & 1 & 0 \\ 0 & 1 & 1 \\ 0 & 0 & 1 \end{bmatrix} \\ 2 & \\ 3 & \end{array}$$

t_4

$$\begin{array}{cccc} & a & c & e \\ 1 & \begin{bmatrix} 1 & 0 & 0 \\ 0 & 1 & 0 \\ 0 & 0 & 1 \end{bmatrix} \\ 2 & \\ 3 & \end{array}$$

t_5

$$\begin{array}{cccc} & a & d & e \\ 1 & \begin{bmatrix} 1 & 0 & 0 \\ 0 & 1 & 0 \\ 0 & 1 & 1 \end{bmatrix} \\ 2 & \\ 3 & \end{array}$$

t_6

$$\begin{array}{cccc} & b & c & d \\ 1 & \begin{bmatrix} 1 & 0 & 0 \\ 1 & 1 & 1 \\ 0 & 0 & 1 \end{bmatrix} \\ 2 & \\ 3 & \end{array}$$

t_7

$$\begin{array}{cccc} & b & c & e \\ 1 & \begin{bmatrix} 1 & 0 & 0 \\ 1 & 1 & 0 \\ 0 & 0 & 1 \end{bmatrix} \\ 2 & \\ 3 & \end{array}$$

t_8

$$\begin{array}{cccc} & b & d & e \\ 1 & \begin{bmatrix} 1 & 0 & 0 \\ 1 & 1 & 0 \\ 0 & 1 & 1 \end{bmatrix} \\ 2 & \\ 3 & \end{array}$$

three maximal connected subgraphs G_1, G_2, and G_3. A subgraph consisting of three maximal connected subgraphs τ_1, τ_2, and τ_3, as shown in Fig. 3-2-3, is a forest because τ_1 is a tree of G_1, τ_2 is a tree of G_2, and τ_3 is a tree of G_3.

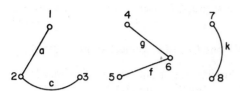

Fig. 3-2-3.   A forest of G.

Although we are concerned only with the properties of trees, we see from Definition 3-2-3 that these properties are also the properties of each maximal connected subgraph of a forest. Hence we should be able to figure out the properties of forest easily.

An important property of a tree is given by the following theorem.

Theorem 3-2-2.   A tree of a connected graph contains no circuits.

Proof.   Suppose a tree contains circuits. Then, by deleting an edge in a circuit, the resultant graph is left to be connected. However, the number of edges will be reduced by one. Thus there will be a connected graph having n_v vertices and $n_v - 2$ edges. Shorting an edge in a connected graph results in a connected graph of one less vertex. Hence, by successively shorting edges, we will have a connected graph having at least two vertices and no edges, which is impossible. Thus a tree cannot have circuits.   QED

From Definition 3-2-2 and Theorem 3-2-2 a tree has the following four properties:

1. It is connected.
2. It contains n_v vertices.
3. It contains no circuits.
4. It contains $n_v - 1$ edges.

We can easily show that any three properties imply the fourth. Furthermore, one pair of these properties is sufficient to define a tree, as shown by the following theorem.

Theorem 3-2-3.   For a connected linear graph of n_v vertices, a subgraph consisting of $n_v - 1$ edges which contains no circuits is a tree.

The proof of this theorem is left to the reader.

There are some simple properties of trees given by the following theorem.

Theorem 3-2-4.   If and only if a linear graph has exactly one path between any two vertices, the linear graph is a tree.

Proof.   Since there exists exactly one path between any two vertices in a linear graph G, it is connected. Moreover, there are obviously no circuits. Hence, by the definition of a tree, linear graph G itself is a tree. The converse can also be proved easily.   QED

Theorem 3-2-5.   There exists at least one tree in a connected linear graph.

Proof.   Since there exists a nonsingular major submatrix of an incidence matrix, the existence of a tree is obvious by Theorem 3-2-1.

Theorem 3-2-6.   In a tree, there exist at least two vertices of degree 1.

Proof.   In Section 1-2, we saw that the sum of the degrees $d(v)$ of all vertices is equal to twice that of the number n_e of edges in a linear graph G, that is,

$$\sum_{v \in G} d(v) = 2n_e \tag{3-2-3}$$

Since G, which we are considering here, is a tree, the number n_e of edges is $n_v - 1$ by Definition 3-2-2 where n_v is the number of vertices in G. Hence

$$\sum_{v \in G} d(v) = 2(n_v - 1) \tag{3-2-4}$$

Suppose the degree $d(v)$ of every vertex in G is equal to or larger than 2. Then

$$\sum_{v \in G} d(v) \geq 2n_v \tag{3-2-5}$$

But Eq. 3-2-4 indicates that $\sum_{v \in G} d(v)$ cannot be larger than $2n_v - 2$. Hence there must be vertices whose degrees are less than 2. Suppose there is only one vertex whose degree is 1 (and all other vertices are of degree at least 2), then

$$\sum_{v \in G} d(v) \geq 2(n_v - 1) + 1$$

which is again larger than $2n_v - 2$.   QED

In Chapter 2 we discussed a τ-graph, which is a subgraph having no circuits. The next theorem shows a condition when a τ-graph is a tree.

Theorem 3-2-7.   For a connected linear graph consisting of n_v vertices, a τ-graph containing $n_v - 1$ edges is a tree.

The proof can be accomplished by using Theorem 3-2-3.

The property that a τ-graph has no circuits means that each τ-graph of a connected linear graph of order n_v can have at most $n_v - 1$ edges. Thus if we know all τ-graphs, we can easily extract all trees by using Theorem 3-2-7. However, it is not easy to obtain all possible τ-graphs of a linear graph unless they are already available to you. Other ways of obtaining all trees, will be studied later.

3-3 CIRCUIT MATRIX

As an incidence matrix shows the location of each edge in a linear graph, a circuit matrix indicates circuits that each edge belongs to. We are now concerned with some properties of circuit matrices, especially the rank of a circuit matrix which is equal to the number of generators in a collection $\{E\}$ of all Euler graphs of a linear graph. Another important property by which we can transform a circuit matrix into an incidence matrix and vice versa will be studied. First, we define an exhaustive circuit matrix as follows.

Definition 3-3-1. The rows of an *exhaustive circuit matrix* B_e of a linear graph represent all possible circuits and edge disjoint union of circuits (Euler graphs) and the columns represent edges. The entry b_{pq} of an exhaustive circuit matrix $B = [b_{pq}]$ is

$$b_{pq} = \begin{cases} 1 & \text{if edge } q \text{ is in circuit (or edge disjoint union} \\ & \text{of circuit) } p \\ 0 & \text{otherwise} \end{cases} \tag{3-3-1}$$

For example, an exhaustive circuit matrix B_e of the linear graph shown in Fig. 3-3-1 is

	a	b	c	d	e	f	g
C_1	1	1	1	0	0	0	0
C_2	1	1	0	1	1	0	0
C_3	1	1	0	1	0	1	1
C_4	0	0	1	1	1	0	0
C_5	0	0	0	0	1	1	1
C_6	0	0	1	1	0	1	1
$C_7 = C_1 \oplus C_5$	1	1	1	0	1	1	1

Since every row in B_e represents a circuit or an edge disjoint union of circuits, the sum (modulo 2) of rows is either a circuit or an edge disjoint union of circuits. For example, sum of the first two rows gives the fourth row of B_e in the previous example.

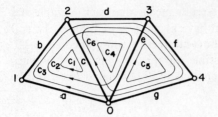

Fig. 3-3-1. A linear graph.

Let A and B_e be an incidence and an exhaustive circuit matrix of a linear graph G consisting of n_e edges and n_v vertices where

$$A = \begin{bmatrix} A_1 \\ A_2 \\ A_3 \\ \vdots \\ A_{n_v-1} \end{bmatrix} \tag{3-3-2}$$

$$B_e = \begin{bmatrix} B_1 \\ B_2 \\ B_3 \\ \vdots \\ B_p \end{bmatrix} \tag{3-3-3}$$

Also let columns of A be arranged so that both kth columns of A and B_e represent the same edge for $k=1, 2, \ldots, n_e$. Then

$$AB_e{}^t = \begin{bmatrix} A_1 B_1{}^t & A_1 B_2{}^t & \cdots & A_1 B_p{}^t \\ A_2 B_1{}^t & A_2 B_2{}^t & \cdots & A_2 B_p{}^t \\ A_{n_v-1} B_1{}^t & A_{n_v-1} B_2{}^t & \cdots & A_{n_v-1} B_p{}^t \end{bmatrix} \tag{3-3-4}$$

where the superscript t means "transpose of" the matrix and

$$A_i B_j{}^t = \sum_{k=1}^{n_e} a_{ik} b_{jk} \tag{3-3-5}$$

Note that $a_{ik} = 1$ means that edge k is incident at vertex i and $b_{jk} = 1$ means that edge k is in circuit (or edge disjoint union of circuits) represented by jth row of B_e. Thus the number of nonzero terms in Eq. 3-3-5 is the number of edges each of which is incident at vertex i and at the same time is in circuit j (or the edge disjoint union of circuits represented by the jth row of B_e). Since every vertex in a circuit or an edge disjoint union of circuits is of even degree, the number of nonzero terms in Eq. 3-3-5 is even. Each nonzero term is 1 and we are using modulo 2 algebra. Thus

$$A_i B_j^t = 0 \qquad (3\text{-}3\text{-}6)$$

for all i and j.

Theorem 3-3-1. Let A and B_e be an incidence and an exhaustive circuit matrix of a linear graph. Then

$$A B_e^t = 0^* \qquad (3\text{-}3\text{-}7)$$

Similarly,

$$B_e A^t = 0 \qquad (3\text{-}3\text{-}8)$$

Example 3-3-1. $A B_e^t$ of the linear graph in Fig. 3-3-1 is

$$
A B_e^t =
\begin{array}{c}
1 \\ 2 \\ 3 \\ 4
\end{array}
\begin{bmatrix}
a & b & c & d & e & f & g \\
1 & 1 & 0 & 0 & 0 & 0 & 0 \\
0 & 1 & 1 & 1 & 0 & 0 & 0 \\
0 & 0 & 0 & 1 & 1 & 1 & 0 \\
0 & 0 & 0 & 0 & 0 & 1 & 1
\end{bmatrix}
\begin{bmatrix}
C_1 & C_2 & C_3 & C_4 & C_5 & C_6 & C_7 \\
1 & 1 & 1 & 0 & 0 & 0 & 1 \\
1 & 1 & 1 & 0 & 0 & 0 & 1 \\
1 & 0 & 0 & 1 & 0 & 1 & 1 \\
0 & 1 & 1 & 1 & 0 & 1 & 0 \\
0 & 1 & 0 & 1 & 1 & 0 & 1 \\
0 & 0 & 1 & 0 & 1 & 1 & 1 \\
0 & 0 & 1 & 0 & 1 & 1 & 1
\end{bmatrix}
\begin{array}{c}
a \\ b \\ c \\ d \\ e \\ f \\ g
\end{array}
= 0
$$

Definition 3-3-2. For a tree (forest) t in a linear graph G, an edge that is not in t is called a *chord*. A set of all edges in G which are not in t is called a *set of chords* with respect to tree (forest) t.

We have learned that for a connected linear graph having n_v vertices, a tree consists of $n_v - 1$ edges. Hence a *set of chords consists of $n_e - v_v + 1$ edges* where n_e is the number of edges in the linear graph.

For example, a set of edges (a, c, e, g) is a tree in a linear graph in Fig. 3-3-1. Thus the set of edges (b, d, f) is a set of chords with respect to tree (a, c, e, g). Note that the number of edges and the number of vertices in the linear graph in Fig. 3-3-1 are $n_e = 7$ and $n_v = 5$, respectively. Hence a set of chords consists of $n_e - n_v + 1 = 3$ edges.

If we insert a chord d to a tree t, the resultant graph $(d) \cup t$ contains exactly one circuit. This can be seen easily once we know that there exists exactly one path between any two vertices in a tree by Theorem 3-2-5. Since we insert each chord to a tree, we have a circuit and there are $n_e - n_v + 1$

---

* Here, 0 indicates a matrix whose entries are all zero.

chords, there are $n_e - n_v + 1$ circuits, each of which contains exactly one chord in the set of chords with respect to tree t. The set of these circuits is called a set of fundamental circuits with respect to a tree t.

Definition 3-3-3. Let $(e_1, e_2, \ldots, e_{n_e - n_v + 1})$ be a set of chords with respect to a tree t of a connected linear graph. Also let C_i be a circuit in $t \cup (e_i)$ for $i = 1, 2, \ldots, n_e - n_v + 1$. Then the collection of circuits $C_1, C_2, \ldots, C_{n_e - n_v + 1}$ is called *a set of fundamental circuits* with respect to a tree t.

From this definition, we can see that each chord in a set of chords with respect to a tree t is in exactly one circuit in a set of fundamental circuits with respect to tree t. Hence the ring sum of any of these fundamental circuits will not become empty. Thus these fundamental circuits are linearly independent. For example, for a tree $t = (a, c, e, g)$ in the linear graph in Fig. 3-3-1, a set of fundamental circuits consists of (a, b, c), (c, d, e), and (e, f, g). Obviously, these are linearly independent.

By rearranging an exhaustive circuit matrix B_e so that the first $n_e - n_v + 1$ columns correspond to all chords and the first $n_e - n_v + 1$ rows represent circuits in the set of fundamental circuits with respect to a tree t, we have

$$
B_e = \begin{array}{c} C_1 \\ C_2 \\ \vdots \\ \vdots \\ \vdots \\ C_{n_e - n_v + 1} \\ \vdots \\ \vdots \end{array}
\begin{array}{c} e_1 e_2 \cdots e_{n_e - n_v + 1} \quad e'_1 \cdots e'_{n_v - 1} \\
\left[\begin{array}{cccc|c}
1 & 0 & & 0 & \cdots \\
0 & 1 & & 0 & \cdots \\
0 & 0 & & 0 & \\
\vdots & \vdots & & \vdots & \\
 & & & 0 & \\
0 & 0 & \cdots & 0 & 1 & \cdots \\
\hline
 & & \cdots & & \cdots \\
 & & \cdots & & \cdots
\end{array} \right] \end{array}
\tag{3-3-9}
$$

We can see that there is a unit matrix at the upper left-hand corner of B_e. This is because each chord is in exactly one of the circuits in the set of fundamental circuits with respect to tree t.

The submatrix obtained by taking the first $n_e - n_v + 1$ rows of B_e in Eq. 3-3-9 is called a fundamental circuit matrix and is symbolized by B_f.

Definition 3-3-4. A *fundamental circuit matrix* B_f of a linear graph is a circuit matrix of the form

$$B_f = [U \quad B_{f_{12}}] \tag{3-3-10}$$

consisting of $n_e - n_v + \rho$ rows where ρ is the number of maximal connected subgraphs in the linear graph, and U is a unit matrix. For example, if we take tree $t = (a, c, e, g)$ in the linear graph shown in Fig. 3-3-1, the set of fundamental circuits with respect to tree (a, c, e, g) are

$$C_1 = (b, a, c)$$
$$C_2 = (d, c, e)$$

and

$$C_3 = (f, e, g)$$

Note that chords are b, d, and f. The fundamental circuit matrix with respect to tree $t = (a, c, e, g)$ is

$$B_f = \begin{array}{c} \\ C_1 \\ C_2 \\ C_3 \end{array} \begin{array}{cccccccc} b & d & f & a & c & e & g \\ \begin{bmatrix} 1 & 0 & 0 & \vdots & 1 & 1 & 0 & 0 \\ 0 & 1 & 0 & \vdots & 0 & 1 & 1 & 0 \\ 0 & 0 & 1 & \vdots & 0 & 0 & 1 & 1 \end{bmatrix} \end{array}$$

By taking a different tree, we will have a different fundamental circuit matrix.

Theorem 3-3-2. In a fundamental circuit matrix $[U \quad B_{f_{12}}]$ of a connected linear graph, edges corresponding to columns of $B_{f_{12}}$ form a tree.

Proof. Let an incidence matrix of a connected linear graph be

$$A = [A_{11} \quad A_{12}]$$

where the ith column of A and the ith column of a fundamental circuit matrix $B_f = [U \quad B_{f_{12}}]$ represent the same edge for all $i = 1, 2, \ldots, n_e$. From Theorem 3-3-1,

$$AB_f^t = A_{11} + A_{12}B_{f_{12}}^t = 0$$

or

$$A_{11} = A_{12}B_{f_{12}}^t$$

Hence

$$A = [A_{12}B_{f_{12}}^t \quad A_{12}] = A_{12}[B_{f_{12}}^t \quad U]$$

Since the rank of A is the same as the number of rows of A, A_{12} must be nonsingular. Thus by Theorem 3-2-1, the edges corresponding to the columns of A_{12} form a tree. The columns of A_{12} and the columns of $B_{f_{12}}$ correspond to the same edges. Hence the edges corresponding to the columns of $B_{f_{12}}$ form a tree. QED

For a separated linear graph G consisting of ρ maximal connected subgraphs $G_1, G_2, \ldots, G_\rho$, a fundamental circuit matrix can be partitioned as

$$B_f = \begin{bmatrix} B_{f1} & & & \\ & B_{f2} & & 0 \\ & & \ddots & \\ 0 & & & B_{f\rho} \end{bmatrix} \tag{3-3-11}$$

where each B_{fp} is a fundamental circuit matrix of G_p. The reason is that any edge in G_p cannot be in a circuit in G_q for $1 \leq p \leq q \leq \rho$. Since each B_{fp} is of the form $[U_p \quad B_{fp_{12}}]$, we can rearrange B_f:

$$B_f = \begin{bmatrix} U_1 & & & & B_{f1_{12}} & & & \\ & U_2 & & 0 & & B_{f2_{12}} & & 0 \\ & & \ddots & & & & \ddots & \\ 0 & & & U_\rho & 0 & & & B_{f\rho_{12}} \end{bmatrix} = [U \quad B_{f_{12}}] \tag{3-3-12}$$

Hence to obtain B_f of a separated graph, we use a forest rather than a tree. In other words, we first pick a forest, say t_f of a given separated graph. Then we arrange the columns of B_f in such a way that the first $n_e - n_v + \rho$ columns represent edges which are not in forest t_f. Now we insert the edge e_1 represented by the first column to forest t_f. The resultant graph $t_f \cup (e_1)$ has exactly one circuit, which obviously contains edge e_1. We represent this circuit by the first row of B_f. Thus there will be 1 at the $(1, 1)$ entry of B_f followed by at least $n_e - n_v + \rho - 1$ zeros. Next we insert edge e_2, represented by the second column of B_f, to forest t_f. The resultant graph $t_f \cup (e_2)$ contains exactly one circuit. This circuit will be represented by the second row of B_f, and so on. The resultant matrix B_f will have the form $[U \quad B_{f_{12}}]$. This matrix is called *a fundamental circuit matrix with respect to a forest*.

A fundamental circuit matrix $B_f = [U \quad B_{f_{12}}]$ of a separated linear graph has the same property as that of a connected graph given by Theorem 3-3-2.

Theorem 3-3-3. In a fundamental circuit matrix $[U \quad B_{f_{12}}]$ of a separated linear graph, edges corresponding to columns of $B_{f_{12}}$ form a forest.

The proof is left to the reader.

Since there is a unit matrix in a fundamental circuit matrix, and a fundamental circuit matrix is a submatrix of a circuit matrix B_e, the rank of exhaustive circuit matrix B_e is at least $n_e - n_v + \rho$ where n_e is the number of edges, n_v is the number of vertices, and ρ is the number of maximal connected subgraphs in a linear graph.

Now we need the Sylvester law of nullity to show that the rank of B_e is $n_e - n_v + 1$.

Theorem 3-3-4 (Sylvester). Let $H = [h_{ij}]_{p,e}$ and $K = [k_{ij}]_{e,q}$ be two matrices where $p \le e, q \le e$, and rank of H is p. Suppose

$$HK = 0 \qquad (3\text{-}3\text{-}13)$$

Then the rank of K is at most $e - p$.

Proof. Without the loss of generality, let

$$H = [H_{11} \quad U] \qquad (3\text{-}3\text{-}14)$$

where U is a unit matrix of order p. By partitioning K as

$$K = \begin{bmatrix} K_{11} \\ K_{12} \end{bmatrix} \qquad (3\text{-}3\text{-}15)$$

we have

$$HK = H_{11}K_{11} + K_{12} = 0 \qquad (3\text{-}3\text{-}16)$$

$$K_{12} = -H_{11}K_{11} \qquad (3\text{-}3\text{-}17)$$

This shows that any row of K_{12} can be obtained by a linear combination of rows of K_{11}. This combination is given by a row of $-H_{11}$. Thus the rank of K is at most equal to the number of rows in K_{11}. Since H_{11} is p by $e - p$ matrix, K_{11} must be $e - p$ by q matrix. Thus the rank of K cannot be larger than $e - p$. QED

Let A and B_e be an incidence and an exhaustive circuit matrix of a linear graph consisting of ρ maximal connected subgraphs. By Theorem 3-3-1 we know that

$$AB_e{}^t = 0 \qquad (3\text{-}3\text{-}18)$$

Also by Theorem 3-1-2, the rank of A is $n_v - \rho$. Thus by the Sylvester theorem, the rank of B_e cannot be larger than $n_e - n_v + \rho$. However, we have found that the rank of B_e is at least $n_e - n_v + \rho$.

Theorem 3-3-5. The rank of an exhaustive circuit matrix of a linear graph consisting of ρ maximal connected subgraphs is $n_e - n_v + \rho$ where n_e is the number of edges and n_v is the number of vertices in the linear graph.

Since the rank of an exhaustive circuit matrix B_e is $n_e - n_v + \rho$, a submatrix of B_e whose rank is $n_e - n_v + \rho$ is sufficient to specify all circuits in a linear graph.

Definition 3-3-5. A submatrix B consisting of $n_e - n_v + \rho$ rows of an exhaustive circuit matrix B_e is called a *circuit matrix* if the rank of B is $n_e - n_v + \rho$.

Note that a fundamental circuit matrix consists of $n_e - n_v + \rho$ of linearly independent rows and is obviously a submatrix of an exhaustive circuit matrix B_e. Furthermore, all edges in a linear graph will be represented by columns of the fundamental circuit matrix. Hence a fundamental circuit matrix is a circuit matrix of a linear graph. Also note that if $n_e - n_v + \rho$ circuits are linearly independent (Definition 2-4-1), then the submatrix of an exhaustive circuit matrix whose rows represent these circuits is a circuit matrix. In Section 1-3, we saw that there are sets of generators which produce all members in $\{E\}$ where $\{E\}$ is a collection of all possible circuits, edge disjoint unions of circuits, and the empty set. Since rows in an exhaustive circuit matrix represent all members in $\{E\}$ except the empty set, the circuits represented by the rows in a circuit matrix will form a set of generators which produce all members in $\{E\}$.

Let B_{11} be a nonsingular major submatrix of a circuit matrix B. Then we can express B as

$$B = [B_{11} \quad B_{12}] \tag{3-3-19}$$

Since B_{11} is nonsingular, there exists the inverse. Let B_{11}^{-1} be the inverse of B_{11}. Then by premultiplying B_{11}^{-1} to Eq. 3-3-19, we have

$$B_{11}^{-1}B = [U \quad B_{11}^{-1}B_{12}] \tag{3-3-20}$$

By Definition 3-3-4, a fundamental circuit matrix B_f is a circuit matrix which can be expressed as

$$B_f = [U \quad B_{f_{12}}] \tag{3-3-21}$$

Thus $B_{11}^{-1}B$ is a fundamental circuit matrix. Furthermore, for any nonsingular major submatrix we choose, we can make B a fundamental circuit matrix by multiplying the inverse of the major submatrix. Hence the columns of any nonsingular major submatrix of B can become the columns of U in a fundamental circuit matrix. We know that the edges represented by the columns of U of a fundamental circuit matrix form a set of chords by Theorems 3-3-2 and 3-3-3. Thus the edges corresponding to the columns of any nonsingular major submatrix of B form a set of chords with respect to a forest.

Converse is also true. That is, for any forest we choose, we can form a fundamental circuit matrix in which the columns of the unit matrix represent all chords with respect to the forest. Since any row in an exhaustive circuit matrix B_e can be obtained by linear combination of the rows in a fundamental circuit matrix, and since circuit matrix B is a submatrix of B_e, the columns in B corresponding to the chords with respect to the forest must be the columns that will become the unit matrix in the fundamental circuit matrix with respect to the forest. Thus the columns in B corresponding to the chords with respect to any forest form a nonsingular major submatrix.

Theorem 3-3-4. A major submatrix of a circuit matrix is nonsingular if and only if the columns correspond to all chords with respect to a forest. When $\rho = 1$, a major submatrix of a circuit matrix is nonsingular if and only if the columns correspond to all chords with respect to a tree.

As an example, Table 3-3-1 shows the nonsingular major submatrices of B and the corresponding sets of chords of the linear graph in Fig. 3-2-1 where

$$B = \begin{matrix} a & b & c & d & e \\ \begin{bmatrix} 1 & 1 & 1 & 0 & 0 \\ 0 & 0 & 1 & 1 & 1 \end{bmatrix} \end{matrix}$$

For convenience, we define "branches" as follows.

Definition 3-3-6. Edges in a chosen tree (or a chosen forest when $\rho > 1$) are called *tree-branches* (forest-branches) or simply *branches*.

Because of the relationship between an incidence matrix and a circuit matrix given by Theorem 3-3-1, we can obtain a circuit matrix from an incidence matrix. To see this, suppose t is a forest (tree when $\rho = 1$) of a linear graph G. We arrange the columns of an incidence matrix A of G such that we can partition A:

$$A = [A_{11} \quad A_{12}] \tag{3-3-22}$$

where the columns of A_{11} correspond to the chords with respect to t. Hence the columns of A_{12} represent all branches of t. We also arrange the columns of the fundamental circuit matrix B_f with respect to t such that the ith column of A and B_f represent the same edge for all $i = 1, 2, \ldots, n_e$. Let this fundamental circuit matrix be

$$B_f = [U \quad B_{f_{12}}] \tag{3-3-23}$$

Since

$$B_f A^t = [U \quad B_{f_{12}}] \begin{bmatrix} A_{11}{}^t \\ A_{12}{}^t \end{bmatrix} = 0 \tag{3-3-24}$$

we have

$$A_{11}{}^t + B_{f_{12}} A_{12}{}^t = 0 \tag{3-3-25}$$

or

$$B_{f_{12}} = A_{11}{}^t [A_{12}{}^t]^{-1} \tag{3-3-26}$$

Matrix Representation of Linear Graph and Trees

Table 3-3-1   Nonsingular Major Submatrices

	Nonsingular Major Submatrix	Set of Chords
$\bar{i}_1$	$\begin{matrix} a & c \\ \begin{bmatrix} 1 & 1 \\ 0 & 1 \end{bmatrix} \end{matrix}$	
$\bar{i}_2$	$\begin{matrix} a & d \\ \begin{bmatrix} 1 & 0 \\ 0 & 1 \end{bmatrix} \end{matrix}$	
$\bar{i}_3$	$\begin{matrix} a & e \\ \begin{bmatrix} 1 & 0 \\ 0 & 1 \end{bmatrix} \end{matrix}$	
$\bar{i}_4$	$\begin{matrix} b & c \\ \begin{bmatrix} 1 & 1 \\ 0 & 1 \end{bmatrix} \end{matrix}$	
$\bar{i}_5$	$\begin{matrix} b & d \\ \begin{bmatrix} 1 & 0 \\ 0 & 1 \end{bmatrix} \end{matrix}$	
$\bar{i}_6$	$\begin{matrix} b & e \\ \begin{bmatrix} 1 & 0 \\ 0 & 1 \end{bmatrix} \end{matrix}$	
$\bar{i}_7$	$\begin{matrix} c & d \\ \begin{bmatrix} 1 & 0 \\ 1 & 1 \end{bmatrix} \end{matrix}$	
$\bar{i}_8$	$\begin{matrix} c & e \\ \begin{bmatrix} 1 & 0 \\ 1 & 1 \end{bmatrix} \end{matrix}$	

Note that A_{12} consists of columns corresponding to edges in forest (tree) t. Hence A_{12} is nonsingular. Thus $A_{12}{}^t$ has the inverse which is indicated by $[A_{12}{}^t]^{-1}$.

From Eq. 3-3-26, we can see that by knowing an incidence matrix, a fundamental circuit matrix can be obtained by

$$B_f = [U \quad A_{11}{}^t[A_{12}{}^t]^{-1}] \tag{3-3-27}$$

For example, the fundamental circuit matrix with respect to tree $t = (a, b, d)$

of the linear graph in Fig. 3-2-1 can be obtained from an incidence matrix as follows.

From

$$A = [A_{11} \quad A_{12}] = \begin{matrix} 1 \\ 2 \\ 3 \end{matrix} \begin{bmatrix} \begin{matrix} c & e \end{matrix} & \begin{matrix} a & b & d \end{matrix} \\ 0 & 0 & 1 & 1 & 0 \\ 1 & 0 & 0 & 1 & 1 \\ 0 & 1 & 0 & 0 & 1 \end{bmatrix}$$

$A_{12}{}^t$ is

$$A_{12}{}^t = \begin{bmatrix} 1 & 0 & 0 \\ 1 & 1 & 0 \\ 0 & 1 & 1 \end{bmatrix}$$

Hence $[A_{12}{}^t]^{-1}$ is

$$[A_{12}{}^t]^{-1} = \begin{bmatrix} 1 & 0 & 0 \\ 1 & 1 & 0 \\ 1 & 1 & 1 \end{bmatrix}$$

Now $A_{11}{}^t[A_{12}{}^t]^{-1}$ is simply

$$A_{11}{}^t[A_{12}{}^t]^{-1} = \begin{bmatrix} 0 & 1 & 0 \\ 0 & 0 & 1 \end{bmatrix} \begin{bmatrix} 1 & 0 & 0 \\ 1 & 1 & 0 \\ 1 & 1 & 1 \end{bmatrix} = \begin{bmatrix} 1 & 1 & 0 \\ 1 & 1 & 1 \end{bmatrix}$$

Thus B_f is

$$B_f = \begin{bmatrix} \begin{matrix} c & e \end{matrix} & \begin{matrix} a & b & d \end{matrix} \\ 1 & 0 & 1 & 1 & 0 \\ 0 & 1 & 1 & 1 & 1 \end{bmatrix}$$

3-4 CUT-SET MATRIX

Consider an incidence matrix A. Suppose we add (modulo 2) one row to another row of A; the resultant row will, in general, no longer represent an incidence set but will represent either a cut-set or an edge disjoint union of cut-sets. Hence a matrix obtained by adding rows of A to each other can be called a "cut-set matrix." In fact, a cut-set matrix is a matrix in which each row represents either a cut-set or an edge disjoint union of cut-sets.

Definition 3-4-1. The rows of an exhaustive cut-set matrix, symbolized by Q_e, represent all cut-sets and edge disjoint unions of cut-sets of a linear graph. Each entry q_{ij} of Q_e is

$$q_{ij} = \begin{cases} 1 & \text{if edge } j \text{ is in a cut-set } i \text{ (or an} \\ & \text{edge disjoint union of cut-sets } i) \\ 0 & \text{otherwise} \end{cases} \quad (3\text{-}4\text{-}1)$$

Since an incidence set is either a cut-set or an edge disjoint union of cut-sets, an incidence matrix is a submatrix of Q_e. Thus the rank of Q_e is at least $n_v - \rho$.

By Theorem 2-4-1, we know that any cut-set can be obtained by the ring sum of incidence sets. Moreover, addition of rows (modulo 2) is the same as the ring sum of sets corresponding to the rows.

Theorem 3-4-1. The rank of an exhaustive cut-set matrix Q_e is equal to $n_v - \rho$.

This theorem leads into the following definition.

Definition 3-4-2. A submatrix Q consisting of $n_v - \rho$ rows of an exhaustive cut-set matrix Q_e is called a *cut-set matrix* if the rank of Q is $n_v - \rho$.

For example, an exhaustive cut-set matrix Q_e of the linear graph in Fig. 3-2-1 is

$$Q_e = \begin{array}{c} \\ S(1) \\ S(2) \\ S(3) \\ S_4 \\ S_5 \\ S_6 \\ S_7 \end{array} \begin{array}{ccccc} a & b & c & d & e \\ \begin{bmatrix} 1 & 1 & 0 & 0 & 0 \\ 0 & 1 & 1 & 1 & 0 \\ 0 & 0 & 0 & 1 & 1 \\ 1 & 0 & 1 & 1 & 0 \\ 1 & 1 & 0 & 1 & 1 \\ 0 & 1 & 1 & 0 & 1 \\ 1 & 0 & 1 & 0 & 1 \end{bmatrix} \end{array}$$

where cut-sets S_4, S_5, S_6, and S_7 are

$$S_4 = S(1) \oplus S(2)$$
$$S_5 = S(1) \oplus S(3)$$
$$S_6 = S(2) \oplus S(3)$$

and

$$S_7 = S(1) \oplus S(2) \oplus S(3)$$

By taking rows 4, 5, and 7, we have

$$
Q = \begin{array}{c} \\ S_4 \\ S_5 \\ S_7 \end{array}
\begin{array}{ccccc} a & b & c & d & e \end{array}
\begin{bmatrix} 1 & 0 & 1 & 1 & 0 \\ 1 & 1 & 0 & 1 & 1 \\ 1 & 0 & 1 & 0 & 1 \end{bmatrix}
$$

which is a cut-set matrix because the rank of Q is equal to that of Q_e. Note that S_5 is an edge disjoint union of cut-sets.

By the definition of cut-set matrices, any incidence matrix A is a cut-set matrix. Thus any cut-set matrix Q can be expressed as

$$Q = DA \tag{3-4-2}$$

where D is a nonsingular matrix of order $n_v - \rho$. (Whenever we say that the rank of a matrix is $n_v - \rho$, we are considering a linear graph which consists of ρ maximal connected subgraphs including the case $\rho = 1$.)

By Eq. 3-3-7, we know that

$$AB_e{}^t = 0 \tag{3-4-3}$$

Premultiplying a nonsingular matrix D, we have

$$DAB_e{}^t = QB_e{}^t = 0 \tag{3-4-4}$$

by Eq. 3-4-2. Furthermore, for any cut-set matrix there exists a nonsingular matrix D such that DA is the cut-set matrix. Hence Eq. 3-4-4 can be written as

$$Q_e B_e{}^t = 0 \tag{3-4-5}$$

This is also true when we change Q_e and B_e by a cut-set matrix Q and a circuit matrix B.

Theorem 3-4-2.   Let Q and B be a cut-set and circuit matrices of a linear graph, respectively, such that column p of Q and B represents the same edge for all p. Then

$$QB^t = 0 \tag{3-4-6}$$

and

$$BQ^t = 0 \tag{3-4-7}$$

To show that the columns of a nonsingular major submatrix of a cut-set matrix corresponds to branches of a forest, consider a fundamental circuit matrix

$$B_f = [U \quad B_{f_{12}}] \tag{3-4-8}$$

with respect to a forest t. (Note that when $\rho = 1$, t is a tree.) Suppose we arrange the columns of a cut-set matrix Q such that the pth column of Q and that of B_f represent the same edge for all p. Let Q be partitioned as

$$Q = [Q_{11} \quad Q_{12}] \tag{3-4-9}$$

By Eq. 3-4-5,

$$QB_f{}^t = [Q_{11} \quad Q_{12}]\begin{bmatrix} U \\ B_{f_{12}}^t \end{bmatrix} = Q_{11} + Q_{12}B_{f_{12}}^t = 0 \tag{3-4-10}$$

Hence

$$Q_{11} = Q_{12}B_{f_{12}}^t \tag{3-4-11}$$

or

$$Q = [Q_{12}B_{f_{12}}^t \quad Q_{12}] = Q_{12}[B_{f_{12}}^t \quad U] \tag{3-4-12}$$

Since the rank of Q is $n_v - \rho$, and the number of rows in Q is also $n_v - \rho$, Q_{12} must be nonsingular in order that there exists a nonsingular major submatrix of Q. Since the edges corresponding to the columns of $B_{f_{12}}$ form a forest, the edges corresponding to the columns of Q_{12} must form a forest.

Theorem 3-4-3. If the columns of a major submatrix of a cut-set matrix correspond to the branches of a forest, then it is nonsingular.

Consider a cut-set matrix Q which is partitioned as

$$Q = [Q_{11} \quad Q_{12}] \tag{3-4-13}$$

Suppose Q_{12} is nonsingular. By premultiplying the inverse $Q_{12}{}^{-1}$, Eq. 3-4-13 becomes

$$Q_{12}{}^{-1}Q = [Q_{12}{}^{-1}Q_{11} \quad U] \tag{3-4-14}$$

Since $Q_{12}{}^{-1}$ is a nonsingular matrix, $Q_{12}{}^{-1}Q$ is another cut-set matrix. Let this cut-set matrix be Q', that is,

$$Q' = Q_{12}{}^{-1}Q = [Q'_{11} \quad U] \tag{3-4-15}$$

Let a circuit matrix B be partitioned as

$$B = [B_{11} \quad B_{12}] \tag{3-4-16}$$

where ith column of B and Q' represent the same edge for all i. Then

$$BQ'^t = B_{11}Q'^t_{11} + B_{12} = 0 \tag{3-4-17}$$

Hence

$$B_{12} = B_{11}Q'^t_{11} \tag{3-4-18}$$

or

$$B = [B_{11} \quad B_{11}Q'^t_{11}] = B_{11}[U \quad Q'^t_{11}] \tag{3-4-19}$$

B is a circuit matrix. Thus, the number of rows in B is equal to the rank of B. This means that B_{11} must be nonsingular. By Theorem 3-3-6, the edges corresponding to the columns of B_{11} form a set of chords with respect to a forest. Hence the edges corresponding to the columns of B_{12} form a forest. We have arranged B so that the columns of B_{12} and Q_{12} correspond to the same edges. Thus the edges corresponding to the columns of Q_{12} form a forest. With Theorem 3-4-3, we have the following theorem.

Theorem 3-4-4. A major submatrix of a cut-set matrix Q is nonsingular if and only if the columns correspond to the branches of a forest. When $\rho = 1$, a major submatrix of Q is nonsingular if and only if the columns of the major submatrix correspond to the branches of a tree.

As we have fundamental circuit matrices, we have fundamental cut-set matrices.

Definition 3-4-3. A cut-set matrix Q_f of the form

$$Q_f = [Q_{f_{11}} \quad U] \tag{3-4-20}$$

is called a *fundamental cut-set matrix.*

Note that the edges corresponding to the columns of U form a forest (a tree when $\rho = 1$).

The cut-sets corresponding to the rows of a fundamental cut-set matrix are called fundamental cut-sets.

Definition 3-4-4. Let Q_f be a fundamental cut-set matrix with respect to a forest t (a tree when $\rho = 1$). Then a set of cut-sets represented by the rows of Q_f is called a *set of fundamental cut-sets* with respect to t.

We will develop a method of obtaining a set of fundamental cut-sets with respect to a forest directly from a linear graph next. Let t be a forest by which a fundamental cut-set matrix in Eq. 3-4-20 is obtained. Also let

$$B_f = [U \quad B_{f_{12}}] \tag{3-4-21}$$

be the fundamental circuit matrix with respect to the same forest t with the arrangement of columns so that the ith columns of B_f and Q_f in Eq. 3-4-20 represent the same edge for all $i = 1, 2, \ldots, n_e$. Then

$$Q_f B_f^t = Q_{f_{11}} + B_{f_{12}}^t = 0 \tag{3-4-22}$$

Hence

$$Q_{f_{11}} = B_{f_{12}}^t \tag{3-4-23}$$

or

$$Q_f = [B_{f_{12}}^t \quad U] \tag{3-4-24}$$

Let the ith columns of Q_f represent edge e_i' for $i = 1, 2, \ldots, n_e - n_v + \rho$, and $(n_e - n_v + \rho + p)$th columns of Q_f represent edges e_p for $p = 1, 2, \ldots,$ $n_v - \rho$ as follows:

$$Q_f = \begin{array}{c} \begin{array}{cc} e_1'e_2'\cdots e_{n_e-n_v+\rho}' & e_1 e_2 \cdots e_{n_v-\rho} \end{array} \\ \left[\begin{array}{c|ccc} & 1 & & \\ & & 1 & \\ B_{f_{12}}^t & & & 0 \\ & & & \ddots \\ & 0 & & 1 \end{array}\right] \end{array} \qquad (3\text{-}4\text{-}25)$$

Thus the set of chords consists of $e_1', e_2', \ldots,$ and $e_{n_e-n_v+\rho}'$ and branches of t are $e_1, e_2, \ldots, e_{n_v-\rho}$. We can see that the cut-set S_j represented by row j of Q_f consists of one branch e_j and the chords corresponding to the columns of $B_{f_{12}}^t$ which have 1 at row j. In B_f row j of $B_{f_{12}}^t$ is the column corresponding to branch e_j as shown in Eq. 3-4-26:

$$B_f = \begin{array}{c} \begin{array}{cc} e_1'e_2'\cdots e_{n_e-n_v+\rho}' & e_1 e_2 \cdots e_j \cdots e_{n_v-\rho} \end{array} \\ \left[\begin{array}{ccc|c} 1 & & 0 & \\ & 1 & & \\ & & \ddots & B_{f_{12}} \\ 0 & & 1 & \end{array}\right] \end{array} \qquad (3\text{-}4\text{-}26)$$

Suppose the rth row of $B_{f_{12}}$ has 1 at column j. Note that column j on $B_{f_{12}}$ corresponds to branch e_j, and 1 at the rth row means that in the fundamental circuit C_r, represented by the rth row, contains branch e_j. We obtain circuit C_r by inserting chord e_r' to forest t. It is clear that C_r contains chord e_r'. Thus 1 at the intersection of the rth row and column j of $B_{f_{12}}$ means that the fundamental circuit C_r contains both e_r' and e_j.

On the other hand, 1 at the intersection of the rth row and column j of $B_{f_{12}}$ means the rth column of $B_{f_{12}}^t$ has 1 at row j. Thus from Eq. 3-4-25, fundamental cut-set S_j, represented by row j, contains edge e_r'. Note that fundamental cut-set S_j contains edge e_j because of the unit matrix in Q_f. Therefore both e_r' and e_j are in S_j.

Suppose the rth row of $B_{f_{12}}$ has 0 at column j. Then we can see that fundamental circuit C_r, represented by the rth row of B_f, does not contain edge e_j. Furthermore, if row j has 0 at the rth column of $B_{f_{12}}^t$, fundamental cut-set S_j, represented by row j of Q_f, does not contain edge e_r'.

Theorem 3-4-5. Chord e_r' is in fundamental cut-set S_j, which contains branch e_j, if and only if branch e_j is in fundamental circuit C_r, which contains chord e'_r.

With this theorem, we can obtain a set of fundamental cut-sets as follows. Let t be a forest in a linear graph. Also let S_j be a fundamental cut-set with respect to t. Suppose branch e_j is in S_j. Then to obtain all chords that are in S_j, we insert every chord e'_p one at a time to forest t to form a linear graph $(e'_p) \cup t$. If a circuit in $(e'_p) \cup t$ contains branch e_j, then e'_p is in S_j. Otherwise, chord e'_p is not in S_j.

Example 3-4-1. For the linear graph in Fig. 3-3-1, suppose we choose a tree $t = (a, c, e, g)$. Then fundamental cut-sets S_a, S_c, S_e, and S_g where $a \in S_a$, $c \in S_c$, $e \in S_e$, and $g \in S_g$ can be obtained as follows.

For S_a, first we insert edge b (a chord) to tree t as shown in Fig. 3-4-1. Since the circuit in the resultant graph contains branch a and chord b, chord b is in S_a. We insert chord d to t as shown in Fig. 3-4-2. The circuit

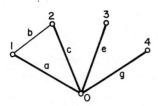

Fig. 3-4-1.   Tree t and chord b.

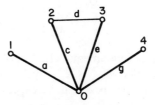

Fig. 3-4-2.   Tree t and chord d.

in the resultant graph does not contain both branch a and chord d. Thus chord d is not in S_a. Similarly, we insert chord f to t as shown in Fig. 3-4-3.

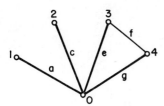

Fig. 3-4-3.   Tree t and chord f.

The circuit in the resultant graph does not contain both branch a and chord f. Hence chord f is not in S_a. Thus we can conclude that fundamental cut-set S_a is

$$S_a = (a, b)$$

For S_c, the insertion of chord b in t gives a circuit containing both chord b and branch c. Hence chord b is in S_c. Similarly, chord d is in S_c. However, chord f is not in S_c because the insertion of chord f into tree t will not produce a circuit which contains both chord f and branch c. Hence

$$S_c = (b, c, d)$$

Similarly, we have

$$S_e = (d, e, f)$$

and

$$S_g = (f, g)$$

Thus

$$
Q_f = \begin{array}{c}
\\
S_a \\
S_c \\
S_e \\
S_g
\end{array}
\begin{array}{cccccccc}
b & d & f & a & c & e & g \\
\left[\begin{array}{ccc|cccc}
1 & 0 & 0 & 1 & 0 & 0 & 0 \\
1 & 1 & 0 & 0 & 1 & 0 & 0 \\
0 & 1 & 1 & 0 & 0 & 1 & 0 \\
0 & 0 & 1 & 0 & 0 & 0 & 1
\end{array}\right]
\end{array}
$$

To obtain a fundamental circuit matrix from a fundamental cut-set matrix Q_f, we note that $Q_{f_{11}}$ is $B_{f_{12}}^t$ by Eq. 3-4-23. Then, by Eq. 3-4-21, a fundamental circuit matrix B_f can be expressed as

$$B_f = [U \quad Q_{f_{11}}^t] \tag{3-4-27}$$

For example, $Q_{f_{11}}$ of the preceding example is

$$
Q_{f_{11}} = \begin{array}{c}
\\
\\
\\
\\
\end{array}
\begin{array}{ccc}
b & d & f \\
\left[\begin{array}{ccc}
1 & 0 & 0 \\
1 & 1 & 0 \\
0 & 1 & 1 \\
0 & 0 & 1
\end{array}\right]
\end{array}
$$

Thus B_f is

$$
B_f = [U \quad Q_{f_{11}}^t] = \begin{array}{ccccccc}
b & d & f & a & c & e & g \\
\left[\begin{array}{ccc|cccc}
1 & 0 & 0 & 1 & 1 & 0 & 0 \\
0 & 1 & 0 & 0 & 1 & 1 & 0 \\
0 & 0 & 1 & 0 & 0 & 1 & 1
\end{array}\right]
\end{array}
$$

This is the fundamental circuit matrix with respect to the same tree

$$t = (a, c, e, g)$$

3-5 REALIZABILITY OF CUT-SET MATRIX (PART 1)

We have learned that circuit and cut-set matrices consist of entries of 1 and 0. Here we will see that not every matrix whose entries are 1 and 0 is either a cut-set matrix or a circuit matrix of a linear graph. In other words, there are some conditions that a matrix should satisfy in order that the matrix be either a cut-set matrix or a circuit matrix of a linear graph. For convenience, we use the following terminology.

Definition 3-5-1. A matrix R is said to be *realizable* as a cut-set (circuit) matrix if there exists a linear graph whose cut-set (circuit) matrix is R.

Since we can obtain a circuit matrix from a cut-set matrix by using Eq. 3-4-19, it is not necessary to study the realizability conditions for circuit matrices. Furthermore, a cut-set matrix can always be expressed as a fundamental cut-set matrix; it is necessary only to find conditions so that a matrix of a form $[R_{11} \quad U]$ is a fundamental cut-set matrix.

First we will study properties of a particular submatrix of a matrix $[R_{11} \quad U]$ called an H-submatrix under the assumption that a linear graph whose cut-set matrix is $[R_{11} \quad U]$ exists.

Definition 3-5-2. Let R be of a form $[R_{11} \quad U]$. An *H-submatrix* with respect to row i of R, symbolized by $H_{(i)}$, is a submatrix of R obtained by (1) deleting all columns which have a nonzero at row i and (2) deleting row i.

For example, suppose R is

$$R = \begin{array}{c} \\ 1 \\ 2 \\ 3 \\ 4 \end{array} \begin{array}{c} b \quad d \quad f \quad a \quad c \quad e \quad g \\ \begin{bmatrix} 1 & 0 & 0 & 1 & 0 & 0 & 0 \\ 1 & 1 & 0 & 0 & 1 & 0 & 0 \\ 0 & 1 & 1 & 0 & 0 & 1 & 0 \\ 0 & 0 & 1 & 0 & 0 & 0 & 1 \end{bmatrix} \end{array}$$

Then we obtain H-submatrix $H_{(1)}$ with respect to row 1 by deleting columns b and a and deleting row 1 as

$$H_{(1)} = \begin{array}{c} \\ 2 \\ 3 \\ 4 \end{array} \begin{array}{c} d \quad f \quad c \quad e \quad g \\ \begin{bmatrix} 1 & 0 & 1 & 0 & 0 \\ 1 & 1 & 0 & 1 & 0 \\ 0 & 1 & 0 & 0 & 1 \end{bmatrix} \end{array}$$

Similarly, $H_{(2)}$ is

$$
H_{(2)} = \begin{array}{c} \\ 1 \\ 3 \\ 4 \end{array}
\begin{array}{cccc}
f & a & e & g \\
\end{array}
\left[
\begin{array}{cccc}
0 & 1 & 0 & 0 \\
1 & 0 & 1 & 0 \\
1 & 0 & 0 & 1
\end{array}
\right]
$$

In order to show that an H-submatrix $H_{(i)}$ with respect to row i is a cut-set matrix of a particular subgraph, we will study the effect of deleting a cut-set on the remaining cut-sets of a linear graph. Let S_i be a fundamental cut-set represented by row i of a fundamental cut-set matrix Q_f of a linear graph G. Then, deleting all columns which have 1 at row i is equivalent to deleting all edges belonging to cut-set S_i from G. As in Eq. 2-2-5, let S_i be expressed as

$$
\mathcal{E}(\Omega \times \Omega) - S_i = \mathcal{E}(\Omega_i \times \Omega_i) \cup \mathcal{E}(\bar{\Omega}_i \times \bar{\Omega}_i) \tag{3-5-1}
$$

where Ω is a set of all vertices in G and cut-set $S_i = \mathcal{E}(\Omega_i \times \bar{\Omega}_i)$. Then by deleting all edges in S_i, we have two subgraphs, $\mathcal{E}(\Omega_i \times \Omega_i)$ and $\mathcal{E}(\bar{\Omega}_i \times \bar{\Omega}_i)$, as shown in Fig. 3-5-1.

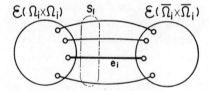

Fig. 3-5-1.   Linear graph G and subgraphs.

Let S_i and S_j be fundamental cut-sets of linear graph G. When all edges in S_i are deleted, it is clear that S_j may not be a fundamental cut-set of the resultant graph G_i. Suppose common edges in S_i and S_j are $e_1, e_2, \ldots, e_k$. Then is a set $S_j - (e_1 e_2 \cdots e_k)$ a fundamental cut-set of G_i? Note that each row of $H_{(i)}$ represents such a set. The answer to this question is given by the following theorem.

Theorem 3-5-1.   Let $S_i = \mathcal{E}(\Omega_i \times \bar{\Omega}_i)$ be a fundamental cut-set with respect to tree (forest) t in a linear graph G and branch e_i be in S_i. Let S_j be another fundamental cut-set containing branch e_j of t. Suppose branch e_j is in subgraph $\mathcal{E}(\Omega_i \times \Omega_i)$. Then a fundamental cut-set S_j' with respect to tree $t_1 = t \cap \mathcal{E}(\Omega_i \times \Omega_i)$ of linear graph $\mathcal{E}(\Omega_i \times \Omega_i)$, which contains branch e_j, will be

$$
S_j' = S_j - S_i \tag{3-5-2}
$$

Note that branches are the edges in a tree (Definition 3-3-6). Also note that $t \cap \mathcal{E}(\Omega_i \times \Omega_i)$ is a portion of tree t, which is in subgraph $\mathcal{E}(\Omega_i \times \Omega_i)$.

Proof. Linear graph $\mathcal{E}(\Omega_i \times \Omega_i) \cup \mathcal{E}(\bar{\Omega}_i \times \bar{\Omega}_i)$ is obviously separated. Hence inserting any chord which is subgraph $\mathcal{E}(\bar{\Omega}_i \times \bar{\Omega}_i)$ into tree t will produce a circuit C which consists only of edges in $\mathcal{E}(\bar{\Omega}_i \times \bar{\Omega}_i)$. This means that branch e_j is not in circuit C. Thus cut-set S_j does not contain any chord in $\mathcal{E}(\bar{\Omega}_i \times \bar{\Omega}_i)$. It is obvious from the definition of a fundamental cut-set that S_j does not contain any branches of t which belong to $\mathcal{E}(\bar{\Omega}_i \times \bar{\Omega}_i)$. Hence S_j does not contain any edge in $\mathcal{E}(\bar{\Omega}_i \times \bar{\Omega}_i)$.

We know that if and only if chord e_p in $\mathcal{E}(\Omega_i \times \Omega_i)$ and tree t_1 produces a circuit which contains branch e_j, chord e_p is in S_j'. However, chord e_p and tree t will produce the same circuit. Thus if chord e_p is in S_j', it is also in S_j, or $S_j' \subset S_j$.

Since S_j does not contain any edge in $\mathcal{E}(\bar{\Omega}_i \times \bar{\Omega}_i)$, any chord which is in S_j but not in S_j' must be a chord in S_i. Thus $S_j' = S_j - S_i$. This argument clearly holds when $\mathcal{E}(\bar{\Omega}_i \times \bar{\Omega}_i)$ is empty. QED

Example 3-5-1. Consider the linear graph in Fig. 3-5-2. Suppose we choose

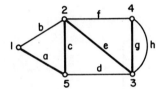

Fig. 3-5-2. A linear graph G.

tree $t = (a, c, e, g)$ and fundamental cut-set $S_i = \mathcal{E}(\Omega_i \times \bar{\Omega}_i) = (d, e, f)$ where $\Omega_i = (1, 2, 5)$ and $\bar{\Omega}_i = (3, 4)$. Linear graph $\mathcal{E}(\Omega_i \times \Omega_i) \cup \mathcal{E}(\bar{\Omega}_i \times \bar{\Omega}_i)$ obtained from G by deleting all edges in S_i is shown in Fig. 3-5-3. We can see that $t_1 = t \cap \mathcal{E}(\Omega_i \times \Omega_i) = (a, c)$ is a tree of subgraph $\mathcal{E}(\Omega_i \times \Omega_i)$.

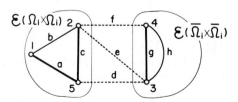

Fig. 3-5-3. Linear graph $\mathcal{E}(\Omega_i \times \Omega_i) \cup \mathcal{E}(\bar{\Omega}_i \times \bar{\Omega}_i)$.

Let $S_j = \mathscr{E}(\Omega_j \times \overline{\Omega}_j) = (b, c, d)$ where $\Omega_j = (1, 5)$ and $\overline{\Omega}_j = (2, 3, 4)$. Note that branch c of t is in S_j. There is only one chord with respect to tree t_1 in $\mathscr{E}(\Omega_i \times \Omega_i)$. This chord b with t_1 will produce a circuit $C = (a, b, c)$, which contains branch c. Hence fundamental cut-set S_j' with respect to tree t_1 in $\mathscr{E}(\Omega_i \times \Omega_i)$ which contains branch c is $S_j' = (b, c)$, and we can see that $S_j' = S_j - S_i$. Now it will be easily shown that H-submatrix $H_{(i)}$ is a fundamental cut-set matrix of subgraph $\mathscr{E}(\Omega_i \times \Omega_i) \cup \mathscr{E}(\overline{\Omega}_i \times \overline{\Omega}_i)$ as follows.

Theorem 3-5-2. Let Q_f be a fundamental cut-set matrix with respect to a tree t of a linear graph G. Let row i of Q_f represent fundamental cut-set $S_i = \mathscr{E}(\Omega_i \times \overline{\Omega}_i)$. Then H-submatrix $H_{(i)}$ with respect to row i of Q_f is a fundamental cut-set matrix with respect to forest $t_1 \cup t_2$ of a linear graph $\mathscr{E}(\Omega_i \times \Omega_i) \cup \mathscr{E}(\overline{\Omega}_i \times \overline{\Omega}_i)$ where $t_1 = t \cap \mathscr{E}(\Omega_i \times \Omega_i)$ and $t_2 = t \cap \mathscr{E}(\overline{\Omega}_i \times \overline{\Omega}_i)$.

Proof. From Theorem 3-5-1, every fundamental cut-set S_j' with respect to tree t_1 in $\mathscr{E}(\Omega_i \times \Omega_i)$ can be expressed as $S_j - S_i$. Similarly, every fundamental cut-set S_k' with respect to tree t_2 in $\mathscr{E}(\overline{\Omega}_i \times \overline{\Omega}_i)$ can be written as $S_k' = S_k - S_i$. An H-submatrix $H_{(i)}$ of Q_f with respect to row i is obtained by deleting all columns that correspond to chords in S_i and by deleting row i, which corresponds to a branch in S_i. Thus every row in $H_{(i)}$ represents $S_p - S_i$ and every row except row i of Q_f is in $H_{(i)}$.   QED

Example 3-5-2. A fundamental cut-set matrix Q_f with respect to tree $t = (a, c, e, g)$ of linear graph G in Fig. 3-5-2 is

$$
Q_f = \begin{array}{c} \\ i \\ 1 \\ 2 \\ 3 \end{array}
\begin{array}{c} \begin{array}{cccccccc} b & d & f & h & e & c & a & g \end{array} \\
\left[\begin{array}{cccc|cccc}
0 & 1 & 1 & 0 & 1 & 0 & 0 & 0 \\
1 & 1 & 0 & 0 & 0 & 1 & 0 & 0 \\
1 & 0 & 0 & 0 & 0 & 0 & 1 & 0 \\
0 & 0 & 1 & 1 & 0 & 0 & 0 & 1
\end{array} \right] \end{array}
$$

An H-submatrix $H_{(i)}$ of Q_f with respect to row i is

$$
H_{(i)} = \begin{array}{c} \\ 1 \\ 2 \\ 3 \end{array}
\begin{array}{c} \begin{array}{ccccc} b & h & c & a & g \end{array} \\
\left[\begin{array}{cc|ccc}
1 & 0 & 1 & 0 & 0 \\
1 & 0 & 0 & 1 & 0 \\
0 & 1 & 0 & 0 & 1
\end{array} \right] \end{array}
$$

which is a fundamental cut-set matrix with respect to forest $t_1 \cup t_2 = (c, a, g)$ of linear graph $\mathscr{E}(\Omega_i \times \Omega_i) \cup \mathscr{E}(\overline{\Omega}_i \times \overline{\Omega}_i)$ in Fig. 3-5-3.

Suppose both $\mathscr{E}(\Omega_i \times \Omega_i)$ and $\mathscr{E}(\overline{\Omega}_i \times \overline{\Omega}_i)$ are nonempty. Then because linear graph $\mathscr{E}(\Omega_i \times \Omega_i) \cup \mathscr{E}(\overline{\Omega}_i \times \overline{\Omega}_i)$ is separated, fundamental cut-sets of $\mathscr{E}(\Omega_i \times \Omega_i) \cup \mathscr{E}(\overline{\Omega}_i \times \overline{\Omega}_i)$ with respect to forest $t_1 \cup t_2$ can be divided into two collections, one of which consists of all fundamental cut-sets with respect to t_1 and the other consists of all fundamental cut-sets with respect to t_2. It can be seen that no fundamental cut-set in one collection has edges in common with any fundamental cut-set in the other collection. Thus H-submatrix $H_{(i)}$, whose rows represent these fundamental cut-sets, can be partitioned as

$$H_{(i)} = \left[\begin{array}{c|c} H_{(i)1} & 0 \\ \hline 0 & H_{(i)2} \end{array}\right] \tag{3-5-3}$$

such that $H_{(i)1}$ represents a fundamental cut-set matrix of linear graph $\mathscr{E}(\Omega_i \times \Omega_i)$ and $H_{(i)2}$ is a fundamental cut-set matrix of linear graph $\mathscr{E}(\overline{\Omega}_i \times \overline{\Omega}_i)$. When $\mathscr{E}(\overline{\Omega}_i \times \overline{\Omega}_i)$ is empty, then $H_{(i)2}$ is also empty.

On the other hand, an arbitrary partition of $H_{(i)}$ into the form in Eq. 3-5-3 may not give the situation that $H_{(i)1}$ and $H_{(i)2}$ are fundamental cut-set matrices of subgraphs $\mathscr{E}(\Omega_i \times \Omega_i)$ and $\mathscr{E}(\overline{\Omega}_i \times \overline{\Omega}_i)$, respectively.

Example 3-5-3. The fundamental cut-set matrix Q_f of the linear graph G in Fig. 3-5-4 with respect to tree $t = (e_1, e_2, e_3, e_4, e_5)$ is

$$Q_f = \begin{array}{c} \\ 1 \\ 2 \\ 3 \\ 4 \\ 5 \end{array} \begin{array}{cccccccccc} a & b & c & d & e_1 & e_2 & e_3 & e_4 & e_5 \\ \left[\begin{array}{ccccccccc} 1 & 0 & 0 & 0 & 1 & 0 & 0 & 0 & 0 \\ 1 & 1 & 0 & 0 & 0 & 1 & 0 & 0 & 0 \\ 1 & 1 & 1 & 0 & 0 & 0 & 1 & 0 & 0 \\ 0 & 1 & 1 & 0 & 0 & 0 & 0 & 1 & 0 \\ 1 & 0 & 0 & 1 & 0 & 0 & 0 & 0 & 1 \end{array}\right] \end{array}$$

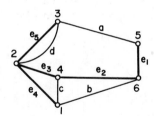

Fig. 3-5-4. A linear graph.

An H-submatrix $H_{(2)}$ is

$$H_{(2)} = \begin{array}{c} \\ 1 \\ 3 \\ 4 \\ 5 \end{array} \begin{array}{cccccc} c & d & e_1 & e_3 & e_4 & e_5 \\ \left[\begin{array}{cccccc} 0 & 0 & 1 & 0 & 0 & 0 \\ 1 & 0 & 0 & 1 & 0 & 0 \\ 1 & 0 & 0 & 0 & 1 & 0 \\ 0 & 1 & 0 & 0 & 0 & 1 \end{array}\right] \end{array}$$

The linear graph $\mathscr{E}(\Omega_2 \times \Omega_2) \cup \mathscr{E}(\overline{\Omega}_2 \times \overline{\Omega}_2)$ obtained from G by deleting all edges in fundamental cut-set $S_2 = (a, b, e_2)$ is shown in Fig. 3-5-5 where $\Omega_2 = (1, 2, 3, 4)$ and $\overline{\Omega}_2 = (5, 6)$.

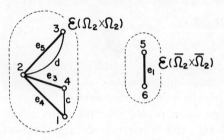

Fig. 3-5-5. Subgraph $\mathscr{E}(\Omega_2 \times \Omega_2) \cup \mathscr{E}(\overline{\Omega}_2 \times \overline{\Omega}_2)$.

By partitioning $H_{(2)}$ as

$$
H_{(2)} = \begin{array}{c} \\ 3 \\ 4 \\ 5 \\ 1 \end{array}
\begin{array}{c} c \quad d \quad e_3 \quad e_4 \quad e_5 \quad e_1 \\
\left[\begin{array}{ccccc|c}
1 & 0 & 1 & 0 & 0 & \\
1 & 0 & 0 & 1 & 0 & 0 \\
0 & 1 & 0 & 0 & 1 & \\
\hline
& & 0 & & & 1
\end{array}\right]
\end{array}
= \left[\begin{array}{c|c} H_{(2)1} & 0 \\ \hline 0 & H_{(2)2} \end{array}\right]
$$

we can see that $H_{(2)1}$ is the fundamental cut-set matrix of $\mathscr{E}(\Omega_2 \times \Omega_2)$ with respect to tree $t_1 = (e_3, e_4, e_5)$ and $H_{(2)2}$ is the fundamental cut-set matrix of $\mathscr{E}(\overline{\Omega}_2 \times \overline{\Omega}_2)$ with respect to tree $t_2 = (e_1)$. On the other hand, we can partition $H_{(2)}$ as

$$
H_{(2)} = \begin{array}{c} \\ 5 \\ 1 \\ 3 \\ 4 \end{array}
\begin{array}{c} d \quad e_5 \quad c \quad e_1 \quad e_3 \quad e_4 \\
\left[\begin{array}{cc|cccc}
1 & 1 & & & 0 & \\
\hline
& & 0 & 1 & 0 & 0 \\
& 0 & 1 & 0 & 1 & 0 \\
& & 1 & 0 & 0 & 1
\end{array}\right]
\end{array}
= \left[\begin{array}{c|c} H_{(2)1} & 0 \\ \hline 0 & H_{(2)2} \end{array}\right]
$$

However, neither $H_{(2)1}$ nor $H_{(2)2}$ of the foregoing partition is the fundamental cut-set matrix of $\mathscr{E}(\Omega_2 \times \Omega_2)$ with respect to tree $t_1 = (e_3, e_4, e_5)$.

Let us next consider how to find out the effect on a fundamental cut-set matrix with respect to a tree t when a branch of t is shorted. We soon will see that the extension of this result and the properties of H-submatrices together give an algorithm for a realizability test of matrices.

Theorem 3-5-3. Let Q_f be a fundamental cut-set matrix with respect to a tree t in a linear graph. Let S_i be a fundamental cut-set represented by row i of Q_f. Let e_i be the branch in S_i. Also let $Q(\bar{i})_f$ be a submatrix of Q_f obtained by (1) deleting row i and (2) deleting all columns of zeros. Then $Q(\bar{i})_f$ is a fundamental cut-set matrix of $G(\bar{e}_i)$ which is obtained from G by shorting branch e_i.

Proof. It is clear that $t - (e_i)$ is a tree of $G(\bar{e}_i)$. Let e_j be a branch of t which is different from e_i. Suppose inserting a chord to tree t produces a circuit G which contains e_j. This circuit will be a circuit containing e_j when branch e_i is shorted. Thus all fundamental cut-sets of G with respect to tree t will be the fundamental cut-sets with respect to tree $t - (e_i)$ except S_i when e_i is shortened.   QED

Suppose we want to obtain a fundamental cut-set matrix M of linear graph G' as shown in Fig. 3-5-6, which is obtained from G in Fig. 3-5-1

$\mathcal{E}(\Omega_i \times \Omega_i)$

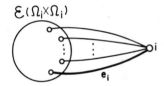

Fig. 3-5-6. Linear graph G'.

making all vertices in subgraph $\mathcal{E}(\bar{\Omega}_i \times \bar{\Omega}_i)$ coincide. Recall that $\mathcal{E}(\bar{\Omega}_i \times \bar{\Omega}_i)$ is one of two maximal connected subgraphs obtained from a linear graph G by deleting all edges in a fundamental cut-set $S_i = \mathcal{E}(\Omega_i \times \bar{\Omega}_i)$ with respect to tree t. We also know that $t_2 = t \cap \mathcal{E}(\bar{\Omega}_i \times \bar{\Omega}_i)$ is a tree of linear graph $\mathcal{E}(\bar{\Omega}_i \times \bar{\Omega}_i)$. Since t_2 contains all vertices in $\mathcal{E}(\bar{\Omega}_i \times \bar{\Omega}_i)$, all vertices in $\mathcal{E}(\bar{\Omega}_i \times \bar{\Omega}_i)$ can be made to coincide by shorting all branches of t_2. Thus by Theorem 3-5-3 a fundamental cut-set matrix M of a linear graph G' with respect to tree $t - t_2$ can be obtained from fundamental cut-set matrix Q_f by deleting all rows that represent fundamental cut-sets containing branches of t_2 and deleting all columns of zeros.

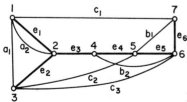

Fig. 3-5-7. A linear graph G.

Example 3-5-4. Consider the linear graph G in Fig. 3-5-7, whose fundamental cut-set matrix Q_f with respect to tree $t = (e_1, e_2, e_3, e_4, e_5, e_6)$ is

$$Q_f = \begin{array}{c} \begin{array}{cccccccccccccc} a_1 & a_2 & c_1 & c_2 & c_3 & b_1 & b_2 & e_1 & e_2 & e_3 & e_4 & e_5 & e_6 \end{array} \\ \left[\begin{array}{ccccccc|cccccc} 1 & 1 & 1 & 0 & 0 & 0 & 0 & 1 & 0 & 0 & 0 & 0 & 0 \\ 1 & 0 & 0 & 1 & 1 & 0 & 0 & 0 & 1 & 0 & 0 & 0 & 0 \\ 0 & 0 & 1 & 1 & 1 & 0 & 0 & 0 & 0 & 1 & 0 & 0 & 0 \\ 0 & 0 & 1 & 1 & 1 & 0 & 1 & 0 & 0 & 0 & 1 & 0 & 0 \\ 0 & 0 & 1 & 0 & 1 & 1 & 1 & 0 & 0 & 0 & 0 & 1 & 0 \\ 0 & 0 & 1 & 0 & 0 & 1 & 0 & 0 & 0 & 0 & 0 & 0 & 1 \end{array} \right] \end{array}$$

The set of fundamental cut-sets with respect to tree t are S_{e_1}, S_{e_2}, S_{e_3}, S_{e_4}, S_{e_5}, and S_{e_6} represented by rows in Q_f.

Let S_i be $S_{e_3} = \mathscr{E}(\Omega_i \times \bar{\Omega}_i) = (c_1, c_2, c_3, e_3)$. Hence $\Omega_i = (1, 2, 3)$ and $\bar{\Omega}_i = (4, 5, 6, 7)$. The corresponding subgraphs $\mathscr{E}(\Omega_i \times \Omega_i)$ and $\mathscr{E}(\bar{\Omega}_i \times \bar{\Omega}_i)$ of S_{e_3} are shown in Fig. 3-5-8. From these subgraphs we can see that $t_1 = (e_1, e_2)$ is a tree of $\mathscr{E}(\Omega_i \times \Omega_i)$ and $t_2 = (e_4, e_5, e_6)$ is a tree of $\mathscr{E}(\bar{\Omega}_i \times \bar{\Omega}_i)$.

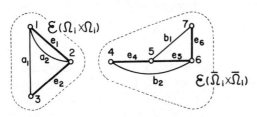

Fig. 3-5-8. Subgraphs $\mathscr{E}(\Omega_i \times \Omega_i)$ and $\mathscr{E}(\bar{\Omega}_i \times \bar{\Omega}_i)$.

The linear graph G' obtained from G by having all vertices in $\mathscr{E}(\bar{\Omega}_i \times \bar{\Omega}_i)$ coincide is shown in Fig. 3-5-9. We can see that $t - t_2 = (e_1, e_2, e_3)$ is a

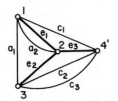

Fig. 3-5-9. Linear graph G'.

tree of G' and the fundamental cut-sets with respect to $t - t_2$ are S_{e_1}, S_{e_2}, and S_{e_3}.

To obtain a fundamental cut-set matrix M of G' with respect to tree $t - t_2$ from Q_f, we delete the rows corresponding to the fundamental cut-sets S_{e_4}, S_{e_5}, and S_{e_6} first. The resultant matrix is

$$
\begin{array}{c}
\\
1\\
2\\
3
\end{array}
\begin{array}{cccccccccccc}
a_1 & a_2 & c_1 & c_2 & c_3 & b_1 & b_2 & e_1 & e_2 & e_3 & e_4 & e_5 & e_6
\end{array}
\begin{bmatrix}
1 & 1 & 1 & 0 & 0 & 0 & 0 & 1 & 0 & 0 & 0 & 0 & 0 \\
1 & 0 & 0 & 1 & 1 & 0 & 0 & 0 & 1 & 0 & 0 & 0 & 0 \\
0 & 0 & 1 & 1 & 1 & 0 & 0 & 0 & 0 & 1 & 0 & 0 & 0
\end{bmatrix}
$$

From this resultant matrix, we delete all columns of zeros to make a fundamental cut-set matrix M of G' with respect to $t - t_2$:

$$
M =
\begin{array}{c}
1\\
2\\
3
\end{array}
\begin{array}{cccccccc}
a_1 & a_2 & c_1 & c_2 & c_3 & e_1 & e_2 & e_3
\end{array}
\begin{bmatrix}
1 & 1 & 1 & 0 & 0 & 1 & 0 & 0 \\
1 & 0 & 0 & 1 & 1 & 0 & 1 & 0 \\
0 & 0 & 1 & 1 & 1 & 0 & 0 & 1
\end{bmatrix}
$$

From Example 3-5-4, it is clear that we can obtain a fundamental cut-set matrix M of G' by (1) deleting all rows corresponding to fundamental cut-sets $S_{e'_1}, S_{e'_2}, \ldots, S_{e'_k}$ where $e'_1, e'_2, \ldots, e'_k$ are branches of tree $t_2 = t \cap \mathscr{E}(\bar{\Omega}_i \times \bar{\Omega}_i)$ of subgraph $\mathscr{E}(\bar{\Omega}_i \times \bar{\Omega}_i)$ and (2) deleting all columns of zeros. Hence we need to know only which rows of Q_f correspond to cut-sets $S_{e'_1}, S_{e'_2}, \ldots, S_{e'_k}$. Now we are ready to use the properties of H-submatrices. Recall that a fundamental cut-set matrix of $\mathscr{E}(\bar{\Omega}_i \times \bar{\Omega}_i)$ is $H_{(i)2}$ in the partition given by Eq. 3-5-3. If there are several ways of partitioning $H_{(i)}$ into the form in Eq. 3-5-3, then one of these partitions will give $H_{(i)2}$, which corresponds to a fundamental cut-set matrix of $\mathscr{E}(\bar{\Omega}_i \times \bar{\Omega}_i)$. Thus the rows that we must delete to obtain a fundamental cut-set matrix M of G' are the rows of $H_{(i)2}$. The columns of zeros produced by deleting these rows obviously correspond to edges in $\mathscr{E}(\bar{\Omega}_i \times \bar{\Omega}_i)$. Thus we can say that M of G' can be obtained from Q_f by deleting all rows and columns of $H_{(i)2}$.

Similarly, a fundamental cut-set matrix of linear graph G'' with respect to tree $t - t_1$ obtained from G in Fig. 3-5-1 by coinciding all vertices of $\mathscr{E}(\Omega_i \times \Omega_i)$ (as shown in Fig. 3-5-10) can be obtained from Q_f by deleting

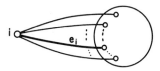

Fig. 3-5-10.   Linear graph G''.

the rows and the columns of $H_{(i)1}$. These fundamental cut-set matrices of G' and G'' are called a pair of M-submatrices.

Definition 3-5-3. Let R be of the form $[R_{11} \quad U]$. For a fixed partition of H-submatrix $H_{(i)}$ of R with respect to row i as in Eq. 3-5-3, a *pair of M-submatrices of R* is a pair of submatrices $M(i)_1$ and $M(i)_2$ where $M(i)_1$ is obtained from R by deleting all rows and columns of $H_{(i)2}$ and $M(i)_2$ is obtained from R by deleting the rows and the columns which belong to $H_{(i)1}$.

Note that when $H_{(i)1}$ is empty, $M(i)_2$ is obtained from R by deleting no rows and no columns. Hence $M(i)_2 = R$. Similarly, if $H_{(i)2}$ is empty, $M(i)_1 = R$.

Example 3-5-5. In Example 3-5-3, we can obtain $H_{(2)1}$ and $H_{(2)2}$ as

$$
H_{(2)1} = \begin{array}{c} \\ 3 \\ 4 \\ 5 \end{array}
\begin{array}{cccccc}
c & d & e_3 & e_4 & e_5 \\
\left[\begin{array}{ccccc}
1 & 0 & 1 & 0 & 0 \\
1 & 0 & 0 & 1 & 0 \\
0 & 1 & 0 & 0 & 1
\end{array}\right]
\end{array}
$$

and

$$
H_{(2)2} = \begin{array}{c} \\ 1 \end{array}
\begin{array}{c}
e_1 \\
[1]
\end{array}
$$

The pair of M-submatrices $M(2)_1$ and $M(2)_2$ of Q_f with respect to row 2 can be obtained by the following process. To obtain $M(2)_1$, we delete row 1 and column e_1 [which form $H_{(2)2}$ from Q_f]. The result is

$$
M(2)_1 = \begin{array}{c} \\ 2 \\ 3 \\ 4 \\ 5 \end{array}
\begin{array}{ccccccccc}
a & b & c & d & e_2 & e_3 & e_4 & e_5 \\
\left[\begin{array}{cccccccc}
1 & 1 & 0 & 0 & 1 & 0 & 0 & 0 \\
1 & 1 & 1 & 0 & 0 & 1 & 0 & 0 \\
0 & 1 & 1 & 0 & 0 & 0 & 1 & 0 \\
1 & 0 & 0 & 1 & 0 & 0 & 0 & 1
\end{array}\right]
\end{array}
$$

Since $H_{(2)1}$ consists of rows 3, 4, and 5 and columns c, d, e_3, e_4, and e_5, deleting these rows and columns gives $M(2)_2$ as

$$
M(2)_2 = \begin{array}{c} \\ 1 \\ 2 \end{array}
\begin{array}{cccc}
a & b & e_1 & e_2 \\
\left[\begin{array}{cccc}
1 & 0 & 1 & 0 \\
1 & 1 & 0 & 1
\end{array}\right]
\end{array}
$$

where

$$
Q_f = \begin{array}{c c}
 & \begin{array}{c c c c c c c c c} a & b & c & d & e_1 & e_2 & e_3 & e_4 & e_5 \end{array} \\
\begin{array}{c} 1 \\ 2 \\ 3 \\ 4 \\ 5 \end{array} &
\left[\begin{array}{c c c c c c c c c}
1 & 0 & 0 & 0 & 1 & 0 & 0 & 0 & 0 \\
1 & 1 & 0 & 0 & 0 & 1 & 0 & 0 & 0 \\
1 & 1 & 1 & 0 & 0 & 0 & 1 & 0 & 0 \\
0 & 1 & 1 & 0 & 0 & 0 & 0 & 1 & 0 \\
1 & 0 & 0 & 1 & 0 & 0 & 0 & 0 & 1
\end{array}\right]
\end{array}
$$

The subgraphs G' and G'' are shown in Fig. 3-5-11.

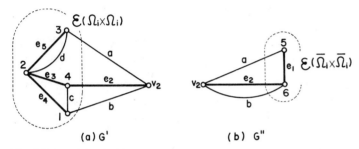

(a) G' (b) G''

Fig. 3-5-11. (a) G'; (b) G''.

We can see that a pair of M-submatrices $M(i)_1$ and $M(i)_2$ of a fundamental cut-set matrix Q_f with respect to row i has the following properties:

1. $M(i)_1$ and $M(i)_2$ are fundamental cut-set matrices of linear graphs.
2. Any row in Q_f except row i is in either $M(i)_1$ or $M(i)_2$ but not in both $M(i)_1$ and $M(i)_2$. Row i of Q_f is in both $M(i)_1$ and $M(i)_2$.
3. Any column in Q_f except columns that have 1 at row i is in either $M(i)_1$ or $M(i)_2$ but not in both $M(i)_1$ and $M(i)_2$. All columns in Q_f that have 1 at row i are in both $M(i)_1$ and $M(i)_2$.
4. In linear graph G' [whose fundamental cut-set matrix is $M(i)_1$ and which is obtained by making vertices coincide in $\mathscr{E}(\bar{\Omega}_i \times \bar{\Omega}_i)$], the fundamental cut-set represented by row i of $M(i)_1$ is an incidence set. Similarly, in linear graph G'' [obtained by making all vertices coincide in $\mathscr{E}(\Omega_i \times \Omega_i)$], the fundamental cut-set represented by row i or $M(i)_2$ is an incidence set.

Properties 1, 2, and 3 are obtained directly from the process of forming $M(i)_1$ and $M(i)_2$ from a fundamental cut-set matrix Q_f. From property 4, we recall that G' is obtained by having all vertices in $\mathscr{E}(\bar{\Omega}_i \times \bar{\Omega}_i)$ coincide.

Hence the fundamental cut-set represented by row i of Q_f becomes an incidence set in G'. Similarly, G'' is obtained by making all vertices in $\mathscr{E}(\Omega_i \times \Omega_i)$ coincide. Hence the fundamental cut-set represented by row i of Q_f becomes an incidence set in G''.

Since we can always obtain a pair of M-submatrices from Q_f and corresponding linear graphs G' and G'' by the procedure discussed previously, we have Theorem 3-5-4.

Theorem 3-5-4. If Q_f is a fundamental cut-set matrix, then there exists a pair of M-submatrices $M(i)_1$ and $M(i)_2$ with respect to a row i of Q_f which have the following two properties:

1. $M(i)_1$ is a fundamental cut-set matrix of a linear graph. Furthermore, the fundamental cut-set represented by row i of $M(i)_1$ is an incidence set in the linear graph.

2. $M(i)_2$ is a fundamental cut-set matrix of a linear graph in which the fundamental cut-set represented by row i of $M(i)_2$ is an incidence set.

Consider linear graph G in Fig. 3-5-1 where fundamental cut-set S_i consists of edges $e_1', e_2', \ldots, e_k'$. Suppose we insert a vertex to the middle point of every edge in S_i as shown in Fig. 3-5-12. After inserting vertices, suppose we

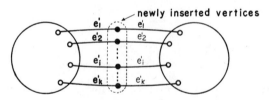

Fig. 3-5-12. Insertion of vertices.

have these vertices coincide to make one vertex i as shown in Fig. 3-5-13. Note that vertex i is a cut-vertex in the resultant linear graph. Now we cut

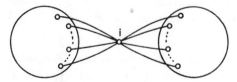

Fig. 3-5-13. Cut vertex i.

vertex i into two so that the resultant linear graph separates as shown in Fig. 3-5-14. We can see that these two graphs G' and G'' are obtained by having all vertices in $\mathscr{E}(\bar{\Omega}_i \times \bar{\Omega}_i)$ and in $\mathscr{E}(\Omega_i \times \Omega_i)$, respectively, coincide.

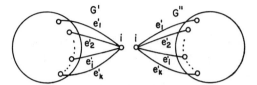

Fig. 3-5-14. Splitting vertex i to form G' and G''.

Suppose G' and G'' in Fig. 3-5-14 are *any* linear graphs which have the properties that (1) $M(i)_1$ and $M(i)_2$ are fundamental cut-set matrices of G' and G'', respectively, and (2) the fundamental cut-set represented by row i of $M(i)_1$ in G' is an incidence set and the fundamental cut-set in G'' represented by row i of $M(i)_2$ is an incidence set. Note that under these conditions, edges that are in both G' and G'' are those connected to one vertex corresponding to the incidence set represented by row i of $M(i)_1$ and $M(i)_2$. Let this vertex be vertex i. That is, vertex i in G' is the vertex corresponding to the incidence set represented by row i of $M(i)_1$, and vertex i in G'' is the vertex corresponding to the incidence set which is represented by row i of $M(i)_2$. Let edges connected to vertex i in G' and G'' be $e_1', e_2', \ldots, e_k'$. Also let vertex v_{1p} in G' be the vertex on which edge e_p' is connected, and vertex v_{2p} be the vertex in G'' on which edge e_p' is connected for $p = 1, 2, \ldots, k$, as shown in Fig. 3-5-15.

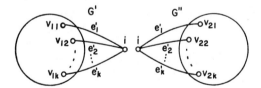

Fig. 3-5-15. Two linear graphs G' and G''.

Then by replacing two e_p' in G' and G'' by one which is connected between vertices v_{1p} and v_{2p} for all $p = 1, 2, \ldots, k$, we have a linear graph G as shown in Fig. 3-5-16. We can see that this linear graph is one whose fundamental cut-set matrix is Q_f. Hence with Theorem 3-5-3, we arrive at Theorem 3-5-5.

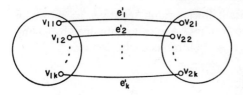

Fig. 3-5-16. Linear graph obtained from G' and G''.

Theorem 3-5-5. A matrix Q_f is realizable as a fundamental cut-set matrix of a linear graph if and only if a pair of M-submatrices $M(i)_1$ and $M(i)_2$ with respect to a row i of matrix Q_f are realizable as fundamental cut-set matrices of linear graphs G' and G'' such that row i of $M(i)_1$ and row i of $M(i)_2$ are incidence sets in G' and G'', respectively.

It is important to note that the names of rows and columns of H-submatrices and M-submatrices are those of corresponding rows and columns of a given matrix Q_f. That is, for example, row i of $M(i)_1$ is the row from row i of Q_f.

As we test a pair of M-submatrices $M(i)_1$ and $M(i)_2$ of a matrix Q_f to see whether Q_f is a fundamental cut-set matrix, we can test a pair of M-submatrices of a matrix $M(i)_1$ to see whether $M(i)_1$ is a fundamental cut-set matrix with the restriction that row i indicates an incidence set. Again, we can take one of this pair as a given matrix and form another pair of M-submatrices to see whether a matrix is a fundamental cut-set matrix with specified rows indicating incidence sets, and so on. To proceed, we employ the following way of indicating the rows representing incidence sets.

Definition 3-5-4. The rows $j_1, j_2, \ldots, j_k$ indicated by the entries inside of the parenthesis of an M-submatrix as $M(j_1 j_2 \cdots j_k)$ are those that must represent incidence sets.

Recall that to obtain a linear graph G from two linear graphs G' and G'' whose fundamental cut-set matrices are $M(i)_1$ and $M(i)_2$, respectively, we only replace edges $e'_p(p = 1, 2, \ldots, k)$ which are connected to vertex i corresponding to the incidence set represented by row i. Hence any incidence set represented by a row in $M(i)_1$ or $M(i)_2$ other than row i will be left to indicate an incidence set. Thus, we have the following theorem.

Theorem 3-5-6. Let $M(j_1 \cdots j_k i)$ and $M(j'_1 \cdots j'_m i)$ be a pair of M-submatrices of a matrix Q_f where $j_1 \neq \cdots \neq j_k \neq j'_1 \neq \cdots \neq j'_m \neq i$. [Note that only i is in both $M(j_1 \cdots j_k i)$ and $M(j'_1 \cdots j'_m i)$.] If and only if $M(j_1 \cdots j_k i)$ and $M(j'_1 \cdots j'_m i)$ are realizable as fundamental cut-set matrices of linear graphs G' and G'' such that (1) the fundamental cut-sets in G' represented by rows $j_1, \ldots, j_k$ and i are incidence sets and (2) the fundamental cut-sets in G''

represented by rows $j'_1, \ldots, j'_m$ and i are incidence sets, then Q_f is realizable as a fundamental cut-set matrix of a linear graph in which fundamental cut-sets represented by rows $j_1, \ldots, j_k, j'_1, \ldots, j'_m$ are incidence sets.

Let $M(i_1)$ be one of a pair of M-submatrices of a fundamental cut-set matrix with respect to row i_1 and $i_2(\neq i_1)$ be a row in $M(i_1)$. Consider a pair of M-submatrices of $M(i_1)$ with respect to row i_2. These two M-submatrices of $M(i_1)$ with respect to row i_2 have the properties that one of these contains row i_1 and row i_2 and the other contains row i_2 but not row i_1. The one that contains both row i_1 and row i_2 is indicated by $M(i_1 i_2)$ and the other, which contains row i_2 but not row i_1, is indicated by $M(i_2)$.

In general, we can obtain a pair of M-submatrices of $M(i_1 i_2 \cdots i_k)$ with respect to row i_{k+1} which is not one of $i_1, i_2, \ldots, i_k$. Let one of these two submatrices contain row $i_1, i_2, \ldots, i_m$ $(m \leq k)$ and, of course, row i_{k+1}. We use the symbol $M(i_1 i_2 \cdots i_m i_{k+1})$ to indicate this submatrix. It is clear that the other submatrix is $M(i_{m+1} \cdots i_k i_{k+1})$. Then, by Theorem 3-5-6, if and only if $M(i_1 \cdots i_m i_{k+1})$ and $M(i_{m+1} \cdots i_k i_{k+1})$ are realizable as fundamental cut-set matrices with the restriction that rows $i_1, i_2, \ldots, i_m, i_{m+1}, \ldots, i_k$, and i_{k+1} represent incidence sets, matrix $M(i_1 i_2 \cdots i_m \cdots i_k)$ is realizable as a fundamental cut-set matrix with rows $i_1, i_2, \ldots, i_m, \ldots, i_k$ representing incidence sets.

We can keep forming a pair of M-submatrices of an M-submatrix $M(i_1 \cdots i_n)$ with respect to row p as long as p is not in the parentheses. When an M-submatrix $M(i_1 i_2 \cdots i_n)$ consists of rows $i_1, i_2, \ldots, i_n$, the M-submatrix is called a minimum M-submatrix.

Definition 3-5-5. An M-submatrix $M(i_1 i_2 \cdots i_n)$ is a *minimum M-submatrix* if $M(i_1 i_2 \cdots i_n)$ consists only of rows $i_1, i_2, \ldots, i_n$.

Since every row in a minimum M-submatrix is in the parentheses, every row must represent an incidence set. Thus a *minimum M-submatrix must be an incidence matrix.*

Since for each row we can form a pair of M-submatrices, from a fundamental cut-set matrix Q_f of a connected linear graph consisting of n_v vertices, we can obtain n_v minimum M-submatrices. Note that Q_f consists of $n_v - 1$ rows. These n_v minimum M-submatrices form a set called a *set of minimum M-submatrices of matrix Q_f.*

Theorem 3-5-7. For a fundamental cut-set matrix Q_f, there exists a set of minimum M-submatrices of Q_f such that each minimum M-submatrix is an incidence matrix.

When an H-submatrix can be partitioned as

$$H = \begin{bmatrix} H_1 & & & \\ & H_2 & & 0 \\ & & \ddots & \\ 0 & & & H_p \end{bmatrix} \qquad (3\text{-}5\text{-}4)$$

then there are many ways of forming a pair of M-submatrices. Thus a set of minimum M-submatrices of a fundamental cut-set matrix Q_f may not be unique which leads to the following theorem.

Theorem 3-5-8. If and only if there exists a set of minimum M-submatrices of a matrix $R = [R_{11} \quad U]$ such that every minimum M-submatrix in the set is an incidence matrix, matrix R is realizable as a fundamental cut-set matrix of a linear graph.

Recall that it is very easy to test whether a matrix is an incidence matrix because every column of an incidence matrix must have at most two 1's. Moreover, if every column of a matrix has at most two 1's, we can construct a linear graph such that the matrix is an incidence matrix of the linear graph.

Example 3-5-6. Suppose a matrix R is given as

$$R = \begin{array}{c} \\ 1 \\ 2 \\ 3 \\ 4 \end{array} \begin{array}{cccccccccc} a & b & c & d & e & f & g & h & i \\ \begin{bmatrix} 1 & 1 & 1 & 0 & 0 & 1 & 0 & 0 & 0 \\ 1 & 0 & 1 & 0 & 0 & 0 & 1 & 0 & 0 \\ 0 & 0 & 1 & 1 & 1 & 0 & 0 & 1 & 0 \\ 0 & 1 & 1 & 0 & 1 & 0 & 0 & 0 & 1 \end{bmatrix} \end{array}$$

H-submatrix $H_{(1)}$, which is obtained by removing every column that has 1 at row 1 and then by removing row 1, can be partitioned as

$$\begin{array}{c} \\ 2 \\ 3 \\ 4 \end{array} \begin{array}{ccccc} g & d & e & h & i \\ \begin{bmatrix} 1 & 0 & 0 & 0 & 0 \\ 0 & 1 & 1 & 1 & 0 \\ 0 & 0 & 1 & 0 & 1 \end{bmatrix} \end{array}$$

Hence the pair of M-submatrices with respect to row 1 are

$$M(1)_1 = \begin{array}{c} \\ 1 \\ 2 \end{array} \begin{array}{ccccc} a & b & c & f & g \\ \begin{bmatrix} 1 & 1 & 1 & 1 & 0 \\ 1 & 0 & 1 & 0 & 1 \end{bmatrix} \end{array}$$

and

$$M(1)_2 = \begin{array}{c} \\ 1 \\ 3 \\ 4 \end{array} \begin{array}{cccccccc} a & b & c & d & e & f & h & i \\ \begin{bmatrix} 1 & 1 & 1 & 0 & 0 & 1 & 0 & 0 \\ 0 & 0 & 1 & 1 & 1 & 0 & 1 & 0 \\ 0 & 1 & 1 & 0 & 1 & 0 & 0 & 1 \end{bmatrix} \end{array}$$

H-submatrix $H_{(2)}$ of $M(1)_1$ is

$$\begin{array}{c} b\ \ f \\ 1[1\ \ 1] \end{array}$$

Thus, by taking $H_{(2)1} = \begin{array}{c} b\ \ f \\ 1[1\ \ 1] \end{array}$ and $H_{(2)2} = \emptyset$, M-submatrix $M(12)$ will be obtained from $M(1)_1$ by deleting no rows and no columns, or

$$M(12) = M(1)_1 \quad \text{(a minimum } M\text{-submatrix)}$$

and M-submatrix $M(2)$ will be obtained from $M(1)_1$ by deleting row 1 and columns b and f as

$$M(2) = \begin{array}{c} a\ \ c\ \ g \\ 2[1\ \ 1\ \ 1] \end{array} \quad \text{(a minimum } M\text{-submatrix)}$$

Note that $M(12) = M(1)_1$ means that every entry of matrix $M(12)$ is equal to the corresponding entry of matrix $M(1)_1$. The symbols $M(12)$ and $M(1)_1$ indicate different restrictions. That is, rows 1 and 2 of $M(12)$ must represent incidence sets. However, only row 1 of $M(1)_1$ need represent an incidence set.

H-submatrix $H_{(3)}$ of $M(1)_2$ is

$$\begin{array}{c} a\ \ b\ \ f\ \ i \\ \begin{bmatrix} 1 & 1 & 1 & 0 \\ 0 & 1 & 0 & 1 \end{bmatrix} \end{array}$$

which cannot be partitioned as in Eq. 3-5-4 with both submatrices being nonempty. Hence, taking $H(3)_2 = \emptyset$, we have

$$M(13) = M(1)_2$$

and

$$M(3) = \begin{array}{c} c\ \ d\ \ e\ \ h \\ 3[1\ \ 1\ \ 1\ \ 1] \end{array} \quad \text{(a minimum } M\text{-submatrix)}$$

Finally, H-submatrix $H_{(4)}$ of $M(13)$, which is obtained from $M(13)$ by deleting every column that has 1 at row 4 and by removing row 4, can be partitioned as

$$
\begin{array}{c}
\begin{array}{cccc} a & f & d & h \end{array} \\
\begin{array}{c} 1 \\ 3 \end{array}
\left[\begin{array}{cc|cc}
1 & 1 & 0 & 0 \\
0 & 0 & 1 & 1
\end{array}\right]
\end{array}
$$

Then the pair of M-submatrices of $M(13)$ with respect to row 4 are

$$
M(14) = \begin{array}{c}
\begin{array}{cccccc} a & b & c & e & f & i \end{array} \\
\begin{array}{c} 1 \\ 4 \end{array}
\left[\begin{array}{cccccc}
1 & 1 & 1 & 0 & 1 & 0 \\
0 & 1 & 1 & 1 & 0 & 1
\end{array}\right]
\end{array} \quad \text{(a minimum } M\text{-submatrix)}
$$

and

$$
M(34) = \begin{array}{c}
\begin{array}{cccccc} b & c & d & e & h & i \end{array} \\
\begin{array}{c} 3 \\ 4 \end{array}
\left[\begin{array}{cccccc}
0 & 1 & 1 & 1 & 1 & 0 \\
1 & 1 & 0 & 1 & 0 & 1
\end{array}\right]
\end{array} \quad \text{(a minimum } M\text{-submatrix)}
$$

Since the set of minimum M-submatrices $M(12)$, $M(2)$, $M(3)$, $M(14)$, and $M(34)$ satisfies Theorem 3-5-8, the given matrix R is a fundamental cut-set matrix of a linear graph.

We can form a linear graph whose fundamental cut-set matrix is Q_f by the following procedure:

1. Form the linear graph whose fundamental cut-set matrix is $M(13)$ from the two linear graphs in Fig. 3-5-17a and b, whose incidence matrices are $M(14)$ and $M(34)$, respectively, as shown in Fig. 3-5-18.

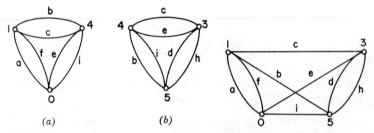

Fig. 3-5-17. Linear graphs. (a) G' for $M(14)$; (b) G'' for $M(34)$.

Fig. 3-5-18. Linear graph for $M(13)$.

2. Construct the linear graph whose fundamental cut-set matrix is Q_f from the two linear graphs of which one is the resultant graph by Step 1 and the

other is that in Fig. 3-5-19, whose incidence matrix is $M(12)$. The resultant graph is shown in Fig. 3-5-20.

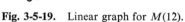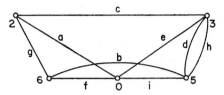

Fig. 3-5-19. Linear graph for $M(12)$.   **Fig. 3-5-20.** Linear graph for Q_f.

Note that it is not necessary to use any linear graphs whose incidence matrix consists of one row to obtain the final linear graph by this procedure. Moreover, it must be noted that Theorem 3-5-8 is also valid for separated linear graphs.

In many cases, we can find that a given matrix is not realizable as a fundamental cut-set matrix before obtaining a set of minimum M-submatrices. One such case is given by the following theorem.

Theorem 3-5-9. Let $M(i_1 i_2 \cdots i_p)$ be an M-submatrix. If there exists a column in which there are at least three 1's in rows $i_1, i_2, \ldots, i_p$, then $M(i_1 i_2 \cdots i_p)$ is not realizable as a fundamental cut-set matrix with rows $i_1, i_2, \ldots, i_p$ representing incidence sets.

Proof. Suppose there are no rows other than $i_1, i_2, \ldots, i_p$ in $M(i_1 i_2 \cdots i_p)$. Then it is obvious that $M(i_1 i_2 \cdots i_p)$ cannot be an incidence matrix. Now we assume that the theorem is true for all matrices of the form $M(\cdots)$ which contain at most k rows $j_1, j_2, \ldots, j_k$ such that these k rows are not in the parentheses. Note that all rows which are in the parentheses must represent incidence sets. Then we will prove the theorem for the matrix $M(i_1 \cdots i_p)$ containing $k + 1$ rows $j_1, j_2, \ldots, j_k, j_{k+1}$ which are not in the parentheses.

Without the loss of generality, suppose column p of $M(i_1 \cdots i_p)$ has three 1's at row $i_1, i_2,$ and i_3.

CASE 1.   Suppose row j_1 has 0 at column p. Also suppose we use row j_1 to form the H-submatrix of $M(i_1 \cdots i_p)$. Then rows $i_1, i_2,$ and i_3 will together be in either H_1 or H_2 when the H-submatrix is partitioned as in Eq. 3-5-3. Thus one of the pair of M-submatrices with respect to row j_1 contains rows $i_1, i_2,$ and i_3. Therefore it contains three 1's at the intersections of column p and rows $i_1, i_2,$ and i_3. Since rows $i_1, i_2,$ and i_3 are in the parentheses, and since this resultant matrix contains at most k rows which are not in the parentheses, the theorem is true for this case.

CASE 2.   Suppose row j_1 has 1 at column p. Let us use row j_1 to form the H-submatrix of $M(i_1 \cdots i_p)$. Then, because column p will be deleted to form the H-submatrix, rows i_1, i_2, and i_3 may not all be in H_1 after partitioning the H-submatrix as in Eq. 3-5-3. If rows i_1, i_2, and i_3 are together in either H_1 or H_2, then the proof is exactly the same as that in Case 1. Suppose row i_1 is in H_1 and rows i_2 and i_3 are in H_2. Then one of the pair of M-submatrices with respect to row j_1 under this partition contains row i_2, i_3, and j_1, all of which have 1 at column p. Since there are at most k rows in this M-submatrix that are not in the parentheses, by assumption, the theorem is true.   QED

The preceding theorem can be expressed another way.

Theorem 3-5-10. If the submatrix of $M(i_1 i_2 \cdots i_p)$ consisting of rows i_1, i_2, $\ldots$, i_p is not an incidence matrix, then $M(i_1 i_2 \cdots i_p)$ is not realizable as a fundamental cut-set matrix with rows i_1, i_2, $\ldots$, i_p being representing incidence sets.

Example 3-5-7.   Consider the matrix

$$
R = \begin{array}{c} \\ 1 \\ 2 \\ 3 \\ 4 \end{array}
\begin{array}{c} a \ b \ c \ d \ e_1 \ e_2 \ e_3 \ e_4 \\ \begin{bmatrix} 1 & 0 & 1 & 1 & 1 & 0 & 0 & 0 \\ 1 & 1 & 0 & 1 & 0 & 1 & 0 & 0 \\ 1 & 1 & 1 & 0 & 0 & 0 & 1 & 0 \\ 1 & 1 & 1 & 1 & 0 & 0 & 0 & 1 \end{bmatrix} \end{array}
$$

By taking row 1, H-submatrix $H_{(1)}$ is

$$
\begin{array}{c} \\ 2 \\ 3 \\ 4 \end{array}
\begin{array}{c} b \ e_2 \ e_3 \ e_4 \\ \begin{bmatrix} 1 & 1 & 0 & 0 \\ 1 & 0 & 1 & 0 \\ 1 & 0 & 0 & 1 \end{bmatrix} \end{array}
$$

Thus the pair of M-submatrices $M_1(1)$ and $M_2(1)$ are

$$M_1(1) = R$$

and

$$
M_2(1) = 1 \begin{array}{c} a \quad c \quad d \quad e_1 \\ [1 \quad 1 \quad 1 \quad 1] \end{array}
$$

By taking row 2 of $M_1(1)$, the H-submatrix is

$$
\begin{array}{c} \\ 1 \\ 3 \\ 4 \end{array}
\begin{array}{c} c \ e_1 \ e_3 \ e_4 \\ \begin{bmatrix} 1 & 1 & 0 & 0 \\ 1 & 0 & 1 & 0 \\ 1 & 0 & 0 & 1 \end{bmatrix} \end{array}
$$

Thus the pair of M-submatrices of $M_1(1)$ with respect to row 2 are

$$M_1(12) = R$$

and

$$M_1(1) = 2 \begin{array}{cccc} a & b & d & e_2 \\ [1 & 1 & 1 & 1] \end{array}$$

Similarly, the pair of M-submatrices of $M_1(12)$ with respect to row 3 are

$$M_1(123) = R$$

and

$$M_1(3) = 3 \begin{array}{cccc} a & b & c & e_3 \\ [1 & 1 & 1 & 1] \end{array}$$

Since $M_1(123)$ contains the submatrix

$$\begin{array}{c} \\ 1 \\ 2 \\ 3 \end{array} \begin{array}{ccccccc} a & b & c & d & e_1 & e_2 & e_3 \\ \begin{bmatrix} 1 & 0 & 1 & 1 & 1 & 0 & 0 \\ 1 & 1 & 0 & 1 & 0 & 1 & 0 \\ 1 & 1 & 1 & 0 & 0 & 0 & 1 \end{bmatrix} \end{array}$$

where every row is in the parentheses (i.e., every row must represent an incidence set), this submatrix must be an incidence matrix. However, column a contains three 1's. Hence this submatrix cannot be an incidence matrix. Thus by Theorem 3-5-10, R cannot be a fundamental cut-set matrix under this set of M-submatrices. Since there is no other set of M-submatrices obtained by using rows 1, 2, and 3, we can conclude that matrix R is not a fundamental cut-set matrix.

The next theorem will be an important one for eliminating unrealizable matrices.

Theorem 3-5-11. For a given matrix $R = [R_{11} \quad U]$, if there exists a submatrix $R' = [R'_{11} \quad U']$ which is not realizable as a fundamental cut-set matrix, then matrix R is not realizable as a fundamental cut-set matrix, where U' is a unit matrix.

Proof. Suppose $R = [R_{11} \quad U]$ is a fundamental cut-set matrix with respect to tree t. Since deleting rows corresponds to shorting branches in t, we can delete rows so that remaining matrix R_a has exactly the same rows as R'. Note that $R_a = [R_{a_{11}} \quad U_a]$ is a fundamental cut-set matrix of a linear graph where U_a is a unit matrix. It can be seen that deleting columns of $R_{a_{11}}$ is equivalent to deleting the chords corresponding to these columns. Hence we can obtain a fundamental cut-set matrix R' of a linear graph by deleting columns of $R_{a_{11}}$ which are not in R'_{11}. This contradicts the assumption that R' is not realizable. Thus R should not be a fundamental cut-set matrix.

QED

Example 3-5-8.   Since the matrix

$$
\begin{array}{ccccccc}
a & b & c & d & e_1 & e_2 & e_3
\end{array}
$$
$$
\begin{bmatrix}
1 & 0 & 1 & 1 & 1 & 0 & 0 \\
1 & 1 & 0 & 1 & 0 & 1 & 0 \\
1 & 1 & 1 & 0 & 0 & 0 & 1
\end{bmatrix}
$$

is not realizable as a fundamental cut-set matrix (see Example 3-5-7), any matrix R of the form $[R_{11} \quad U]$ which contains this matrix as a submatrix is not realizable as a fundamental cut-set matrix.

3-6 TRANSFORMATION FROM A CUT-SET MATRIX TO AN INCIDENCE MATRIX

We have learned that an incidence matrix A consists of rows in an exhaustive cut-set matrix Q_e of a linear graph. A fundamental cut-set matrix Q_f is also formed by the rows of an exhaustive cut-set matrix Q_e. Furthermore, the ranks of both incidence matrix A and a fundamental cut-set matrix Q_f are equal to that of an exhaustive cut-set matrix Q_e. Thus there exists a nonsingular matrix D such that

$$A = DQ_f \tag{3-5-5}$$

We must now determine how to obtain D satisfying Eq. 3-5-5.

Let $\{M_j; \ j = 1, 2, \ldots, n_v - \rho + 1\}$ be a set of minimum M-submatrices of Q_f which has been studied previously. Also let $\{g_j; \ j = 1, 2, \ldots, n_v - \rho + 1\}$ be a set of linear graphs such that an incidence matrix of g_j is M_j for $j = 1, 2, \ldots, n_v - \rho + 1$. It is clear that g_j has a reference vertex v_j which is not represented by any row of incidence matrix M_j. Note that every vertex represented by a row in a matrix M_j in $\{M_j; \ j = 1, 2, \ldots, n_v - \rho + 1\}$ appears in exactly two linear graphs in $\{g_j; \ j = 1, 2, \ldots, n_v - \rho + 1\}$. Moreover, when we combine all linear graphs in $\{g_j; \ j = 1, 2, \ldots, n_v - \rho + 1\}$ to obtain a linear graph G whose fundamental cut-set matrix is a given matrix Q_f, all vertices represented by the rows in any matrix M_j in $\{M_j; \ j = 1, 2, \ldots, n_v - \rho + 1\}$ will be removed. In other words, linear graph G has no vertices represented by rows in $M_j \in \{M_j; \ j = 1, 2, \ldots, n_v - \rho + 1\}$. It is also clear that during the process of forming G from $g_j \in \{g_j = 1, 2, \ldots, n_v - \rho + 1\}$, we will not produce any new vertices. Thus G will consist only of vertices that are the reference vertices of g_j.

Example 3-5-9.   Consider a fundamental cut-set matrix Q_f where

$$
Q_f = \begin{array}{c} \\ 1 \\ 2 \\ 3 \\ 4 \end{array}
\begin{array}{c} \begin{array}{ccccccccc} a & b & c & d & e & f & g & h & i \end{array} \\
\begin{bmatrix}
1 & 1 & 1 & 0 & 0 & 1 & 0 & 0 & 0 \\
1 & 0 & 1 & 0 & 0 & 0 & 1 & 0 & 0 \\
0 & 0 & 1 & 1 & 1 & 0 & 0 & 1 & 0 \\
0 & 1 & 1 & 0 & 1 & 0 & 0 & 0 & 1
\end{bmatrix}
\end{array}
$$

which is equal to R in Example 3-5-6. From the results in Example 3-5-6, a set of minimum M-submatrices of Q_f is

$$\{M_1, M_2, M_3, M_4, M_5\}$$

where $M_1 = M(12)$, $M_2 = M(2)$, $M_3 = M(3)$, $M_4 = M(14)$, and $M_5 = M(34)$. A set of linear graphs $\{g_1, g_2, g_3, g_4, g_5\}$ is given in Fig. 3-5-21. Note

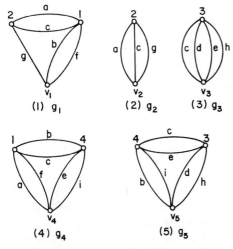

Fig. 3-5-21.   Linear graphs in (1) g_1, (2) g_2, (3) g_3, (4) g_4, and (5) g_5.

that the reference vertex of g_j is symbolized by v_j. By combining all these linear graphs, we will have a linear graph G as shown in Fig. 3-5-22 whose fundamental cut-set matrix is Q_f. Note that the vertices in G are the reference vertices of g_1, g_2, g_3, g_4, and g_5.

We know that the incidence set corresponding to the reference vertex v_j in g_j is equal to the ring sum of all incidence sets corresponding to the rows of M_j. In other words, row j of an incidence matrix A of linear graph G, which

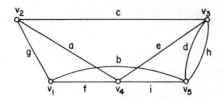

Fig. 3-5-22. Linear graph G.

corresponds to the reference vertex v_j of g_j, can be obtained by the sum of all rows of M_j. Thus the jth row of a nonsingular matrix D must make the summation of the rows which belong to M_j. From DQ_f, we can see that ith column of D corresponds to the ith row of Q_f. In other words, to make a summation of the j_1th, the j_2th, ..., and the j_kth rows of Q_f by multiplying the row i of D, we must give 1 at the entries $(i, j_1), (i, j_2), \ldots, (i, j_k)$ of D.

Theorem 3-5-12. Let Q_f be a fundamental cut-set matrix of a linear graph G consisting of ρ maximal connected subgraphs. Then any matrix D satisfying two conditions:

1. The $n_v - \rho$ rows of D correspond to $n_v - \rho$ of the minimum M-submatrices of Q_f.
2. The entry d_{ij} of $D = [d_{ij}]$ is

$$d_{ij} = \begin{cases} 1 & \text{if the minimum } M\text{-submatrix corresponding to the } i\text{th} \\ & \text{row of } D \text{ contains the } j\text{th row of } Q_f \\ 0 & \text{otherwise} \end{cases}$$

has the property that DQ_f is an incidence matrix of linear graph G.

Suppose there exists a nonsingular matrix $\underline{D}$ such that $\underline{D}Q_f$ is an incidence matrix of a linear graph G. Then from this linear graph G with knowing fundamental cut-sets given by Q_f, we can obtain a set $\{g_j; \ j = 1, 2, \ldots, n_v - \rho + 1\}$ of linear graphs corresponding to a set $\{M_j; \ j = 1, 2, \ldots, n_v - \rho + 1\}$ of minimum M-submatrices of Q_f. Thus we can obtain a nonsingular matrix D by Theorem 3-5-12 such that DQ_f is an incidence matrix of G where row i of DQ_f and row i of DQ_f represent the same vertex in G. Hence

$$\underline{D}Q_f = DQ_f \qquad (3\text{-}5\text{-}6)$$

Since $\underline{D}$ is nonsingular, we can conclude that

$$\underline{D} = D \qquad (3\text{-}5\text{-}7)$$

Theorem 3-5-13. Let Q_f be a fundamental cut-set matrix of a linear graph G. For any nonsingular matrix D which has the property that DQ_f is an incidence matrix, there exists a set of minimum M-submatrices of Q_f such that D satisfies the conditions in Theorem 3-5-12.

Example 3-5-10. Consider a fundamental cut-set matrix Q_f given in Example 3-5-6. Suppose we choose M_1, M_2, M_3, and M_4 of $\{M_1, M_2, M_3, M_4, M_5\}$. Then by Theorem 3-5-12, we can obtain a nonsingular matrix D as

$$D = \begin{array}{c} \\ M_1 \\ M_2 \\ M_3 \\ M_4 \end{array} \begin{array}{cccc} 1 & 2 & 3 & 4 \\ \left[\begin{array}{cccc} 1 & 1 & 0 & 0 \\ 0 & 1 & 0 & 0 \\ 0 & 0 & 1 & 0 \\ 1 & 0 & 0 & 1 \end{array}\right] \end{array}$$

and we have

$$DQ_f = \begin{bmatrix} 1 & 1 & 0 & 0 \\ 0 & 1 & 0 & 0 \\ 0 & 0 & 1 & 0 \\ 1 & 0 & 0 & 1 \end{bmatrix} \begin{bmatrix} 1 & 1 & 1 & 0 & 0 & 1 & 0 & 0 & 0 \\ 1 & 0 & 1 & 0 & 0 & 0 & 1 & 0 & 0 \\ 0 & 0 & 1 & 1 & 1 & 0 & 0 & 1 & 0 \\ 0 & 1 & 1 & 0 & 1 & 0 & 0 & 0 & 1 \end{bmatrix}$$

$$= \begin{bmatrix} 0 & 1 & 0 & 0 & 0 & 1 & 1 & 0 & 0 \\ 1 & 0 & 1 & 0 & 0 & 0 & 1 & 0 & 0 \\ 0 & 0 & 1 & 1 & 1 & 0 & 0 & 1 & 0 \\ 1 & 0 & 0 & 0 & 1 & 1 & 0 & 0 & 1 \end{bmatrix}$$

which is clearly an incidence matrix.

PROBLEMS

1. Write a fundamental circuit matrix and a fundamental cut-set matrix of a linear graph in Fig. P-3-1 with respect to (a) tree $t = (a, b, c, d, e)$ and (b) tree $t = (e, f, g, h, i)$.

Fig. P-3-1.

2. Suppose a connected linear graph g consists of n_v vertices. Prove that (a) if g contains $n_v - 1$ edges, g is a tree, and (b) if g contains no circuits, g is a tree.

3. Prove that any edge in a connected linear graph can be in a tree.

4. Prove or disprove the statement "for any path in a connected linear graph, there exists a tree which contains the path."

5. Prove that the set of chords with respect to any tree in a complete graph forms a connected subgraph.

6. Suppose degree $d(v)$ of a vertex v in a tree t is k. Then prove that there are at least k vertices of degree 1 in t.

7. Prove that the subgraph consisting of all chords with respect to any tree in a complete graph G contains at least one circuit if G contains at least five vertices.

8. Prove that any tree and any cut-set have at least one edge in common.

9. Prove Theorem 3-2-3.

10. Prove Theorem 3-3-3.

11. Let t be a tree in a linear graph G. Let b be a chord. Suppose the circuit in $t \cup (b)$ consists of edges $b, e_1, e_2, \ldots, e_k$. Then show that $t \cup (b) - (e_p)$ is a tree for $1 \le p \le k$. Note that $t \cup (b) - (e_p)$ consists of edge b and all edges in t except e_p.

12. Let e be an edge in a tree t. Also set $S = (e, a_1, a_2, \ldots, a_k)$ be a fundamental cut-set with respect to t. (Hence all edges $a_1, a_2, \ldots, a_k$ are not in t.) Prove that $t \cup (a_p) - (e)$ is a tree for $1 \le p \le k$. Note that $t \cup (a_p) - (e)$ consists of edge a_p and all edges in t except edge e.

13. Test whether the following matrix is realizable as a fundamental cut-set matrix:

$$\begin{bmatrix} 1 & 1 & 0 & 0 & 0 & 1 & 0 & 0 & 0 & 0 & 0 \\ 1 & 0 & 1 & 0 & 0 & 0 & 1 & 0 & 0 & 0 & 0 \\ 1 & 0 & 0 & 1 & 0 & 0 & 0 & 1 & 0 & 0 & 0 \\ 0 & 0 & 0 & 1 & 1 & 0 & 0 & 0 & 1 & 0 & 0 \\ 0 & 0 & 1 & 0 & 1 & 0 & 0 & 0 & 0 & 1 & 0 \\ 0 & 1 & 0 & 0 & 1 & 0 & 0 & 0 & 0 & 0 & 1 \end{bmatrix}$$

14. Is there a matrix consisting of 1 and 0 which can be neither a cut-set matrix nor a circuit matrix? If there is such a matrix, show one.

15. Find a linear graph whose fundamental cut-set matrix is

$$
\begin{array}{cccccccccc}
1 & 2 & 3 & 4 & 5 & 6 & 7 & 8 & 9 & 10
\end{array}
$$
$$
\begin{bmatrix}
1 & 1 & 1 & 0 & 0 & 0 & 1 & 0 & 0 & 0 \\
1 & 0 & 0 & 1 & 1 & 0 & 0 & 1 & 0 & 0 \\
0 & 0 & 0 & 1 & 1 & 1 & 0 & 0 & 1 & 0 \\
0 & 0 & 1 & 0 & 1 & 1 & 0 & 0 & 0 & 1
\end{bmatrix}
$$

16. Is there a linear graph whose fundamental circuit matrix is the matrix in Problem 13?

17. Find a linear graph whose fundamental circuit matrix is

$$
\begin{array}{ccccccccc}
1 & 2 & 3 & 4 & 5 & 6 & 7 & 8 & 9
\end{array}
$$
$$
\begin{bmatrix}
1 & 1 & 1 & 0 & 0 & 1 & 0 & 0 & 0 \\
1 & 1 & 1 & 1 & 1 & 0 & 1 & 0 & 0 \\
0 & 1 & 1 & 1 & 0 & 0 & 0 & 1 & 0 \\
0 & 0 & 1 & 1 & 1 & 0 & 0 & 0 & 1
\end{bmatrix}
$$

18. Find a linear graph whose fundamental circuit matrix is

$$
\begin{array}{ccccccccc}
1 & 2 & 3 & 4 & 5 & 6 & 7 & 8 & 9
\end{array}
$$
$$
\begin{bmatrix}
1 & 0 & 1 & 0 & 1 & 1 & 0 & 0 & 0 \\
0 & 0 & 1 & 1 & 1 & 0 & 1 & 0 & 0 \\
1 & 1 & 0 & 1 & 0 & 0 & 0 & 1 & 0 \\
0 & 1 & 0 & 1 & 1 & 0 & 0 & 0 & 1
\end{bmatrix}
$$

CHAPTER
4
PLANAR GRAPHS

4-1 TWO-ISOMORPHIC GRAPHS

We now turn to the properties of a particular class of linear graphs, planar graphs, not only because of the four-color problem which ignites the research on planar graphs but, more important, because of the increasing demands on planarity in modern technology such as in the area of electronic industries. We are especially concerned with Whitney's duality, Kuratowski's planar graphs, and application of the properties of planar graphs to prove Tutte's realizability condition of cut-set (circuit) matrices. But first, we will study two types of equivalence of linear graphs.

Consider linear graphs G and G' in Fig. 4-1-1a and b. By comparing the

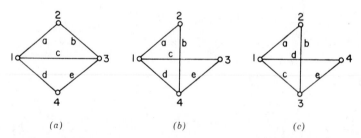

Fig. 4-1-1. Isomorphic graphs. (a) Linear graph G; (b) linear graph G'; (c) remaining vertices and edges of G'.

two endpoints of edge b, we can see that G and G' are different linear graphs. However, if we rename vertices and edges of G' as shown in Fig. 4-1-1c (rename vertex 3 as 4, vertex 4 as 3, edge c as d, and edge d as c), the resultant linear graph becomes identical to G. Hence we could say that G and G' are similar. We call these graphs isomorphic to each other.

138

Definition 4-1-1. Two linear graphs G and G' are *isomorphic* to each other if there exists a $1:1$ correspondence between the edges of G and G' such that the incidence relationships will be preserved.

If instead of preserving the incidence relationship, the correspondence between the circuits of G and G' is preserved, these two linear graphs are said to be 2-isomorphic to each other.

Definition 4-1-2. Linear graphs G and G' are said to be *2-isomorphic* to each other if there exists a $1:1$ correspondence between the edges such that the circuit relationships will be preserved.

For example, linear graphs G and G' in Fig. 4-1-2a and b are 2-isomorphic,

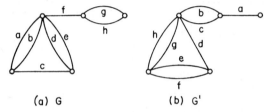

(a) G   (b) G'

Fig. 4-1-2. 2-isomorphic graphs (*a*) G and (*b*) G'.

because by making a $1:1$ correspondence between the edges of G and G' as shown in Table 4-1-1, the circuits in G correspond to the circuits in G' as shown in Table 4-1-2.

Table 4-1-1 One to One Correspondence between the Edges of G and G'

Edges in G	Edges in G'
a	h
b	g
c	d
d	f
e	e
f	a
g	b
h	c

Table 4-1-2 One to One Correspondence between the Circuits of G and G'

Circuits in G	Circuits in G'
ab	hg
de	ef
gh	bc
acd	hdf
ace	hde
bcd	gdf
bce	gde

It can be seen from the definition of isomorphism that if linear graphs G and G' are isomorphic, then by interchanging rows and columns (if necessary), the exhaustive incidence matrix A_e of G becomes identical with the exhaustive incidence matrix A_e' of G'. For example, the exhaustive incidence matrix of G in Fig. 4-1-1a is

$$A_e = \begin{array}{c} \\ 1 \\ 2 \\ 3 \\ 4 \end{array} \begin{array}{ccccc} a & b & c & d & e \\ \left[\begin{array}{ccccc} 1 & 0 & 1 & 1 & 0 \\ 1 & 1 & 0 & 0 & 0 \\ 0 & 1 & 1 & 0 & 1 \\ 0 & 0 & 0 & 1 & 1 \end{array}\right] \end{array}$$

By interchanging the third and fourth rows, we have

$$\begin{array}{c} \\ 1 \\ 2 \\ 3 \\ 4 \end{array} \begin{array}{ccccc} a & b & c & d & e \\ \left[\begin{array}{ccccc} 1 & 0 & 1 & 1 & 0 \\ 1 & 1 & 0 & 0 & 0 \\ 0 & 0 & 0 & 1 & 1 \\ 0 & 1 & 1 & 0 & 1 \end{array}\right] \end{array}$$

Now by interchanging the third and fourth columns, we have

$$\begin{array}{c} \\ 1 \\ 2 \\ 3 \\ 4 \end{array} \begin{array}{ccccc} a & b & c & d & e \\ \left[\begin{array}{ccccc} 1 & 0 & 1 & 1 & 0 \\ 1 & 1 & 0 & 0 & 0 \\ 0 & 0 & 1 & 0 & 1 \\ 0 & 1 & 0 & 1 & 1 \end{array}\right] \end{array}$$

which is the exhaustive incidence matrix A_e' of G' in Fig. 4-1-1b.

On the other hand, if linear graphs G and G' are 2-isomorphic, then by interchanging rows and columns (if necessary), an exhaustive circuit matrix B_e of G becomes the same as that of G'. For example, suppose we use the following matrices B and B' of G and G' in Fig. 4-1-2a and b, respectively:

$$
B = \begin{array}{c} \\ C_1 \\ C_2 \\ C_3 \\ C_4 \end{array}
\begin{array}{cccccccc} a & b & c & d & e & f & g & h \\ \end{array}
\left[\begin{array}{cccccccc}
1 & 1 & 0 & 0 & 0 & 0 & 0 & 0 \\
0 & 0 & 0 & 1 & 1 & 0 & 0 & 0 \\
0 & 0 & 0 & 0 & 0 & 0 & 1 & 1 \\
1 & 0 & 1 & 1 & 0 & 0 & 0 & 0
\end{array}\right]
$$

$$
B' = \begin{array}{c} \\ C_1' \\ C_2' \\ C_3' \\ C_4' \end{array}
\begin{array}{cccccccc} a & b & c & d & e & f & g & h \\ \end{array}
\left[\begin{array}{cccccccc}
0 & 0 & 0 & 0 & 0 & 0 & 1 & 1 \\
0 & 0 & 0 & 0 & 1 & 1 & 0 & 0 \\
0 & 1 & 1 & 0 & 0 & 0 & 0 & 0 \\
0 & 0 & 0 & 1 & 1 & 0 & 0 & 1
\end{array}\right]
$$

Then the exhaustive circuit matrix B_e of G is

$$
B_e = \begin{array}{c} \\
C_1 \\ C_2 \\ C_3 \\ C_4 \\
C_1 \oplus C_2 \\ C_1 \oplus C_3 \\ C_1 \oplus C_4 \\
C_2 \oplus C_3 \\ C_2 \oplus C_4 \\ C_3 \oplus C_4 \\
C_1 \oplus C_2 \oplus C_3 \\ C_1 \oplus C_2 \oplus C_4 \\
C_1 \oplus C_3 \oplus C_4 \\ C_2 \oplus C_3 \oplus C_4 \\
C_1 \oplus C_2 \oplus C_3 \oplus C_4 \end{array}
\begin{array}{cccccccc} a & b & c & d & e & f & g & h \\ \end{array}
\left[\begin{array}{cccccccc}
1 & 1 & 0 & 0 & 0 & 0 & 0 & 0 \\
0 & 0 & 0 & 1 & 1 & 0 & 0 & 0 \\
0 & 0 & 0 & 0 & 0 & 0 & 1 & 1 \\
1 & 0 & 1 & 1 & 0 & 0 & 0 & 0 \\
1 & 1 & 0 & 1 & 1 & 0 & 0 & 0 \\
1 & 1 & 0 & 0 & 0 & 0 & 1 & 1 \\
0 & 1 & 1 & 1 & 0 & 0 & 0 & 0 \\
0 & 0 & 0 & 1 & 1 & 0 & 1 & 1 \\
1 & 0 & 1 & 0 & 1 & 0 & 0 & 0 \\
1 & 0 & 1 & 1 & 0 & 0 & 1 & 1 \\
1 & 1 & 0 & 1 & 1 & 0 & 1 & 1 \\
0 & 1 & 1 & 0 & 1 & 0 & 0 & 0 \\
0 & 1 & 1 & 1 & 0 & 0 & 1 & 1 \\
1 & 0 & 1 & 0 & 1 & 0 & 1 & 1 \\
0 & 1 & 1 & 0 & 1 & 0 & 1 & 1
\end{array}\right]
$$

By interchanging columns of B_e so that the order of columns becomes f, g, h, c, a, b, d, e,

$$
B_e = \begin{array}{r} \\ 1 \\ 2 \\ 3 \\ 4 \\ 5 \\ 6 \\ 7 \\ 8 \\ 9 \\ 10 \\ 11 \\ 12 \\ 13 \\ 14 \\ 15 \end{array}
\begin{array}{cccccccc}
f & g & h & c & a & b & d & e \\
\hline
0 & 0 & 0 & 0 & 1 & 1 & 0 & 0 \\
0 & 0 & 0 & 0 & 0 & 0 & 1 & 1 \\
0 & 1 & 1 & 0 & 0 & 0 & 0 & 0 \\
0 & 0 & 0 & 1 & 1 & 0 & 1 & 0 \\
0 & 0 & 0 & 0 & 1 & 1 & 1 & 1 \\
0 & 1 & 1 & 0 & 1 & 1 & 0 & 0 \\
0 & 0 & 0 & 1 & 0 & 1 & 1 & 0 \\
0 & 1 & 1 & 0 & 0 & 0 & 1 & 1 \\
0 & 0 & 0 & 1 & 1 & 0 & 0 & 1 \\
0 & 1 & 1 & 1 & 1 & 0 & 1 & 0 \\
0 & 1 & 1 & 0 & 1 & 1 & 1 & 1 \\
0 & 0 & 0 & 1 & 0 & 1 & 0 & 1 \\
0 & 1 & 1 & 1 & 0 & 1 & 1 & 0 \\
0 & 1 & 1 & 1 & 1 & 0 & 0 & 1 \\
0 & 1 & 1 & 1 & 0 & 1 & 0 & 1
\end{array}
$$

Then C'_1 of B' is equal to the second row, C'_2 is the first row, C'_3 is the third row, and C'_4 is the ninth row of B_e. Thus by interchanging rows, this matrix becomes the same as the exhaustive circuit matrix B'_e of G'. Note that circuits in B and B' may not correspond to each other even if G and G' are 2-isomorphic to each other. This is the reason why we use the exhaustive circuit matrix B_e in this example.

Since all rows of an exhaustive circuit matrix can be generated from a fundamental circuit matrix, if a fundamental circuit matrix of one linear graph is also a fundamental circuit matrix of another linear graph, then these linear graphs are 2-isomorphic to each other.

Theorem 4-1-1. Linear graphs are 2-isomorphic to each other if and only if they have the same fundamental circuit matrix.

By use of Eq. 3-4-24, we can modify this theorem.

Theorem 4-1-2. Linear graphs are 2-isomorphic to each other if and only if they have the same fundamental cut-set matrix.

These theorems may not be practical for testing whether two linear graphs are 2-isomorphic with each other. However, they are rather important. For example, we can say that all linear graphs that satisfy a given fundamental cut-set matrix are 2-isomorphic to each other, which will be a great help when we design linear graphs in later chapters.

The next method of testing 2-isomorphism is a geometrical one introduced by Whitney. If linear graphs G and G' are 2-isomorphic with each other, then they become isomorphic under successive applications of either or both of the following operations:

1. If there exists a cut vertex, cut it in two so that the number of maximal connected subgraphs will be increased by one.

2. If a linear graph consists of two subgraphs H and $\overline{H}$, which have only two vertices, say p and q, in common, then interchange the name of these two vertices p and q in H. Geometrically, this operation is equivalent to turning subgraph H around at these vertices.

For example, linear graphs G and G' in Fig. 4-1-3a and b are 2-isomorphic because, by the following steps, G becomes isomorphic to G'.

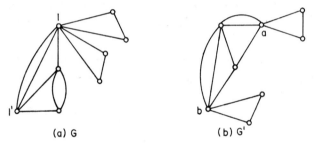

(a) G   (b) G'

Fig. 4-1-3. 2-isomorphic graphs (a) G and (b) G'.

STEP 1. Using operation 1 twice to cut vertex 1 in G, we obtain the linear graph G_s shown in Fig. 4-1-4a. Similarly, using operation 1 to cut vertices a and b in G', we have the linear graph G'_s shown in Fig. 4-1-4b.

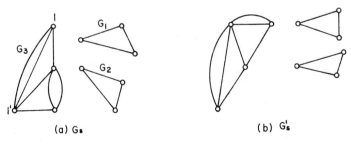

(a) G$_s$ (b) G'$_s$

Fig. 4-1-4. 2-Isomorphic graphs (a) G_s and (b) G'_s.

STEP 2. Since G_3 of G_s in Fig. 4-1-4a consists of two subgraphs H and $\overline{H}$ as shown in Fig. 4-1-5 which are joined by two vertices 1 and 1', we can use operation 2 to obtain linear graph G_{st} as shown in Fig. 4-1-6.

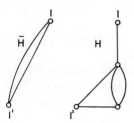

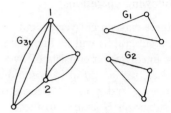

Fig. 4-1-5. Subgraphs H and $\overline{H}$.   **Fig. 4-1-6.**   Linear graph G_{st}.

STEP 3. Similarly, G_{3t} of G_{st} consists of two subgraphs which have exactly two vertices 1 and 2 in common. Hence we can use operation 2 to obtain a linear graph which is isomorphic to one in Fig. 4-1-4b. Thus G and G' are 2- isomorphic.

It is easily seen that the circuits are invariant under these operations. As long as given linear graphs consist of few edges, this method would be suitable for testing 2-isomorphism.

Another method is using a circuit matrix of one linear graph and a cut-set matrix of the other linear graph.

Let

$$B_1 = [B_{1_{11}} \quad B_{1_{12}}] \tag{4-1-1}$$

be a circuit matrix of linear graph G_1 where $B_{1_{11}}$ is nonsingular. Also let

$$Q_2 = [Q_{2_{11}} \quad Q_{2_{12}}] \tag{4-1-2}$$

be a cut-set matrix of linear graph G_2 where $Q_{2_{12}}$ is nonsingular. Suppose

$$B_1 Q_2{}^t = 0 \tag{4-1-3}$$

Then

$$B_{1_{11}} Q_{2_{11}}^t + B_{1_{12}} Q_{2_{12}}^t = 0 \tag{4-1-4}$$

Since, $Q_{2_{12}}$ is nonsingular,

$$B_{1_{12}} = B_{1_{11}} Q_{2_{11}}^t Q_{2_{12}}^{t-1} \tag{4-1-5}$$

Hence

$$\begin{aligned} B_1 &= [B_{1_{11}} \quad B_{1_{11}} Q_{2_{11}}^t Q_{2_{12}}^{t-1}] \\ &= B_{1_{11}}[U \quad Q_{2_{11}}^t Q_{2_{12}}^{t-1}] \end{aligned} \tag{4-1-6}$$

On the other hand, by using a fundamental circuit matrix $B_2 = [U \quad B_{2_{12}}]$ of G_2, we always have

$$B_2 Q_2{}^t = 0 \qquad (4\text{-}1\text{-}7)$$

or

$$Q_{2_{11}}^t + B_{2_{12}} Q_{2_{12}}^t = 0 \qquad (4\text{-}1\text{-}8)$$

Hence

$$B_{2_{12}} = Q_{2_{11}}^t Q_{2_{12}}^{t-1} \qquad (4\text{-}1\text{-}9)$$

or

$$B_2 = [U \quad Q_{2_{11}}^t Q_{2_{12}}^{t-1}] \qquad (4\text{-}1\text{-}10)$$

Since B_2 is a circuit matrix of G_2, by premultiplying any nonsingular matrix whose entries are either 1 or 0 to Eq. 4-1-10, the resultant matrix is also a circuit matrix of G_2. Since $B_{1_{11}}$ is a nonsingular matrix, we use this to make another circuit matrix of G_2:

$$B_{1_{11}} B_2 = B_2' = B_{1_{11}}[U \quad Q_{2_{11}}^t Q_{2_{12}}^{t-1}] \qquad (4\text{-}1\text{-}11)$$

However, this circuit matrix is exactly the same as the circuit matrix of G_1 given by Eq. 4-1-6. Thus, by Theorem 4-1-1, G_1 and G_2 are 2-isomorphic.

Theorem 4-1-3. Let B_1 be a circuit matrix of linear graph G_1, and Q_2' be a cut-set matrix of linear graph G_2. Then G_1 and G_2 are 2-isomorphic if and only if there exists Q_2 obtained by interchanging columns of Q_2' such that

$$Q_2 B_1{}^t = 0 \qquad (4\text{-}1\text{-}12)$$

Example 4-1-1. Let B be a circuit matrix of G in Fig. 4-1-2a where

$$
B = \begin{array}{c}
\begin{array}{cccccccc} a & b & c & d & e & f & g & h \end{array} \\
\left[\begin{array}{cccccccc}
1 & 1 & 0 & 0 & 0 & 0 & 0 & 0 \\
0 & 0 & 0 & 1 & 1 & 0 & 0 & 0 \\
0 & 0 & 0 & 0 & 0 & 0 & 1 & 1 \\
1 & 0 & 1 & 1 & 0 & 0 & 0 & 0
\end{array} \right]
\end{array}
$$

Also let Q_1' be a cut-set matrix of G' in Fig. 4-1-2b where

$$
Q_1' = \begin{array}{c}
\begin{array}{cccccccc} a & b & c & d & e & f & g & h \end{array} \\
\left[\begin{array}{cccccccc}
1 & 0 & 0 & 0 & 0 & 0 & 0 & 0 \\
0 & 1 & 1 & 0 & 0 & 0 & 0 & 0 \\
0 & 0 & 0 & 1 & 1 & 1 & 0 & 0 \\
0 & 0 & 0 & 1 & 0 & 0 & 1 & 1
\end{array} \right]
\end{array}
$$

By interchanging columns of Q_1', we have

$$
Q_1 = \begin{array}{c} \begin{matrix} e & f & d & g & h & a & b & c \end{matrix} \\ \begin{bmatrix} 0 & 0 & 0 & 0 & 0 & 1 & 0 & 0 \\ 0 & 0 & 0 & 0 & 0 & 0 & 1 & 1 \\ 1 & 1 & 1 & 0 & 0 & 0 & 0 & 0 \\ 0 & 0 & 1 & 1 & 1 & 0 & 0 & 0 \end{bmatrix} \end{array}
$$

so that

$$ Q_1 B^t = 0 $$

Thus G and G' in Fig. 4-1-2a and b are 2-isomorphic.

4-2 PLANAR GRAPH

Special linear graphs known as planar graphs are industrially as well as theoretically important. However, at present, only a few properties of planar graphs are known. One which was found by Kuratowski will be studied first.

Definition 4-2-1. A linear graph G is said to be *planar* if it can be drawn on a plane such that no two edges in G are crossing each other except possibly at their endpoints.

For example, linear graph G in Fig. 4-2-1 can be redrawn as shown in Fig. 4-2-2 where no edges are crossing. Hence linear graph G is planar. On the other hand, the linear graph in Fig. 4-2-3 is not planar.

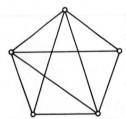

Fig. 4-2-1. A linear graph G.

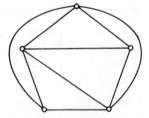

Fig. 4-2-2. Planar graph G.

Consider a sphere S on a plane Z as shown in Fig. 4-2-4. Let N be the topmost point on sphere S. The straight line which passes through N and a point p on sphere S intersects plane Z at point p'. This point p' is called the *corre-*

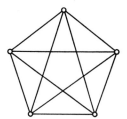

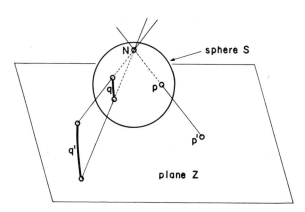

Fig. 4-2-3.   Nonplanar linear graph.

Fig. 4-2-4.   Sphere S and plane Z.

sponding point of point p. We can see that every point except point N on
sphere S corresponds to one and only one point on plane Z. Hence, for a
line-segment q on sphere S, we can obtain the corresponding line segment q'
on plane Z as long as the line segment on sphere S is not passing through
point N.

Suppose there is a linear graph drawn on a sphere S. Then we can obtain
the corresponding linear graph drawn on a plane Z by placing sphere S on
plane Z such that no edges and vertices of the linear graph lie on the topmost
point N on sphere S. Conversely, a linear graph drawn on a plane can be
drawn on a sphere S.

Suppose a linear graph G is drawn on a plane Z such that no two edges are
crossing. Also suppose G has at least one circuit. Then we can consider that
edges of G divides plane Z into regions. For example, in Fig. 4-2-5, a linear
graph G divides a plane into five regions R_0, R_1, R_2, R_3, and R_4. R_0 is called
the *outer region*. That is, the unbounded region is the outer region. The
other regions are called *inner regions*. Note that a linear graph may not give
a unique set of regions. For example, if we draw the same linear graph G as

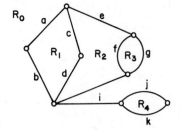

Fig. 4-2-5. A planar graph and regions.

shown in Fig. 4-2-6, we have regions R'_0, R'_1, R'_2, R'_3, and R'_4, whose boundaries (edges which form boundaries of these regions) are not the same as those of R_0, R_1, R_2, R_3, and R_4.

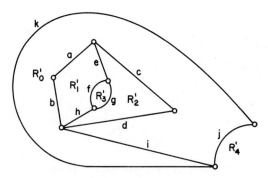

Fig. 4-2-6. Planar graph G and regions.

One interesting property of a planar graph is that *any region obtained by drawing a planar graph G on a plane can be made the outer region* by redrawing G. This can be shown by first choosing any region R. Note that a particular drawing of a given planar graph is assumed to be given. Now we draw this drawing onto sphere S by the procedure discussed previously. Then we rotate sphere S such that the topmost point N will be inside region R. Then we redraw the drawing (on the sphere) back onto a plane. The resultant drawing obviously gives R as the outer region.

Definition 4-2-2. Let endpoints of edge e_1 be u_1 and u_2 and endpoints of edge e_2 be v_1 and v_2 where $u_1 \neq u_2$ and $v_1 \neq v_2$. Then e_1 and e_2 are said to be in *series* if $u_2 = v_1$ and no other edges have u_2 as their endpoints. The term *series-edges* is used to indicate that the edges are in series.

For example, edges e_1 and e_2 in the linear graph in Fig. 4-2-7 are in series, or we say that e_1 and e_2 are series edges.

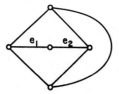

Fig. 4-2-7.   Series-edges e_1 and e_2.

Consider two linear graphs G and G' in Fig. 4-2-8. These are obviously different linear graphs. However, when edge e in G is replaced by series edges e_1 and e_2, the resultant graph becomes isomorphic to G'. Such linear graphs are called homeomorphic.

Fig. 4-2-8.   (*a*) Linear graph G; (*b*) linear graph G'.

Definition 4-2-3.   Two linear graphs G and G' are *homeomorphic* if and only if there exists a linear graph G'' such that each of G and G' becomes isomorphic to G'' by replacing edges by series edges.

For example, linear graphs G_1 and G_2 in Fig. 4-2-9 are homeomorphic.

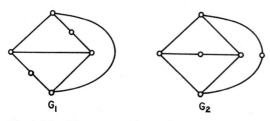

Fig. 4-2-9.   Homeomorphic graphs.

Now we are ready to learn an important property which clearly indicates the difference between planar and nonplanar graphs.

Theorem 4-2-1 (Kuratowski). A finite linear graph is not planar if and only if it contains a subgraph homeomorphic to either one of the two basic nonplanar graphs in Fig. 4-2-10.

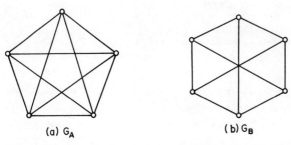

(a) G_A (b) G_B

Fig. 4-2-10. Two basic nonplanar graphs. (a) G_A; (b) G_B.

Note that a finite linear graph is a linear graph containing a finite number of edges.

Example 4-2-1. The linear graph in Fig. 4-2-11a is nonplanar because it contains a subgraph as shown in Fig. 4-2-11b which is homeomorphic to a basic nonplanar graph G_B.

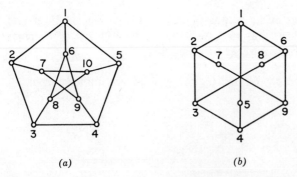

(a)   (b)

Fig. 4-2-11. (a) A nonplanar graph G; (b) subgraph of G.

Proof. From the definition, if a linear graph is planar, then there is a drawing of the linear graph on a plane where no edges are crossing. By removing some edges in the drawing, we have a drawing of a subgraph on a plane where no edges are crossing. Hence if a linear graph is planar, any subgraph of the linear graph is also planar. Thus a planar graph cannot contain a subgraph homeomorphic to a basic nonplanar graph. This proves the first part of the theorem.

To show that a nonplanar graph G' always contains a subgraph homeomorphic to one of the two basic nonplanar graphs, we consider a non-separable subgraph G'' of G' which is also nonplanar, but deleting any one edge from G'' makes it planar. It is easily seen that such a subgraph exists in any nonplanar graph. Now we take a linear graph G which is homeomorphic to G'' in which there are no series edges. It is obvious that if G contains a subgraph homeomorphic to one of two basic nonplanar graphs, G'' and G' also contain a subgraph homeomorphic to one of the two basic nonplanar graphs. Hence we need only to consider a nonseparable linear graph G to prove the second-part of Theorem 4-2-1.

Consider linear graph $G(\bar{e})$, which is a subgraph obtained from G by deleting edge e. Since $G(\bar{e})$ is planar, there exists a drawing of $G(\bar{e})$ on a plane such that no edges are crossing each other. Let endpoints of edge e be u and v. In this drawing, there exists the largest circuit C which contains vertices u and v such that any other circuits containing u and v are inside of C. The circuit C divides $G(\bar{e})$ into two parts G_o and G_i where G_o consists of all edges drawn outside of C and G_i consists of all edges drawn inside of G as shown in Fig. 4-2-12. Hence $G(\bar{e}) = G_o \cup G_i \cup C$ where G_o, G_i, and C are edge disjoint.

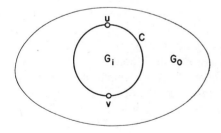

Fig. 4-2-12. Circuit C and subgraphs G_o and G_i.

Suppose when we draw circuit C starting from vertex p in the clockwise direction, we pass vertices p, r_1, r_2, ..., r_k, q, r_{k+1}, ..., r_n in that order and back to p. We divide these vertices into three, one consisting of vertices r_1, r_2, ..., r_k, another of r_{k+1}, ..., r_n, and the third of vertices p and q. For convenience, we use the symbol $[p, q]$ to indicate the first set of the vertices, that is, $[p, q] = (r_1 r_2 \cdots r_k)$. Likewise, the symbol $[q, p] = (r_{k+1} \cdots r_n)$. Note that (p, q) is a set consisting of p and q. With these symbols, all vertices in C can be expressed as $[p, q] \cup [q, p] \cup (p, q)$ and any two of sets $[p, q]$, $[q, p]$, and (p, q) have no vertices in common.

In the drawing of $G(\bar{e})$ in Fig. 4-2-12, we cannot add edge e between u and v outside of C without crossing other edges because G is nonplanar. Hence there must be a path P_o in G_o from a vertex a in $[v, u]$ to a vertex b in $[u, v]$. Similarly, in G_i, there must be a path P_i from a vertex c in $[v, u]$ to a vertex d in $[u, v]$ as shown in Fig. 4-2-13. Depending on the locaton of vertices c and d we have the following cases.

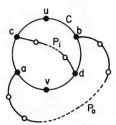

Fig. 4-2-13. Paths P_o and P_i.

CASE 1. Suppose $c \in [a, u]$ and $d \in [b, v]$ as shown in Fig. 4-2-13. Then it is clear that a subgraph consisting of C, P_o, P_i, and edge e is homeomorphic to basic nonplanar graph G_B. This is also true when $c \in [v, a]$ and $d \in [u, b]$.

CASE 2. Suppose $c \in [a, u]$ and $d \in [u, b]$ as shown in Fig. 4-2-14. In order to prevent to have a new drawing of $G(\bar{e})$ in which P_i and P_o will be outside of C without crossing edges, there must be a path P_1 as shown in Fig. 4-2-15a. Note that P_1 is from a vertex e in $[b, a]$ to a vertex f in P_i other than c and d. We can get the same result by inserting a path P_1 as shown in Fig. 4-2-15b. However, the resultant graph has a circuit C' which contains both vertices u and v and is outside of circuit C. This contradicts the assumption that every circuit containing vertices u and v are inside of circuit C. Hence we need consider only the structure shown in Fig. 4-2-15a for this case.

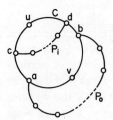

Fig. 4-2-14. $G(\bar{e})$ for Case 2.

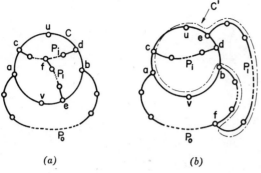

(a)   (b)

Fig. 4-2-15. Location of P_1. (a) Path P_1; (b) circuit C'.

The linear graph in Fig. 4-2-15a with edge e has a subgraph as shown in Fig. 4-2-16 which is homeomorphic to G_B. This is also true when $e = v$ and (or) $c = a$. Because of the symmetry, the situation when $c \in [v, a]$ and $d \in [b, v]$ will give the same result. Note that we have covered the case when either $c = a$ or $b = d$ but not both $c = a$ and $b = d$, which will be the next case.

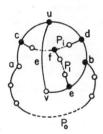

Fig. 4-2-16. Nonplanar subgraph.

CASE 3. The final case needed to complete the proof is when $c = a$ and $b = d$. Under this circumstance, the linear graph in Fig. 4-2-15a can be redrawn so that P_i (and P_o) can be outside of C unless there is another path P_2 as shown in Fig. 4-2-17.

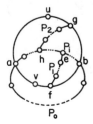

Fig. 4-2-17. Location of P_2.

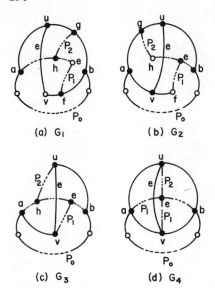

(a) G_1

(b) G_2

(c) G_3

(d) G_4

Fig. 4-2-18. Nonplanar graphs. (a) G_1; (b) G_2; (c) G_3; (d) G_4.

In this linear graph with edge e, one of four linear graphs in Fig. 4-2-18 will be a subgraph depending on the location of P_2. That is, the linear graph in (a) is a subgraph when $g \in [u, b]$ (where f can be equal to v and e can be equal to h), the linear graph in (b) becomes a subgraph when $g \in [a, u]$ (where e can be equal to h and f can be equal to v), the one in (c) is a subgraph when $g = u$ and $h \neq e$, and one in (d) is a subgraph when $g = u$ and $h = e$. The first three linear graphs G_1, G_2, and G_3 are homeomorphic to G_B and the last one G_4 is homeomorphic to G_A. This completes the proof of Theorem 4-2-1.

QED

4-3 DUALITY

Kuratowski's theorem, discussed in Section 4-2, is rather difficult to use for testing whether a linear graph is planar because it requires examination of all subgraphs of order at least five. The property of duality given by Whitney, on the other hand, guarantees the existence of another linear graph when a given linear graph is planar. Hence for several applications, Whitney's theorem may be easier to use for testing planar graphs. To study Whitney's duality theorem, we define the nullity of a linear graph as follows.

Definition 4-3-1. The nullity N of a linear graph G is

$$N = n_e - n_v + \rho \qquad (4\text{-}3\text{-}1)$$

where n_e is the number of edges, n_v is the number of vertices, and ρ is the number of maximal connected subgraphs of G.

Recall that the rank R of a linear graph (by Definition 2-2-1) is

$$R = n_v - \rho \qquad (4\text{-}3\text{-}2)$$

Hence the nullity N can be expressed as

$$N = n_e - R \qquad (4\text{-}3\text{-}3)$$

From the preceding equation, we can see that the sum of the nullity and the rank of a linear graph G is equal to the number of edges in G:

$$N + R = n_e \qquad (4\text{-}3\text{-}4)$$

For example, a linear graph G in Fig. 4-2-11a consists of 15 edges and 10 vertices and is connected. Hence the nullity N of G is 6. Note that the number of linearly independent circuits in a linear graph is also $n_e - n_v + \rho$ by Theorem 3-3-5, which is the rank of a circuit matrix of the linear graph. Hence the nullity of a linear graph is the rank of a circuit matrix of the linear graph.

In the discussion of dual graphs, we must include self-loops as edges in a linear graph.

AGREEMENT.   In this section, linear graphs can have self-loops.

Consider a linear graph G_1 consisting of edges $e_1, e_2, \ldots, e_{n_e}$ and a linear graph G_2 consisting of edges $e'_1, e'_2, \ldots, e'_{n_e}$. Suppose we choose edge e'_p in G_2 as the edge corresponding to edge e_p in G_1 for $p = 1, 2, \ldots, n_e$. Hence we have a 1:1 correspondence between the edges of G_1 and the edges of G_2. Let g_1 be a subgraph of G_1 consisting of edges $e_{i_1}, e_{i_2}, \ldots, e_{i_k}$. Then the subgraph g_2 of G_2 which consists of edges $e'_{i_1}, e'_{i_2}, \ldots, e'_{i_k}$ is said to be the *corresponding subgraph* of g_1. Also, for convenience, the symbol $\bar{g}$ is used to indicate the complement of subgraph g of a linear graph G, that is, $\bar{g}$ consists of all edges in G except those in g and isolated vertices which are not in g.

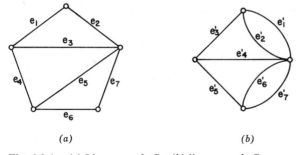

(a)   (b)

Fig. 4-3-1.   (a) Linear graph G_1; (b) linear graph G_2.

Example 4-3-1. Consider two linear graphs G_1 and G_2 in Fig. 4-3-1a and b. Suppose edge e_p in G_1 corresponds to edge e'_p in G_2 for $p = 1, 2, \ldots, 7$. Also suppose we choose a subgraph g_1 in G_1 which consists of edges e_1, e_2, e_3, and e_5 as shown in Fig. 4-3-2a. Then the subgraph g_2 in G_2 corresponding to g_1

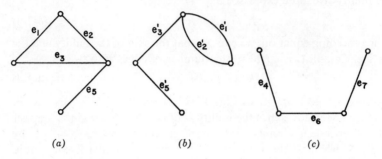

Fig. 4-3-2. (a) Subgraph g_1 of G_1; (b) subgraph g_2 of G_2; (c) complement $\bar{g}_1$.

will be the one shown in Fig. 4-3-2b. Moreover, the complement $\bar{g}_1$ of g_1 will consist of edges e_4, e_6, and e_7 as shown in Fig. 4-3-2c. With these subgraphs, we can define duality of linear graphs as follows.

Definition 4-3-2. A linear graph G_2 is said to be a *dual* of linear graph G_1 if there exists a $1:1$ correspondence between edges of G_1 and G_2 such that for every subgraph g_1 of G_1

$$n_1 = R_2 - \bar{r}_2 \qquad (4\text{-}3\text{-}5)$$

where n_1 is the nullity of g_1, R_2 is the rank of G_2, $\bar{r}_2$ is the rank of $\bar{g}_2$, g_2 is the subgraph of G_2 corresponding to g_1, and $\bar{g}_2$ is the complement of g_2.

With this definition, if $g_2 = G_2$, then $\bar{g}_2$ is a subgraph of G_2 containing no edges. Hence $\bar{r}_2$ is zero, and $g_1 = G_1$, which means that $n_1 = N_1$ (the nullity of G_1). Thus if G_2 is dual of G_1, from Eq. 4-3-5, we have

$$N_1 = R_2 \qquad (4\text{-}3\text{-}6)$$

Since there are the same number n_e of edges in G_1 and G_2 where G_2 is a dual of G_1, Eq. 4-3-6 can be written as

$$n_e - N_1 = n_e - R_2 \qquad (4\text{-}3\text{-}7)$$

By Eq. 4-3-3, the foregoing equation becomes

$$R_1 = N_2 \qquad (4\text{-}3\text{-}8)$$

Thus from Eqs. 4-3-6 and 4-3-8, we can say that if G_1 and G_2 are dual, then $N_1 = R_2$ and $R_1 = N_2$.

Let $\bar{g}_1$ be a subgraph of G_1 consisting of n_{e_1} edges. Let n_{e_2} be

$$n_{e_2} = n_e - n_{e_1} \qquad (4\text{-}3\text{-}9)$$

or

$$n_{e_1} + n_{e_2} = n_e \qquad (4\text{-}3\text{-}10)$$

where n_e is the number of edges in G_1. When G_2 is a dual of G_1, then the number of edges in g_2 of G_2 is n_{e_2} where g_2 is the complement of $\bar{g}_2$ which corresponds to $\bar{g}_1$. Then Eq. 4-3-5 becomes

$$\bar{n}_1 = R_2 - r_2 \qquad (4\text{-}3\text{-}11)$$

where $\bar{n}_1$ is the nullity of $\bar{g}_1$ and r_2 is the rank of g_2. This can be expressed as

$$n_e - \bar{n}_1 = n_e - R_2 + r_2 \qquad (4\text{-}3\text{-}12)$$

or, by Eqs. 4-3-3 and 4-3-10,

$$n_{e_2} + n_{e_1} - \bar{n}_1 = N_2 + r_2 \qquad (4\text{-}3\text{-}13)$$

By moving n_{e_2} to the right-hand side of the equation, we have

$$\bar{r}_1 = N_2 - (n_{e_2} - r_2) \qquad (4\text{-}3\text{-}14)$$

or

$$\bar{r}_1 = N_2 - n_2 \qquad (4\text{-}3\text{-}15)$$

By Eq. 4-3-8, the preceding equation becomes

$$n_2 = R_1 - \bar{r}_1 \qquad (4\text{-}3\text{-}16)$$

This shows that if G_2 is dual of G_1, then G_1 is a dual of G_2.

For a planar graph G_1, we know a way of constructing another planar graph G_2 which is known to be a dual of G_1.

Definition 4-3-3. Let G_1 of n_e edges be a planar graph which has been drawn on a plane such that no edges are crossing each other. Then a process of drawing a linear graph G_2 from G_1, by inserting a vertex to every region of G_1 and giving n_e edges such that exactly one edge of G_1 and one edge of G_2 cross each other, is called the *D-process*. The resultant linear graph G_2 is called the linear graph obtained from G_1 by a *D-process*.

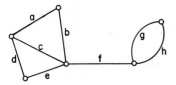

Fig. 4-3-3. A planar graph.

For example, we can construct a planar graph G_2 from G_1 in Fig. 4-3-3 by first inserting vertices 1, 2, 3, and 4 to each region, then giving edge a' to cross edge a, edge b' to cross edge b, and so on, as shown in Fig. 4-3-4.

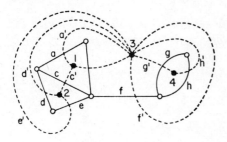

Fig. 4-3-4. *D*-Process.

Now we are ready to show that the linear graph G_2 obtained from a planar graph G_1 by the D-process is a dual of G_1.

It can be seen that for a connected linear graph G_1, if G_2 is obtained from G_1 by the D-process, then G_1 can be obtained from G_2 by the D-process.

To show that linear graph G' obtained from a finite linear graph G by a D-process is connected, we consider a drawing of G on a plane Z. Then we place a point v in any region R and a point v' in any other region R'. Since the number of vertices is finite in G, there exists a line L on Z starting from point v and terminating at point v' such that L does not pass through any vertices of G. Let $e_1, e_2, \ldots, e_p$ be the edges of G that intersect line L. Consider a linear graph g consisting of edges $e_1', e_2', \ldots, e_p'$ of G' which are the corresponding edges of $e_1, e_2, \ldots, e_p$. It is clear that points v and v' correspond to vertices v and v' of g. Because line L will pass through regions one by one, we can draw g starting from vertex v and terminating at vertex v' without lifting a pencil. We may have to trace an edge more than once to do this; however, it is clear that there exists a path between vertices v and v' in g. This is true for any two vertices in G' we choose. Thus by the definition of connected graphs, G' is connected.

Theorem 4-3-1. A linear graph obtained from another by a D-process is connected.

Example 4-3-2. In linear graph G, we place points v and v' in regions R and R', respectively. Then we draw line L without passing any vertices as shown in Fig. 4-3-5. This line L passes edges $e_1, e_2, e_3, e_4,$ and e_5. In linear graph G' obtained from G by a D-process (Fig. 4-3-6), the corresponding edges of $e_1, e_2, e_3, e_4,$ and e_5 are $e_1', e_2', e_3', e_4',$ and e_5', which form subgraph g' as shown

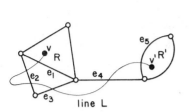

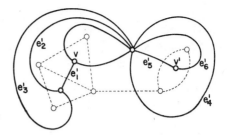

Fig. 4-3-5. A linear graph G and line L.   **Fig. 4-3-6.** Linear graph G'.

in Fig. 4-3-7. This subgraph g' can be drawn without lifting a pencil by starting from vertex v and drawing edges in the order e_1', e_2', e_3', e_2', e_4', e_5'. Note that we trace e_2' twice. It is easily seen that in the sequence $(e_1'e_2'e_3'e_2'e_4'e_5')$ there exists a sequence $(e_1'e_3'e_5')$ which is a path between vertices v and v'.

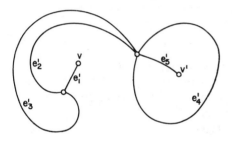

Fig. 4-3-7. A subgraph g'.

Because of a D-process, there are some relationships between circuits of one linear graph and cut-sets of the other.

Theorem 4-3-2. Let $C = (e_1, e_2, \ldots, e_n)$ be a circuit in G_1. Let G_2 be obtained from G_1 by the D-process. Then edges e_1', e_2', $\ldots$, e_n' form a cut-set in G_2 where e_p' is the corresponding edge of e_p for $p = 1, 2, \ldots, n$.

Proof. Note that whenever we say that G_2 is obtained from G_1 by the D-process, G_1 (and G_2) is assumed to be drawn on a plane such that no two edges are crossing each other. Hence regions are clearly specified.

Let $W_1, W_2, \ldots, W_p$ be regions inside of circuit C. Let $v_1, v_2, \ldots, v_p$ be the vertices in G_2 which are inside of regions $W_1, W_2, \ldots, W_p$. Also let $\Omega_1 = (v_1, v_2, \ldots, v_p)$, and $\bar{\Omega}_1 = \Omega - \Omega_1$ where Ω is the set of all vertices in G_2. Then $\mathscr{E}(\Omega_1 \times \bar{\Omega}_1)$ is the set of edges e_1', e_2', $\ldots$, e_n' which correspond to the edges in C. We know that $\mathscr{E}(\Omega_1 \times \bar{\Omega}_1)$ is either a cut-set or an edge disjoint union of cut-sets. If $\mathscr{E}(\Omega_1 \times \bar{\Omega}_1)$ is not a cut-set, then either subgraph $\mathscr{E}(\Omega_1 \times \Omega_1)$ or subgraph $\mathscr{E}(\bar{\Omega}_1 \times \bar{\Omega}_1)$ must be separated (see Sections 2-2 and 2-3).

Consider a subgraph g_1 of G_1 which consists of edges $e_1, e_2, \ldots, e_n$ of circuit C and all edges inside C as shown in Fig. 4-3-8a. Note that the edges

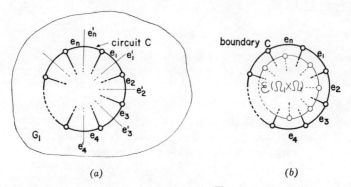

(a)   (b)

Fig. 4-3-8. Circuit C and cut-set $\mathscr{E}(\Omega_1 \times \bar{\Omega}_1)$. (a) Linear graph G_1 and G_2; (b) g_1 and $\mathscr{E}(\Omega_1 \times \Omega_1)$.

in G_2 corresponding to edges in $g_1 - (e_1, e_2, \ldots, e_n)$ form $\mathscr{E}(\Omega_1 \times \Omega_1)$ as shown in Fig. 4-3-8b. Let g_s be a linear graph obtained from g_1 by shorting all edges $e_1, e_2, \ldots, e_n$. We can visualize g_s by considering the boundary (which is circuit C) of g_1 as one vertex. It can be seen that $\mathscr{E}(\Omega_1 \times \Omega_1)$ is also a linear graph obtained from g_s by a D-process. Hence, by Theorem 4-3-1, $\mathscr{E}(\Omega_1 \times \Omega_1)$ is connected. Similarly, a linear graph $\mathscr{E}(\bar{\Omega}_1 \times \bar{\Omega}_1)$ is connected. Thus $\mathscr{E}(\Omega_1 \times \bar{\Omega}_1)$ is a cut-set.   QED

Theorem 4-3-3. Let $S = (e_1, e_2, \ldots, e_n)$ be a cut-set in G_1. Also let G_2 be obtained from G_1 by the D-process. Then edges $e_1', e_2', \ldots, e_n'$ corresponding to edges $e_1, e_2, \ldots, e_n$ of S form a circuit in G_2.

Proof. If G_1 is connected, then it is obvious. Suppose G_1 is separated. Then we can see that G_1' obtained from G_2 by the D-process after G_2 is formed from G_1 by the D-process has cut vertices such that by splitting cut vertices properly, G_1' becomes G_1 as shown in Fig. 4-3-9a and b. Under this circumstance, cut-set of G_1' is also a cut-set of G_1.   QED

From the last two theorems, we can see that

$$R_1 = N_2 \tag{4-3-17}$$

where R_1 is the rank of G_1 and N_2 is the nullity of G_2 which is obtained from G_1 by the D-process.

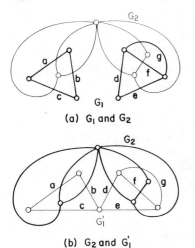

(a) G_1 and G_2

(b) G_2 and G_1' **Fig. 4-3-9.** (*a*) G_1 and G_2; (*b*) G_2 and G_1'.

Theorem 4-3-4. A planar graph G_2 obtained from G_1 by the *D*-process is a dual of G_1.

Proof. Suppose we short one edge e in G_1 to obtain g_1'. Then, by deleting edge e' corresponding to edge e in G_2, we will have the exact same resultant graph as that obtained from g_1' by the *D*-process. In general, suppose $\bar{g}_1$ is a subgraph of G_1 and $\bar{g}_2$ is the corresponding subgraph of G_2. Note that G_1 and G_2 are already drawn on a plane by the result of the *D*-process. Suppose we short all edges in $\bar{g}_1$ to obtain g_1'. Then by deleting all edges in $\bar{g}_2$ from G_2, we have the resultant graph whose nullity is exactly the same as one obtained from g_1' by the *D*-process.

The rank of g_1' is equal to $R_1 - \bar{r}_1$ where R_1 is the rank of G_1 and $\bar{r}_1$ is the rank of $\bar{g}_1$. By Eq. 4-3-17, this rank is equal to the nullity n_2 of g_2 which is obtained from G_2 by deleting all edges in $\bar{g}_2$, or

$$n_2 = R_1 - \bar{r}_1 \qquad (4\text{-}3\text{-}18)$$

This is true for any subgraph $\bar{g}_1$ of G_1. Note that g_2 is the compliment of $\bar{g}_2$. Thus by definition, G_1 is a dual of G_2, and G_2 is a dual of G_1.   QED

We know from Theorem 4-3-4 that a *D*-process gives a dual graph. The next important question is whether any dual graph can be obtained by a *D*-process or not. It has been found by Theorem 4-3-1 that a dual graph obtained by a *D*-process is connected. Hence if there exists a dual graph (of a linear graph) which is separated, then it cannot be obtained by a *D*-process. We know such a linear graph exists. For example, if we choose a separated linear graph G and obtain a linear graph G' by a *D*-process, then G and G'

are dual. Thus G is a dual graph of G' and G is separated. Thus the answer to the above question is no because there are dual graphs that cannot be obtained by a D-process. However, we will find that there is a relationship between a dual graph of a linear graph G obtained by a D-process and any other dual graph of G. But first we consider the effect on the nullity of a linear graph when some edges are shorted.

Theorem 4-3-5. Let C be a circuit in a linear graph. Then shorting some but not all edges in C will not change the nullity and shorting all edges in C will reduce the nullity by exactly one.

Because self-loops are allowed in this chapter, shorting an edge e means to coincide the two endpoints of edge e and to delete only edge e.

Proof. Let G_s be a linear graph obtained from G by shorting edges in C. Also let n_e and n_{se} be the numbers of edges and n_v and n_{sv} be the numbers of vertices in G and G_s, respectively. It should be noted that shorting edges will not change the number of maximal connected subgraphs unless C itself is a maximal connected subgraph. Suppose we short k edges in C where k is less than the number of edges in C. Then in the resultant graph G_s,

$$n_{se} = n_e - k \qquad (4\text{-}3\text{-}19)$$

and

$$n_{sv} = n_v - k \qquad (4\text{-}3\text{-}20)$$

Thus the nullity of G_s is

$$n_{se} - n_{sv} + \rho = n_e - n_v + \rho \qquad (4\text{-}3\text{-}21)$$

which proves the first part of the theorem.

Suppose we short all n edges in C. Then in the resultant graph G_s,

$$n_{se} = n_e - n \qquad (4\text{-}3\text{-}22)$$

and

$$n_{sv} = n_v - n + 1 \qquad (4\text{-}3\text{-}23)$$

for the case when C is not a maximal connected subgraph. Thus the nullity of G_s is

$$n_{se} - n_{sv} + \rho = n_e - n_v + \rho - 1 \qquad (4\text{-}3\text{-}24)$$

which is exactly one less than the nullity of G.

When C itself is a maximal connected subgraph, then shorting all edges in C makes

$$n_{sv} = n_v - n \qquad (4\text{-}3\text{-}25)$$

but the number of maximal connected subgraphs will be reduced by one. Hence the nullity of G_s is

$$n_{se} - n_{sv} + \rho - 1 = n_e - n_v + \rho - 1 \qquad (4\text{-}3\text{-}26)$$

which is again one less than that of G. QED

Theorem 4-3-6. Let G and G' be dual. Let e be an edge in G and e' be the corresponding edge in G'. Also let G_s be obtained from G by shorting edge e and G'_o be obtained from G' by deleting edge e'. Then G_s and G'_o are dual.

Proof. Consider a subgraph g of G consisting only of edge e. Because G and G' are dual, the following equation must be satisfied:

$$n = R' - \bar{r}' \qquad (4\text{-}3\text{-}27)$$

where n is the nullity of g, R' is the rank of G', and $\bar{r}'$ is the rank of G'_o. If e is a self-loop, nullity n of g is one. Hence the rank of G'_o must be one less than that of G' from Eq. 4-3-27. If e is not a self-loop, nullity of g is zero. Hence the rank of G' and that of G'_o are the same. With these results, we can show that G_s and G'_o are dual as follows. Let g be a subgraph of G. Suppose g contains edge e. Let g_s be obtained from g by shorting edge e and g'_o be the corresponding subgraph of g_s in G'_o. We must now determine whether the equation

$$n_s = R'_o - \bar{r}'_o \qquad (4\text{-}3\text{-}28)$$

satisfies or not. Note that n_s is the nullity of g_s, R'_o is the rank of G'_o, and $\bar{r}'_o$ is the rank of $\bar{g}'_o$ which is the complement of g'_o.

Suppose e is a self-loop. Since $\bar{g}'$ is the complement of the corresponding subgraph of g in G', subgraph $\bar{g}'$ in G' is the same as $\bar{g}'_o$ of G'_o. Hence rank $\bar{r}'_o$ is the rank $\bar{r}'$ of $\bar{g}'$. On the other hand, because e is a self-loop, the rank R'_o of G'_o is one less than that of G'. Thus $R'_o - \bar{r}'_o$ is one less than $R' - \bar{r}'$. The nullity n_s of g_s is one less than that of g by Theorem 4-3-5 because of self-loop e. Thus Eq. 4-3-28 holds.

If e is not a self-loop, then from the previous results, the rank R'_o of G'_o is the same as that of G'. Furthermore, the nullity n_s of g_s is the same as that of g by Theorem 4-3-5. Hence Eq. 4-3-28 also holds when e is not a self-loop. We conclude, therefore, Eq. 4-3-28 is true for any g_s obtained from g by shorting edge e in g. However, it is clear that for any subgraph g_s of G_s, there exists a subgraph g of G such that (1) g contains edge e and (2) g becomes g_s by shorting edge e. Hence Eq. 4-3-28 holds for any subgraph g_s of G_s. Thus G_s and G'_o are dual. QED

Theorem 4-3-7. Let G and G' be dual. Then a circuit in G corresponds to a cut-set in G'.

Note that from Theorem 4-3-2 a circuit of G corresponds to a cut-set of G' if G' is obtained from G by a D-process. This theorem emphasizes that the property holds for any two linear graphs which are dual to each other.

Proof. Suppose we take g equal to a circuit C in G. We are going to show that the corresponding subgraph g' of g is a cut-set in G'. Let G_s be obtained from G by shorting all edges in g and G'_o be obtained from G' by deleting all edges in g'. By Theorem 4-3-5,

$$N - 1 = N_s \qquad\qquad (4\text{-}3\text{-}29)$$

where N is the nullity of G and N_s is the nullity of G_s.

By successive application of Theorem 4-3-6, we can easily show that G_s and G'_o are dual. Hence

$$N_s = R'_o \qquad\qquad (4\text{-}3\text{-}30)$$

where R'_o is the rank of G'_o. Since

$$N = R' \qquad\qquad (4\text{-}3\text{-}31)$$

Eq. 4-3-29 indicates that the rank of G'_o is one less than that of G'. This means that deleting all edges in g' will reduce the rank of G' by one.

If we take g as a proper subgraph of circuit C, then the nullity of G_s is the same as that of G by Theorem 4-3-5. Hence Eqs. 4-3-29 and 4-3-30 show that the rank of G' will not be reduced if some but not all edges, which are the corresponding edges of those in circuit C, are deleted. Thus by the definition of cut-sets, the edges corresponding to those in a circuit in G form a cut-set in G'. QED

Now we are ready to learn an important property of linear graphs which are duals of linear graphs.

Theorem 4-3-8. Let G_1 be a dual of G'. Then G_2 is a dual of G' if and only if G_1 and G_2 are 2-isomorphic.

Proof. Suppose G_2 is a dual of G'. Then there exists a $1:1$ correspondence between edges of G_2 and edges of G' such that circuits of G_1 correspond to cut-sets of G'. Since there is a $1:1$ correspondence between edges of G_1 and edges of G' such that circuits of G_1 correspond to cut-sets of G', we can make a $1:1$ correspondence between edges of G_1 and edges of G_2 such that a $1:1$ correspondence between circuits of G_1 and circuits of G_2 can be obtained. This proves the first part.

Next, suppose G_1 and G_2 are 2-isomorphic. Since G_1 and G' are dual, for any subgraph g_1 of G_1, Eq. 4-3-5 should hold, that is,

$$n_1 = R' - \bar{r}' \qquad\qquad (4\text{-}3\text{-}32)$$

where n_1 is the nullity of g_1, R' is the rank of G', and $\bar{r}'$ is the rank of the complement of the corresponding subgraph g' of g, in G'. Since G_1 and G_2 are 2-isomorphic, there exists a 1:1 correspondence between edges of G_1 and edges of G_2 such that circuits of G_1 correspond to circuits of G_2. Hence there exists a subgraph g_2 of G_2 formed by edges which correspond to edges in g_1. It is clear that the nullity n_2 of g_2 is equal to the nullity n_1 of g_1. It is also obvious that g' is the corresponding subgraph of g_2. Thus we have

$$n_2 = R' - \bar{r}' \qquad (4\text{-}3\text{-}33)$$

which proves that G_2 and G' are dual. QED

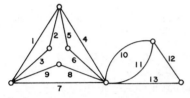

Fig. 4-3-10. A linear graph G.

Example 4-3-3. Consider a linear graph G as shown in Fig. 4-3-10. The dual graph G_1 of G obtained by a D-process, shown in Fig. 4-3-11a, can be redrawn

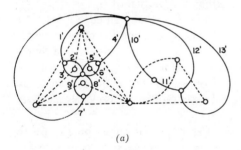

(a)

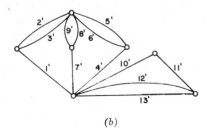

(b)

Fig. 4-3-11. A dual graph G_1 of G. (a) G_1 by a D-process; (b) redrawing of G_1.

as shown in Fig. 4-3-11*b*. A linear graph G_2 shown in Fig. 4-3-12 is also a dual of *G*. We can see that G_1 and G_2 are 2-isomorphic.

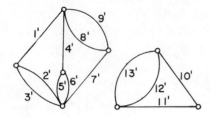

Fig. 4-3-12. A dual graph G_2 of *G*.

By modifying Theorem 4-3-7, we will have an important theorem about the existence of a dual graph.

Theorem 4.3.9. Let *B* be a circuit matrix of a linear graph *G*. Then *G* has a dual if and only if there exists a linear graph *G'* such that there is 1:1 correspondence between edges of *G* and *G'* such that a cut-set matrix of $G' = B$.

Proof. If there exists a linear graph *G'* which is a dual of *G*, then this theorem is true by Theorem 4-3-7.

Suppose *G'* is a linear graph such that there is a 1:1 correspondence between edges of *G* and edges of *G'* so that a cut-set matrix of *G'* is *B*. With this assumption, we are going to prove that *G'* is a dual of *G*.

Let G_e be a linear graph obtained from *G* by shorting an edge *e*. Also let B_e be obtained from a circuit matrix *B* of *G* by deleting the column representing edge *e* and deleting a row of zeros if such a row exists. Note that a row of zeros will be produced by deleting a column if edge *e* is a self-loop. We can see that B_e is a circuit matrix of G_e. Let G'_e be obtained from *G'* by deleting edge *e'* which corresponds to edge *e* in *G*. Then B_e is clearly a cut-set matrix of G'_e.

Instead of shorting an edge, if we short all edges in a subgraph *g* of *G*, we will have the following. Let G_g be a linear graph obtained from *G* by shorting all edges in subgraph *g*. Also let B_g be obtained from *B* by deleting the columns representing all edges in *g* and deleting the rows of zeros. Then we can see that B_g is a circuit matrix of G_g. Let G'_g be obtained from *G'* by deleting all edges in *g'*, which is the corresponding subgraph *g* in *G'*. Then B_g is a cut-set matrix of G'_g. Since the number of rows of a circuit matrix is the nullity of a corresponding linear graph, if *n* is the nullity of *g*, then $N - n$ is the nullity of G_g where *N* is the nullity of *G*. Moreover, because B_g is a cut-set matrix, $N - n$ is the rank of $\bar{g}'$. That is,

$$N - n = \bar{r}' \qquad\qquad (4\text{-}3\text{-}34)$$

Furthermore, the rank R' of G' is equal to the nullity N of G. Hence we have

$$R' - n = \bar{r}' \qquad (4\text{-}3\text{-}35)$$

from which Eq. 4-3-18 can be obtained. This is true for any subgraph g of G. Hence G and G' are dual. QED

Theorem 4-3-9 leads to the following important theorem which characterizes planar graphs.

Theorem 4-3-10 (Whitney). A linear graph has a dual if and only if it is planar.

Proof. The first part is obvious because we can use a D-process.
The proof of the second part can be accomplished by showing that a nonplanar graph has no dual as follows.
Suppose a linear graph G contains the basic nonplanar graph in Fig. 4-3-13 as a subgraph. Then a fundamental circuit matrix B of linear graph

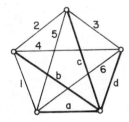

Fig. 4-3-13. One of two basic nonplanar graph and tree $t = (a, b, c, d)$.

G can be chosen such that the fundamental circuit matrix B_1 of the nonplanar graph with respect to tree $t = (a, b, c, d)$ is a submatrix of B where

$$
B_1 = \begin{array}{c}
\begin{array}{c} \\ 1 \\ 2 \\ 3 \\ 4 \\ 5 \\ 6 \end{array}
\begin{array}{c}
\begin{array}{cccccccccc} 1 & 2 & 3 & 4 & 5 & 6 & a & b & c & d \end{array} \\
\left[\begin{array}{cccccc|cccc}
1 & 0 & 0 & 0 & 0 & 0 & 1 & 1 & 0 & 0 \\
0 & 1 & 0 & 0 & 0 & 0 & 0 & 1 & 1 & 0 \\
0 & 0 & 1 & 0 & 0 & 0 & 0 & 0 & 1 & 1 \\
0 & 0 & 0 & 1 & 0 & 0 & 0 & 1 & 0 & 1 \\
0 & 0 & 0 & 0 & 1 & 0 & 1 & 0 & 1 & 0 \\
0 & 0 & 0 & 0 & 0 & 1 & 1 & 0 & 0 & 1
\end{array}\right]
\end{array}
\qquad (4\text{-}3\text{-}36)
$$

It can be seen that *even if each edge in this nonplanar graph is replaced by the series connection of edges, B_1 will be a submatrix of a fundamental circuit matrix of such a nonplanar graph.* Consider B_1 as a fundamental cut-set matrix. For convenience, we rearrange columns so that we have the form of a fundamental cut-set matrix:

$$
B_1' = \begin{array}{c} \\ 1 \\ 2 \\ 3 \\ 4 \\ 5 \\ 6 \end{array}
\begin{array}{c} \begin{array}{ccccccccccc} a & b & c & d & 1 & 2 & 3 & 4 & 5 & 6 \end{array} \\
\left[\begin{array}{cccc|cccccc}
1 & 1 & 0 & 0 & 1 & 0 & 0 & 0 & 0 & 0 \\
0 & 1 & 1 & 0 & 0 & 1 & 0 & 0 & 0 & 0 \\
0 & 0 & 1 & 1 & 0 & 0 & 1 & 0 & 0 & 0 \\
0 & 1 & 0 & 1 & 0 & 0 & 0 & 1 & 0 & 0 \\
1 & 0 & 1 & 0 & 0 & 0 & 0 & 0 & 1 & 0 \\
1 & 0 & 0 & 1 & 0 & 0 & 0 & 0 & 0 & 1
\end{array}\right]
\end{array}
\qquad (4\text{-}3\text{-}37)
$$

The H-submatrix $H_{(1)}$ of B_1' with respect to row 1 is

$$
H_{(1)} = \begin{array}{c} \\ 2 \\ 3 \\ 4 \\ 5 \\ 6 \end{array}
\begin{array}{c} \begin{array}{ccccccc} c & d & 2 & 3 & 4 & 5 & 6 \end{array} \\
\left[\begin{array}{ccccccc}
1 & 0 & 1 & 0 & 0 & 0 & 0 \\
1 & 1 & 0 & 1 & 0 & 0 & 0 \\
0 & 1 & 0 & 0 & 1 & 0 & 0 \\
1 & 0 & 0 & 0 & 0 & 1 & 0 \\
0 & 1 & 0 & 0 & 0 & 0 & 1
\end{array}\right]
\end{array}
\qquad (4\text{-}3\text{-}38)
$$

which cannot be partitioned as in Eq. 3-5-3 with both H_1 and H_2 being nonempty. Thus one of two M-submatrices of B_1' with respect to row 1 will be

$$ M(1) = B_1' \qquad (4\text{-}3\text{-}39) $$

Similarly, one of two M-submatrices of $M(1)$ with respect to row 2 will be

$$ M(12) = B_1' \qquad (4\text{-}3\text{-}40) $$

Furthermore, one of the pair of M-submatrices of $M(12)$ with respect to row 4 is

$$ M(124) = B_1' \qquad (4\text{-}3\text{-}41) $$

Since column b of $M(124)$ has three 1's at rows 1, 2, and 4, $M(124) = B_1'$ is not a fundamental cut-set matrix with rows 1, 2, and 4 being representing incidence sets by Theorem 3-5-10. Furthermore, using rows 1, 2, and 4, we always obtain matrix $M(124)$ as an M-submatrix. Thus we can say that B_1' is not a fundamental cut-set matrix. Hence by Theorem 3-5-11, B cannot be a fundamental cut-set matrix. Thus G has no dual by Theorem 4-3-9.

Continuing the proof, suppose linear graph G contains the other basic nonplanar graph G_2 shown in Fig. 4-3-14 as a subgraph. Then a fundamental

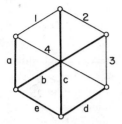

Fig. 4-3-14. A basic nonplanar graph and tree (a, b, c, d, e).

circuit matrix B of G can be chosen such that the fundamental circuit matrix B_1 of G_2 with respect to tree $t = (a, b, c, d, e)$ is a submatrix of B where

$$
B_1 =
\begin{array}{c}
\\
1 \\
2 \\
3 \\
4
\end{array}
\begin{array}{cccccc}
a & b & c & d & e & 1 \ 2 \ 3 \ 4 \\
\end{array}
\left[
\begin{array}{ccccc|cccc}
1 & 0 & 1 & 0 & 1 & 1 & 0 & 0 & 0 \\
0 & 1 & 1 & 0 & 1 & 0 & 1 & 0 & 0 \\
0 & 1 & 0 & 1 & 1 & 0 & 0 & 1 & 0 \\
1 & 0 & 0 & 1 & 1 & 0 & 0 & 0 & 1
\end{array}
\right]
\qquad (4\text{-}3\text{-}42)
$$

Note that the unit matrix is placed at the right-hand side for convenience. As before, if B_1 can be a cut-set matrix of a linear graph, then it is possible that G may have a dual. (However, even if B_1 can be a cut-set matrix, there is no guarantee that B can be a cut-set matrix.)

With row 1, one of two M-submatrices of B_1 is

$$M(1) = B_1 \qquad (4\text{-}3\text{-}43)$$

One of two M-submatrices of $M(1)$ with respect to row 2 is

$$M(12) = B_1 \qquad (4\text{-}3\text{-}47)$$

Furthermore, one of two M-submatrices of $M(12)$ with respect to row 3 is

$$M(123) = B_1 \qquad (4\text{-}3\text{-}45)$$

Since column e of $M(123)$ has three 1's in rows 1, 2, and 3, by Theorem 3-5-10, B_1 cannot be a fundamental cut-set matrix with rows 1, 2, and 3 representing incidence sets. Also using rows 1, 2, and 3, we will get $M(123) = B_1$ as one of M-submatrices of B_1. Thus we can say that B_1 cannot be a fundamental cut-set matrix. Since by Theorem 3-5-11, B cannot be a fundamental cut-set matrix, G has no dual. By Theorem 4-2-1, which is if and only if G is a nonplanar, G contains a subgraph which is homeomorphic to either of these two basic nonplanar graphs, we can conclude that the theorem is true.   QED

By knowing that only a planar graph has a dual, we can obtain interesting properties of planar graphs. Let G' be a dual of a connected linear graph G.

Suppose G is drawn on a plane such that no two edges are crossing each other. Also suppose G' is obtained from G by D-process. Then we know that each vertex of G' is in each region of G.

By Theorem 2-4-6, we know that there exists $n'_v - 1$ linearly independent incidence sets in G' where n'_v is the number of vertices in G'. Let n_e be the number of edges in G. (Note that the number of edges in G and the number of edges in G' are the same.) Since the number of linearly independent circuits in G is $n_e - n_v + 1$ by Theorem 3-3-5 where n_v is the number of vertices in G, and since circuits of G correspond to cut-sets of G', we have

$$n'_v - 1 = n_e - n_v + 1 \qquad (4\text{-}3\text{-}46)$$

Thus

$$n'_v = n_e - n_v + 2 \qquad (4\text{-}3\text{-}47)$$

We say that a linear graph G is a *maximal planar graph* if (1) there is no parallel edges in G and (2) by adding one edge between any two vertices in G, G is no longer a planar. Consider a maximal planar graph G. Then we can see that the circuit corresponding to the boundary of every region (including the outer region) consists of exactly three edges. If this were not the case, we could insert one edge to make a region whose boundary consists of more than three edges into two regions without crossing edges.) This means that in G', which is obtained from G by a D-process, there are exactly three edges connected to every vertex. Thus the number of edges must be equal to $3n'_v/2$. Hence by Eq. 4-3-47

$$n_e = \frac{3n'_v}{2} = \frac{3(n_e - n_v + 2)}{2} \qquad (4\text{-}3\text{-}48)$$

or

$$n_e = 3(n_v - 2) \qquad (4\text{-}3\text{-}49)$$

Theorem 4-3-11. Any planar graph containing no parallel edges has at most $3(n_v - 2)$ edges where n_v is the number of vertices in the planar graph.

Suppose a planar graph G contains no parallel edges. If the degree of every vertex in G is at least six, then the number n_e of edges is at least

$$\frac{6n_v}{2} = 3n_v \qquad (4\text{-}3\text{-}50)$$

However, by the previous theorem, the number of edges cannot be more than $3(n_v - 2)$, which is obviously smaller than $3n_v$.

Theorem 4-3-12. A planar graph containing no parallel edges has at least one vertex of degree 5 or less.

Suppose every vertex of a planar graph (containing no parallel edges) is of degree 5. Then the number of edges in the graph will be $5n_v/2$. By Theorem 4-3-11,

$$\frac{5n_v}{2} \le 3(n_v - 2) \tag{4-3-51}$$

or

$$n_v \ge 12 \tag{4-3-52}$$

There exists a maximal planar graph consisting of exactly 12 vertices in which every vertex is of degree 5 as shown in Fig. 4-3-15.

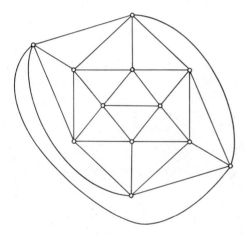

Fig. 4-3-15. A maximal planar graph in which every vertex is of degree 5.

4-4 REALIZABILITY OF CUT-SET MATRIX (PART II)

A testing algorithm for realizability of cut-set matrices in Section 3-5 requires operation on a given matrix. Hence it may not be convenient to employ this algorithm when we need to build a fundamental cut-set matrix under some restrictions. Here we study properties of a cut-set matrix introduced by Tutte in 1959. But first we will study a relationship between a cut-set matrix and a circuit matrix of a planar graph.

Suppose $[U \quad B_{f_{12}}]$ is a fundamental circuit matrix of a planar graph G. Then by Theorem 4-3-7, matrix $[B_{f_{12}} \quad U]$ is a fundamental cut-set matrix of a graph G' which is a dual of G. Similarly, if $[Q_{11} \quad U]$ is a cut-set matrix of a planar graph G, then matrix $[U \quad Q_{11}]$ is a circuit matrix of a dual graph G' of G. On the other hand, suppose a matrix $[N \quad U]$ is a circuit matrix of a *nonplanar* graph. Then it is impossible that $[N \quad U]$ is a fundamental cut-set matrix because a nonplanar graph cannot have a dual by Theorem 4-3-10. Also Theorem 3-5-11 says that if a matrix $[N \quad U]$ contains a submatrix which is not a cut-set matrix, then the matrix cannot be a fundamental cut-set matrix.

Theorem 4-4-1. A matrix $[N \quad U]$ cannot be a fundamental cut-set matrix if there is a submatrix which is a fundamental circuit matrix of either of two basic nonplanar graphs of Kuratowski.

Note that a nonplanar graph has either of Kuratowski's two basic nonplanar graphs by Theorem 4-2-1.

Let D be a nonsingular matrix of order equal to the number of rows in $[N \quad U]$ and whose entries are either 1 or 0. Suppose rearranging columns of $D[N \quad U]$, we can obtain a new matrix $[N' \quad U]$. In Section 3-4, we saw that if we chose a different tree, we would have a different fundamental cut-set matrix. Hence, if $[N \quad U]$ is a fundamental cut-set matrix of a linear graph, then $[N' \quad U]$ is also a fundamental cut-set matrix of the same linear graph but with respect to a different tree. Theorem 4-4-1 must hold for any fundamental cut-set matrix. Thus we can extend Theorem 4-4-1 if we define a normal form of a matrix as follows.

Definition 4-4-1. A matrix containing a unit matrix is referred to as a matrix in normal form.

Now Theorem 4-4-1 can be extended.

Theorem 4-4-2. A matrix $[N \quad U]$ is not a fundamental cut-set matrix if any normal form of $[N \quad U]$ contains a submatrix which is a fundamental circuit matrix of either of Kuratowski's two basic nonplanar graphs.

There is one interesting matrix, symbolized by N_o, which has a peculiar property where

$$N_o = \begin{bmatrix} 1 & 1 & 1 & 0 \\ 1 & 1 & 0 & 1 \\ 1 & 0 & 1 & 1 \end{bmatrix} \tag{4-4-1}$$

Lemma 4-4-1. Neither $[N_o \quad U_o]$ nor $[N_o{}^t \quad U_o']$ is a fundamental cut-set matrix where U_o and U_o' are unit matrices.

The proof of this lemma is left to the reader. With this lemma, Theorem 4-4-2 can be extended.

Theorem 4-4-3. A matrix $[N \quad U]$ is not a fundamental cut-set matrix if any normal form of $[N \quad U]$ contains any of (1) $[N_o \quad U_o]$ (2) $[N_o{}^t \quad U_o']$, and (3) a fundamental circuit matrix of either of Kuratowski's two basic non-planar graphs.

The conditions given in Theorem 4-4-3 are also sufficient for a matrix to be a fundamental cut-set matrix. In order to show this, we use a particular matrix called a minimum nonrealizable matrix defined as follows.

Definition 4-4-2. A matrix $[N \quad U]$ is a *minimum nonrealizable matrix* if it satisfies the following four conditions:

CONDITION 1. No normal form of $[N \quad U]$ contains a circuit matrix of either of two basic nonplanar graphs of Kuratowski.

CONDITION 2. No normal form of $[N \quad U]$ contains either $[N_o \quad U_o]$ or $[N_o{}^t \quad U_o']$ where U_o and U_o' are unit matrices and N_o is a matrix given by Eq. 4-4-1.

CONDITION 3. $[N \quad U]$ is not a cut-set matrix.

CONDITION 4. If we delete any row r from any formal form $[N' \quad U]$ of $[N \quad U]$, the resultant matrix $[N' \quad U]_{-r}$ is a cut-set matrix. Furthermore, if we delete any column c from N', the resultant matrix $[N'_{-c} \quad U]$ is a cut-set matrix.

First we investigate some properties of minimum nonrealizable matrices under the assumption that such matrices exist. For simplicity, we define the following symbols.

Definition 4-4-3. Let r be a row in a minimum nonrealizable matrix $[N \quad U]$. Let c be a column in N. Then $G(-r)$, $G(-c)$, $G(-r-c)$ and $G(-c-r)$ are defined by

$G(-r)$:   A linear graph whose cut-set matrix is $[N \quad U]_{-r}$.
$G(-c)$:   A linear graph whose cut-set matrix is $[N_{-c} \quad U]$.
$G(-r-c)$:   A linear graph obtained from $G(-r)$ by deleting the edge corresponding to column c of $[N \quad U]_{-r}$.
$G(-c-r)$:   A linear graph obtained from $G(-c)$ by shorting the edge (coinciding two endpoints of the edge and deleting the edge) corresponding to the column in U which has 1 at row r of $[N_{-c} \quad U]$

where the symbol $[N \quad U]_{-r}$ indicates the matrix obtained by deleting row r from a minimum nonrealizable matrix $[N \quad U]$ and the symbol $[N_{-c} \quad U]$ indicates the matrix obtained by deleting column c of N.

First consider some properties of linear graphs whose cut-set matrices are the submatrices of $[N \quad U]$.

Lemma 4-4-2. Let $G_0(-r - c)$ be a linear graph whose cut-set matrix is $[N_{-c} \quad U]_{-r}$ which is obtained from a minimum nonrealizable matrix $[N \quad U]$ by deleting column c in N and row r. Then $G_0(-r - c)$ is 2-isomorphic to $G(-r - c)$ and $G(-c - r)$.

Proof. If we delete the chord in $G(-r)$ corresponding to column c, we have a linear graph $G(-r - c)$ whose cut-set matrix is $[N_{-c} \quad U]_{-r}$. Hence by Theorem 4-1-2, $G_0(-r - c)$ and $G(-r - c)$ are 2-isomorphic to each other. Similarly, if we short the branch in $G(-c)$ corresponding to the column in U which has 1 at row r of $[N_{-c} \quad U]$, we have a linear graph $G(-c - r)$ whose cut-set matrix is $[N_{-c} \quad U]_{-r}$. Thus $G(-c - r)$ and $G_0(-r - c)$ are 2-isomorphic to each other by Theorem 4-1-2. QED

The next lemma shows some properties of linear graph $G(-c)$.

Lemma 4-4-3. Let $r_1, r_2, \ldots, r_k$ be the rows in $[N \quad U]$ which have 1 at column c. Let branch r_p be the edge corresponding to column in U which has 1 at row r_p for $p = 1, 2, \ldots, k$. Then linear graph $G(-c)$ has the following properties:

1. When any one of branches $r_1, r_2, \ldots, r_k$ is shorted, the remaining branches will either be a path or become a path by 2-isomorphic operations.

2. Branches $r_1, r_2, \ldots, r_k$ can neither be a path nor become a path by any 2-isomorphic operations. Note that branches $r_1, r_2, \ldots, r_k$ are edges in the tree corresponding to the columns of U of $[N_{-c} \quad U]$. The 2-isomorphic operations are the geometric operations to obtain 2-isomorphic graphs introduced in Section 4-1.

Proof. Branches $r_2, r_3, \ldots, r_k$ and chord c must form a circuit in $G'(-r_1)$ in order that $[N \quad U]_{-r_1}$ be a fundamental cut-set matrix of $G(-r_1)$ where $G'(-r_1)$ is a linear graph 2-isomorphic to $G(-r_1)$. Hence branches $r_2, r_3, \ldots, r_k$ form a path in $G'(-r_1 - c)$ which is 2-isomorphic to $G(-r_1 - c)$. By Lemma 4-4-2, $G(-r_1 - c)$ and $G(-c - r_1)$ are 2-isomorphic to each other. Hence $G'(-r_1 - c)$ and $G(-c - r_1)$ are 2-isomorphic to each other. Thus there exists a path consisting of $r_2, r_3, \ldots, r_k$ in a linear graph $G'(-c - r_1)$ obtained from $G'(-c)$ by shorting r_1 where $G'(-c)$ is 2-isomorphic to $G(-c)$. Instead of r_1, we can use $r_p(1 < p \le k)$ to reach the same conclusion. Hence property 1 holds in $G(-c)$.

Suppose branches $r_1, r_2, \ldots, r_k$ form a path in $G'(-c)$ which is 2-isomorphic to $G(-c)$. Then we can obtain a linear graph G by inserting chord c to $G'(-c)$ so that the chord c forms a circuit with branches $r_1, r_2, \ldots, r_k$. Then a fundamental cut-set matrix of G will be $[N \quad U]$ because a fundamental cut-set matrix of $G'(-c)$ is $[N_{-c} \quad U]$. However, $[N \quad U]$ is a minimum nonrealizable matrix. Hence property 2 must hold in $G(-c)$. QED

One of the important properties of $[N \quad U]$ is given by the next theorem.

Theorem 4-4-4. Let $[N \quad U]$ be a minimum nonrealizable matrix. Then any column of N' of any normal form $[N' \quad U]$ of $[N \quad U]$ has either two nonzero entries or three nonzero entries.

Proof. It is necessary only to prove that k in Lemma 4-4-3 is not larger than 3. To show this, we will investigate all possible structures of $G(-c)$ with $k \geq 3$ which have the properties in Lemma 4-4-3. Recall that in $G(-c)$, branches $r_1, r_2, \ldots, r_k$ can neither be a path nor become a path by 2-isomorphic operations (property 2). However, if we short branch $r_p(1 \leq p \leq k)$, the remaining branches will either be a path or become a path by 2-isomorphic operations (property 1). Hence by choosing $p = 1$, we have the following two classes of structures which will be the entire candidates for $G(-c)$:

CASE 1. Without shorting r_1, either $r_2, \ldots, r_k$ are already a path or these branches become a path by 2-isomorphic operations.

CASE 2. Without shorting r_1, neither $r_2, \ldots, r_k$ forms a path nor do these branches become a path by 2-isomorphic operations.

The candidates of $G(-c)$ belonging to Class 1 are of two types, depending on whether or not branch r_1 is attached to a path formed by $r_2, \ldots, r_k$. Thus we can see that all structures for the candidates for $G(-c)$ in Class 1 will be either G_d or G_e in Fig. 4-4-1d and e. Note that there are the candidates for $G(-c)$ in Class 1, which becomes the same structure as G_d or G_e by 2-isomorphic operations. However, for the proof of Theorem 4-4-4, we do not need to consider them.

The candidates for $G(-c)$ belonging to Class 2 can be divided into two types depending on the kind of 2-isomorphic operations necessary in order to make branches $r_2, \ldots, r_k$ into a path after shorting branch r_1. As we have seen in Section 4-1, there are two types of geometrical operations for 2-isomorphism. For convenience, we will call them Type 1 and Type 2.

Definition 4-4-4. A *2-isomorphic operation of Type 1* is an operation to split a cut-vertex into two so that the number of maximal connected subgraphs will be increased by one. The reverse of this operation, also called a 2-isomorphic operation of Type 1, is to connect two maximal connected subgraphs g_r and g_s by coinciding a vertex in g_r and a vertex in g_s. A *2-isomorphic operation of Type 2* is the geometrical operation of turning around one of two subgraphs, which are connected by two vertices, at these vertices.

The candidates for $G(-c)$ belonging to Class 2 that require the 2-isomorphic operation of Type 1 can be represented by structure G_c in Fig.

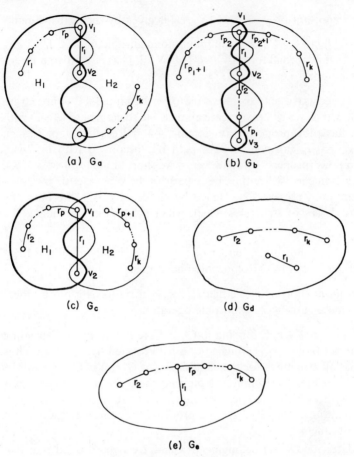

Fig. 4-4-1. Candidates for $G(-c)$. (a) G_a; (b) G_b; (c) G_c; (d) G_d; (e) G_e.

4-4-1c. Of course, all structures that can be obtained from G_c by 2-isomorphic operations before shorting r_1 belong to this type, but we need not consider them here. For the candidates for $G(-c)$ that use the 2-isomorphic operation of Type 2, there are two structures depending on how branches $r_2, r_3, \ldots, r_k$ are connected to become a path by the 2-isomorphic operation. These are G_a and G_b in Fig. 4-4-1a and b. Again, all structures that can be obtained from G_a and G_b by 2-isomorphic operations before shorting branch r_1 can be neglected. Thus in order to prove Theorem 4-4-4, we need only investigate five structures in Fig. 4-4-1.

STRUCTURE A.   Consider linear graph G_a in Fig. 4-4-1a. If $p < 2$, that is, all $r_2, r_3, \ldots, r_k$ are in subgraph H_2, G_a is the same structure as G_d in Fig. 4-4-1d, which will be studied later. So we assume that $p \geq 2$. Note that when branch r_1 is shorted, vertices v_1 and v_2 become one vertex. This new vertex and vertex v_3 become a pair of vertices of a 2-isomorphic operation of Type 2 by which $r_2, r_3, \ldots, r_p, r_{p+1}, \ldots, r_k$ become one path. Instead of shorting r_1, suppose we short r_2. If a 2-isomorphic operation which becomes possible by shorting r_2 is that of Type 1, the original graph G_a must be one shown in Fig. 4-4-2. Hence it is clear that the remaining branches $r_1, r_3, \ldots, r_k$ cannot be a path by 2-isomorphic operations.

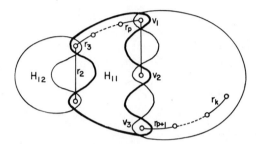

Fig. 4-4-2. G_a with H_{12} and H_{22}.

If a 2-isomorphic operation which becomes possible by shorting r_2 is that of Type 2, then the use of this operation will not change the structure of H_2 or the location of r_1. Thus in order to make $r_1, r_3, \ldots, r_k$ a path, some of branches $r_3, \ldots, r_p$ must be shifted to form a path between v_2 and v_3, which is impossible unless the path was there to begin with. Since the path cannot be there by property 2, we can conclude that G_a with $p \geq 2$ is not $G(-c)$.

STRUCTURE B. Consider G_b in Fig. 4-4-1b. If $p_1 < 2$, it becomes G_a so $p_1 \geq 2$. If $p_2 = p_1$, which means that none of $r_1, r_2, \ldots, r_k$ is in H_1 and $r_1, r_2, \ldots, r_k$ form a path. Hence p_2 must be larger than p_1. Furthermore, it is clear that $k \geq p_2 + 1$. Suppose $p_1 > 2$, then shorting r_2, we have the resultant graph which is identical to G_b except that the number of series edges which form a path between v_2 and v_3 is reduced by 1. Thus $r_1, r_3, \ldots, r_k$ cannot become a path by 2-isomorphic operations. This means that $p_1 = 2$.

With $p_1 = 2$, suppose $p_2 > 3$. Since there exist at least two branches r_3 and r_4 in H_1 for this case, shorting r_3 will give the same results as for G_a. Hence G_b is not $G(-c)$ for this case.

The only remaining case is when $p_2 = 3$ and $k = 4$ as in Fig. 4-4-3. In order that this structure is valid, there must be paths P_1, P_2, P_3, and P_4 as shown in the figure. This linear graph obviously has a subgraph which is

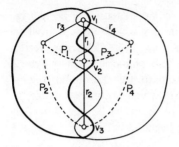

Fig. 4-4-3. Paths P_1, P_2, P_3, and P_4.

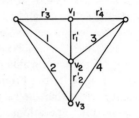

Fig. 4-4-4. G_3.

homeomorphic to linear graph G_3 in Fig. 4-4-4. Hence there is a fundamental cut-set matrix Q_f of G_b which contains a fundamental cut-set matrix Q_f of G_3 as its submatrix. Consider a fundamental cut-set matrix Q_f of G_3 where

$$Q_f = \begin{array}{c} \begin{array}{cccccccc} 1 & 2 & 3 & 4 & r_1' & r_2' & r_3' & r_4' \end{array} \\ \left[\begin{array}{cccc|cccc} 1 & 1 & 1 & 1 & 1 & 0 & 0 & 0 \\ 0 & 1 & 0 & 1 & 0 & 1 & 0 & 0 \\ 1 & 1 & 0 & 0 & 0 & 0 & 1 & 0 \\ 0 & 0 & 1 & 1 & 0 & 0 & 0 & 1 \end{array}\right] \end{array}$$

If G_b is $G(-c)$, then a normal form of $[N \quad U]$ must contain the following matrix $[\underline{N} \quad \underline{U}]$:

$$[\underline{N} \quad \underline{U}] = \begin{bmatrix} 1 & & & \\ & 1 & & \\ & & 1 & \\ & & & 1 \end{bmatrix} \begin{array}{c} Q_f \end{array} = \begin{array}{c} \begin{array}{ccccccccc} c & 1 & 2 & 3 & 4 & r_1' & r_2' & r_3' & r_4' \end{array} \\ \left[\begin{array}{ccccc|cccc} 1 & 1 & 1 & 1 & 1 & 1 & 0 & 0 & 0 \\ 1 & 0 & 1 & 0 & 1 & 0 & 1 & 0 & 0 \\ 1 & 1 & 1 & 0 & 0 & 0 & 0 & 1 & 0 \\ 1 & 0 & 0 & 1 & 1 & 0 & 0 & 0 & 1 \end{array}\right] \end{array}$$

However, this matrix is a fundamental circuit matrix of one of the two basic nonplanar graphs of Kuratowski. Thus $[N \quad U]$ does not satisfy Condition 1. Hence $[N \quad U]$ is not a minimum nonrealizable matrix, which is a contradiction. Thus G_b is not $G(-c)$.

STRUCTURE C. Consider G_c in Fig. 4-4-1c. If $p < 2$ (i.e., no branches of $r_2, \ldots, r_k$ are in H_1), G_c becomes G_d, which will be discussed later. Hence we assume that $p \geq 2$. By shorting r_2 rather than r_1, we can easily see that the remaining branches $r_1, r_3, \ldots, r_k$ cannot become a path by 2-isomorphic operations. Hence G_c is not $G(-c)$.

STRUCTURE D.   Consider G_d in Fig. 4-4-1d. If we short r_2 instead of shorting r_1, the structure of G_d will be either G_a or G_c in order that G_d be a candidate for $G(-c)$. Thus G_d is not $G(-c)$.

STRUCTURE E.   Consider G_e in Fig. 4-4-1e. For $k > 3$, if we short any branch other than r_1, the structure of G_e becomes one of the others in Fig. 4-4-1 in order that G_e be a candidate for $G(-c)$. However, we found that these cannot be $G(-c)$. Thus G_e can be $G(-c)$ only when $k = 3$, which gives the structure shown in Fig. 4-4-5. We can now conclude that $k \leq 3$. Note

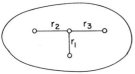

Fig. 4-4-5. G_e with $k = 3$.

that k is the number of nonzeros in column c of $[N \quad U]$. Since this must be true for taking any column of any normal form of $[N \quad U]$, we can conclude that any column in N' has either two or three nonzeros where $[N' \quad U]$ is any normal form of $[N \quad U]$, which proves the theorem.   QED

We have the same property for the rows of N as that for columns given by the previous theorem.

Theorem 4-4-5.   Let $[N \quad U]$ be a minimum nonrealizable matrix. Then any row of N' of any normal form $[N' \quad U]$ of $[N \quad U]$ has either two or three nonzero entries.

Proof.   We need only show that $[N^t \quad U]$ is a minimum nonrealizable matrix when $[N \quad U]$ is a minimum nonrealizable matrix where U is a unit matrix and N^t is the transpose of N. First, we will show that $[N^t \quad U]$ satisfies Condition 4. Suppose $[N \quad U]_{-r}$ and $[N_{-c} \quad U]$ are cut-set matrices of nonplanar graphs. Then $[N \quad U]_{-r}$ and $[N_{-c} \quad U]$ contain a cut-set matrix of either of two basic nonplanar graphs of Kuratowski. This means that there are normal forms of $[N \quad U]_{-r}$ and $[N_{-c} \quad U]$ which have a column of more than three nonzero entries because we can choose trees as shown in Fig. 4-4-6a and b. This violates Theorem 4-4-4. Thus $[N \quad U]_{-r}$ and $[N_{-c} \quad U]$ are cut-set matrices of planar graphs. Hence $[N^t_{-r} \quad U]$ and $[N^t \quad U]_{-c}$ are cut-set matrices. This indicates that $[N^t \quad U]$ satisfies Condition 4. For Condition 1, suppose $[N^t \quad U]$ contains a circuit matrix of either of two basic nonplanar graphs. Since $[N^t \quad U]_{-c}$ is a cut-set matrix, $[N^t \quad U]$ is a circuit matrix of one of two basic nonplanar graphs itself. Then $[N \quad U]$ must be a cut-set matrix, which is a contradiction.

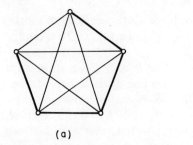
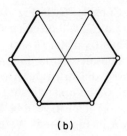

(a)   (b)

Fig. 4-4-6. Basic nonplanar graphs and chosen trees.

$[N^t \quad \underline{U}]$ must satisfy Condition 2 because if $[N_o \quad U_o]$ is in a normal form of $[N^t \quad \underline{U}]$, then $[N_o^t \quad U_o]$ is in a normal form of $[N \quad U]$ where U_o is a unit matrix. This is also true for $[N_o^t \quad U_o]$. Finally, $[N^t \quad \underline{U}]$ cannot be a cut-set matrix because if it is, $[N \quad U]$ is a circuit matrix of a nonplanar graph, which violates Condition 1. Hence $[N^t \quad \underline{U}]$ satisfies Condition 3. Thus $[N^t \quad \underline{U}]$ is a minimum nonrealizable matrix. Hence by Theorem 4-4-4, this theorem is true.   QED

The next theorem and Theorem 4-4-3 give the realizability condition introduced by Tutte.

Theorem 4-4-6. Minimum nonrealizable matrices do not exist.

Proof. If a matrix is not a cut-set matrix, there must be at least one column having at least three 1's. Furthermore, any row of N has either two or three 1's by Theorem 4-4-5. Hence submatrix N of a minimum nonrealizable matrix $[N \quad U]$ can be assumed to have the following configuration:

$$N = \begin{bmatrix} 1 & 1 & a_{13} & 0 & \cdots \\ 1 & a_{22} & a_{23} & a_{24} & \cdots \\ 1 & a_{32} & a_{33} & a_{34} & \cdots \\ 0 & a_{42} & a_{43} & a_{44} & \cdots \\ \vdots & \vdots & \vdots & \vdots & \end{bmatrix}$$

First we will consider the case when $a_{13} = 1$. $[N \quad U]$ for this case will be

$$[N \quad U] = \begin{bmatrix} 1 & 1 & 1 & 0 & \cdots & 1 & 0 & 0 & & \\ 1 & a_{22} & a_{23} & a_{24} & \cdots & 0 & 1 & 0 & & 0 \\ 1 & a_{32} & a_{33} & a_{34} & \cdots & 0 & 0 & 1 & & \\ 0 & & & & & 0 & 0 & 0 & 1 & \\ \vdots & \vdots & \vdots & \vdots & & \vdots & \vdots & \vdots & & 0 \\ 0 & & & & & 0 & 0 & 0 & & 1 \end{bmatrix}$$

If $a_{22} = a_{32}$, then $a_{22} = a_{32} = 1$ makes the first and second columns of N identical because there are at most three 1's in every column of N by Theorem 4-4-4. However, deleting one of these columns cannot make the remaining matrix realizable as a fundamental cut-set matrix. This means that $[N \quad U]$ is not a minimum nonrealizable matrix. Thus $a_{22} = a_{32} = 1$ is not possible. When $a_{22} = a_{32} = 0$, we have

$$
[N \quad U] =
\begin{array}{c c c c c c c c c c}
 & 1 & 2 & 3 & 4 & & 1' & 2' & 3' & \\
\left[\begin{array}{c c c c c c c c c c}\right.
 & 1 & 1 & 1 & 0 & \cdots & 1 & 0 & 0 & \\
 & 1 & 0 & a_{23} & a_{24} & \cdots & 0 & 1 & 0 & & 0 \\
 & 1 & 0 & a_{33} & a_{34} & \cdots & 0 & 0 & 1 & \\
 & 0 & a_{42} & a_{43} & a_{44} & \cdots & 0 & 0 & 0 & 1 \\
 & \vdots & \vdots & \vdots & \vdots & & \vdots & \vdots & \vdots & 0 & \ddots \\
 & 0 & & & & & & & & & 1 \left.\right]
\end{array}
$$

By interchanging columns 1 and 1', we have

$$
\begin{array}{c c c c c c c c}
 1' & 2 & 3 & 4 & & 1 & 2' & 3' \\
\left[\begin{array}{c c c c c c c c}\right.
 1 & 1 & 1 & 0 & \cdots & 1 & 0 & 0 \\
 0 & 0 & a_{23} & a_{24} & \cdots & 1 & 1 & 0 & \quad 0 \\
 0 & 0 & a_{33} & a_{34} & \cdots & 1 & 0 & 1 \\
 0 & a_{42} & a_{43} & a_{44} & \cdots & 0 & 0 & 0 & 1 \\
 \vdots & \vdots & \vdots & \vdots & & \vdots & \vdots & \vdots & 0 & \ddots \\
 0 & & & & & & & & & 1 \left.\right]
\end{array}
$$

In order to make this a normal form, we must add (modular 2) the first row to the second and third rows, which gives

$$
\begin{array}{c c c c c c c c}
 1' & 2 & 3 & 4 & & 1 & 2' & 3' \\
\left[\begin{array}{c c c c c c c c}\right.
 1 & 1 & 1 & 0 & \cdots & 1 & 0 & 0 \\
 1 & 1 & 1 + a_{23} & a_{24} & \cdots & 0 & 1 & 0 \\
 1 & 1 & 1 + a_{33} & a_{34} & \cdots & 0 & 0 & 1 & \quad 0 \\
 0 & a_{42} & a_{43} & a_{44} & \cdots & 0 & 0 & 0 & 1 \\
 \vdots & \vdots & \vdots & \vdots & & \vdots & \vdots & \vdots & 0 & \ddots \\
 0 & & & & & & & & & 1 \left.\right]
\end{array}
$$

By Theorem 4-4-4, column 2 must be identical to column 1'. Hence $[N \quad U]$

is not a minimum nonrealizable matrix. Thus $a_{22} \neq a_{32}$. Without the loss of generality, let $a_{22} = 1$ and $a_{32} = 0$:

$$[N \quad U] = \begin{bmatrix} 1 & 1 & 1 & 0 & \cdots & 1 & 0 & 0 & & & \\ 1 & 1 & a_{23} & a_{24} & \cdots & 0 & 1 & 0 & & & \\ 1 & 0 & a_{33} & a_{34} & \cdots & 0 & 0 & 1 & & & 0 \\ 0 & a_{42} & a_{43} & a_{44} & \cdots & 0 & 0 & 0 & 1 & & \\ \vdots & \vdots & \vdots & \vdots & & \vdots & \vdots & \vdots & & 0 & \ddots \\ 0 & & & & & & & & & & 1 \end{bmatrix}$$

Similarly, $a_{23} \neq a_{33}$. It is also true for $[N^t \quad U]$ by Theorem 4-4-5. Hence $a_{22} \neq a_{23}$ and $a_{32} \neq a_{33}$. Thus

$$\begin{array}{c} \begin{array}{ccccccccc} & 1 & 2 & 3 & 4 & & 1' & 2' & 3' \end{array} \\ \begin{array}{c} 1 \\ 2 \\ 3 \\ 4 \\ \vdots \\ \end{array} \begin{bmatrix} 1 & 1 & 1 & 0 & \cdots & 1 & 0 & 0 & & & \\ 1 & 1 & 0 & a_{24} & \cdots & 0 & 1 & 0 & & & \\ 1 & 0 & 1 & a_{34} & \cdots & 0 & 0 & 1 & & & 0 \\ 0 & a_{42} & a_{43} & a_{44} & \cdots & 0 & 0 & 0 & 1 & & \\ \vdots & \vdots & \vdots & \vdots & & \vdots & \vdots & \vdots & & 0 & \ddots \\ 0 & & & & & & & & & & 1 \end{bmatrix} \end{array}$$

Suppose $a_{42} \neq a_{43}$. Without the loss of generality, let $a_{42} = 1$ and $a_{43} = 0$. Then by interchanging columns 2 and $1'$ and making the resultant matrix a normal form by adding row 1 to rows 2, 4, and so forth, we will produce at least four 1's in column 3. Thus by Theorem 4-4-4, $[N \quad U]$ is not a minimum nonrealizable matrix. Hence $a_{42} = a_{43}$.

If $a_{42} = a_{43} = 1$, we have

$$[N \quad U] = \begin{bmatrix} 1 & 1 & 1 & 0 & \cdots & 1 & 0 & 0 & & & \\ 1 & 1 & 0 & a_{24} & \cdots & 0 & 1 & 0 & & & \\ 1 & 0 & 1 & a_{34} & \cdots & 0 & 0 & 1 & & & 0 \\ 0 & 1 & 1 & a_{44} & \cdots & 0 & 0 & 0 & 1 & & \\ \vdots & \vdots & \vdots & \vdots & & \vdots & \vdots & \vdots & & 0 & \ddots \\ & & 0 & & & & & & & & 1 \end{bmatrix}$$

in which $N_o{}^t$ is a submatrix. Thus $[N \quad U]$ is not a minimum nonrealizable matrix. Thus $a_{42} = a_{43} = 0$. If any $(r, 2)$ entry for $r > 4$ is 1, we can interchange row r and row 4 then interchange columns belonging to U to make the resultant matrix a normal form. Thus $a_{42} = a_{43} = 0$ means $a_{r2} = a_{r3} = 0$ for $r = 4, 5, \ldots$. This is also true for a_{2s} and a_{3s} by considering $[N^t \quad U]$ for $s \geq 4$. Thus we have

$$[N \quad U] = \begin{bmatrix} 1 & 1 & 1 & 0 & \cdots & 0 & 1 & 0 & 0 & & \\ 1 & 1 & 0 & 0 & \cdots & 0 & 0 & 1 & 0 & & 0 \\ 1 & 0 & 1 & 0 & \cdots & 0 & 0 & 0 & 1 & & \\ 0 & 0 & 0 & & & & 0 & 0 & 0 & & \\ \vdots & \vdots & \vdots & & N' & & \vdots & \vdots & \vdots & U' & \\ 0 & 0 & 0 & & & & 0 & 0 & 0 & & \end{bmatrix}$$

If $[N \quad U]$ is not a cut-set matrix, we can see that $[N' \quad U']$ must not be a cut-set matrix. Thus $[N \quad U]$ with $a_{13} = 1$ is not a minimum nonrealizable matrix.

For $a_{13} = 0$, $[N \quad U]$ becomes

$$[N \quad U] = \begin{bmatrix} 1 & 1 & 0 & 0 & \cdots & 0 & & \\ 1 & a_{22} & a_{23} & a_{24} & \cdots & & & \\ 1 & a_{32} & a_{33} & a_{34} & \cdots & & U & \\ 0 & & & & & & & \\ \vdots & & & & & & & \\ 0 & & & & & & & \end{bmatrix}$$

$a_{22} = a_{32}$ cannot make $[N \quad U]$ a minimum nonrealizable matrix. Thus $a_{22} = 1$ and $a_{32} = 0$, or

$$[N \quad U] = \begin{matrix} 1 \\ 2 \\ 3 \\ 4 \\ \\ \\ \end{matrix} \begin{bmatrix} 1 & 1 & 0 & 0 & \cdots & 0 & & \\ 1 & 1 & a_{23} & & & & & \\ 1 & 0 & a_{33} & & & & U & \\ 0 & & & & & & & \\ \vdots & & & & & & & \\ 0 & & & & & & & \end{bmatrix}$$

If $a_{23} = 1$, we have $[N \quad U]$, which is the same form as one with $a_{13} = 1$. Furthermore, if $(2, p)$ for $p > 3$ in N is 1, we can shift it to the $(2, 3)$ position. Thus $a_{23} = a_{24} = \cdots = 0$ in N. But this situation makes row 1 and row 2 of N identical. Thus $[N \quad U]$ with $a_{13} = 0$ is not a minimum nonrealizable matrix. We can now conclude that minimum nonrealizable matrices do not exist, which proves Theorem 4-4-6. QED

We can see that any matrix which is not a fundamental cut-set matrix satisfying Conditions 1 and 2 can be reduced to a matrix satisfying Condition 4. Hence Theorem 4-4-6 means that any matrix which is not a fundamental cut-set matrix does not satisfy Conditions 1 and 2. This is the converse of Theorem 4-4-3.

Theorem 4-4-7 (Tutte). A matrix $[N \quad U]$ is a fundamental cut-set matrix if and only if no normal form of $[N \quad U]$ contains any of (1) $[N_o \quad U_o]$ (2) $[N_o^t \quad U_o']$, and (3) a fundamental circuit matrix of either of Kuratowski's two basic nonplanar graphs.

By the relationship between a cut-set matrix and a circuit matrix given by Eq. 3-4-24, Theorem 4-4-7 can be modified.

Theorem 4-4-8. A matrix $[N \quad U]$ is a fundamental circuit matrix if and only if no normal form of $[N \quad U]$ contains any of (1) $[N_o \quad U_o]$, (2) $[N_o^t \quad U_o']$, and (3) a fundamental cut-set matrix of either of Kuratowski's two basic nonplanar graphs.

PROBLEMS

1. Suppose Q_1 and Q_2 are cut-set matrices of linear graphs G_1 and G_2, respectively. Suppose $Q_1 = DQ_2$ where D is a nonsingular matrix consisting of 1 and 0. Can we say that G_1 and G_2 are 2-isomorphic to each other?

2. Are there several linear graphs having the same fundamental cut-set matrix?

3. Test whether the two linear graphs in Fig. P-4-3 are 2-isomorphic to each other.

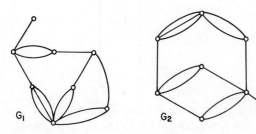

G_1 G_2

Fig. P-4-3.

4. How can we test whether a linear graph is planar without drawing it?

5. Is linear graph G_3 in Fig. P-4-5 planar?

Fig. P-4-5.

6. State a simple method to test whether two linear graphs are homeomorphic.

7. Test whether linear graphs G and G' in Fig. P-4-7 are homeomorphic.

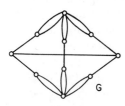 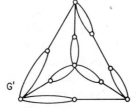

Fig. P-4-7.

8. Let G_1 and G_2 be linear graphs whose fundamental circuit matrices are B_1 and B_2, respectively. Suppose B_1 (nonempty) can be obtained by deleting rows of B_2 and deleting columns of zeros. What can we say about G_1 and G_2?

9. A degree-3 graph is a linear graph in which every vertex is of degree 3. There exists a degree-3 graph which is one of two basic nonplanar graphs of Kuratowski as shown in Fig. 4-3-14. Let this graph be symbolized by N_3. Prove or disprove that if a cut-set matrix of degree-3 graph G does not contain a cut-set matrix of N_3 as a submatrix, then G is planar.

10. Prove Lemma 4-4-1.

5

SPECIAL CUT-SET AND PSEUDO-CUT

5-1 CUT-SET SEPARATING TWO SPECIFIED VERTICES

Suppose we want to destroy all paths between two vertices i and j in a connected linear graph G by deleting edges. Let H be a set of edges such that the deletion of all edges in H from G will destroy all paths in $\{P_{ij}\}$. Thus any path between i and j in G has at least one edge in H. Note that $\{P_{ij}\}$ is a collection of all possible paths between vertices i and j (Definition 1-4-3).

Let us consider the resultant graph G' after deleting all edges in H. G' must have at least two maximal connected subgraphs which may be just isolated vertices, such that one of these contains vertex i and another contains vertex j. Thus the rank of G' is at most one less than that of G. This means that set H has a subset which is a cut-set. We have found (Eq. 2-2-5) that the deletion of all edges in a cut-set $S = \mathcal{E}(\Omega_1 \times \overline{\Omega}_1)$ from a linear graph G produces two subgraphs $\mathcal{E}(\Omega_1 \times \Omega_1)$ and $\mathcal{E}(\overline{\Omega}_1 \times \overline{\Omega}_1)$. If $\mathcal{E}(\Omega_1 \times \Omega_1)$ contains vertex i and $\mathcal{E}(\overline{\Omega}_1 \times \overline{\Omega}_1)$ contains vertex j, then the deletion of all edges in S will destroy all paths between i and j. Thus in order that a set H be a minimum set of edges by deletion of which will destroy all paths in $\{P_{ij}\}$, H must be a particular cut-set $\mathcal{E}(\Omega_1 \times \overline{\Omega}_1)$ such that $i \in \mathcal{E}(\Omega_1 \times \Omega_1)$ and $j \in \mathcal{E}(\overline{\Omega}_1 \times \overline{\Omega}_1)$. For convenience, we use the following definition.

Definition 5-1-1. A cut-set $S = \mathcal{E}(\Omega_1 \times \overline{\Omega}_1)$ is said to *separate vertices i and j* if $i \in \Omega_1$ and $j \in \overline{\Omega}_1$.

With this definition, we can say that set H must be a cut-set separating vertices i and j in order that set H be a minimum set.

Theorem 5-1-1. If deleting all edges in a set S will destroy all paths between vertices i and j in a connected linear graph G, and if deleting some but not all edges in S will not destroy all paths between i and j, then S is a cut-set separating i and j. Conversely, if all edges in cut-set $S = \mathscr{E}(\Omega_1 \times \bar{\Omega}_1)$ are deleted from a linear graph G, all paths from a vertex in Ω_1 to a vertex in $\bar{\Omega}_1$ will be destroyed.

In order to indicate whether a cut-set separates two specified vertices, we use the symbols $S(i; j)$ and $S(ij; .)$ as follows.

Definition 5-1-2. The symbol $S(i; j)$ indicates that cut-set S separates vertices i and j and the symbol $S(ij; .)$ shows that S does not separate vertices i and j.

Example 5-1-1. Consider the linear graph in Fig. 5-1-1. Let $\Omega_1 = (1, \ 2)$.

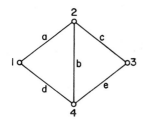

Fig. 5-1-1. A linear graph.

Then the cut-set $S = \mathscr{E}(\Omega_1 \times \bar{\Omega}_1)$ consists of edges b, c, and d. Since vertex 1 is in Ω_1 and vertex 3 is in $\bar{\Omega}_1$, the cut-set S separates vertices 1 and 3. We use the symbol $S(1; 3)$ to indicate this fact. On the other hand, if we are interesting vertices 1 and 2, the cut-set S will not separate them. We use the symbol $S(12; .)$ to indicate this. Let $S_1 = (a, b, c)$. Then we can see that S_1 will not separate vertices 1 and 3. Hence we use the symbol $S_1(13; .)$ to show this fact.

We frequently would like to find all possible cut-sets that separate two specified vertices. Instead of finding all such cut-sets directly from a linear graph, it may be simpler if we can generate such cut-sets from a set of linearly independent cut-sets.

In Section 2-3, we learned that the collection $\{S\}$ is an Abelian group under the ring sum. Moreover, there are $n_v - \rho$ generators by which all sets in $\{S\}$ can be generated by the ring sum operation. These cut-sets separating vertices i and j are obviously in $\{S\}$. However, in order to obtain only those cut-sets separating i and j, we must know some things about cut-sets of types $S(i; j)$ and $S(ij; .)$ such as:

1. What type of cut-set will we have if we ring sum two cut-sets of type $S(i; j)$?
2. What will $S(i; j) \oplus S(ij; .)$ be?

We also know that the ring sum of two cut-sets may or may not be a cut-set. However, we will study the case when the ring sum of two cut-sets is a cut-set first.

Theorem 5-1-2. If the ring sum of two cut-sets $S_1(i; j) \oplus S_2(i; j)$ is a cut-set, it is a cut-set that does not separate vertices i and j. If $S_1(i; j) \oplus S_2(ij; .)$ is a cut-set, it separates vertices i and j. Finally, if the ring sum of two cut-sets $S_3(ij; .) \oplus S_4(ij; .)$ is a cut-set, it does not separate vertices i and j.

Proof. For the proof of the first part, consider the linear graph shown in Fig. 2-3-1 where vertex i is in Ω_{11} and vertex j is in Ω_{22}. Hence $S_1(i; j)$ and $S_2(i; j)$ are given by Eqs. 2-3-13 and 2-3-14 where $\Omega_1 = \Omega_{11} \cup \Omega_{12}$ and $\Omega_2 = \Omega_{11} \cup \Omega_{21}$. Then by Eq. 2-3-17 (with Ω instead of Ω_c),

$$S_1(i; j) \oplus S_2(i; j) = \mathscr{E}(\Omega_3 \times \bar{\Omega}_3) \qquad (5\text{-}1\text{-}1)$$

where $\Omega_3 = \Omega_{11} \cup \Omega_{22}$. Since vertices i and j are in Ω_3, the first part of the theorem is correct. The second and the third parts can easily be proved the same way.   QED

Consider the linear graph in Fig. 5-1-2. It can be seen that cut-set $S_1 = (a, d)$ does not separate vertices i and j, but cut-set $S_2 = (b, d)$ separates i and j. Also $S_1 \oplus S_2$ is a cut-set. Hence by Theorem 5-1-2, $S_1 \oplus S_2 = S_a$ separates i and j.

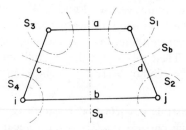

Fig. 5-1-2. A linear graph.

We have seen that a collection $\{S\}$ of all possible cut-sets, edge disjoint unions of cut-sets, and the empty set is an Abelian group. We know that there are $n_v - \rho$ generators (linearly independent cut-sets) in $\{S\}$ which can generate all members in $\{S\}$. For example, with cut-sets S_1, S_2, and S_3 of the linear graph in Fig. 5-1-2, we can express all members in $\{S\}$ as

$$\{S\} = \{\emptyset, S_1, S_2, S_3, S_1 \oplus S_2, S_1 \oplus S_3, S_2 \oplus S_3, S_1 \oplus S_2 \oplus S_3\}$$

We can see that S_2, $S_1 \oplus S_2$, and $S_1 \oplus S_2 \oplus S_3$ are the cut-sets separating vertices i and j. If we include $S_2 \oplus S_3$, the collection

$$\{S(i; j)\} = \{S_2, S_1 \oplus S_2, S_2 \oplus S_3, S_1 \oplus S_2 \oplus S_3\}$$

has the property that every member contains S_2. Note that only S_2 in the chosen linearly independent cut-sets separates i and j. Then the collection of all cut-sets separating i and j is given by min $\{S(i; j)\}$ (see Definition 1-6-1). However, Theorem 5-1-2 alone does not indicate that we should consider $S_2 \oplus S_3$ and $S_1 \oplus S_2 \oplus S_3$. Hence we should study the additional cases which are (1) the ring sum of two cut-sets when the result is not a cut-set and (2) the ring sum of more than two cut-sets. The next few theorems give the properties of the ring sum of two cut-sets when the result is not a cut-set.

Theorem 5-1-3. If the ring sum of two cut-sets $S_1(i; j) \oplus S_2(ij; .)$ is not a cut-set, it can be expressed as an edge disjoint union of cut-sets $S'_1, S'_2, \ldots, S'_p$ such that exactly one of $S'_1, S'_2, \ldots, S'_p$ separates vertices i and j.

Proof. Suppose $S_1(i; j)$ and $S_2(ij; .)$ are given by Eqs. 2-3-13 and 2-3-14. Suppose further that vertex i is in Ω_{11} and vertex j is in Ω_{21} where $\Omega_1 = \Omega_{11} \cup \Omega_{12}$ and $\Omega_2 = \Omega_{11} \cup \Omega_{21}$. Then by Eq. 2-3-17 with Ω rather than Ω_c, the resultant graph obtained by deleting all edges in $S_1(i; j) \oplus S_2(ij; .)$ consists of $\mathcal{E}(\Omega_3 \times \Omega_3)$ and $\mathcal{E}(\bar{\Omega}_3 \times \bar{\Omega}_3)$ where $\Omega_3 = \Omega_{11} \cup \Omega_{22}$. Note that $i \in \Omega_3$ and $j \in \bar{\Omega}_3$. Suppose $\mathcal{E}(\Omega_3 \times \Omega_3)$ consists of k maximal connected subgraphs and $\mathcal{E}(\bar{\Omega}_3 \times \bar{\Omega}_3)$ consists of m maximal connected subgraphs as shown in Fig. 2-3-7. Without loss of generality, let $i \in g_{a1}$ and $j \in g_{b1}$. Furthermore, in Fig. 2-3-8, g_{c1} contains g_{b1}. Thus vertex j is in g_{c1}. Then it can be seen that in Fig. 2-3-8, one of the cut-sets in $\mathcal{E}(\Omega_{a1} \times \bar{\Omega}_{a1})$ separates vertices i and j. Let this cut-set be S'_1. It is clear that S'_1 is only the cut-set in the edge disjoint union of cut-sets in $S_1(i; j) \oplus S_2(ij; .)$ which separates vertices i and j when we choose S'_1 as one of cut-sets in the edge disjoint union of cut-sets.
$$\text{QED}$$

Consider the linear graph in Fig. 5-1-2. Let $S_a(i; j) = (a, b)$ and $S_b(ij; .) = (c, d)$. Then $S_a \oplus S_b = (a, b, c, d)$, which is an edge disjoint union of cut-sets. The question is which cut-sets form this edge disjoint union of cut-sets. One set of cut-sets which form the edge disjoint union of cut-sets clearly consists of $S_a(i; j)$ and $S_b(ij; .)$. Another set consists of $S_1 = (a, d)$ and $S_4 = (b, c)$ and the other set consists of $S_2 = (b, d)$ and $S_3 = (a, c)$. In other words, this edge disjoint union of cut-sets (a, b, c, d) can be considered as an edge disjoint union of (1) S_a and S_b, (2) S_1 and S_4, or (3) S_2 and S_3. The preceding theorem shows that *one* of these sets consists of cut-sets in which exactly one separates i and j. The next theorem says that *any* of these sets contains exactly one cut-set which separates i and j.

Theorem 5-1-4. Suppose there exists a path between vertices i and j in a linear graph G. If the ring sum of two cut-sets $S_1(i; j) \oplus S_2(ij; .)$ is not a cut-set, then any representation of $S_1(i; j) \oplus S_2(ij; .)$ into an edge disjoint union of cut-sets contains exactly one cut-set which separates vertices i and j.

Proof. With the same assumption as in the proof of Theorem 5-1-3, we have the linear graph shown in Figs. 2-3-7 and 2-3-8. Note that vertex j is in g_{b1}. The set of edges $\mathscr{E}(\Omega_{a1} \times \Omega_{b1})$ must be in exactly one cut-set that forms an edge disjoint union of cut-sets, because those edges are in $S_1(i;j) \oplus S_2(ij; .)$ and these cut-sets have no edges in common. On the other hand, in order to separate vertices i and j, the cut-set must contain the set $\mathscr{E}(\Omega_{a1} \times \Omega_{b1})$.   QED

For the ring sum of two cut-sets $S_1(i;j) \oplus S_2(i;j)$ and two cut-sets $S_1(ij; .) \oplus S_2(ij; .)$, we have Theorem 5-1-5.

Theorem 5-1-5. If the ring sum of two cut-sets $S_1(i;j) \oplus S_2(i;j)$ is not a cut-set, it can be expressed as an edge disjoint union of cut-sets $S_1', S_2', \ldots, S_p'$ such that either exactly two of $S_1', S_2', \ldots, S_p'$ separate vertices i and j or none of $S_1', S_2', \ldots, S_p'$ separates vertices i and j.

Proof. We use the same assumptions as those in the proof of Theorem 5-1-3 except $j \in \Omega_{22}$ this time. Therefore there are two possible locations of vertices i and j in the linear graph in Fig. 2-3-7. Suppose vertices i and j are both in g_{a1}. Then it is clear that none of $S_1', S_2', \ldots, S_p'$ separates vertices i and j. Suppose vertex i is in g_{a1} and vertex j is in g_{a2}. Then it is also clear from the proof of Theorem 5-1-3 that exactly one cut-set in $\mathscr{E}(\Omega_{a1} \times \overline{\Omega}_{a1})$ separates vertices i and j, and exactly one cut-set in $\mathscr{E}(\Omega_{a2} \times \overline{\Omega}_{a2})$ separates vertices i and j. Thus by choosing these cut-sets to form an edge disjoint union of cut-sets, there are exactly two cut-sets which separates vertices i and j.   QED

Theorem 5-1-6. If the ring sum of two cut-sets $S_1(i;j) \oplus S_2(i;j)$ is not a cut-set, then any representation of $S_1(i;j) \oplus S_2(i;j)$ into an edge disjoint union of cut-sets contains either exactly two cut-sets which separate vertices i and j or no cut-sets separating vertices i and j.

Proof. We use the same assumptions as those in the proof of Theorem 5-1-5. When vertices i and j are in g_{a1}, it is obvious that none of cut-sets separates vertices i and j. Thus we assume that vertex $i \in g_{a1}$ and vertex j is in g_{a2}. Since g_{a2} must be in one of g_{cr} for $r = 1, 2, \ldots, n$ in Fig. 2-3-9, we assume that g_{a2} is in g_{c1} as shown in Fig. 5-1-3.

Subgraphs $g_{d1}, g_{d2}, \ldots, g_{du}$ with g_{a2} are the maximal connected subgraphs in the resultant graph obtained from g_{c1} by deleting all edges connected from vertices in g_{a1} and g_{a2} to vertices in g_{c1} other than in g_{a2}. In other words, these subgraphs become maximal connected subgraphs in g_{c1} when all edges in $\mathscr{E}(\Omega_{a1} \times \Omega_{d1}), \mathscr{E}(\Omega_{a1} \times \Omega_{d2}), \ldots, \mathscr{E}(\Omega_{a1} \times \Omega_{dt}), \mathscr{E}(\Omega_{d1} \times \Omega_{a2}), \mathscr{E}(\Omega_{d2} \times \Omega_{a2}), \ldots, \mathscr{E}(\Omega_{du} \times \Omega_{a2})$ are deleted where Ω_{dr} is the set of all vertices in g_{dr} for $r = 1, 2, \ldots, u$. Note that edges from vertices in g_{a1} to vertices in g_{c1} are those connected to vertices in $g_{d1}, g_{d2}, \ldots, g_{dt}$. On the other hand, edges connected from vertices in g_{a2} to vertices in the remaining g_{c1} are those connected to the vertices in subgraphs $g_{d1}, g_{d2}, \ldots, g_{dt}, g_{dt+1}, \ldots, g_{du}$ because g_{c1} is connected by assumption.

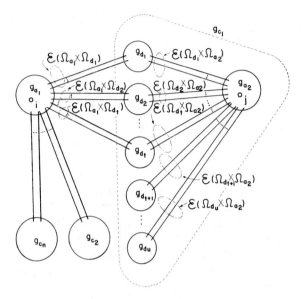

Fig. 5-1-3.   Maximal connected subgraphs $g_{d1}, g_{d2}, \ldots, g_{du}$.

We can see that any choice of edge disjoint union of cut-sets consisting of edges in $\mathcal{E}(\Omega_{a1} \times \Omega_{d1})$, $\mathcal{E}(\Omega_{a1} \times \Omega_{d2}), \ldots, \mathcal{E}(\Omega_{a1} \times \Omega_{dt})$, $\mathcal{E}(\Omega_{a2} \times \Omega_{d1})$, $\mathcal{E}(\Omega_{a2} \times \Omega_{d2}), \ldots, \mathcal{E}(\Omega_{a2} \times \Omega_{dt})$ will contain either two cut-sets separating i and j or no cut-sets separating i and j. We can see also that any choice of edge disjoint union of cut-sets consisting of remaining edges cannot have cut-sets separating i and j. Thus the theorem is true.   QED

Theorem 5-1-7.   If the ring sum of two cut-sets $S_1(ij; .) \oplus S_2(ij; .)$ is not a cut-set, then any representation of $S_1(ij; .) \oplus S_2(ij; .)$ into an edge disjoint union of cut-sets contains either exactly two cut-sets that separate vertices i and j or cut-sets that separate vertices i and j.

The proof of this theorem is similar to that of Theorem 5-1-6.

In Fig. 5-1-4, there are two linear graphs. Figure 5-1-4a is an example of $S_1(i; j) \oplus S_2(i; j)$ which shows two possible choices of sets of cut-sets to represent $S_1(i; j) \oplus S_2(i; j)$. That is, we can choose S_{a1} and S_{a2} to make an

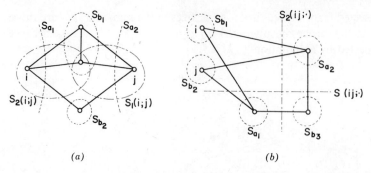

Fig. 5-1-4. Examples of (a) $S_1(i; j) \oplus S_2(i; j)$ and (b) $S_1(ij; \cdot) \oplus S_2(ij; \cdot)$.

edge disjoint union of cut-sets or S_{b1} and S_{b2} to make an edge disjoint union of cut-sets for $S_1(i; j) \oplus S_2(i; j)$. Note that both S_{a1} and S_{a2} separate vertices i and j; however, both S_{b1} and S_{b2} do not separate vertices i and j.

In Fig. 5-1-4b, there are three different ways of choosing cut-sets to represent $S_1(ij; \cdot) \oplus S_2(ij; \cdot)$ as an edge disjoint union of cut-sets. Since $S_1(ij; \cdot)$ and $S_2(ij; \cdot)$ have no edges in common, these two can be the cut-sets to form an edge disjoint union of cut-sets. By choosing S_{a1} and S_{a2}, both of which do not separate vertices i and j, we have a representation of $S_1(ij; \cdot) \oplus S_2(ij; \cdot)$. By using S_{b1}, S_{b2}, and S_{b3} to represent $S_1(ij; \cdot) \oplus S_2(ij; \cdot)$, we have the case where exactly two cut-sets separate vertices i and j.

As we discussed earlier, each cut-set can be expressed as the ring sum of generators. Obviously a cut-set separating two specified vertices can also be expressed by the same way. Hence, in order to generate all possible cut-sets that separate two specified vertices by the ring sum of generators, we should know about the ring sum of more than two cut-sets which is equivalent to knowing about the ring sum of two edge disjoint unions of cut-sets. For this, we extend the definition of the symbols $S(i; j)$ and $S(ij; \cdot)$ as follows.

Definition 5-1-3. If set $S = \mathscr{E}(\Omega_1 \times \overline{\Omega}_1)$ has the property that $i \in \Omega_1$ and $j \in \overline{\Omega}_1$, then S is said to separate i and j and is symbolized by $S(i; j)$. If $S = \mathscr{E}(\Omega_1 \times \overline{\Omega}_1)$ has the property that both i and j are together in Ω_1 (or $\overline{\Omega}_1$), then S is not separating i and j and is symbolized by $S(ij; \cdot)$. Note that S is either a cut-set or an edge disjoint union of cut-sets.

With these definitions, Theorems 5-1-2 to 5-1-7 can be extended.

Theorem 5-1-8. The ring sum of $S_1(i; j) \oplus S_2(i; j)$ and $S_1'(ij; \cdot) \oplus S_2'(ij; \cdot)$ will not separate vertices i and j, and $S_1''(i; j) \oplus S_2''(ij; \cdot)$ will separate vertices i and j.

The proof for this theorem is the same as those of previous theorems.

Suppose each vertex in a linear graph G indicates an island and each edge represents a bridge between two islands. The problem is to find the minimum number of bridges that must be destroyed in order to have no paths between two specified islands i and j. To answer this problem suppose we define the value of set S, symbolized by $V[S]$, as the number of edges in the set. Also suppose $\{S^o(i;j)\}$ is the set of all possible cut-sets that separate vertices i and j and $\{S(i;j)\}$ is the set of all possible cut-sets and edge disjoint unions of cut-sets that separate vertices i and j. Then the minimum number N of bridges to be destroyed would be

$$N = \min\{V[S]; \quad S \in \{S^o(i;j)\}\} \tag{5-1-2}$$

For any edge disjoint union S of cut-sets that separates vertices i and j, there exists a cut-set S' which is a proper subset of S and which separates vertices i and j. Thus

$$\min\{V[S]; \quad S \in \{S^o(i;j)\}\} = \min\{V[S]; \quad S \in \{S(i;j)\}\} \tag{5-1-3}$$

Thus we need only generate all sets in $\{S(i;j)\}$ without knowing which of them are cut-sets. As an example, consider G in Fig. 5-1-5. By Theorem 3-1-4,

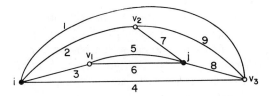

Fig. 5-1-5.   Linear graph G with vertices i and j.

we only need to know $n_v - 1$ linearly independent incidence sets in order to generate all possible cut-sets and edge disjoint unions of cut-sets where $n_v = 5$ for this example. Suppose we choose the following incidence sets as $n_v - 1$ linearly independent sets:

$$S(i) = (1, 2, 3, 4)$$
$$S(j) = (5, 6, 7, 8)$$
$$S(v_1) = (3, 5, 6)$$

and

$$S(v_2) = (2, 7, 9)$$

By Theorem 5-1-8, any set in $\{S(i;j)\}$ must contain either $S(i)$ or $S(j)$ but not both of these cut-sets. Note that only $S(i)$ and $S(j)$ separate vertices i and j in these linearly independent sets. Thus we can obtain $\{S(i;j)\}$ by taking either $S(i)$ or $S(j)$ and all possible combinations of the remaining incidence sets $S(v_1)$ and $S(v_2)$ as

$$S(i) = (1, 2, 3, 4)$$
$$S(i) \oplus S(v_1) = (1, 2, 4, 5, 6)$$
$$S(i) \oplus S(v_2) = (1, 3, 4, 7, 9)$$
$$S(i) \oplus S(v_1) \oplus S(v_2) = (1, 4, 5, 6, 7, 9)$$
$$S(j) = (5, 6, 7, 8)$$
$$S(j) \oplus S(v_1) = (3, 7, 8)$$
$$S(j) \oplus S(v_2) = (2, 5, 6, 8, 9)$$

and

$$S(j) \oplus S(v_1) \oplus S(v_2) = (2, 3, 8, 9)$$

Thus it can be seen that $N = 3$ and the bridges which should be destroyed are 3, 7, and 8.

One of the important applications of cut-sets of type $S(i; j)$ is in the field of communication nets where the capacity between two vertices is directly related to a cut-set of type $S(i; j)$. This topic will be studied in a later chapter.

5-2 PSEUDO-CUT

Consider two linear graphs G_1 and G_2 which are dual to each other. In Section 4-3, we found that any circuit in G_1 corresponds to a cut-set in G_2. Here we will introduce a subgraph called a pseudo-cut which has properties similar to those of paths. Furthermore, we will see that any path in G_1 corresponds to a pseudo-cut in a dual graph G_2. We begin by studying a subgraph of G_2 corresponding to a path in G_1.

Let e be an edge in G_1 whose endpoints are vertices i and j. Let e' be the edge corresponding to edge e in G_2 where G_1 and G_2 are dual to each other. Consider a path p ($\neq e$) between vertices i and j in G_1. Then we know that $(e) \cup p$ is a circuit in G_1 and there exists a cut-set S in G_2 corresponding to this circuit $(e) \cup p$ in G_1. Thus the subgraph in G_2 corresponding to path p in G_1 is $S \oplus (e')$. Clearly, this set $S \oplus (e')$ is not a cut-set. We call such a set a *pseudo-cut* with respect to edge e'.

Consider two distinct paths p_1 and p_2 between vertices i and j in G_1. Suppose neither p_1 nor p_2 consists of edge e'. Then there are two pseudo-cuts $S_1 \oplus (e')$ and $S_2 \oplus (e')$ in G_2 where S_1 and S_2 in G_2 are the cut-sets corresponding to circuits $p_1 \cup (e)$ and $p_2 \cup (e)$ in G_1, respectively. We know that $p_1 \oplus p_2$ is either a circuit or an edge disjoint union of circuits. Hence the subgraph in G_2 corresponding to $p_1 \oplus p_2$ must be either a cut-set or an edge disjoint union of cut-sets. From

$$[S_1 \oplus (e')] \oplus [S_2 \oplus (e')] = S_1 \oplus S_2 \qquad (5\text{-}2\text{-}1)$$

we can see that the corresponding subgraph in G_2 is indeed either a cut-set or an edge disjoint union of cut-sets by Theorem 2-3-2. In other words, the ring sum of two pseudo-cuts with respect to an edge is either a cut-set or an edge disjoint union of cut-sets. However, as we have defined paths independent of whether a linear graph is planar or not, we will define pseudo-cuts with respect to an edge as follows.

Definition 5-2-1. A *pseudo-cut* U with respect to an edge e' is a set of edges equal to

$$U = S \oplus (e') \tag{5-2-2}$$

where S is either a cut-set containing edge e' or the empty set.

With this definition, it is clear that the ring sum of two pseudo-cuts with respect to the same edge is either a cut-set or an edge disjoint union of cut-sets.

For convenience, we define the symbols $\{S\}$ and $\{S^\circ\}$ as follows.

Definition 5-2-2. The symbol $\{S\}$ indicates the collection of all cut-sets, edge disjoint unions of cut-sets, and the empty set. The symbol $\{S^\circ\}$ is the collection of all cut-sets and the empty set. That is

$$\{S^\circ\} = \min\{S\} \tag{5-2-3}$$

(See Definition 1-6-1.)

We now further generalize the definition of pseudo-cuts.

Definition 5-2-3. Let U_1 be a proper subset of a set in $\{S\}$. Then collection $\{U\}_{U_1}$ is defined as

$$\{U\}_{U_1} = \min\{U_1 \oplus S; \quad S \in \{S\}\} \tag{5-2-4}$$

where each set in $\{U\}_{U_1}$ is called a *pseudo-cut with respect to U_1* or a *pseudo-cut in collection* $\{U\}_{U_1}$.

It is clear from the definition that a pseudo-cut is not a cut-set. However, the ring sum of two pseudo-cuts with respect to U_1 is either a cut-set or an edge disjoint union of cut-sets. To show this, let U_a and U_b be in $\{U\}_{U_1}$. Then U_a and U_b can be expressed as

$$U_a = U_1 \oplus S_1 \tag{5-2-5}$$

and

$$U_b = U_1 \oplus S_2 \tag{5-2-6}$$

where S_1 and S_2 are in $\{S\}$. Hence

$$U_a \oplus U_b = U_1 \oplus S_1 \oplus U_1 \oplus S_2 = S_1 \oplus S_2 \tag{5-2-7}$$

which is clearly either a cut-set or an edge disjoint union of cut-sets. Suppose $U_a = U_1$; then

$$U_a \oplus U_b = U_1 \oplus U_1 \oplus S_2 = S_2 \qquad (5\text{-}2\text{-}8)$$

which is again a cut-set.

Theorem 5-2-1. The ring sum of any two pseudo-cuts with respect to U_1 is either a cut-set or an edge disjoint union of cut-sets.

This theorem leads directly to the next.

Theorem 5-2-2.

$$\{U\}_{U_1} \otimes \{U\}_{U_1} = \{S^o\} \qquad (5\text{-}2\text{-}9)$$

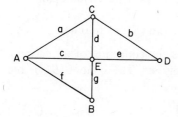

Fig. 5-2-1. A linear graph.

Example 5-2-1. Consider the linear graph in Fig. 5-2-1. Let $U_1 = (a)$. Then

$$\{U\}_a = \{(a), (c, f), (b, d), (d, e), (c, g)\}$$

and

$$\begin{aligned}\{U\}_a \otimes \{U\}_a = \{ &\emptyset, (a, c, f), (a, b, d), (a, d, e), (a, c, g), (b, c, d, f), \\ &(c, d, e, f), (f, g), (b, e), (b, c, d, g), (c, d, e, g)\}\end{aligned}$$

which is $\{S^o\}$. Note that

$$\begin{aligned}\{S\} = \{ &\emptyset, (a, c, f), (a, b, d), (a, d, e), (a, c, g), (b, e), (f, g), (c, d, e, f), \\ &(a, b, c, e, f), (a, d, e, f, g), (b, e, f, g), (b, c, d, f), (c, d, e, g), \\ &(a, b, c, e, g), (a, b, d, f, g), (b, c, d, g)\}\end{aligned}$$

as in Example 2-3-3.

In Definition 5-2-3 collection $\{U\}_{U_1}$ is given by $\min\{U_1 \oplus S; \quad S \in \{S\}\}$. If U_1 is a cut-set in S, then $\min\{S \oplus S'; \quad S' \in \{S\}\}$ is clearly equal to $\{S^o\}$ by Theorem 2-3-2.

Theorem 5-2-3. Let S be a set in $\{S\}$. Then

$$\{S^o\} = \min\{S \oplus S'; \quad S' \in \{S\}\} \qquad (5\text{-}2\text{-}10)$$

Note that if we replace $\{S\}$ by $\{S^o\}$ in the right-hand side of Eq. 5-2-10, we may not have equality, that is, we only can state that

$$\{S^o\} \supset \min \{S \oplus S'; \quad S' \in \{S^o\}\} \tag{5-2-11}$$

The following theorems should be obvious.

Theorem 5-2-4.

$$\{S^o\} \otimes \{S^o\} = \{S^o\} \tag{5-2-12}$$

Theorem 5-2-5.

$$\{U\}_{U_1} \otimes \{S^o\} = \{U\}_{U_1} \tag{5-2-13}$$

Recall that if paths P_1 and P_2 are in $\{P_{ij}\}$ (Definition 1-4-3), then

$$\begin{aligned}\{P_{ij}\} &= \min \{P_1 \oplus E; \quad E \in \{E\}\} \\ &= \min \{P_2 \oplus E; \quad E \in \{E\}\}\end{aligned}$$

(Theorem 1-6-12). We now see that $\{U\}_{U_1}$ has a similar property.

Theorem 5-2-6. Let U_2 be in $\{U\}_{U_1}$. Then

$$\{U\}_{U_2} = \{U\}_{U_1} \tag{5-2-14}$$

Proof. Since U_2 is in $\{U\}_{U_1}$, there exists a cut-set S' such that

$$U_2 = U_1 \oplus S' \tag{5-2-15}$$

Thus

$$\begin{aligned}\{U\}_{U_2} &= \min \{U_2 \oplus S; \quad S \in \{S\}\} \\ &= \min \{U_1 \oplus S' \oplus S; \quad S \in \{S\}\} \\ &= \min \{U_1 \oplus S''; \quad S'' \in \{S\}\} \end{aligned} \tag{5-2-16}$$
$$\text{QED}$$

For example, in Example 5-2-1, (c, f) is in $\{U\}_a$. By taking $U_1 = (c, f)$, we have

$$\begin{aligned}\{U\}_{(cf)} &= \min \{(c, f) \oplus S; \quad S \in \{S\}\} \\ &= \{(c, f), (a), (c, g), (d, e), (b, d)\}\end{aligned}$$

which is the same as $\{U\}_a$.

The following theorem is very similar to Theorem 1-6-9 concerning collections of paths.

Theorem 5-2-7. Let S be a cut-set in $\{S\}$. Also let $U_1, U_2, \ldots, U_k$ be proper subsets of S such that any pair of these sets $U_1, U_2, \ldots, U_k$ have no edges in common and the union of all these sets $U_1, U_2, \ldots, U_k$ is S. Then

$$\{U\}_{U_1} \otimes \{U\}_{U_2} \otimes \cdots \otimes \{U\}_{U_k} = \{S^o\} \tag{5-2-17}$$

We can prove this theorem by induction if we can prove the following theorem.

Theorem 5-2-8. Let U_0 be a proper subset of a cut-set. Also let U_1 and U_2 be proper subsets of U_0 such that U_1 and U_2 have no edges in common and the union of U_1 and U_2 is U_0. Then

$$\{U\}_{U_1} \otimes \{U\}_{U_2} = \{U\}_{U_0} \tag{5-2-18}$$

Proof. We need to prove that (1) for any $U_0' \in \{U\}_{U_0}$, there exists $U_1' \in \{U\}_{U_1}$ and $U_2' \in \{U\}_{U_2}$ such that $U_1' \oplus U_2' = U_0'$ and (2) for any $U_1' \in \{U\}_{U_1}$ and $U_2' \in \{U\}_{U_2}$, $U_1' \oplus U_2'$ is either in $\{U\}_{U_0}$ or a proper subset of $U_1' \oplus U_2'$ is in $\{U\}_{U_0}$.

For (1), let

$$U_0' = U_0 \oplus S' \tag{5-2-19}$$

where $S' \in \{S\}$. Since

$$U_1 \oplus U_2 = U_0 \tag{5-2-20}$$

we have

$$U_1 \oplus U_2 \oplus S' = U_0' \tag{5-2-21}$$

If $U_1 \oplus S'$ is in $\{U\}_{U_1}$, then since $U_2 \in \{U\}_{U_2}$, these two pseudo-cuts $U_1 \oplus S'$ and U_2 are the two desired sets. This is also true when $U_2 \oplus S'$ is in $\{U\}_{U_2}$. Thus we only have to consider when neither $U_1 \oplus S' \in \{U\}_{U_1}$ nor $U_2 \oplus S' \in \{U\}_{U_2}$. Under this circumstance, there exist sets S_1 and S_2 in $\{S\}$ such that

$$U_1 \oplus S_1 \subsetneqq U_1 \oplus S' \tag{5-2-22}$$

$$U_2 \oplus S_2 \subsetneqq U_2 \oplus S' \tag{5-2-23}$$

where $U_1 \oplus S_1$ is in $\{U\}_{U_1}$ and $U_2 \oplus S_2$ is in $\{U\}_{U_2}$ (because of Eq. 5-2-4). Since $U_1 \oplus U_2 \oplus S'$ is in $\{U\}_{U_0}$, we have

$$U_1 \oplus U_2 \oplus S_1 \supsetneqq U_1 \oplus U_2 \oplus S' \tag{5-2-24}$$

Note that $U_1 \oplus U_2 \oplus S_1$ is not in $\{U\}_{U_0}$. Otherwise, this case becomes the same as the previous case because of the assumption that $U_1 \oplus S_1 \in \{U\}_{U_1}$.

From Eq. 5-2-24,

$$S_1 - (U_1 \cup U_2) \supset S' - (U_1 \cup U_2) \tag{5-2-25}$$

Similarly from Eqs. 5-2-22 and 5-2-23, we have

$$S' - U_1 \supset S_1 - U_1 \tag{5-2-26}$$

and

$$S' - U_2 \supset S_2 - U_2 \tag{5-2-27}$$

These two equations give

$$(S' - U_1) \cup (S' - U_2) \supset (S_1 - U_1) \cup (S_2 - U_2)$$
$$\supsetneqq S_1 - U_1$$
$$\supsetneqq S_1 - (U_1 \cup U_2) \qquad (5\text{-}2\text{-}28)$$

because $U_2 \neq S_2$. Since the left-hand side of this equation is equal to $S' - (U_1 \cup U_2)$, we have

$$S' - (U_1 \cup U_2) \supsetneqq S_1 - (U_1 \cup U_2) \qquad (5\text{-}2\text{-}29)$$

However, from Eq. 5-2-25, $S' - (U_1 \cup U_2)$ is a subset of $S_1 - (U_1 \cup U_2)$, which is a contradiction. This is the situation that neither $U_1 \oplus S' \in \{U\}_{U_1}$ nor $U_2 \oplus S' \in \{U\}_{U_2}$ will exist, which proves the first part.

For (2), let $U_1' = U_1 \oplus S_1'$ and $U_2' = U_2 \oplus S_2'$ where S_1' and S_2' are in $\{S\}$. Then

$$U_1' \oplus U_2' = U_1 \oplus S_1' \oplus U_2 \oplus S_2' = U_0 \oplus S'' \qquad (5\text{-}2\text{-}30)$$

which must be in $\{U\}_{U_0}$ unless there exists a proper subset of $U_0 \oplus S''$ in $\{U\}_{U_0}$. QED

Example 5-2-2. $\{S\}$ of the linear graph in Fig. 5-2-1 is

$$\{S\} = \{\emptyset, (a, c, f), (a, b, d), (a, d, e), (a, c, g), (b, e), (f, g),$$
$$(c, d, e, f), (a, b, c, e, f), (a, d, e, f, g), (b, e, f, g), (b, c, d, f),$$
$$(c, d, e, g), (a, b, c, e, g), (a, b, d, f, g), (b, c, d, g)\}$$

With $U_1 = (a)$, we have

$$\{U\}_a = \min \{(a) \oplus S; \quad S \in \{S\}\}$$
$$= \{(a), (c, f), (b, d), (d, e), (c, g)\}$$

With $U_1 = (c)$, we have

$$\{U\}_c = \min \{(c) \oplus S; \quad S \in \{S\}\}$$
$$= \{(c), (a, f), (a, g), (d, e, f), (b, d, f), (d, e, g), (b, d, g)\}$$

Then

$$\{U\}_a \oplus \{U\}_c = \{(a, c), (f), (g), (c, d, e), (b, c, d)\} = \{U\}_{(ac)}$$

Also note that (f) is in $\{U\}_{(ac)}$. Thus

$$\{U\}_{(ac)} = \{U\}_f$$

by Theorem 5-2-6. Indeed,

$$\{U\}_f = \min \{(f) \oplus S; \quad S \in \{S\}\}$$

gives $\{U\}_{(ac)}$.

When a path consists of one edge e, then by Theorem 1-6-10, we have

$$\{P(e)\} = \min\{(e) \oplus C; \quad C \in \{C\}\} \tag{5-2-31}$$

Note that $\{C\}$ is the collection of all circuits and the empty set. We have the same situation when U_1 consists of one edge.

Theorem 5-2-9. For an edge e in a linear graph,

$$\{U\}_e = \min\{(e) \oplus S; \quad S \in \{S^o\}\} \tag{5-2-32}$$

Note that the number of members in $\{S^o\}$ is, in general, much smaller than that in $\{S\}$.

Proof. Let $U \in \{U\}_e$. Suppose

$$U = (e) \oplus S$$

where S is an edge disjoint union of cut-sets $S_1, S_2, \ldots, S_p$. Since one of $S_1, S_2, \ldots, S_p$ must contain edge e in order that U be in $\{U\}_e$. Let $e \in S_1$. Then $(e) \oplus S_1$ rather than $(e) \oplus S$ will be in $\{U\}_e$. Thus S of $U = (e) \oplus S$ must be a cut-set.   QED

Note that if U_1 contains more than one edge

$$\{U\}_{U_1} \supset \min\{U_1 \oplus S; \quad S \in \{S^o\}\} \tag{5-2-33}$$

If we have $\{S^o\}$ rather than $\{S\}$, the following theorem gives $\{U\}_{U_1}$ for $U_1 = \{e_1, e_2, \ldots, e_k\}$.

Theorem 5-2-10. Let $U_1 = \{e_1, e_2, \ldots, e_k\}$. Then

$$\{U\}_{U_1} = \{U\}_{e_1} \otimes \{U\}_{e_2} \otimes \cdots \otimes \{U\}_{e_k} \tag{5-2-34}$$

This theorem can be proved easily by using Theorem 5-2-8.
In Example 5-2-2,

$$\{S^o\} = \{\emptyset, (a, c, f), (a, b, d), (a, d, e), (a, c, g), (b, e),$$
$$(f, g), (c, d, e, f), (b, c, d, f), (c, d, e, g), (b, c, d, g)\}$$

To obtain $\{U\}_{(cd)}$, for example, instead of using Eq. 5-2-4, we can obtain $\{U\}_c$ and $\{U\}_d$ as follows:

$$\{U\}_c = \min\{(c) \oplus S; \quad S \in \{S^o\}\}$$
$$= \{(c), (a, f), (a, g), (d, e, f), (b, d, f), (d, e, g), (b, d, g)\}$$

and

$$\{U\}_d = \min\{(d) \oplus S; \quad S \in \{S^o\}\}$$
$$= \{(d), (a, b), (a, e), (c, e, f), (b, c, f), (c, e, g), (b, c, g)\}$$

Then

$$\{U\}_{(cd)} = \{U\}_c \otimes \{U\}_d$$
$$= \{(c, d), (a, b, c), (a, c, e), (e, f), (b, f), (e, g), (b, g), (a, d, f), (a, d, g)\}$$

Suppose a linear graph is mapped on the surface of a doughnut. If we only need to break all circuits each of which circles around the center hole of the doughnut by deleting edges, then these edges form a pseudo-cut unless a proper subset of the set of these edges does the job. For example, consider this situation as shown in Fig. 5-2-2. The set $\{S\}$ of all cut-sets, edge

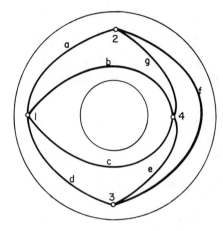

Fig. 5-2-2. A linear graph on a doughnut.

disjoint unions of cut-sets, and the empty set can be obtained from $n_v - 1$ linearly independent incidence sets. Thus we need only the following incidence sets to obtain a set of pseudo-cuts with respect to U_1:

$$S(1) = (a, b, c, d)$$
$$S(2) = (a, f, g)$$
$$S(3) = (d, e, f)$$

By taking $U_1 = (c, d)$, which obviously breaks all circuits circling around the center hole, we have

$$U_2 = (c, d) \oplus S(1) = (a, b)$$
$$U_3 = (c, d) \oplus S(2) = (a, c, d, f, g)$$
$$U_4 = (c, d) \oplus S(3) = (c, e, f)$$
$$U_5 = (c, d) \oplus S(1) \oplus S(2) = (b, f, g)$$
$$U_6 = (c, d) \oplus S(1) \oplus S(3) = (a, b, d, e, f)$$
$$U_7 = (c, d) \oplus S(2) \oplus S(3) = (a, c, e, g)$$

and

$$U_8 = (c, d) \oplus S(1) \oplus S(2) \oplus S(3) = (b, d, e, g)$$

Thus sets of edges each of which breaks all circuits circling around the center hole and no subset of which does the job would be U_1, U_2, U_4, U_5, U_7, and U_8, which obviously form $\{U\}_{U_1}$.

Consider the ring sum of two pseudo-cuts U_a and U_b where $U_a \in \{U\}_{U_1}$ and $U_b \in \{U\}_{U_2}$:

$$U_a \oplus U_b = (S_a \oplus U_1) \oplus (S_b \oplus U_2) = S_a \oplus S_b \oplus (U_1 \oplus U_2) = S_c \oplus U$$
$$(5\text{-}2\text{-}35)$$

where S_a, S_b, and S_c are in $\{S\}$ and

$$U = U_1 \oplus U_2 \qquad\qquad (5\text{-}2\text{-}36)$$

Since there is no guarantee that U is a proper subset of a cut-set in $\{S\}$, $S_c \oplus U$ may not be a pseudo-cut by the previous definition. However, any two sets, $S_c \oplus U$ and $S_d \oplus U$, satisfy

$$(S_c \oplus U) \oplus (S_d \oplus U) = S_e$$

where S_c, S_d, and S_e are in $\{S\}$, that is, the ring sum of any two such sets is in $\{S\}$. Hence we extend the definition of pseudo-cuts to include such sets.

Definition 5-2-4. For a set $\mathscr{E}$ of edges, the set $\{U\}_{\mathscr{E}}$ of pseudo-cuts with respect to $\mathscr{E}$ is defined as

$$\{U\}_{\mathscr{E}} = \min \{\mathscr{E} \oplus S; \quad S \in \{S\}\} \qquad (5\text{-}2\text{-}37)$$

With this definition, we have the next theorem.

Theorem 5-2-11. Let $\{U\}_{U_1}$ and $\{U\}_{U_2}$ be sets of pseudo-cuts with respect to U_1 and U_2, respectively. Then

$$\{U\}_{U_1} \otimes \{U\}_{U_2} = \{U\}_{U_1 \oplus U_2} \qquad (5\text{-}2\text{-}38)$$

Proof. By the definition of operation $\otimes$,

$$\begin{aligned}
\{U\}_{U_1} \otimes \{U\}_{U_2} &= \min \{U_a \oplus U_b; \quad U_a \in \{U\}_{U_1}, U_b \in \{U\}_{U_2}\} \\
&= \min \{S_a \oplus U_1 \oplus S_b \oplus U_2; \quad S_a, S_b \in \{S\}\} \\
&= \min \{S \oplus (U_1 \oplus U_2); \quad S \in \{S\}\} \qquad (5\text{-}2\text{-}39)
\end{aligned}$$

which is the definition of $\{U\}_{U_1 \oplus U_2}$.   QED

Consider a collection $[\{U\}, \{S^o\}]$ of sets of pseudo-cuts with respect to all possible sets of edges. Note that if the sets of edges is the empty set $\emptyset$, then $\{U\}_{\emptyset} = \{S^o\}$. Thus collection $[\{U\}, \{S^o\}]$ contains set $\{S^o\}$ which is the set of all possible cut-sets and the empty set. For any sets R_a, R_b, and R_c in $[\{U\}, \{S^o\}]$, we have the following:

1. $R_a \otimes R_b = R_b \otimes R_a \in [\{U\}, \{S^o\}]$.
2. There exists $\{S^o\}$ in $[\{U\}, \{S^o\}]$ such that $R_a \oplus \{S^o\} = R_a$.
3. $R_a \otimes R_a = \{S^o\}$.

4. $R_a \otimes (R_b \otimes R_c) = (R_a \otimes R_b) \otimes R_c.$

Thus $[\{U\}, \{S^o\}]$ is an Abelian group under the operation $\otimes$.

Example 5-2-3. Consider the linear graph in Fig. 5-2-3. To obtain $\{S\}$, we

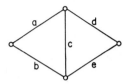

Fig. 5-2-3. A linear graph G.

choose linearly independent cut-sets $S_1 = (a, b)$, $S_2 = (a, c, e)$ and $S_3 = (d, e)$. Then we perform

$$S_1 \oplus S_2 = (b, c, e) = S_4$$
$$S_1 \oplus S_3 = (a, b, d, e) = S_5$$
$$S_2 \oplus S_3 = (a, c, d) = S_6$$

and

$$S_1 \oplus S_2 \oplus S_3 = (b, c, d) = S_7$$

Hence

$$S = \{S_1, S_2, S_3, S_4, S_5, S_6, S_7, \emptyset\}$$

We can see that

$$S^o = \{S_1, S_2, S_3, S_4, S_6, S_7, \emptyset\}$$

Let $U_1 = (a)$. Then

$$\{U\}_{U_1} = \min \{U_1 \oplus S_i; \quad S_i \in \{S\}\}$$
$$= \{(a), (b), (c, e), (c, d)\}$$

Note that

$$\{U\}_{U_1} \otimes \{U\}_{U_1} = \{\emptyset, (a, b), (a, c, e), (a, c, d), (b, c, e), (b, c, d), (d, e)\}$$
$$= \{S^o\}$$

Let $U_2 = (e)$. Then

$$\{U\}_{U_2} = \{(e), (a, c), (d), (b, c)\}$$

Let $U_3 = (c)$. Note that

$$U_1 \oplus U_2 \oplus U_3 = S_2$$

and

$$U_p \cap U_q = \emptyset \qquad \text{for} \quad 1 \le p < q \le 3$$

Now

$$\{U\}_{U_1} \otimes \{U\}_{U_2} = \min \{U_a \oplus U_b; \quad U_a \in \{U\}_{U_1}, U_b \in \{U\}_{U_2}\}$$
$$= \{(a, e), (c), (a, d), (b, e), (b, d)\} = \{U\}_{U_3}$$

It can easily be seen that

$$\{U\}_{U_1} \otimes \{U\}_{U_2} \otimes \{U\}_{U_3} = \{U\}_{U_3} \otimes \{U\}_{U_3} = \{S^o\}$$

Let $U_4 = (b, c)$. Then

$$\{U\}_{U_4} = \{(b, c), (a, c), (e), (d)\}$$

Now

$$\{U\}_{U_4} \otimes \{U\}_{U_1} = \{(c), (b, e), (b, d), (a, e), (a, d)\} = \{U\}_{U_3}$$

This is because

$$\{U\}_{U_4} \otimes \{U\}_{U_1} = \{U\}_{U_4 \oplus U_1} = \{U\}_{(abc)} = \{U\}_{S_1 \oplus (c)}$$
$$= \{U\}_{(c)} = \{U\}_{U_3}$$

When we studied paths, we saw a set which is an edge disjoint union of a path and circuits. We have a similar set with pseudo-cuts and cut-sets.

Definition 5-2-4. A set Y is called an edge disjoint union of a pseudo-cut and cut-sets if

$$Y = U \oplus S \tag{5-2-40}$$

and

$$U \cap S = \emptyset \tag{5-2-41}$$

where U is a pseudo-cut and S is either a cut-set or an edge disjoint union of cut-sets.

Definition 5-2-5. Let $\{U\}_U$ be a collection of all possible pseudo-cuts with respect to a set U of edges. Then the symbol $\{Y_U\}$ is a collection of all pseudo-cuts in $\{U\}_U$, edge disjoint unions of a pseudo-cut U and cut-sets, and the empty set.

It can be seen that

$$\{Y_U\} = \{U \oplus S; \quad S \in \{S\}\} \tag{5-2-42}$$

Furthermore, from the definition of a collection of pseudo-cuts, we can see that

$$\min \{Y_U\} = \{U\}_U \tag{5-2-43}$$

Example 5-2-4. In Example 5-2-3 we found that

$$\{U\}_{U_1} = \{(a), (b), (c, e), (c, d)\}$$

Also we know that

$$\{S\} = \{\emptyset, (a, b), (a, c, e), (d, e), (b, c, e), (a, b, d, e), (a, c, d), (b, c, d)\}$$

Thus

$$\{Y_{U_1}\} = \{U \oplus S; \quad U \in \{U\}_{U_1}, S \in \{S\}\}$$
$$= \{(a), (b), (c, e), (c, d), (a, b, c, e), (a, b, c, d), (a, d, e), (b, d, e)\}$$

The same collection can be obtained by

$$\{Y_{U_1}\} = \{U_1 \oplus S; \quad S \in \{S\}\}$$
$$= \{(a), (b), (c, e), (a, d, e), (a, b, c, e), (b, d, e), (c, d), (a, b, c, d)\}$$

An interesting property of $\{Y_U\}$ is given by the following theorem.

Theorem 5-2-12. The ring sum of odd number of sets in $\{Y_U\}$ is a set in $\{Y_U\}$ and the ring sum of even number of sets in $\{Y_U\}$ is a set in $\{S\}$.

The proof is obvious from Eq. 5-2-42. Note that all pseudo-cuts with respect to U are in $\{Y_U\}$.

This theorem suggests that if we collect all sets in $\{Y_U\}$ and all sets in $\{S\}$, we will have a group under the ring sum. In other words, by defining $[Y_U, S]$ as the collection of all sets in $\{Y_U\}$ and $\{S\}$, we can state that $[Y_U, S]$ is an Abelian group under the ring sum. Recall that $\{S\}$ is the set of all cut-sets, edge disjoint unions of cut-sets, and the empty set.

5-3 ABELIAN GROUPS

We have studied the following collections of subgraphs which are Abelian groups:

$\{E\}$:   All possible circuits, edge disjoint unions of circuits, and the empty set.

$\{E, M_{ij}\}$:   All possible sets in $\{M_{ij}\}$ and $\{E\}$ where $\{M_{ij}\}$ is the set of all possible M-graphs of type M_{ij} which is either a path or an edge disjoint union of a path and circuits.

$[\{\tau\}, \{C\}]$:   Collection of all collections of the form $\{\tau_{(i_1 i_2 \cdots i_r)}\}$ and $\{C\}$ where $\{C\}$ is the collection of all circuits and the empty set and $\{\tau_{(i_1 \cdots i_r)}\}$ is the set of all possible τ-graph of type $\tau_{(i_1 \cdots i_r)}$.

$\{S\}$:   All possible cut-sets, edge disjoint unions of cut-sets, and the empty set.

$\{S, Y_U\}$:   Collection of all sets in $\{Y_U\}$ and $\{S\}$ where $\{Y_U\}$ is the set of all possible pseudo-cuts with respect to U and edge disjoint unions of a pseudo-cut with respect to U and cut-sets. (U is a set of edges.)

$[\{U\}, \{S^o\}]$:   Collection of all collections of the form $\{U\}_{U_p}$ and $\{S^o\}$ where $\{S^o\}$ is the collection of all cut-sets and the empty set.

We also saw that there are $n_e - n_v + \rho$ linearly independent sets in $\{E\}$ by which all sets in $\{E\}$ can be generated. Similarly, there are $n_v - \rho$ linearly independent sets in $\{S\}$ by which all sets in $\{S\}$ can be generated. Recall that n_e is the number of edges, n_v is the number of vertices, and ρ is the number of maximal connected subgraphs in a linear graph. Variables are the number of linearly independent sets in $\{E, M_{ij}\}$, $[\{\tau\}, \{C\}]$, $\{S, Y_U\}$, and $[\{U\}, \{S^o\}]$.

Theorem 5-3-1.   There exist $n_e - n_v + \rho + 1$ linearly independent sets in $\{E, M_{ij}\}$.

Proof.   From the definition of M_{ij}, we can see that

$$\{M_{ij}\} = \{M \oplus E; \quad E \in \{E\}\} \tag{5-3-1}$$

where M is a set in $\{M_{ij}\}$. It is clear that $n_e - n_v + \rho$ linearly independent sets in $\{E\}$ are linearly independent sets in $\{E, M_{ij}\}$. Furthermore, any set M in $\{M_{ij}\}$ and any linearly independent sets in $\{E\}$ are linearly independent by Eq. 5-3-1. Thus there are at least $n_e - n_v + \rho + 1$ linearly independent sets in $\{E, M_{ij}\}$.

Since any set in $\{M_{ij}\}$ can be expressed as the ring sum of M and $E \in \{E\}$, and any set in $\{E\}$ can be expressed as the ring sum of linearly independent sets in $\{E\}$, we can easily see that there are no more than $n_e - n_v + \rho + 1$ linearly independent sets in $\{E, M_{ij}\}$.   QED

Similarly, by knowing that there are $n_v - \rho$ linearly independent sets in $\{S\}$, we have the next theorem.

Theorem 5-3-2.   There exist exactly $n_v - \rho + 1$ linearly independent sets in $\{S, Y_U\}$ where $\{Y_U\} \neq \{S\}$.

Let $1, 2, \ldots, n_v$ be the vertices in a nonseparable linear graph. Then it is easily seen that $\{\tau_{(12)}\}, \{\tau_{(13)}\}, \ldots, \{\tau_{(1n_v)}\}$ are linearly independent. For any pair of vertices i and j where $1 < i \neq j$, $\{\tau_{(ij)}\}$ can be expressed as

$$\{\tau_{(ij)}\} = \{\tau_{(i1)}\} \otimes \{\tau_{(1j)}\} \tag{5-3-2}$$

by Theorem 1-7-8. Also any set $\{\tau_{(i_1 \cdots i_r)}\}$ can be expressed as

$$\{\tau_{(i_1 \cdots i_r)}\} = \{\tau_{(1i_1)}\} \otimes \{\tau_{(1i_2)}\} \otimes \cdots \otimes \{\tau_{(1i_r)}\} \tag{5-3-3}$$

Furthermore,

$$\{\tau_{(i_1 \cdots i_r)}\} \otimes \{\tau_{(i_1 \cdots i_r)}\} = \{C\} \tag{5-3-4}$$

by Theorems 1-6-9 and 1-7-6.

Theorem 5-3-3. For a nonseparable linear graph, there exist $n_v - 1$ linearly independent sets in $[\{\tau\}, \{C\}]$.

For $[\{U\}, \{S^o\}]$, let $t = (e_1, e_2, \ldots, e_n)$ be a tree (a forest, in the case of a separated linear graph) and $(b_1, b_2, \ldots, b_m)$ be the set of chords corresponding to t where $n = n_v - \rho$ and $n + m = n_e$. We will show that sets $\{U\}_{b_1}, \{U\}_{b_2}, \ldots, \{U\}_{b_m}$ are linearly independent and adding any other set $\{U\}_{\mathscr{E}}$, where $\mathscr{E} \neq (b_p)$ for $1 \leq \rho \leq m$, make them dependent.

Since there are no cut-sets consisting only of edges in $(b_1, b_2, \ldots, b_m)$, the ring product of any combination of $\{U\}_{b_1}, \{U\}_{b_2}, \ldots, \{U\}_{b_m}$ will not be equal to $\{S^o\}$. Thus these sets are linearly independent.

Suppose $\mathscr{E} = (e_r)$, $1 \leq r \leq n$. We know that there exists a fundamental cut-set S_{e_r} with respect to t such that S_{e_r} contains edge e_r. Let this cut-set be $(b_{i_1}, b_{i_2}, \ldots, b_{i_k}, e_r)$ where $1 \leq i_1 < \cdots < i_k \leq m$. Then by Theorem 5-2-7

$$\{U\}_{b_{i_1}} \otimes \{U\}_{b_{i_2}} \otimes \cdots \otimes \{U\}_{b_{i_k}} \otimes \{U\}_{e_r} = \{S^o\} \tag{5-3-5}$$

or

$$\{U\}_{b_{i_1}} \otimes \{U\}_{b_{i_2}} \otimes \cdots \otimes \{U\}_{b_{i_k}} = \{U\}_{e_r} \tag{5-3-6}$$

Thus $\{U\}_{b_1}, \{U\}_{b_2}, \ldots, \{U\}_{b_m}, \{U\}_{e_r}$, are linearly dependent.

Let $\mathscr{E}$ be a set of more than one edge, say

$$\mathscr{E} = (b_{i_1}, \ldots, b_{i_p}, e_{j_1}, \ldots, e_{j_q})$$

Then by Theorem 5-2-8

$$\{U\}_{\mathscr{E}} = \{U\}_{b_{i_1}} \otimes \cdots \otimes \{U\}_{b_{i_p}} \otimes \{U\}_{e_{j_1}} \otimes \cdots \otimes \{U\}_{e_{j_q}} \tag{5-3-7}$$

By Eq. 5-3-6, every set $\{U\}_{e_{j_r}}$ for $r = 1, \ldots, q$, can be replaced by the ring product of the sets in $\{U\}_{b_1}, \{U\}_{b_2}, \ldots, \{U\}_{b_m}$. Thus $\{U\}_{b_1}, \ldots, \{U\}_{b_m}$, and $\{U\}_{\mathscr{E}}$ are linearly dependent. Also it is clear that

$$\{U\}_{\mathscr{E}} \otimes \{U\}_{\mathscr{E}} = \{S^o\} \tag{5-3-8}$$

Hence we have Theorem 5-3-4.

Theorem 5-3-4. There are exactly $n_e - n_v + \rho$ linearly independent sets in $[\{U\}, \{S^o\}]$.

There are several applications of these collections. For example, collections of type $\{E, M_{ij}\}$ are related to switching networks. An interesting application of $[\{\tau\}, \{C\}]$ is to obtain all possible trees in a linear graph. Because a tree is a set of $n_v - 1$ edges having exactly one path between any pair of vertices, any set in $\{\tau_{(i_1 i_2 \cdots i_r)}\}$ which consists of $n_v - 1$ edges is a tree. Thus by knowing $[\{\tau\}, \{C\}]$, we can obtain all possible trees.

PROBLEMS

1. Find all possible cut-sets which separate vertices p and q from $S_1 = (a, b, c)$, $S_2 = (c, d, f, h)$, $S_3 = (e, f, h)$, and $S_4 = (b, d, f, g)$ of linear graph G in Fig. P-5-1.

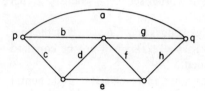

Fig. P-5-1.

2. Let S be a cut-set which does not separate vertices i and j. Prove that there exists a path P from i to j such that P and S have no edges in common.

3. Let S be a cut-set which separates vertices i and j. Prove that any path from i to j contains an edge in S.

4. Let $\{P_{ij}\}$ be a set of all paths from i to j. Also let S be a set of edges. Then if

$$S \cap P \neq \emptyset \qquad \text{for } all \quad P \in \{P_{ij}\}$$

and if any proper subset S' of S has the property that

$$S' \cap P = \emptyset \qquad \text{for } some \quad P \in \{P_{ij}\}$$

Prove S is a cut-set.

5. Suppose $S(i; j) \oplus S(j; k)$ is a cut-set. Prove or disprove that

$$S(i; j) \oplus S(j; k) = S(i; k)$$

6. Obtain all possible pseudo-cuts with respect to $u = (e, f)$ of a linear graph in Fig. P-5-1.

7. What kinds of linear graphs and sets u's of edges satisfy that $U_1 \oplus U_2$ is a cut-set for all U_1 and U_2 in $\{U\}_u$?

8. Is it possible to have

$$\{U\}_{u_1} \otimes \{U\}_{u_2} = \{S\}$$

where $\{U\}_{u_1} \neq \{U\}_{u_2}$?

9. Let $e_1, e_2, \ldots$, and e_n all be edges in a linear graph. Is

$$\{U\}_{(e_1)} \otimes \{U\}_{(e_2)} \otimes \cdots \otimes \{U\}_{(e_n)}$$

a set of pseudo-cuts or $\{S\}$?

10. Prove Theorem 5-1-8.
11. Prove Theorem 5-2-3.
12. Prove Theorem 5-2-4.
13. Prove Theorem 5-2-7.

CHAPTER
6
ORIENTED LINEAR GRAPH

6-1 INCIDENCE AND CIRCUIT MATRICES OF ORIENTED GRAPHS

When the endpoints of an edge are ordered, the edge is said to be *oriented* or is called an *oriented edge*. To indicate the direction of an oriented edge, an arrow on a line segment is usually used as shown in Fig. 6-1-1a. A linear

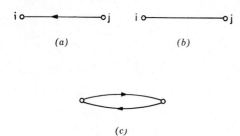

(a)   (b)

(c)

Fig. 6-1-1. (a) Oriented edge; (b) non-oriented edge; (c) two oriented edges.

graph in which all edges are oriented is said to be *oriented* or is called an *oriented linear graph* (or simply an *oriented graph*). The linear graph in Fig. 6-1-2 is an oriented graph. In order to define circuits, cut-sets, and so forth, we define the corresponding nonoriented graph as follows.

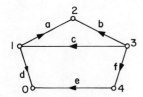

Fig. 6-1-2. Oriented graph.

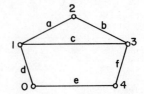

Fig. 6-1-3. Corresponding nonoriented graph.

210

Definition 6-1-1. A linear graph G_n is the *corresponding nonoriented graph* of an oriented graph G_o. G_n can be obtained from G_o by replacing every oriented edge by a nonoriented edge.

As an example, the linear graph in Fig. 6-1-3 is the corresponding non-oriented graph of the oriented graph in Fig. 6-1-2.

Definition 6-1-2. An oriented graph is said to be *connected* if the corresponding nonoriented graph is connected.

For example, the oriented graph in Fig. 6-1-2 is connected because the corresponding nonoriented graph in Fig. 6-1-3 is connected. Similar way, we define separable graphs, separated graphs, and maximal connected subgraphs.

Since the difference between oriented graphs and nonoriented graphs occur whether or not edges have orientation, by properly defining circuits, cut-sets, paths, and so on, in oriented graphs, as we will see later, all properties of nonoriented graphs will be carried over to oriented graphs. On the other hand, some special subgraphs—directed circuits, directed paths, and semicuts in oriented graphs, which are different from circuits, paths, and cut-sets—are very important for applications of oriented graphs. These will be studied later in this chapter.

It is important to note that there are cases where oriented edges can be replaced by nonoriented edges and vice versa. For example, we will see that for lumped linear bilateral electrical networks, each oriented edge can be replaced by a nonoriented edge. If we are studying a flow from a vertex to another vertex, two oriented edges in Fig. 6-1-1c can be replaced by one nonoriented edge in Fig. 6-1-1b. On the other hand, there are cases where oriented edges and nonoriented edges have completely different properties so that the interchange of these edges is almost impossible. As an example, if we are studying simultaneous flows in a communication net, all the configurations in Fig. 6-1-1 are different, as will be seen in later chapters. First, we will study similarities between oriented graphs and nonoriented graphs.

One way of indicating the location and the orientations of each edge in an oriented graph is by using an exhaustive incidence matrix, defined as follows.

Definition 6-1-3. An *exhaustive incidence matrix* of an oriented linear graph of n_v vertices and n_e edges is a matrix $A_e = [a_{ij}]$ of order n_v by n_e where

$$a_{ij} = \begin{cases} 1 & \text{if edge } j \text{ is incident at vertex } i \text{ and is oriented away} \\ & \text{from vertex } i \\ -1 & \text{if edge } j \text{ is incident at vertex } i \text{ and is oriented toward} \\ & \text{vertex } i \\ 0 & \text{otherwise} \end{cases}$$

For example, exhaustive incidence matrix A_e of the oriented graph in Fig. 6-1-2 is as follows:

$$A_e = \begin{array}{c} \\ 1 \\ 2 \\ 3 \\ 4 \\ 0 \end{array} \begin{array}{c} \begin{array}{cccccc} a & b & c & d & e & f \end{array} \\ \left[\begin{array}{cccccc} 1 & 0 & -1 & 1 & 0 & 0 \\ -1 & -1 & 0 & 0 & 0 & 0 \\ 0 & 1 & 1 & 0 & 0 & 1 \\ 0 & 0 & 0 & 0 & 1 & -1 \\ 0 & 0 & 0 & -1 & -1 & 0 \end{array} \right] \end{array}$$

Note that there exist exactly two nonzeros (one is $+1$ and the other is -1) in each column of A_e for an oriented linear graph. Recall that A_e of a nonoriented linear graph has exactly two 1's in each column.

We saw previously that the rank of an exhaustive incidence matrix of a nonoriented graph is $n_v - \rho$. This is also true for oriented graphs.

Theorem 6-1-1. The rank of an exhaustive incidence matrix of an oriented graph is $n_v - \rho$ where n_v is the number of vertices and ρ is the number of maximal connected subgraphs in the linear graph.

The proof can be accomplished in the same way as that for nonoriented graphs, except that entries are treated as real integers.

Definition 6-1-4. A matrix A obtained from an exhaustive incidence matrix A_e of a *connected* oriented graph by deleting one row is called an *incidence matrix* and the vertex corresponding to the row in A_e which has been deleted is called the *reference vertex*.

Recall that we defined an incidence matrix of a nonoriented graph in the same way. When $\rho > 1$, we define an incidence matrix as follows.

Definition 6-1-5. A matrix A obtained by deleting ρ rows of A_e of an oriented linear graph is called an *incidence matrix* if the rank of A is $n_v - \rho$ where ρ is the maximal connected subgraphs in the oriented graph.

This definition of an incidence matrix of an oriented graph is again the same as that of a nonoriented graph. Recall that in the case of nonoriented graphs, the determinant of any nonsingular major submatrix of an incidence matrix is 1. However, the determinant of a nonsingular major submatrix of an incidence matrix of an oriented graph may not be 1.

Let A_s be a nonsingular submatrix of order $n_v - \rho$ of an incidence matrix A of an oriented linear graph consisting of ρ maximal connected subgraphs and containing n_v vertices. Then there exists at least one column in A_s containing only one nonzero element. Because if no columns containing only one nonzero element exist, then every column of A_s has two nonzero elements one of which is $+1$ and the other -1. Hence the determinant of A_s is zero, which contradicts the assumption that A_s is nonsingular. Let $a_{ik} = \pm 1$ and all elements except a_{ik} in the kth column of A_s be zeros. By the expansion of $|A_s|$ in terms of elements in the kth column, we have

$$|A_s| = \pm(-1)^{i+k}|A_{s_{ik}}| \tag{6-1-1}$$

where $A_{s_{ik}}$ is the square submatrix obtained by deleting the ith row and the kth column of A_s. If there are no columns which have exactly one nonzero entry, then $|A_{s_{ik}}| = 0$, which gives $|A_s| = 0$ by Eq. 6-1-1. However, $|A_s| \neq 0$ by assumption. Hence there must be at least one column in $A_{s_{ik}}$ in which there is only one nonzero entry.

Let the (p, q) entry of $A_{s_{ik}}$ be ± 1. Then

$$|A_s| = \pm(-1)^{i+k}(\pm 1)(-1)^{p+q}|A_{s_{ik},pq}| \tag{6-1-2}$$

where $A_{s_{ik},pq}$ is the square submatrix obtained from $A_{s_{ik}}$ by deleting row p and column q. We now can see that the determinant of A_s is either $+1$ or -1.

Consider a subgraph T representing a tree of a connected oriented linear graph. An incidence matrix A_T of linear graph T is the square matrix of order $n_v - 1$ where n_v is the number of vertices in T. Since an incidence matrix of a connected oriented linear graph is of rank $n_v - 1$, A_T is nonsingular. Hence the submatrix of order $n_v - 1$ of an incidence matrix of a connected oriented linear graph whose columns correspond to branches of a tree is nonsingular. Conversely, if a square submatrix A_s of order $n_v - 1$ of an incidence matrix of a connected oriented linear graph G is nonsingular, the subgraph corresponding to A_s (i.e., the subgraph consisting of edges corresponding to the columns of A_s) must be connected, must contain $n_v - 1$ edges, and must contain n_v vertices. Thus the subgraph corresponding to a nonsingular submatrix A_s of order $n_v - 1$ is a tree of G. We may now generalize.

Theorem 6-1-2. Let A be an incidence matrix of an oriented graph consisting of ρ maximal connected subgraphs. If and only if a subgraph is a forest (a tree when $\rho = 1$), the major submatrix of A corresponding to the subgraph is nonsingular. The determinant of such a nonsingular major submatrix is either 1 or -1.

Example 6-1-1. An incidence matrix A of the linear graph in Fig. 6-1-2 is

$$A = \begin{array}{c} \\ 1 \\ 2 \\ 3 \\ 4 \end{array} \begin{array}{c} \begin{array}{cccccc} a & b & c & d & e & f \end{array} \\ \left[\begin{array}{cccccc} 1 & 0 & -1 & 1 & 0 & 0 \\ -1 & -1 & 0 & 0 & 0 & 0 \\ 0 & 1 & 1 & 0 & 0 & 1 \\ 0 & 0 & 0 & 0 & 1 & -1 \end{array} \right] \end{array}$$

where vertex 0 is the reference vertex. If we take a square submatrix A_s of A

$$A_s = \begin{array}{c} \\ 1 \\ 2 \\ 3 \\ 4 \end{array} \begin{array}{cccc} a & b & d & e \\ \left[\begin{array}{cccc} 1 & 0 & 1 & 0 \\ -1 & -1 & 0 & 0 \\ 0 & 1 & 0 & 0 \\ 0 & 0 & 0 & 1 \end{array} \right] \end{array}$$

then

$$A_{s_{44}} = \begin{array}{c} 1 \\ 2 \\ 3 \end{array} \begin{array}{ccc} a & b & d \\ \left[\begin{array}{ccc} 1 & 0 & 1 \\ -1 & -1 & 0 \\ 0 & 1 & 0 \end{array} \right] \end{array}$$

and

$$A_{s_{44,32}} = \begin{array}{c} 1 \\ 2 \end{array} \begin{array}{cc} a & d \\ \left[\begin{array}{cc} 1 & 1 \\ -1 & 0 \end{array} \right] \end{array}$$

Finally,

$$A_{s_{44,32,12}} = \begin{array}{c} \\ 2 \end{array} \begin{array}{c} a \\ [-1] \end{array}$$

Hence

$$|A_s| = |A_{s_{44}}| = -|A_{s_{44,32}}| = |A_{s_{44,32,12}}| = -1$$

Note that edges a, b, d, and e form a tree.

If we take A_s of A as

$$A_s = \begin{array}{c} \\ 1 \\ 2 \\ 3 \\ 4 \end{array} \begin{array}{cccc} a & b & d & f \\ \left[\begin{array}{cccc} 1 & 0 & 1 & 0 \\ -1 & -1 & 0 & 0 \\ 0 & 1 & 0 & 1 \\ 0 & 0 & 0 & -1 \end{array} \right] \end{array}$$

we can find that

$$|A_s| = 1$$

where edges a, b, d, and f form a tree.

Note that if we choose A_s as

$$
A_s = \begin{array}{c c} & \begin{array}{cccc} a & b & c & d \end{array} \\ \begin{array}{c} 1 \\ 2 \\ 3 \\ 4 \end{array} & \left[\begin{array}{cccc} 1 & 0 & -1 & 1 \\ -1 & -1 & 0 & 0 \\ 0 & 1 & 1 & 0 \\ 0 & 0 & 0 & 0 \end{array} \right] \end{array}
$$

the determinant of A_s is zero and edges a, b, c, and d do not form a tree.

In order that we may have the same properties among circuits in oriented graphs as in nonoriented graphs, we would like to define a circuit in an oriented graph as we defined that in a nonoriented graph. Since each row of a circuit matrix of a nonoriented graph represents a circuit, we will indicate a circuit by each row of a circuit matrix of an oriented graph, too. We use the following definitions.

Definition 6-1-6. A circuit in an oriented graph is a circuit in the corresponding nonoriented graph with an orientation assigned to it by a cyclic ordering of vertices. An arrow will be employed to indicate the orientation of a circuit.

For example, C_1, C_2, and C_3 in Fig. 6-1-4 represent circuits. We define an edge disjoint union of circuits similarly.

Definition 6-1-7. An edge disjoint union of circuits in an oriented graph is a union of circuits no two of which have edges in common.

Definition 6-1-8. The element b_{ij} of an exhaustive circuit matrix $B_e = [b_{ij}]$ of an oriented graph is defined as

$$
b_{ij} = \begin{cases} 1 & \text{if edge } j \text{ is in circuit } i \text{ (or edge disjoint union of circuits } i) \text{ and the orientation of the edge and that of the circuit (edge disjoint union of circuits) coincide} \\ -1 & \text{if edge } j \text{ is in circuit } i \text{ (or edge disjoint union of circuits } i) \text{ and the orientation of the edge and that of the circuits (edge disjoint union of circuits) do not coincide} \\ 0 & \text{otherwise} \end{cases}
$$

For example, an exhaustive circuit matrix of the oriented linear graph in Fig. 6-1-4 is

$$
B_e = \begin{array}{c c} & \begin{array}{ccccc} a & b & d & e & f \end{array} \\ \begin{array}{c} C_1 \\ C_2 \\ C_3 \end{array} & \left[\begin{array}{ccccc} 1 & 1 & 0 & 0 & -1 \\ 0 & 0 & -1 & 1 & 1 \\ 1 & 1 & -1 & 1 & 0 \end{array} \right] \end{array}
$$

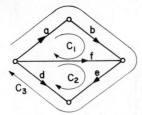

Fig. 6-1-4. Representation of oriented circuits.

As in the case of nonoriented graphs, we can define a set of fundamental circuits and a fundamental circuit matrix of an oriented graph as follows.

Definition 6-1-9. Let G be a connected oriented graph containing n_v vertices and n_e edges. Also let t be a tree in G and e_p be a chord for $p = 1$, $2, \ldots, n_e - n_v + 1$. Then a *fundamental circuit* C_p with respect to tree t is a circuit in $t \cup (e_p)$ whose orientation coincides with the orientation of chord e_p. A set of all fundamental circuits $C_1, C_2, \ldots, C_{n_e - n_v + 1}$ is called a *set of fundamental circuits with respect to tree t*.

Example 6-1-2. Consider the connected oriented graph in Fig. 6-1-5. Let a

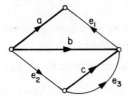

Fig. 6-1-5. An oriented graph.

tree t be (a, b, c). Then circuit C_1 in subgraph $t \cup (e_1)$ with the orientation which coincides with the orientation of chord e_1 as shown in Fig. 6-1-6a is

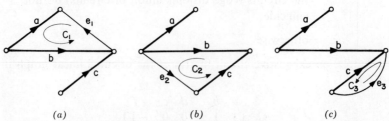

(a)   (b)   (c)

Fig. 6-1-6. A set of fundamental circuits with respect to $t = (a, b, c)$. (a) $t \cup (e_1)$ and C_1; (b) $t \cup (e_2)$ and C_2; (c) $t \cup (e_3)$ and C_3.

in a set of fundamental circuits with respect to t. We can see that the circuits C_1, C_2, and C_3 in Fig. 6-1-6 form a set of fundamental circuits with respect to tree t.

For a separated oriented graph, a set of fundamental circuits with respect to a forest is defined as follows.

Definition 6-1-10. Let G be an oriented graph consisting of n_v vertices, n_e edges, and ρ maximal connected subgraphs. Let t be a forest and e_p a chord for $p = 1, 2, \ldots, n_e - n_v + \rho$. Then a *fundamental circuit C_p with respect to forest t* is a circuit in $t \cup (e_p)$ whose orientation agrees with the orientation of chord e_p. A set of all fundamental circuits C_p ($p = 1, 2, \ldots, n_e - n_v + \rho$) is called a *set of fundamental circuits with respect to forest t.*

We choose the orientation of a chord as the orientation of the fundamental circuit containing the chord so that we have a unit matrix when we form a fundamental circuit matrix, as we shall see next.

Definition 6-1-11. Let G be an oriented graph consisting of n_v vertices, n_e edges, and ρ maximal connected subgraphs. Let $(e_1, e_2, \ldots, e_{n_e - n_v + \rho})$ be a set of chords with respect to a forest t (which will be a tree if $\rho = 1$). Also let B_e be an exhaustive circuit matrix of G whose columns are arranged so that the first $n_e - n_v + \rho$ columns correspond to chords $e_1, e_2, \ldots, e_{n_e - n_v + \rho}$. Then a submatrix B_f of B_e is called a *fundamental circuit matrix with respect to forest t* if row p of B_f corresponds to a fundamental circuit C_p in $t \cup (e_p)$ for $p = 1, 2, \ldots, n_e - n_v + \rho$.

It can easily be seen from this definition that a fundamental circuit matrix B_f can be expressed as

$$B_f = [U \quad B_{f_{12}}] \tag{6-1-3}$$

where U is a unit matrix.

Example 6-1-3. A fundamental circuit matrix B_f of the oriented graph in Fig. 6-1-7 with respect to tree $t = (e, f, g, h)$ is

$$B_f = \begin{array}{c} \begin{array}{cccccccc} a & b & c & d & & e & f & g & h \end{array} \\ \left[\begin{array}{cccc|cccc} 1 & 0 & 0 & 0 & 1 & 1 & 0 & 0 \\ 0 & 1 & 0 & 0 & -1 & -1 & 0 & 0 \\ 0 & 0 & 1 & 0 & 0 & -1 & 1 & 0 \\ 0 & 0 & 0 & 1 & 0 & 0 & 1 & -1 \end{array} \right] \end{array}$$

Since a fundamental circuit matrix B_f is a submatrix of an exhaustive circuit matrix B_e of an oriented linear graph, the rank of B_e is at least $n_v - n_v + \rho$ where n_e is the number of edges, n_v is the number of vertices,

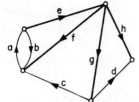

Fig. 6-1-7. An oriented graph.

and ρ is the number of maximal connected subgraphs in the oriented linear graph.

To show the rank of an exhaustive circuit matrix, we use the orthogonal property of an incidence and a circuit matrix of an oriented linear graph. Note that for a nonoriented graph, $B_e A_e{}^t = 0$ was established with modulo-2 algebra. However, we are using real integers here.

Let

$$A_e = \begin{bmatrix} A_1 \\ A_2 \\ \vdots \\ A_{n_v} \end{bmatrix} \quad \text{and} \quad B_e = \begin{bmatrix} B_1 \\ B_2 \\ \vdots \\ B_n \end{bmatrix}$$

be the incidence and the circuit matrices of an oriented linear graph. Then the product $A_e B_e{}^t$ is

$$\begin{bmatrix} A_1 \\ A_2 \\ \vdots \\ A_{n_v} \end{bmatrix} [B_1{}^t B_2{}^t \cdots B_n{}^t] = \begin{bmatrix} A_1 B_1{}^t & A_1 B_2{}^t & \cdots & A_1 B_n{}^t \\ A_2 B_1{}^t & A_2 B_2{}^t & \cdots & A_2 B_n{}^t \\ & & \cdots & \\ A_{n_v} B_1{}^t & A_{n_v} B_2{}^t & \cdots & A_{n_v} B_n{}^t \end{bmatrix} \tag{6-1-4}$$

where

$$A_i B_j{}^t = \sum_{k=1}^{n_e} a_{ik} b_{jk} \tag{6-1-5}$$

It can be seen that $a_{ip} b_{jp}$ is nonzero only if edge p is incident at vertex i and also is in circuit j (or an edge disjoint union of circuits j). Suppose edge p is incident at vertex i and is also in circuit j. Then there exists at least one edge other than p which is incident at vertex i and also is in circuit j. Let q be such an edge as shown in Fig. 6-1-8. The sign of a_{ip} (also a_{iq}) is determined by whether the arrow direction of edge p (edge q) is toward or away from vertex i. On the other hand, the sign of b_{jp} (b_{jp}) is determined by whether or not the arrow direction of edge p (edge q) and the arrow direction of circuit j

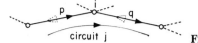

Fig. 6-1-8.   Circuit j and edges p and q.

(or an edge disjoint union of circuits j) coincide. Hence, if the sign of a_{ip} is the same as that of b_{jp}, then the sign of a_{iq} must be opposite to the sign of b_{jq}. Moreover, the degree of each vertex in a circuit or edge disjoint union of circuits is even. Thus if there exist nonzero terms in Eq. 6-1-5, they appear as pairs, one of which is $+1$ and the other is -1. Hence

$$A_e B_e^t = 0 \qquad\qquad (6\text{-}1\text{-}6)$$

and

$$B_e A_e^t = 0 \qquad\qquad (6\text{-}1\text{-}7)$$

For example, an exhaustive incidence matrix of an oriented graph G in Fig. 6-1-4 is

$$
A_e =
\begin{array}{c}
\begin{array}{ccccc} a & b & d & e & f \end{array} \\
\left[
\begin{array}{ccccc}
1 & 0 & 1 & 0 & 1 \\
-1 & 1 & 0 & 0 & 0 \\
0 & -1 & 0 & 1 & -1 \\
0 & 0 & -1 & -1 & 0
\end{array}
\right]
\end{array}
$$

An exhaustive circuit matrix of G is

$$
B_e =
\begin{array}{c}
\begin{array}{ccccc} a & b & d & e & f \end{array} \\
\left[
\begin{array}{ccccc}
1 & 1 & 0 & 0 & -1 \\
0 & 0 & -1 & 1 & 1 \\
1 & 1 & -1 & 1 & 0
\end{array}
\right]
\end{array}
$$

Hence

$$
A_e B_e^t =
\left[
\begin{array}{ccccc}
1 & 0 & 1 & 0 & 1 \\
-1 & 1 & 0 & 0 & 0 \\
0 & -1 & 0 & 1 & -1 \\
0 & 0 & -1 & -1 & 0
\end{array}
\right]
\left[
\begin{array}{ccc}
1 & 0 & 1 \\
1 & 0 & 1 \\
0 & -1 & -1 \\
0 & 1 & 1 \\
-1 & 1 & 0
\end{array}
\right]
= 0
$$

(Note that 0 in the right-hand side of the preceding equation is a matrix whose entries are all 0.)

We know that the rank of an exhaustive incidence matrix A_e is $n_v - p$. By Eq. 6-1-7 and the Sylvester's law of nullity (Theorem 3-3-4), we can see that the rank of an exhaustive circuit matrix B_e is at most $n_e - n_v + \rho$. We also know that a fundamental circuit matrix B_f is of rank $n_e - n_v + \rho$ and is a submatrix of B_e.

Theorem 6-1-3. The rank of an exhaustive circuit matrix B_e of an oriented graph is $n_e - n_v + \rho$ where n_e is the number of edges, n_v the number of vertices, and ρ the number of maximal connected subgraphs in the oriented graph.

Since the rank of an exhaustive circuit matrix B_e is $n_e - n_v + \rho$, there are $n_e - n_v + \rho$ rows of B_e such that all other rows can be obtained from these rows. This means that such $n_e - n_v + \rho$ rows are sufficient to know all about circuits in an oriented graph. Hence we define a circuit matrix B which consists of these $n_e - n_v + \rho$ rows of B_e as follows.

Definition 6-1-12. The symbol B will be used for representing a circuit matrix of order $n_e - n_v + \rho$ by n_e which is formed by taking $n_e - n_v + \rho$ rows of B_e so that the rank of B is $n_e - n_v + \rho$.

With this definition, we have the next important property of a circuit matrix of an oriented graph.

Theorem 6-1-4. A major submatrix of a circuit matrix B of an oriented graph is nonsingular if and only if the columns correspond to all chords with respect to a forest (a tree when $\rho = 1$).

Recall that we have exactly the same property in the case of nonoriented graphs.

Proof. Let an incidence matrix A of an oriented graph G be partitioned as

$$A = [A_{11} \quad A_{12}] \tag{6-1-8}$$

where A_{12} corresponds to a forest. Let a circuit matrix B be partitioned as

$$B = [B_{11} \quad B_{12}] \tag{6-1-9}$$

where the ith column of B and the ith column of A correspond to the same edge in G for all $i = 1, 2, \ldots, n_e$. By Eq. 6-1-7, we have

$$BA^t = [B_{11} \quad B_{12}]\begin{bmatrix} A_{11}{}^t \\ A_{12}{}^t \end{bmatrix} = B_{11}A_{11}{}^t + B_{12}A_{12}{}^t = 0 \tag{6-1-10}$$

By Theorem 6-1-2, A_{12} is nonsingular. Thus we can multiply A_{12}^{t-1} by Eq. 6-1-10 to obtain

$$B_{11}A_{11}{}^t(A_{12}{}^t)^{-1} + B_{12} = 0 \tag{6-1-11}$$

or

$$B_{12} = -B_{11}A_{11}{}^t(A_{12}{}^t)^{-1} \tag{6-1-12}$$

Thus B can be expressed as

$$B = B_{11}[U \quad -A_{11}{}^t(A_{12}{}^t)^{-1}] \qquad (6\text{-}1\text{-}13)$$

Since B is a circuit matrix, B_{11} must be nonsingular in order that the rank of B be equal to the number of rows in B. On the other hand, the columns of B_{11} correspond to the columns of A_{11} and the columns of A_{11} correspond to all chords with respect to a forest. Thus if the columns of a major submatrix of B correspond to all chords, the major submatrix is nonsingular.

Next suppose a circuit matrix B is partitioned as

$$B = [B_{11} \quad B_{12}] \qquad (6\text{-}1\text{-}14)$$

where B_{11} is nonsingular. Now we partition an incidence matrix A as

$$A = [A_{11} \quad A_{12}] \qquad (6\text{-}1\text{-}15)$$

such that the ith column of A and the ith column of B correspond to the same edge in G for all $i = 1, 2, \ldots, n_e$. By Eq. 6-1-5, we have

$$AB^t = A_{11}B_{11}{}^t + A_{12}B_{12}{}^t = 0 \qquad (6\text{-}1\text{-}16)$$

Hence

$$A_{11} = -A_{12}B_{12}{}^t(B_{11}{}^t)^{-1} \qquad (6\text{-}1\text{-}17)$$

which gives

$$A = A_{12}[-B_{12}{}^t(B_{11}{}^t)^{-1} \quad U] \qquad (6\text{-}1\text{-}18)$$

Since A is an incidence matrix, A_{12} must be nonsingular. Thus by Theorem 6-1-2, the columns of A_{12} correspond to a forest. Thus the columns of B_{11} correspond to all chords with respect to a forest.   QED

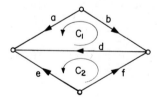

Fig. 6-1-9. Oriented graph.

Example 6-1-4. Consider a circuit matrix B of the oriented graph in Fig. 6-1-9:

$$B = \begin{array}{c} \\ \\ \end{array}\overset{\displaystyle a \quad\ b \quad\ \ d \quad\ \ e \quad f}{\begin{bmatrix} 1 & -1 & -1 & 0 & 0 \\ 0 & 0 & 1 & -1 & 1 \end{bmatrix}}$$

A major submatrix consisting of column a and b is singular and edges a and b do not form a set of chords with respect to any tree. On the other hand, a nonsingular major submatrix consists of columns a and d, that is,

$$\begin{matrix} a & d \end{matrix}$$
$$\begin{vmatrix} 1 & -1 \\ 0 & 1 \end{vmatrix} = 1$$

and edges a and d form a set of chords with respect to tree $t = (b, e, f)$. Another nonsingular major submatrix consists of columns a and e whose determinant is

$$\begin{matrix} a & e \end{matrix}$$
$$\begin{vmatrix} 1 & 0 \\ 0 & -1 \end{vmatrix} = -1$$

and edges a and e form a set of chords with respect to tree (b, d, f). Note that the determinant is -1 in this case.

6-2 ELEMENTARY TREE TRANSFORMATION

Here we are going to discuss a general property of trees of a linear graph which is independent of whether a linear graph is oriented or not. However, we need this property for the coming discussion of the invariant property of major determinants of a circuit and a cut-set matrices of an oriented graph.

Definition 6-2-1. Let t_1 and t_2 be trees in a linear graph G. The distance between two trees t_1 and t_2, symbolized by $d(t_1, t_2)$ is defined as

$$d(t_1, t_2) = \text{the number of edges in } t_1 \text{ but not in } t_2 \qquad (6\text{-}2\text{-}1)$$

For example, suppose

$$t_1 - t_2 = (e_1, e_2, \ldots, e_k) \qquad (6\text{-}2\text{-}2)$$

Then

$$d(t_1, t_2) = k \qquad (6\text{-}2\text{-}3)$$

Since the number of edges in a tree is $n_v - 1$,

$$d(t_1, t_2) = d(t_2, t_1) \qquad (6\text{-}2\text{-}4)$$

Suppose the distance between t_1 and t_2 is one, that is,

$$d(t_1, t_2) = 1 \qquad (6\text{-}2\text{-}5)$$

Let

$$t_1 - t_2 = (e) \qquad (6\text{-}2\text{-}6)$$

and

$$t_2 - t_1 = (e') \qquad (6\text{-}2\text{-}7)$$

Then it is clear that

$$t_1 \oplus (e, e') = t_2 \qquad (6\text{-}2\text{-}8)$$

The transformation from tree t_1 to tree t_2 by Eq. 6-2-8 is called an elementary tree transformation.

Definition 6-2-2. A transformation from a tree t_1 to another tree t_2 by

$$t_1 \oplus \mathscr{E} = t_2 \qquad (6\text{-}2\text{-}9)$$

is an *elementary tree transformation* if set $\mathscr{E}$ of edges consists of exactly two edges.

Note that set $\mathscr{E}$ must be as follows:

$$\mathscr{E} = t_1 \oplus t_2 = (t_1 - t_2) \cup (t_2 - t_1) \qquad (6\text{-}2\text{-}10)$$

Hence, in order that $\mathscr{E}$ consists of two edges, the distance between these two trees $d(t_1, t_2)$ must be one. Further, if two trees t_1 and t_2 are distance one from each other, then there is an elementary tree transformation which gives t_2 from t_1. It is also clear that if t_2 can be obtained from t_1 by an elementary tree transformation, then t_1 can be obtained from t_2 by an elementary tree transformation. For example, consider two trees t_1 and t_2 in the linear graph in Fig. 6-1-7 where

$$t_1 = (e, f, g, h)$$

and

$$t_2 = (d, e, f, g)$$

Since the distance between t_1 and t_2 is $d(t_1, t_2) = 1$, we can obtain t_2 from t_1 by an elementary tree transformation:

$$t_2 = t_1 \oplus (d, h)$$

Similarly,

$$t_1 = t_2 \oplus (d, h)$$

Suppose t_1 and t_k are trees of a linear graph G. Let

$$t_1 - t_k = (e_1, e_2, \ldots, e_k) \qquad (6\text{-}2\text{-}11)$$

and

$$t_k - t_1 = (a_1, a_2, \ldots, a_k) \qquad (6\text{-}2\text{-}12)$$

Then the distance between t_1 and t_k is k or

$$d(t_1, t_k) = k \qquad (6\text{-}2\text{-}13)$$

Suppose we insert edge a_1 to tree t_1 as shown in Fig. 6-2-1. Let C be the

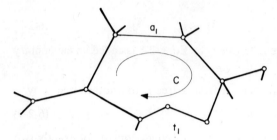

Fig. 6-2-1. A tree t_1 and edge a_1.

fundamental circuit formed by edge a_1 with tree t_1. Tree t_k cannot contain all edges in C because a tree cannot have a circuit by definition. Thus there exists at least one edge in C which is not in t_k (this edge is not edge a_1 because edge a_1 is in t_k), or

$$C - t_k \subseteq t_1 - t_k \qquad (6\text{-}2\text{-}14)$$

Let $e_1 \in (C - t_k)$. Then deleting edge e_1 from $t_1 \cup (a_1)$ will produce a tree because $e_1 \in C$ and the deletion of edge e_1 from $t_1 \cup (a_1)$ destroys only the circuit C in $t_1 \cup (a_1)$. Let this tree be t_2. Then

$$t_2 = t_1 \oplus (a_1, e_1) \qquad (6\text{-}2\text{-}15)$$

is an elementary tree transformation. Furthermore,

$$t_2 - t_k = (e_2, \ldots, e_k) \qquad (6\text{-}2\text{-}16)$$

$$t_k - t_2 = (a_2, \ldots, a_k) \qquad (6\text{-}2\text{-}17)$$

and

$$d(t_2, t_k) = k - 1 \qquad (6\text{-}2\text{-}18)$$

We can see from this that there exists a sequence of trees $t_1 t_2 \cdots t_{k-1} t_k$ such that all adjacent trees are of distance one. In other words, t_k can be obtained from t_1 by successive elementary tree transformations.

Theorem 6-2-1. Let t_1 and t_k be trees of a connected linear graph. If $d(t_k, t_1)$ is k, then t_k can be obtained from t_1 by exactly k successive elementary tree transformations.

Example 6-2-1. Consider the linear graph in Fig. 6-1-7. Suppose we choose $t_1 = (e, f, g, h)$ and $t_k = (a, c, d, h)$. Then

$$d(t_1, t_k) = 3$$

and

$$t_k - t_1 = (a, c, d)$$

By inserting edge a to tree t_1 as shown in Fig. 6-2-2, we can see that there is a

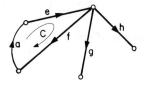

Fig. 6-2-2. Tree t_1 and edge a.

fundamental circuit $C = (a, e, f)$. By deleting edge e from $t_1 \cup (a)$, we will have a tree $t_2 = (a, f, g, h)$. Hence

$$t_2 = t_1 \oplus (a, e)$$

is an elementary tree transformation. Now we have

$$t_k - t_2 = (c, d)$$

By inserting edge c to tree t_2 as shown in Fig. 6-2-3, fundamental circuit C' with respect to tree t_2 can be obtained where

$$C' = (c, f, g)$$

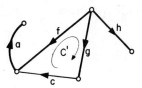

Fig. 6-2-3. Tree t_2 and edge c.

Thus by deleting edge f from $t_2 \cup (c)$, we will have a tree t_3 which consists of edges a, c, g, and h. Hence

$$t_3 = t_2 \oplus (c, f)$$

is an elementary tree transformation. With t_3, we have

$$t_k - t_3 = (d)$$

Now a similar process gives

$$t_k = t_3 \oplus (d, g)$$

which is clearly an elementary tree transformation. Thus we have a sequence of trees t_1, t_2, t_3, t_k such that every adjacent tree is of distance one. We have shown that t_k can be obtained from t_1 by three successive elementary tree transformations where the distance between t_1 and t_k is three. It can be seen that for a separated linear graph, we can define an elementary forest transformation and show that any forest can be obtained from any other forest by the successive elementary forest transformations.

6-3 VALUES OF NONZERO MAJOR DETERMINANTS OF A CIRCUIT MATRIX

In Section 6-1 we found that the determinant of any nonsingular major submatrix of an incidence matrix is always either $+1$ or -1. However, a nonsingular major determinant of a circuit matrix may be neither $+1$ nor -1. For example, a circuit matrix of the oriented graph in Fig. 6-3-1 is

$$B = \begin{array}{c} \\ C_1 \\ C_2 \end{array} \begin{array}{cccc} a & b & d & e \\ \begin{bmatrix} 1 & -1 & -1 & 1 \\ 1 & -1 & 1 & -1 \end{bmatrix} \end{array}$$

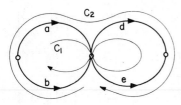

Fig. 6-3-1.   Oriented graph and circuits C_1 and C_2.

The determinant of a major submatrix consisting of columns a and d is

$$\begin{array}{cc} a & d \\ \begin{vmatrix} 1 & -1 \\ 1 & 1 \end{vmatrix} \end{array} = 2$$

However, we will see that the absolute value of the determinant of all nonsingular major submatrices of a circuit matrix are the same, which is a very important property for many applications.

Let

$$B_f = [U \quad B_{f_{12}}] \tag{6-3-1}$$

be a fundamental circuit matrix with respect to a tree t_1 of an oriented graph G. Recall that the columns of $B_{f_{12}}$ correspond to the branches of tree t_1. Let B_{11} be a nonsingular major submatrix of B_f. Then by rearranging columns of B_f, we obtain a circuit matrix B_2:

$$B_2 = [B_{11} \quad B_{12}] \qquad (6\text{-}3\text{-}2)$$

If we can show that B_2 becomes a fundamental circuit matrix just by adding and (or) subtracting rows of B_2 and multiplying -1 (if necessary) to rows of the resultant matrix, then we know that the determinant of B_{11} is either $+1$ or -1.

Since B_{11} is assumed to be nonsingular, the columns of B_{12} correspond to branches of a tree by Theorem 6-1-4. Let this tree be t_2. Suppose

$$d(t_1, t_2) = 1 \qquad (6\text{-}3\text{-}3)$$

Then we know that there are edges e_1 and a_1 where e_1 is in t_1 and a_1 is in t_2 such that

$$t_2 = t_1 \oplus (a_1, e_1) \qquad (6\text{-}3\text{-}4)$$

which is an elementary tree transformation.

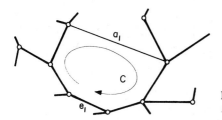

Fig. 6-3-2.   Tree t_1 and edge a_1 [subgraph $t_1 \cup (a_1)$].

Consider subgraph $t_1 \cup (a_1)$ where C is the circuit in the subgraph as shown in Fig. 6-3-2. Note that

$$t_1 \cup (a_1) = t_2 \cup (e_1) \qquad (6\text{-}3\text{-}5)$$

by Eq. 6-3-4.

Let $\{C_1\}$ be the set of fundamental circuits with respect to t_1. Note that rows of B_f represent all circuits in $\{C_1\}$. Let $\{C_2\}$ be the set of fundamental circuits with respect to t_2. It is clear that circuit C is in both $\{C_1\}$ and $\{C_2\}$.

Let $C_{12}, C_{13}, \ldots, C_{1k}$ be those circuits in $\{C_1\}$ that do not contain edge e_1 and $C_{1k+1}, \ldots, C_{1m}$ be those circuits in $\{C_1\}$ which contain edge e_1 where m is the number of fundamental circuits in linear graph G. Then we can see easily that $C_{12}, C_{13}, \ldots, C_{1k}$ are also in $\{C_2\}$.

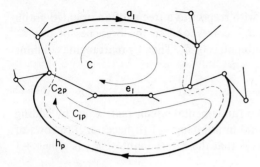

Fig. 6-3-3. Circuit C_{1p}.

Consider circuit C_{1p} ($k + 1 \leq p \leq m$) in $\{C_1\}$ shown in Fig. 6-3-3. We can see that C_{1p} is not a fundamental circuit with respect to t_2 because it contains two chords e_1 and h_p. However, there is a corresponding fundamental circuit C_{2p} in $\{C_2\}$ which contains the same chord h_p. Note that h_p is a chord with respect to both t_1 and t_2. Furthermore, we can see that

$$C_{2p} = C_{1p} \oplus C \tag{6-3-6}$$

Now consider matrix B_2. For convenience, let the first row of B_2 represent circuit C and the pth row represent circuit C_{1p} for $p = 2, 3, \ldots, m$. The orientation of C is assigned to agree with the orientation of a_1 in B_f because a_1 is a chord with respect to t_1. To make B_2 a fundamental circuit matrix, e_1 rather than a_1 is a chord with respect to t_2. Hence the orientation of C must agree with that of e_1 in order that the first row of B_2 represent fundamental circuit C with respect to t_2. Thus we multiply -1 to the first row if the orientation of C with respect to t_2 disagrees with that of C with respect to t_1.

Since the fundamental circuit C_{1p} for $2 \leq p \leq k$ with respect to t_1 is also the fundamental circuits with respect to t_2, these rows of B_2 represent fundamental circuits C_{2p} with respect to t_2 for $2 \leq p \leq k$.

Finally, circuit C_{1p} for $k + 1 \leq p \leq m$ represented by the pth row of B_2 contains edge e_1. Hence C_{1p} is not a fundamental circuit with respect to t_2, so we must change C_{1p} to a corresponding circuit C_{2p}. Since the orientation of both C_{1p} and C_{2p} agree with that of chord h_p, Eq. 6-3-6 indicates that by either adding or subtracting the first row (representing circuit C) from the pth row, we will have a new pth row which represents circuit C_{2p}. Hence circuit matrix B_2 in Eq. 6-3-2 can be changed to a fundamental circuit matrix

$$B_{2_f} = [U \quad B_{2_{f_{12}}}] \tag{6-3-7}$$

by addition or subtraction of a row from others, multiplication of -1 to one row (if necessary), and rearranging rows. These operations will not change the absolute value of the determinant of any nonsingular major submatrix of B_f. A nonsingular major submatrix of B_f and the corresponding nonsingular major submatrix of B_{2_f} (whose columns represent the same set of edges) thus have the property that the determinants are the same but the signs are different. Furthermore, since the major $\underline{B}_{11}$ becomes U in B_{2_f}, we can say that the determinant $\underline{B}_{11}$ is either $+1$ or -1. In general, any major determinant of B_f that corresponds to a set of chords with respect to a tree of distance one from t_1 is either $+1$ or -1.

By rearranging rows and columns of B_{2_f}, we can obtain

$$B_3 = [\underline{B}_{11} \quad \underline{B}_{12}] \tag{6-3-8}$$

where $\underline{B}_{11}$ is nonsingular. Suppose the columns corresponding to $\underline{B}_{12}$ is a tree t_3 which is distance one from t_2. Then we can see that the determinant of $\underline{B}_{11}$ is either $+1$ or -1. Hence the major determinant of B_{2_f} corresponding to the set of chords with respect to tree t_3 (which is distance 2 from t_1) is either $+1$ or -1. In general, any major determinant of B_{2_f} corresponding to a set of chords with respect to a tree of distance one from t_2 can be chosen to be $\underline{B}_{11}$ in Eq. 6-3-8. We now can say that any nonzero major determinant of B_f corresponding to the set of chords of a tree of distance two from t_1 is either $+1$ or -1.

The fact that any tree can be obtained from t_1 by successive elementary tree transformation brings us to the next theorem.

Theorem 6-3-1. The determinant of any nonsingular major submatrix of a fundamental circuit matrix is either $+1$ or -1.

Example 6-3-1. Fundamental circuit matrix B_f with respect to tree $t_1 = (g, h, i, j)$ of the oriented linear graph in Fig. 6-3-4 is

$$B_f = \begin{array}{c} \\ \\ \\ \\ \end{array} \begin{array}{cccccccc} a & b & d & f & g & h & i & j \\ \left[\begin{array}{cccccccc} 1 & 0 & 0 & 0 & 1 & -1 & 1 & 0 \\ 0 & 1 & 0 & 0 & -1 & 1 & 0 & 0 \\ 0 & 0 & 1 & 0 & 0 & 1 & -1 & 1 \\ 0 & 0 & 0 & 1 & 0 & -1 & 1 & 0 \end{array}\right] \end{array}$$

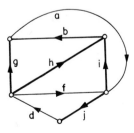

Fig. 6-3-4. Oriented linear graph.

The fundamental circuit matrix B_{2_f} with respect to tree $t_2 = (g, h, d, j)$ can be obtained from B_f by the following procedure:

1. Interchange columns d and i and multiply the third row by -1:

$$
\begin{array}{cccccccc}
a & b & i & f & g & h & d & j \\
\end{array}
$$
$$
\begin{bmatrix}
1 & 0 & 1 & 0 & 1 & -1 & 0 & 0 \\
0 & 1 & 0 & 0 & -1 & 1 & 0 & 0 \\
0 & 0 & 1 & 0 & 0 & -1 & -1 & -1 \\
0 & 0 & 1 & 1 & 0 & -1 & 0 & 0
\end{bmatrix}
$$

2. Since $t_2 = t_1 \oplus (i, d)$, the third row corresponds to the fundamental circuit with respect to both t_1 and t_2. Using this circuit, we must change circuits represented by rows in the foregoing matrix (except the third row) which contain edge i to those which do not contain edge i. This can be done by adding -1 times the third row to the first row and to the fourth row. The resultant matrix shown below is the fundamental circuit matrix with respect to t_2:

$$
\begin{array}{cccccccc}
& a & b & i & f & g & h & d & j \\
\end{array}
$$
$$
B_{2_f} = \begin{bmatrix}
1 & 0 & 0 & 0 & 1 & 0 & 1 & 1 \\
0 & 1 & 0 & 0 & -1 & 1 & 0 & 0 \\
0 & 0 & 1 & 0 & 0 & -1 & -1 & -1 \\
0 & 0 & 0 & 1 & 0 & 0 & 1 & 1
\end{bmatrix}
$$

We know that any circuit corresponding to a row in an exhaustive circuit matrix B_3 can be obtained by a linear combination of rows in a fundamental circuit matrix B_f. Hence any circuit matrix B of rank $n_e - n_v + \rho$ of an oriented graph can be expressed as

$$B = DB_f \tag{6-3-9}$$

where D is a nonsingular matrix. Thus the determinant of any nonsingular major submatrix of B is equal to $\pm |D|$.

Theorem 6-3-2. Let B be a circuit matrix. Also let $|B_{11}| = k$ where B_{11} is a nonsingular major submatrix of B. Then the determinant of any nonsingular major submatrix of B is $\pm k$.

For example, a circuit matrix of the oriented graph in Fig. 6-3-1 is

$$
\begin{array}{cccc}
a & b & d & e \\
\end{array}
$$
$$
B = \begin{bmatrix}
1 & -1 & -1 & 1 \\
1 & -1 & 1 & -1
\end{bmatrix}
$$

The absolute values of the determinant of nonsingular submatrices are

$$
\begin{array}{cc} a & d \\ \left\| \begin{array}{cc} 1 & -1 \\ 1 & 1 \end{array} \right\| \end{array} =
\begin{array}{cc} a & e \\ \left\| \begin{array}{cc} 1 & 1 \\ 1 & -1 \end{array} \right\| \end{array} =
\begin{array}{cc} b & d \\ \left\| \begin{array}{cc} -1 & -1 \\ -1 & 1 \end{array} \right\| \end{array} =
\begin{array}{cc} b & e \\ \left\| \begin{array}{cc} -1 & 1 \\ -1 & -1 \end{array} \right\| \end{array} = 2
$$

The fundamental circuit matrix B_f with respect to $t = (b, e)$ is

$$
B_f = \begin{array}{cccc} a & d & b & e \\ \begin{bmatrix} 1 & 0 & -1 & 0 \\ 0 & 1 & 0 & -1 \end{bmatrix} \end{array}
$$

By choosing matrix D as

$$
\begin{bmatrix} 1 & -1 \\ 1 & 1 \end{bmatrix}
$$

we have

$$
DB_f = \begin{bmatrix} 1 & -1 \\ 1 & 1 \end{bmatrix} \begin{array}{cccc} a & d & b & e \\ \begin{bmatrix} 1 & 0 & -1 & 0 \\ 0 & 1 & 0 & -1 \end{bmatrix} \end{array} = \begin{array}{cccc} a & d & b & e \\ \begin{bmatrix} 1 & -1 & -1 & 1 \\ 1 & 1 & -1 & -1 \end{bmatrix} \end{array} = B
$$

with columns b and d being interchanged. Note that $|D| = 2$.

6-4 CUT-SET MATRIX

We have used the symbol $\mathscr{E}(\Omega_1 \times \Omega_2)$ for a set of edges connected from a vertex in Ω_1 to a vertex in Ω_2. In the case of nonoriented graphs,

$$
\mathscr{E}(\Omega_1 \times \Omega_2) = \mathscr{E}(\Omega_2 \times \Omega_1) \tag{6-4-1}
$$

because edges have no orientation. However, in the case of oriented graphs,

$$
\mathscr{E}(\Omega_1 \times \Omega_2) \neq \mathscr{E}(\Omega_2 \times \Omega_1) \tag{6-4-2}
$$

For example, if $\Omega_1 = (1, 2)$ and $\Omega_2 = \bar{\Omega}_1$, then $\mathscr{E}(\Omega_1 \times \bar{\Omega}_1)$ of the oriented graph in Fig. 6-4-1 is (b, d, e). On the other hand, $\mathscr{E}(\bar{\Omega}_1 \times \Omega_1)$ is empty.

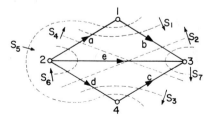

Fig. 6-4-1.   Cut-sets in an oriented linear graph.

Suppose by deleting all edges in a cut-set S of a linear graph G, we have two connected subgraphs G_1 and G_2. Let Ω_1 be the set of all vertices in G_1 and $\overline{\Omega}_1$ be the set of all vertices in G_2. If G is nonoriented, we know that $\mathscr{E}(\Omega_1 \times \overline{\Omega}_1)$ is cut-set S. However, if G is oriented, $\mathscr{E}(\Omega_1 \times \overline{\Omega}_1)$ may not be cut-set S. We can see that cut-set S in an oriented graph can be expressed as

$$S = \mathscr{E}(\Omega_1 \times \overline{\Omega}) \cup \mathscr{E}(\overline{\Omega}_1 \times \Omega_1) \qquad (6\text{-}4\text{-}3)$$

Definition 6-4-1.　When one (but not both) of $\mathscr{E}(\Omega_1 \times \overline{\Omega}_1)$ and $\mathscr{E}(\overline{\Omega}_1 \times \Omega_1)$ of a cut-set S in Eq. 6-4-3 is the empty set, then S is called a *directed cut-set*.

Since edges have orientations, cut-sets also possess orientations. Consider a cut-set $S = \mathscr{E}(\Omega_1 \times \overline{\Omega}_1) \cup \mathscr{E}(\overline{\Omega}_1 \times \Omega_1)$. The orientation of S is either from Ω_1 to $\overline{\Omega}_1$ or from $\overline{\Omega}_1$ to Ω_1. For convenience, an arrow will be used for specifying the orientation of a cut-set. For example, the arrow of cut-set S shown in Fig. 6-4-2 indicates that the orientation of S is from Ω_1 to $\overline{\Omega}_1$ where Ω_1 is the set of all vertices in G_1. Even if $S = \mathscr{E}(\Omega_1 \times \overline{\Omega}_1) \cup \mathscr{E}(\overline{\Omega}_1 \times \Omega_1)$ is an edge disjoint union of cut-sets, only one orientation will be given to S, either from Ω_1 to $\overline{\Omega}_1$ or from $\overline{\Omega}_1$ to Ω_1.

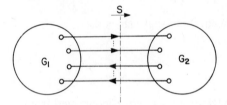

Fig. 6-4-2.　Cut-set and its orientation.

Instead of using Ω_1 and $\overline{\Omega}_1$ to indicate the orientation of cut-set (or edge disjoint union of cut-sets) S, we can use two vertices p and q if $p \in \Omega_1$ and $q \in \overline{\Omega}_1$. That is, we can say that *the orientation of cut-set S is from vertex p to vertex q, which tacitly indicates that p and q are separated by S.*

Example 6-4-1.　The cut-sets and the edge disjoint unions of cut-sets of the oriented graph in Fig. 6-4-1 are

$$S_1 = \mathscr{E}((1) \times (\overline{1})) \cup \mathscr{E}((\overline{1}) \times (1)) = (a, b)$$
$$S_2 = \mathscr{E}((1, 2) \times (\overline{1, 2})) \cup \mathscr{E}((\overline{1, 2}) \times (1, 2)) = (b, d, e)$$
$$S_3 = \mathscr{E}((4) \times (\overline{4})) \cup \mathscr{E}((\overline{4}) \times (4)) = (c, d)$$
$$S_4 = \mathscr{E}((1, 3) \times (\overline{1, 3})) \cup \mathscr{E}((\overline{1, 3}) \times (1, 3)) = (a, c, e)$$
$$S_5 = \mathscr{E}((1, 4) \times (\overline{1, 4})) \cup \mathscr{E}((\overline{1, 4}) \times (1, 4)) = (a, b, c, d)$$
$$S_6 = \mathscr{E}((2) \times (\overline{2})) \cup \mathscr{E}((\overline{2}) \times (2)) = (a, d, e)$$

and

$$S_7 = \mathscr{E}((3) \times (\overline{3})) \cup \mathscr{E}((\overline{3}) \times (3)) = (b, c, e)$$

S_2, S_4, S_6, and S_7 are directed cut-sets and S_5 is an edge disjoint union of cut-sets. The orientations of these cut-sets can be chosen as those indicated in Fig. 6-4-1. For example, the orientation of S_5 is from $(1, 4)$ to $(2, 3)$. If it is more convenient to use vertices 1 and 2 rather than sets of vertices $(1, 4)$ and $(2, 3)$, we can say that the orientation of S_5 is from 1 to 2.

Now we can define an exhaustive cut-set matrix as follows.

Definition 6-4-2. An exhaustive cut-set matrix Q_e of an oriented graph G is defined as follows:

$$Q_e = [q_{ij}] \tag{6-4-4}$$

where

$$q_{ij} = \begin{cases} 1 & \text{if edge } j \text{ is in cut-set (or edge disjoint union of cut-sets) } i \text{ and the orientation of edge } j \text{ agrees with that of cut-set (or edge disjoint union of cut-sets) } i \\ -1 & \text{if edge } j \text{ is in cut-set (or edge disjoint union of cut-sets) } i \text{ and the orientation of edge } j \text{ disagrees with that of cut-set (or edge disjoint union of cut-sets) } i \\ 0 & \text{otherwise} \end{cases} \tag{6-4-5}$$

such that every cut-set and every edge disjoint union of cut-sets in G will be represented by a row in Q_e.

For example, an exhaustive cut-set matrix Q_e of the oriented graph in Fig. 6-4-1 is

$$Q_e = \begin{array}{c} \\ S_1 \\ S_2 \\ S_3 \\ S_4 \\ S_5 \\ S_6 \\ S_7 \end{array} \begin{array}{ccccc} a & b & c & d & e \\ \left[\begin{array}{ccccc} -1 & 1 & 0 & 0 & 0 \\ 0 & -1 & 0 & -1 & -1 \\ 0 & 0 & -1 & 1 & 0 \\ 1 & 0 & 1 & 0 & 1 \\ -1 & 1 & 1 & -1 & 0 \\ -1 & 0 & 0 & -1 & -1 \\ 0 & -1 & -1 & 0 & -1 \end{array}\right] \end{array}$$

From the definition of an exhaustive cut-set matrix Q_e, we can see that each row represents a cut-set or an edge disjoint union of cut-sets. For convenience, we use the symbol $R(S)$ to indicate a row in Q_e which represents a cut-set (or an edge disjoint union of cut-sets) S. Let a cut-set (or edge disjoint union of cut-sets) be

$$S_c = \mathscr{E}(\Omega_1 \times \bar{\Omega}_1) \cup \mathscr{E}(\bar{\Omega}_1 \times \Omega_1) \tag{6-4-6}$$

where $\Omega_1 = (v_1, v_2, \ldots, v_n)$. Then we know from Theorem 2-4-1 that

$$S_c = S_1 \oplus S_2 \oplus \cdots \oplus S_n \tag{6-4-7}$$

where

$$S_r = \mathscr{E}((v_r) \times (\overline{v_r})) \cup \mathscr{E}((\overline{v_r}) \times (v_r)) \tag{6-4-8}$$

is the incidence set corresponding to vertex v_r for $r = 1, 2, \ldots, n$. We want to see whether

$$R(S_c) = \sum_{r=1}^{n} R(S_r) \tag{6-4-9}$$

where the orientation of S_c is from Ω_1 to $\overline{\Omega}_1$ and the orientation of S_r is from (v_r) to $(\overline{v_r})$ for $r = 1, 2, \ldots, n$. One way is by testing whether each column of $R(S_c)$ is equal to the sum of corresponding columns of $R(S_r)$. Since each column represents an edge, we need consider only two types of edges, one the edges in S_c and the other the edges not in S_c but in S_r.

TYPE 1.   Let edge e be an edge in S_c. Also let v_p and v_q be the two endpoints of edge e and the orientation of edge e be from v_p to v_q. Then we have 1 or -1 at column e in $R(S_c)$ depending on whether $v_p \in \Omega_1$ or $v_q \in \Omega_1$. Suppose $v_p \in \Omega_1$. Then $1 \leq p \leq n$ and $q > n$. Since only one incidence set S_p among $S_1, S_2, \ldots, S_n$ contains edge e, and in $R(S_p)$ we have 1 at column e, Eq. 6-4-9 is true for this case. We can show easily that Eq. 6-4-9 is true for the case when v_q is in Ω_1.

TYPE 2.   Suppose edge e is not in S_c. It is clear that if edge e is not in any of $S_1, S_2, \ldots, S_n$, Eq. 6-4-9 is true. Hence we assume that edge e is in at least one of these incidence sets. Then the two endpoints of edge e must be in Ω_1. Let v_p and v_q be the two end points of edge e and the orientation of edge e be from v_p to v_q. Without the loss of generality, let $1 \leq p < q \leq n$. Then there are exactly two incidence sets S_p and S_q among $S_1, S_2, \ldots, S_n$ which contain edge e. Furthermore, column e of $R(S_p)$ is $+1$ and column e of $R(S_q)$ is -1. Thus Eq. 6-4-9 is true for this case. We have assumed that the orientation of incidence set S_r is from (v_r) to $(\overline{v_r})$. If we do not assume this, we must modify Eq. 6-4-9 as follows:

$$R(S_c) = \sum_{r=1}^{n} k_r R(S_r) \tag{6-4-10}$$

where

$$k_r = \begin{cases} 1 & \text{if the orientation of } S_r \text{ is from } (\overline{v_r}) \text{ to } (v_r) \\ -1 & \text{if the orientation of } S_r \text{ is from } (v_r) \text{ to } (\overline{v_r}) \end{cases} \tag{6-4-11}$$

This is because by the definition of Q_e, the multiplication of (-1) to $R(S_r)$ is equivalent to the change of the orientation of S_r. This result indicates that any row in an exhaustive cut-set matrix Q_e can be obtained by a linear combination of rows in an exhaustive incidence matrix A_e.

Theorem 6-4-1. The rank of an exhaustive cut-set matrix of an oriented graph G is $n_v - \rho$ where n_v is the number of vertices and ρ is the number of maximal connected subgraphs in G.

From this theorem, we can see that any submatrix Q of Q_e whose rank is $n_v - \rho$ can be expressed as

$$Q = DA \tag{6-4-12}$$

where D is a nonsingular matrix.

Since

$$AB^t = 0 \tag{6-4-13}$$

by Eq. 6-1-6, we have

$$QB^t = DAB^t = 0 \tag{6-4-14}$$

Definition 6-4-2. Let G be an oriented graph consisting of ρ maximal connected subgraphs. A submatrix Q of an exhaustive cut-set matrix Q_e of G is called a *cut-set matrix* of G if Q consists of $n_v - \rho$ rows of Q_e and the rank of Q is $n_v - \rho$.

Theorem 6-4-2. Let Q and B be a cut-set and a circuit matrix of an oriented graph, respectively. Then

$$QB^t = 0 \tag{6-4-15}$$

and

$$BQ^t = 0 \tag{6-4-16}$$

When a cut-set matrix is of a normal form, we call it a fundamental cut-set matrix of an oriented graph.

Definition 6-4-3. A cut-set matrix Q_f which has the form

$$Q_f = [Q_{f_{11}} \quad U] \tag{6-4-17}$$

is called a *fundamental cut-set matrix*.

A fundamental cut-set matrix of an oriented graph is obtained in exactly the same way as is that of a nonoriented graph except an orientation is assigned to each fundamental cut-set, which was not required for nonoriented graphs. In other words, for a given tree t (or forest when $\rho > 1$), we can find the set of fundamental cut-sets of an oriented graph in exactly the same way as we found that of a nonoriented graph. Then we choose the orientation of each fundamental cut-set so that it agrees with the orientation of the branch of t which is in the fundamental cut-set. Using this set of oriented fundamental cut-sets, we can obtain a fundamental cut-set matrix as

$$Q_f = [Q_{11} \quad U] \tag{6-4-18}$$

where the columns of unit matrix U correspond to the branches of t. For example, if we choose a tree t as $t = (a, d, e)$, the set of fundamental cut-sets of the oriented linear graph in Fig. 6-4-1 consists of

$$S_a = \mathcal{E}((2, 3, 4) \times (1)) \cup \mathcal{E}((1) \times (2, 3, 4))$$
$$S_d = \mathcal{E}((1, 2, 3) \times (4)) \cup \mathcal{E}((4) \times (1, 2, 3))$$

and

$$S_e = \mathcal{E}((1, 2, 4) \times (3)) \cup \mathcal{E}((3) \times (1, 2, 4))$$

where the orientation of S_2 is from $(2,3,4)$ to (1), that of S_d is from $(1,2,3)$ to (4), and that of S_e is from $(1,2,4)$ to (3). Hence the fundamental cut-set matrix Q_f with respect to tree t will be

$$Q_f = \begin{array}{c} \\ S_a \\ S_d \\ S_e \end{array} \overset{\begin{array}{ccccc} b & c & a & d & e \end{array}}{\begin{bmatrix} -1 & 0 & 1 & 0 & 0 \\ 0 & -1 & 0 & 1 & 0 \\ 1 & 1 & 0 & 0 & 1 \end{bmatrix}}$$

Using a fundamental cut-set matrix and a fundamental circuit matrix with respect to the same tree (forest), Theorem 6-4-2 gives

$$Q_f B_c^{\,t} = [Q_{11} \quad U] \begin{bmatrix} U \\ B_{c_{12}}^t \end{bmatrix} = 0 \tag{6-4-19}$$

Thus

$$Q_{11} = -B_{c_{12}}^t \tag{6-4-20}$$

or we can express

$$Q_f = [-B_{c_{12}}^t \quad U] \tag{6-4-21}$$

Similarly, we can express fundamental circuit matrix as

$$B_c = [U \quad -Q_{11}{}^t] \tag{6-4-22}$$

For example, from the fundamental cut-set matrix with respect to tree $t = (a, d, e)$ of the oriented linear graph in Fig. 6-4-1, we have

$$Q_{11} = \overset{\begin{array}{cc} b & c \end{array}}{\begin{bmatrix} -1 & 0 \\ 0 & -1 \\ 1 & 1 \end{bmatrix}}$$

Hence the fundamental circuit matrix with respect to tree t of the oriented linear graph is

$$B_c = \begin{array}{cccccc} & b & c & a & d & e \\ & \begin{bmatrix} 1 & 0 & \vdots & 1 & 0 & -1 \\ 0 & 1 & \vdots & 0 & 1 & -1 \end{bmatrix} \end{array}$$

6-5 REALIZABILITY OF FUNDAMENTAL CUT-SET MATRICES

In Section 3-5, we used M-submatrices for testing whether a matrix is a fundamental cut-set matrix of a nonoriented graph. Here we will see that the same method can be used for testing a fundamental cut-set matrix of an oriented graph. However, we will study a property of a fundamental cut-set matrix of an oriented graph first.

Theorem 6-5-1. If a matrix $Q_f = [Q_{11} \quad U]$ is a fundamental cut-set matrix of an oriented graph G, then Q_f is also a fundamental cut-set matrix of another oriented graph G' which is obtained by reversing the orientations of all edges in G.

Proof. When the orientation of every edge is reversed, we have the same cut-set matrix by reversing the orientation of cut-sets.   QED

The next theorem shows an important property of a pair of M-submatrices which leads to a realizability condition of a fundamental cut-set matrix of an oriented graph.

Theorem 6-5-2. Let a pair of M-submatrices of a matrix $M_{1+2} = [M_{11} \quad U]$ with respect to row p be M_1 and M_2. Suppose there exists oriented graph g_k ($k = 1, 2$) such that (1) the fundamental cut-set matrix of g_k is M_k for $k = 1, 2$, and (2) there exists vertex p in g_k such that either row p in M_k or (-1) times row p in M_k represents an incidence set with respect to vertex p for $k = 1, 2$. Then there exists a linear graph g_{1+2} such that (1) M_{1+2} is a fundamental cut-set matrix of g_{1+2} and (2) for every row q other than row p in M_k if either row q or (-1) times row q represents an incidence set in g_k, then either row q or (-1) times row q in M_{1+2} represents an incidence set in g_{1+2} for $k = 1, 2$.

Proof. We prove this theorem by constructing an oriented graph g_{1+2} which satisfies the conditions in the theorem.

Since M_k is an M-submatrix of M_{1+2} with respect to row p, there exists a vertex p in g_k such that either row p or (-1) times row p in M_k represents an incidence set with respect to vertex p for $k = 1, 2$. In other words, the edges connected to vertex p in g_1 and the edges connected to vertex p in g_2 are the same. Let $(e_1, e_2, \ldots, e_m)$ be a set of all edges connected to vertex p in g_1 (and g_2). For convenience, we will say that the orientations of edge e in g_1 and in g_2 "agree" if the orientation of e in g_1 with respect to vertex p agrees with the orientation of e in g_2 with respect to vertex p. Otherwise, we will say that the orientations of edge e in g_1 and in g_2 "disagree." For example, if the orientation of edge e in g_1 is away from vertex p and that of e in g_2 is toward vertex p, then the orientations of e in g_1 and g_2 disagree. On the other hand, if the orientation of e in g_1 is away from vertex p and that of e in g_2 is also away from vertex p, then the orientations of e in g_1 and g_2 agree.

Since either row p or (-1) times row p of M_k indicates an incidence set in g_k for $k = 1, 2$, if the orientations of one edge which is connected to a vertex p in g_1 and g_2 agree, the orientations of every edge which is connected to a vertex p in g_1 and g_2 agree. On the other hand, if the orientations of one edge which is connected to vertex p in g_1 and in g_2 disagree, then the orientations of every edge which is connected to vertex p in g_1 and in g_2 disagree.

CASE 1.   Suppose the orientations of an edge which is connected to vertex p in g_1 and in g_2 disagree as shown in Fig. 6-5-1. Let v_{1r} and p in g_1 be the

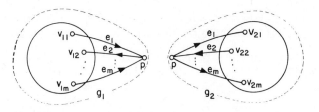

Fig. 6-5-1.   Linear graphs g_1 and g_2.

endpoints of edge e_r and v_{2r} and p in g_2 the endpoints of edge e_r for $r = 1, 2, \ldots, m$. Then we can obtain g_{1+2} by deleting edges $e_1, e_2, \ldots, e_m$ from g_1 and g_2 and then inserting edge e_r between v_{1r} and v_{2r} with the orientation which agrees with that of e_r in g_1 with respect to v_{1r} for $r = 1, 2, \ldots, m$ as shown in Fig. 6-5-2. It is apparent that the resultant graph g_{1+2} satisfies the conditions in the theorem.

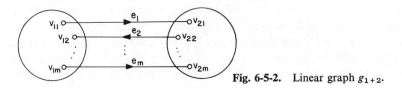

Fig. 6-5-2.   Linear graph g_{1+2}.

CASE 2. Suppose the orientations of edge e_r in g_1 and in g_2 agree as shown in Fig. 6-5-3. Then by Theorem 6-5-1, we can reverse the orientations of all

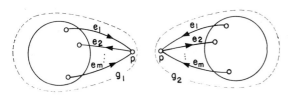

Fig. 6-5-3. Linear graphs g_1 and g_2.

edges in g_2 to obtain a new linear graph g_2' whose fundamental cut-set matrix is M_2. Note that if a row q in M_2 represents an incidence set in g_2 with respect to vertex q, then (-1) times row q represents an incidence set in g_2' with respect to vertex q. Similarly, if (-1) times row q in M_2 represents an incidence set in g_2 with respect to vertex q, then row q in M_2 represents an incidence set in g_2' with respect to vertex q. Now considering g_1 and g_2', this case becomes the previous case. Hence we can obtain a desired oriented graph g_{1+2}.   QED

Theorem 6-5-3 is similar to Theorem 3-5-8, which is a realizability condition for nonoriented graphs.

Theorem 6-5-3. A matrix $R = [R_{11} \quad U]$ is a fundamental cut-set matrix of an oriented graph if and only if there exists a set of minimum M-submatrices of R such that every minimum M-submatrix in the set becomes an incidence matrix by multiplying some rows by (-1).

Note that a matrix which consists of $+1$, -1, and 0 is an incidence matrix if and only if every column has either one nonzero or two nonzeros with opposite signs. We know that for an oriented graph, an incidence set with respect to a vertex v can be expressed as $\mathscr{E}((v) \times (\bar{v})) \cup \mathscr{E}((\bar{v}) \times (v))$, whose orientation is from (v) to $(\bar{v})$. We also know that for a vertex p, a set $\mathscr{E}((p) \times (\bar{p})) \cup \mathscr{E}((\bar{p}) \times (p))$ is either a cut-set or an edge disjoint union of cut-sets. Suppose the orientation of $\mathscr{E}((p) \times (\bar{p})) \cup \mathscr{E}((\bar{p}) \times (p))$ is from $(\bar{p})$ to (p). Then set $\mathscr{E}((p) \times (\bar{p})) \cup \mathscr{E}((\bar{p}) \times (p))$ is not an incidence set because of the opposite orientation. Suppose the set is represented by a row p of a cut-set matrix. We know that multiplying (-1) times row p is equivalent to reversing the orientation of the set. Hence by multiplying row p by (-1), set $\mathscr{E}((p) \times (\bar{p})) \cup \mathscr{E}((\bar{p}) \times (p))$ represented by row p becomes an incidence set. By knowing this fact and using Theorem 6-5-2 we can prove Theorem 6-5-3 (which is left to the reader).

Example 6-5-1.   Consider a matrix R:

$$R = \begin{array}{c} \\ 1 \\ 2 \\ 3 \\ 4 \end{array} \begin{array}{cccccccc} a & b & c & d & e & f & g & h & i \\ \left[\begin{array}{ccccc|cccc} -1 & 1 & 1 & 0 & 0 & 1 & 0 & 0 & 0 \\ 1 & 0 & -1 & 0 & 0 & 0 & 1 & 0 & 0 \\ 0 & 0 & 1 & -1 & -1 & 0 & 0 & 1 & 0 \\ 0 & 1 & 1 & 0 & -1 & 0 & 0 & 0 & 1 \end{array}\right] \end{array}$$

A set of minimum M-submatrices will be

$$M_1 = \begin{array}{c} 1 \\ 2 \end{array} \begin{array}{ccccc} a & b & c & f & g \\ \left[\begin{array}{ccccc} -1 & 1 & 1 & 1 & 0 \\ 1 & 0 & -1 & 0 & 1 \end{array}\right] \end{array}$$

$$M_2 = 2 \begin{array}{c} a & c & g \\ \left[1 & -1 & 1 \right] \end{array}$$

$$M_3 = 3 \begin{array}{c} c & d & e & h \\ \left[1 & -1 & -1 & 1 \right] \end{array}$$

$$M_4 = \begin{array}{c} 1 \\ 4 \end{array} \begin{array}{cccccc} a & b & c & e & f & i \\ \left[\begin{array}{cccccc} -1 & 1 & 1 & 0 & 1 & 0 \\ 0 & 1 & 1 & -1 & 0 & 1 \end{array}\right] \end{array}$$

and

$$M_5 = \begin{array}{c} 3 \\ 4 \end{array} \begin{array}{cccccc} b & c & d & e & h & i \\ \left[\begin{array}{cccccc} 0 & 1 & -1 & -1 & 1 & 0 \\ 1 & 1 & 0 & -1 & 0 & 1 \end{array}\right] \end{array}$$

By multiplying row 4 of M_4 and M_5 by (-1), we have

$$M_4' = \begin{array}{c} 1 \\ 4 \end{array} \begin{array}{cccccc} a & b & c & e & f & i \\ \left[\begin{array}{cccccc} -1 & 1 & 1 & 0 & 1 & 0 \\ 0 & -1 & -1 & 1 & 0 & -1 \end{array}\right] \end{array}$$

and

$$M_5' = \begin{array}{c} 3 \\ 4 \end{array} \begin{array}{cccccc} b & c & d & e & h & i \\ \left[\begin{array}{cccccc} 0 & 1 & -1 & -1 & 1 & 0 \\ -1 & -1 & 0 & 1 & 0 & -1 \end{array}\right] \end{array}$$

Since M_1, M_2, M_3, M_4', and M_5' are incidence matrices, matrix R is a fundamental cut-set matrix of an oriented graph by Theorem 6-5-3.

6-6   DIRECTED SUBGRAPHS

In the previous section, we found that the cut-set and circuit matrices of oriented graphs have the same properties as those of nonoriented graphs. In other words, the subgraphs such as cut-sets, circuits, and paths in an oriented graph are the same as those in the corresponding nonoriented graph except that we assign orientations to them so that the orientations of edges in a subgraph can be compared to the orientation of the subgraph. Thus almost every property we have studied about nonoriented graphs in the previous chapters can become the properties of oriented graphs without major modifications. On the other hand, there are important subgraphs of oriented graphs whose particular properties cannot be obtained from the corresponding nonoriented graphs. These are so called directed circuits, directed paths, and directed trees. We need certain definitions in order to study these subgraphs.

Definition 6-6-1.   A *directed edge train* is an edge train having the orientation from the initial vertex to the final vertex and satisfying the condition that the orientation of every edge in the edge train coincides with the orientation of the edge train.

For example, in the oriented graph in Fig. 6-6-1, (*abcde*) is a directed edge train but (*abf*) is not a directed edge train.

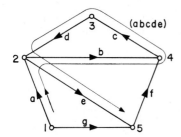

Fig. 6-6-1.   An oriented graph.

Definition 6-6-2.   We say that a directed edge train is closed if the initial vertex and the final vertex are the same, and is open otherwise.

For example, (*bcd*) in Fig. 6-6-1 is a closed directed edge train, and (*abcde*) is an open directed edge train.

Similar to the degree of a vertex in a nonoriented graph, we define the outgoing degree and incoming degree of a vertex in an oriented graph.

Definition 6-6-3.   The *outgoing degree*, $d^+(v)$ of a vertex v is the number of edges that are connected to vertex v whose orientations are away from vertex v. The *incoming degree*, $d^-(v)$ of a vertex v is the number of edges that are connected to vertex v and whose orientations are toward vertex v.

For example, the outgoing degree $d^+(2)$ and the incoming degree $d^-(2)$ of vertex 2 in Fig. 6-6-1 are

$$d^+(2) = 2$$

and

$$d^-(2) = 2$$

The outgoing degree $d^+(5)$ and the incoming degree $d^-(5)$ of vertex 5 are

$$d^+(5) = 1$$

and

$$d^-(5) = 2$$

By this definition, we can see that

$$d(v) = d^+(v) + d^-(v) \tag{6-6-1}$$

where $d(v)$ is the degree of vertex v.

Definition 6-6-4. An oriented graph is called a *directed Euler graph* if for every vertex v, the outgoing degree $d^+(v)$ is equal to the incoming degree $d^-(v)$.

For example, the oriented graph in Fig. 6-6-2 is a directed Euler graph but that in Fig. 6-6-3 is not.

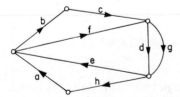

Fig. 6-6-2.   A directed Euler graph.

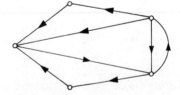

Fig. 6-6-3.   An Euler graph.

We now can easily show the following theorem.

Theorem 6-6-1. A connected oriented graph is a directed Euler graph if and only if there exists a closed directed edge train which contains all edges in the graph.

For example, a closed edge train $(abcdefgh)$ of the directed Euler graph in Fig. 6-6-2 contains all edges in the graph.

Definition 6-6-5.   When a connected directed Euler graph has the property that every vertex is degree 2 (i.e., there are exactly two edges connected to every vertex), then it is called a *directed circuit*.

For example, all directed circuits in Fig. 6-6-2 are (a, b, c, d, h), (a, b, c, g, h), (a, d, f, h), (a, f, g, h), (b, c, d, e), (b, c, e, g), (d, e, f), and (e, f, g).

It can be seen readily that if C_1 and C_2 are directed circuits and $C_1 \cap C_2 \neq \emptyset$, then $C_1 \oplus C_2$ is not a directed circuit. In general, the ring sum of two directed Euler graphs may not be a directed Euler graph. However, we have the following situation.

Theorem 6-6-2.   Let G be a directed Euler graph. Then for any directed circuit that is a proper subgraph of G, there exists subgraphs E_1 and E_2 of G both of which are nonempty directed Euler graphs such that

$$E_1 \oplus E_2 = C \qquad\qquad (6\text{-}6\text{-}2)$$

Before proving this theorem, let us study another theorem.

Theorem 6-6-3.   Let E be a directed Euler graph and E_1 be a proper subgraph of E which is also a directed Euler graph. Then

$$E \oplus E_1 = E - E_1 \qquad\qquad (6\text{-}6\text{-}3)$$

is a directed Euler graph.

This theorem can be proved by considering the incoming degree and the outgoing degree of every vertex in $E - E_1$. By Theorem 6-6-3 with $E_1 = E - C$ and $E_2 = E$, we have the proof of Theorem 6-6-2.

Theorem 6-6-4.   A directed Euler graph contains at least one directed circuit.

The proof is easily accomplished by constructing a closed directed edge train. The next theorem is rather obvious but very important.

Theorem 6-6-5.   A directed Euler graph is either a directed circuit or an edge disjoint union of directed circuits.

Proof.   By Theorem 6-6-4, there exists at least one directed circuit in a directed Euler graph G. Let this directed circuit be C_1. Then by Theorem 6-6-3, $G - C_1$ is a directed Euler graph. Hence there exists at least one directed circuit in $G - C_1$, and so on. Finally, $G - C_1 - C_2 - \cdots - C_k$ becomes empty. Thus we can conclude that G is the edge disjoint union of directed circuits $C_1, C_2, \ldots, C_k$. When $k = 1$, G is obviously a directed circuit.   QED

For example, the directed Euler graph in Fig. 6-6-2 can be considered an edge disjoint union of directed circuits (a, d, f, h) and (b, c, e, g).

For nonoriented graphs, we have studied an M-graph, which is an edge disjoint union of a path and circuits. Here we introduce directed M-graphs and directed paths. We next show that a directed M-graph is an edge disjoint union of a directed path and directed circuits.

Definition 6-6-6.   An oriented graph is called a *directed M-graph* of type $M(i \times j)$ if it satisfies the following three conditions:
1. $d^+(v) = d^-(v)$ for all vertices except i and j.
2. $d^+(i) = d^-(i) + 1$.
3. $d^-(j) = d^+(j) + 1$.

Let G consist of all edges in an open directed edge train whose initial vertex is i and whose final vertex is j. Then G is a directed M-graph of type $M(i \times j)$. For example, the oriented graph in Fig. 6-6-4 is a directed M-graph of type $M(i \times j)$.

Fig. 6-6-4.   An M-graph of type $M(i \times j)$.

When a directed M-graph of type $M(i \times j)$ is not connected, then one of a maximal connected subgraph must be a directed M-graph of type $M(i \times j)$ and others must be directed Euler graphs. For example, the oriented graph in Fig. 6-6-5 is a separated directed M-graph of type $M(i \times j)$. Note that all maximal connected subgraphs except one are directed Euler graphs.

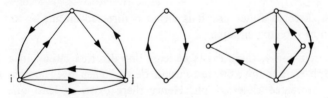

Fig. 6-6-5.   A separated directed M-graph of type $M(i \times j)$.

A particular directed M-graph is a directed path.

Definition 6-6-7. Let P be a connected directed M-graph of type $M(i \times j)$. If $d^+(v) = 1$ for all vertices except j and $d^+(j) = 0$, then P is called a *directed path from i to j.*

By this definition, we have Theorem 6-6-6.

Theorem 6-6-6. A directed M-graph of type $M(i \times j)$ is either a directed path from i to j or an edge disjoint union of a directed path from i to j and directed circuits.

The proof can be accomplished by using a directed edge train. We next find a property similar to the property of M-graphs in the nonoriented case which states that the sum of two M-graphs having the same terminal vertices is an Euler graph.

Theorem 6-6-7. Let $M(i \times j)$ be a directed M-graph of type $M(i \times j)$ and $M(j \times i)$ be a directed M-graph of type $M(j \times i)$. Suppose

$$M(i \times j) \cap M(j \times i) = \emptyset \qquad (6\text{-}6\text{-}4)$$

Then

$$M(i \times j) \oplus M(j \times i) = E \qquad (6\text{-}6\text{-}5)$$

is a directed Euler graph.

The proof of this theorem can easily be accomplished by checking the incoming degree and the outgoing degree of every vertex in E.

We now can generalize directed M-graphs.

Definition 6-6-8. A directed M-graph of type $M[i(k) \times j(k)]$ is defined as an oriented graph satisfying the following conditions:
1. $d^+(v) = d^-(v)$ for all vertices except i and j.
2. $d^+(i) = d^-(i) + k$.
3. $d^-(j) = d^+(j) + k$.

Fig. 6-6-6. A directed M-graph of type $M[i(2) \times j(2)]$.

For example, the oriented graph in Fig. 6-6-6 is a directed M-graph of type $M[i(2) \times j(2)]$.

Theorem 6-6-8. A directed M-graph of type $M[i(k) \times j(k)]$ is either an edge disjoint union of k directed paths from i to j or an edge disjoint union of k directed paths from i to j and directed circuits.

The proof can be accomplished by using mathematical induction on k and directed edge trains.

It is rather important to know whether or not an oriented graph contains directed circuits. The next theorem gives a sufficient condition for the existence of directed circuits.

Theorem 6-6-9. If the outgoing degree of every vertex of a finite connected oriented graph G is one, then G has exactly one directed circuit.

Proof. Let G be an oriented graph in which $d^+(v) = 1$ for all vertices. First, we will prove that G has at most one directed circuit. Shorting an edge in G, the resultant graph G_1 obviously satisfies $d^+(v) = 1$ for all vertices. We can keep shorting edges as long as we do not produce any self-loops. Note that we have no more than two edges in parallel in any directed graph satisfying $d^+(v) = 1$ for all vertices. Also note that only when two edges are in parallel, as shown in Fig. 6-6-7, do we form a self-loop by shorting one of

Fig. 6-6-7. Oriented graph g_p.

these edges. Let a graph consisting of two parallel edges be denoted by g_p. Consider a resultant graph G_k after shorting all possible edges without producing self-loops. Since the original graph G is connected and shorting edges will not produce separated graphs, G_k must be isomorphic to g_p. We also know that shorting edges will not decrease the number of directed circuits. Hence the original graph G has at most one directed circuit.

Now we will show that the original graph G has at least one directed circuit. Starting from a vertex v, we can obtain a directed edge train. Because $d^+(v) = 1$ for all vertices, this edge train will be terminated only when the final vertex is one that has previously been passed by the edge train. Hence if the directed edge train is terminated, there will be at least one directed circuit. Because G is finite, the edge train must be terminated. QED

Theorem 6-6-10. Let G be a connected oriented graph such that $d^-(v) = 1$ for all vertices. Then G contains exactly one directed circuit.

Proof. By reversing the orientation of every edge in G, the resultant oriented graph will satisfy the condition in Theorem 6-6-9. Thus G has exactly one directed circuit.   QED

For example, connected oriented graph G in Fig. 6-6-8 satisfies the condition in Theorem 6-6-10. Thus there exists exactly one directed circuit.

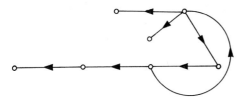

Fig. 6-6-8.   A connected oriented graph.

Theorem 6-6-11.   An oriented graph which satisfies

$$d^+(v) > 0$$

and

$$d^-(v) > 0$$

for all vertices has at least one directed circuit.

This can be proven easily by using a directed edge train.
Consider the oriented graph in Fig. 6-6-9. It can be seen that this oriented

Fig. 6-6-9.   An oriented graph containing a directed circuit.

graph does not satisfy the conditions in Theorem 6-6-11. However, this oriented graph contains a directed circuit. This example suggests a way of modifying the preceding theorem so that it becomes a necessary and sufficient condition.

Theorem 6-6-12.   A necessary and sufficient condition that an oriented graph contains at least one directed circuit is that there exists a subgraph which satisfies

$$d^+(v) > 0$$

and

$$d^-(v) > 0$$

for all vertices in the subgraph.

Although this theorem gives a necessary and sufficient condition, to test whether or not an oriented graph contains a directed circuit by investigating all possible subgraphs is not an efficient method. In order to test the existence of directed circuits more effectively, we define a W-process of eliminating vertices as follows.

Definition 6-6-9. Let v be a vertex in an oriented graph G. Suppose either $d^+(v) = 0$ or $d^-(v) = 0$. Then deleting all edges connected on vertex v is called the W-process of eliminating vertex v.

Note that there will be no directed paths passing through a vertex v if this vertex v can be eliminated by a W-process. In other words, if a vertex v is in a directed circuit, then we cannot apply a W-process to eliminate vertex v. With this process, we have the next theorem.

Theorem 6-6-13. If and only if an oriented graph G has no directed circuit, then the W-process can be employed successively to eliminate all edges in G.

The proof can be accomplished by using Theorem 6-6-11. This theorem gives an algorithm of testing the existence of directed circuits which can be seen readily in the next example.

Example 6-6-1. Consider oriented graph G in Fig. 6-6-10. Since $d^+(1) = 0$ and $d^-(2) = 0$, we can use a W-process to eliminate vertices 1 and 2. The resultant graph G_1 is shown in Fig. 6-6-11a. In G_1, $d^+(3) = 0$ and $d^-(4) = 0$.

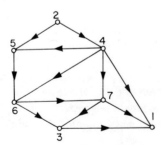

Fig. 6-6-10. An oriented graph.

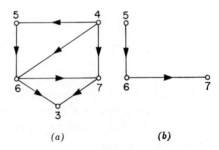

(a)   (b)

Fig. 6-6-11. Resultant graphs by the W-process. (a) G_1; (b) G_2.

Thus we can use a W-process to eliminate vertices 3 and 4. The resultant graph G_2 is shown in Fig. 6-6-11b. In G_2, $d^-(5) = 0$ and $d^+(7) = 0$. Thus by W-process to remove v vertices 5 and 7, we remove all edges; there are now no edges left. Hence the original graph G has no directed circuits.

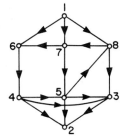

Fig. 6-6-12. An oriented graph G_o.

Consider oriented graph G_o in Fig. 6-6-12. It can be seen that $d^-(1) = 0$ and $d^+(2) = 0$. Hence by W-process to eliminate vertices 1 and 2, we have the resultant graph G_o' as shown in Fig. 6-6-13a. In G_o', $d^+(3) = 0$. Hence by

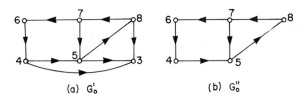

(a) G_o' (b) G_o''

Fig. 6-6-13. Resultant graphs by the W-process. (a) G_o'; (b) G_o''.

W-process to eliminate vertex 3, we have the resultant graph G_o'' shown in Fig. 6-6-13b. Since there are no more vertices in G_o'' whose either incoming degree or the outgoing degree is zero, we cannot use W-process to eliminate vertices and there are edges in G_o''. Thus the given oriented graph G_o has at least one directed circuit.

PROBLEMS

1. Prove Theorem 6-1-1.

2. Write an incidence matrix A of an oriented graph in Fig. P-6-2 with vertex v_0 being the reference vertex.

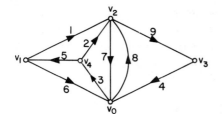

Fig. P-6-2.

3. Write a fundamental cut-set matrix with respect to tree $t = (1, 2, 3, 4)$ of the oriented graph in Fig. P-6-2.

4. Write a fundamental circuit matrix with respect to tree $t = (1, 2, 3, 4)$ of the oriented graph in Fig. P-6-2.

5. Is it possible to pick linearly independent circuits in an oriented graph so that the circuit matrix, whose rows correspond to these linearly independent circuits, has a nonzero major determinant which is not ± 1? Try it with (a) the oriented graph in Fig. P-6-5a and (b) the oriented graph in Fig. P-6-5b.

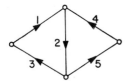

Fig. P-6-5-a.

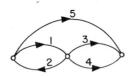

Fig. P-6-5b.

6. Let S_e be a fundamental cut-set with respect to a tree where edge e is in t. Prove that

$$t \oplus (e, b)$$

is a tree for every edge b ($\neq e$) in S_e.

7. Let C_b be a fundamental circuit with respect to a tree t where edge b is not in t. Prove that

$$t \oplus (b, e)$$

is a tree for every edge e ($\neq b$) in C_b.

8. Prove Theorem 6-5-3.

9. Prove Theorem 6-6-1.

10. Prove Theorem 6-6-3.

11. Prove Theorem 6-6-8.
12. Prove Theorem 6-6-11.
13. Prove Theorem 6-6-12.
14. Prove Theorem 6-6-13.
15. Test whether the oriented graph in Fig. P-6-15 has a directed circuit.

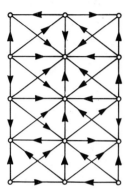
Fig. P-6-15.

CHAPTER
7
TOPOLOGICAL ANALYSIS OF PASSIVE NETWORKS

7-1 KIRCHHOFF'S LAWS

Kirchhoff's laws are very important in characterizing physical and non-physical systems, especially when these systems are represented by linear graphs. In other words, we can classify systems by whether the systems satisfy (1) Kirchhoff's current law, (2) Kirchhoff's voltage law, (3) both laws, or (4) neither law. For example, a communication net satisfies Kirchhoff's current law when we modify it slightly. An electrical network satisfies both Kirchhoff's laws. Hence, before studying electrical networks, we will study properties of Kirchhoff's laws when systems are represented by linear graphs.

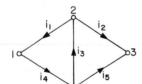

Fig. 7-1-1. Weighted oriented linear graph.

Consider the oriented linear graph in Fig. 7-1-1 in which the symbol i_k associated with each edge represents a current (a flow) known as a *branch current* or an *edge flow* through the edge whose direction of the branch current agrees with the orientation of the edge. A weight such as a branch current associated with an edge is called an *edge weight*.

Definition 7-1-1. When an edge weight (or edge weights) is given to every edge, a linear graph is called a *weighted linear graph*.

252

Consider a weighted oriented linear graph G in which each edge has a branch current as its edge weight. Let A be an incidence matrix of G whose pth column represents edge e_p for $p = 1, 2, \ldots, n_e$. Also let i_p be the branch current of edge e_p. Since each row of an incidence matrix A indicates all edges incident at a vertex with their orientations, the product AI_e of incidence matrix A and a column matrix

$$I_e = \begin{bmatrix} i_1 \\ i_2 \\ \vdots \\ i_{n_e} \end{bmatrix} \tag{7-1-1}$$

gives the set of equations each of which indicates the sum of branch currents incident at a vertex. By setting these equations to zero, we have the Kirchhoff's current law:

$$AI_e = 0 \tag{7-1-2}$$

For example, from the weighted linear graph in Fig. 7-1-1, we have

$$
\begin{array}{c}
\begin{array}{ccccc} i_1 & i_2 & i_3 & i_4 & i_5 \end{array} \\
\begin{array}{c} 1 \\ 2 \\ 3 \end{array}
\begin{bmatrix} -1 & 0 & 0 & 1 & 0 \\ 1 & 1 & -1 & 0 & 0 \\ 0 & -1 & 0 & 0 & -1 \end{bmatrix}
\end{array}
\begin{bmatrix} i_1 \\ i_2 \\ i_3 \\ i_4 \\ i_5 \end{bmatrix}
= \begin{array}{c} 1 \\ 2 \\ 3 \end{array}
\begin{bmatrix} -i_1 + i_4 \\ i_1 + i_2 - i_3 \\ -i_2 - i_5 \end{bmatrix} = 0
$$

Instead of using a branch current as a weight of each edge, if we use a branch voltage v_k as a weight of each edge in an oriented linear graph, we will have the Kirchhoff's voltage law:

$$BV_e = 0 \tag{7-1-3}$$

where B is a circuit matrix of an oriented graph and

$$V_e = \begin{bmatrix} v_1 \\ v_2 \\ \vdots \\ v_{n_e} \end{bmatrix} \tag{7-1-4}$$

with the assumption that the positive polarity of each branch voltage v_k is at the tail of the arrow of the corresponding edge. From Eqs. 7-1-2 and 7-1-3, we can see that $n_v - \rho$ of Kirchhoff's current equations are linearly independent and $n_e - n_v + \rho$ of Kirchhoff's voltage equations are linearly independent where n_e is the number of edges, n_v is the number of vertices, and ρ is the number of maximal connected subgraphs of a weighted oriented linear graph.

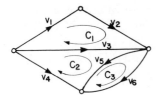

Fig. 7-1-2. Weighted linear graph.

Example 7-7-1. Consider the weighted linear graph in Fig. 7-1-2. By using circuits C_1, C_2, and C_3, we have the circuit matrix

$$
\begin{array}{c}
\phantom{B={}}\begin{array}{cccccc} v_1 & v_2 & v_3 & v_4 & v_5 & v_6 \end{array} \\
B = \begin{array}{c} C_1 \\ C_2 \\ C_3 \end{array}\left[\begin{array}{cccccc} 1 & 1 & -1 & 0 & 0 & 0 \\ 0 & 0 & 1 & -1 & 1 & 0 \\ 0 & 0 & 0 & 0 & -1 & 1 \end{array}\right]
\end{array}
$$

Then from Eq. 7-1-3, we have

$$
BV_e = \begin{array}{c}\begin{array}{cccccc} v_1 & v_2 & v_3 & v_4 & v_5 & v_6 \end{array}\\ \left[\begin{array}{cccccc} 1 & 1 & -1 & 0 & 0 & 0 \\ 0 & 0 & 1 & -1 & 1 & 0 \\ 0 & 0 & 0 & 0 & -1 & 1 \end{array}\right]\end{array}\begin{bmatrix} v_1 \\ v_2 \\ v_3 \\ v_4 \\ v_5 \\ v_6 \end{bmatrix} = \begin{bmatrix} v_1 + v_2 - v_3 \\ v_3 - v_4 + v_5 \\ -v_5 + v_6 \end{bmatrix} = 0
$$

We must note that instead of using an incidence matrix, we can use a cut-set matrix in Eq. 7-1-2 where

$$
QI_e = 0 \tag{7-1-5}
$$

because promultiplication of a nonsingular matrix to Eq. 7-1-2 will give this equation.

7-2 MESH TRANSFORMATION

When a system satisfies Kirchhoff's current law, branch currents cannot all be independent. One way of expressing all branch currents as a function of chosen independent branch currents is the *mesh transformation*.

Suppose we partition column matrix I_e in Eq. 7-1-1:

$$
I_e = \begin{bmatrix} I_c \\ I_t \end{bmatrix} \tag{7-2-1}
$$

such that I_c is a column matrix of all branch currents corresponding to the chords with respect to a forest (tree) t and I_t is a column matrix of all branch currents corresponding to the branches of forest (tree) t. Hence the number of entries in I_c is $n_e - n_v + \rho$ and the number of entries in I_t is $n_v - \rho$ where n_e is the number of edges, n_v is the number of vertices, and ρ is the number of maximal connected subgraphs of a weighted oriented linear graph. We also partition incidence matrix A:

$$A = [A_{11} \quad A_{12}] \tag{7-2-2}$$

where A_{11} consists of $n_e - n_v + \rho$ columns corresponding to the chords with respect to forest (tree) t and A_{12} consists of $n_v - \rho$ columns corresponding to the branches of forest (tree) t. Then, by Eq. 7-1-2,

$$AI_e = [A_{11} \quad A_{12}]\begin{bmatrix} I_c \\ I_t \end{bmatrix} = 0 \tag{7-2-3}$$

or

$$I_t = -A_{12}^{-1}A_{11}I_c \tag{7-2-4}$$

Thus

$$I_e = \begin{bmatrix} U \\ -A_{12}^{-1}A_{11} \end{bmatrix} I_c \tag{7-2-5}$$

Since we know that from Eq. 6-1-13 that with $B_{11} = U$,

$$B_f = [U \quad -(A_{12}^{-1}A_{11})^t] \tag{7-2-6}$$

we have

$$I_e = B_f{}^t I_c \tag{7-2-7}$$

Moreover, any circuit matrix B can be expressed as

$$B = DB_f \tag{7-2-8}$$

where D is a nonsingular matrix of order $n_e - n_v + \rho$, that is,

$$B = [B_{11} \quad B_{12}] = D[U \quad B_{f_{12}}] \tag{7-2-9}$$

Hence

$$B_{11} = D \tag{7-2-10}$$

is nonsingular. Thus

$$B_f = D^{-1}B = B_{11}^{-1}B \tag{7-2-11}$$

This, with Eq. 7-2-7, gives

$$I_e = B^t(B_{11}^{-1})^t I_c \qquad (7\text{-}2\text{-}12)$$

or

$$I_e = B^t M(I) \qquad (7\text{-}2\text{-}13)$$

where

$$M(I) = (B_{11}^{-1})^t I_c \qquad (7\text{-}2\text{-}14)$$

By substituting I_e of Eq. 7-2-13 in Eq. 7-1-2, we have

$$AI_e = AB^t M(I) = 0 \qquad (7\text{-}2\text{-}15)$$

Since AB^t is always zero, the Kirchhoff's current law satisfies even if $M(I)$ is a column matrix of an arbitrary set of $n_e - n_v + \rho$ functions. However, once $M(I)$ is chosen, I_e is fixed by Eq. 7-2-13 which is known as the *mesh transformation*. Because by transposing the matrices of both sides of Eq. 7-2-13, we have

$$I_e^{\ t} = [m_1 \quad m_2 \quad \cdots \quad m_{n_e - n_v + \rho}]B \qquad (7\text{-}2\text{-}16)$$

where

$$M(I)^t = [m_1 \quad m_2 \quad \cdots \quad m_{n_e - n_v + \rho}] \qquad (7\text{-}2\text{-}17)$$

We can see from this equation that element m_p (for $p = 1, 2, \ldots, n_e - n_v + \rho$) is multiplied by every entry of pth row of B. Hence m_p can be considered a weight given to the circuit corresponding to the pth row of circuit matrix B. Thus $M(I)$ can be considered a column matrix of so-called loop

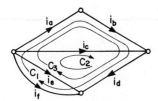

Fig. 7-2-1.   A weighted linear graph.

currents. For example, I_e of the weighted oriented linear graph in Fig. 7-2-1 can be obtained from $M(I)$ as follows where

$$M(I) = \begin{bmatrix} i_1 \\ i_2 \\ i_3 \end{bmatrix}$$

By choosing circuits C_1, C_2, and C_3 as shown in Fig. 7-2-1, we have

$$
B = \begin{array}{c} \\ C_1 \\ C_2 \\ C_3 \end{array}
\begin{array}{cccccc} a & b & c & d & e & f \\ \left[\begin{array}{cccccc} 1 & 1 & 0 & 1 & 0 & -1 \\ 0 & 0 & 1 & 1 & -1 & 0 \\ 1 & 1 & 0 & 1 & -1 & 0 \end{array}\right] \end{array}
$$

Then by using Eq. 7-2-13, we have

$$
I_e = B^t M(I) = \begin{bmatrix} 1 & 0 & 1 \\ 1 & 0 & 1 \\ 0 & 1 & 0 \\ 1 & 1 & 1 \\ 0 & -1 & -1 \\ -1 & 0 & 0 \end{bmatrix} \begin{bmatrix} i_1 \\ i_2 \\ i_3 \end{bmatrix} = \begin{array}{c} i_a \\ i_b \\ i_c \\ i_d \\ i_e \\ i_f \end{array} \begin{bmatrix} i_1 + i_3 \\ i_1 + i_3 \\ i_2 \\ i_1 + i_2 + i_3 \\ -i_2 - i_3 \\ -i_1 \end{bmatrix}
$$

It is important to note that if we choose $n_e - n_v + \rho$ loop currents corresponding to the linearly independent circuits indicated by the rows in a circuit matrix and use Eq. 7-2-13 to express branch currents, a system automatically satisfies Kirchhoff's current law.

7-3 NODE TRANSFORMATION

When a system satisfies Kirchhoff's voltage law, branch voltages are related to each other. To see this, let B_f be a fundamental circuit matrix of a weighted oriented linear graph with respect to a forest (tree) t. Also let column matrix V_e be partitioned as follows:

$$
V_e = \begin{bmatrix} V_c \\ V_t \end{bmatrix} \tag{7-3-1}
$$

where V_c consists of branch voltages which are the weights of chords with respect to forest (tree) t and V_t consists of branch voltages corresponding to branches of forest (tree) t. Thus V_c consists of $n_e - n_v + \rho$ entries and V_t consists of $n_v - \rho$. Then by Kirchhoff's voltage law (Eq. 7-1-3), we have

$$
B_c V_e = \begin{bmatrix} U & B_{c_{12}} \end{bmatrix} \begin{bmatrix} V_c \\ V_t \end{bmatrix} = 0 \tag{7-3-2}
$$

or

$$
V_c = -B_{c_{12}} V_t \tag{7-3-3}
$$

Hence

$$V_e = \begin{bmatrix} -B_{c_{12}} \\ U \end{bmatrix} V_t \tag{7-3-4}$$

Since from Eq. 6-1-12 with $B_{11} = U$ we know

$$B_{c_{12}} = -(A_{12}^{-1}A_{11})^t \tag{7-3-5}$$

we have

$$V_e = \begin{bmatrix} (A_{12}^{-1}A_{11})^t \\ U \end{bmatrix} V_t \tag{7-3-6}$$

which is equal to

$$V_e = \begin{bmatrix} A_{11}{}^t(A_{12}{}^{-1})^t \\ A_{12}{}^t(A_{12}{}^{-1})^t \end{bmatrix} V_t = \begin{bmatrix} A_{11}{}^t \\ A_{12}{}^t \end{bmatrix} (A_{12}{}^{-1})^t V_t$$

$$= A^t(A_{12}{}^{-1})^t V_t \tag{7-3-7}$$

or

$$V_e = A^t N(V) \tag{7-3-8}$$

where

$$N(V) = (A_{12}{}^{-1})^t V_t \tag{7-3-9}$$

Equation 7-3-8 is called the *node transformation*. It can be seen from

$$BV_e = BA^t N(V) = 0 \tag{7-3-10}$$

that (since $BA^t = 0$ for a linear graph) Kirchhoff's voltage law is satisfied even if $N(V)$ is the column matrix of any arbitrary set of $n_v - \rho$ functions. However, we must note that once $N(V)$ is chosen, V_e will be fixed by Eq. 7-3-8.

By transposing both sides of Eq. 7-3-8, we have

$$V_e{}^t = [n_1 \quad n_2 \quad \cdots \quad n_{n_v - \rho}]A \tag{7-3-11}$$

where

$$N(V)^t = [n_1 \quad n_2 \quad \cdots \quad n_{n_v - \rho}] \tag{7-3-12}$$

Equation 7-3-11 indicates that n_p will be multiplied by every entry of the pth row of A. Hence n_p can be considered a weight to the vertex corresponding to the pth row of A. Since the reference vertex is not represented by a row of A, n_p indicates a voltage from the vertex corresponding to the pth row of A to the reference vertex, which is known as a *node voltage*. For example,

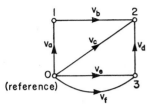

Fig. 7-3-1.   A weighted linear graph.

V_e of the weighted oriented graph in Fig. 7-3-1 can be obtained from $N(V)$ as follows where

$$N(V) = \begin{bmatrix} v_1 \\ v_2 \\ v_3 \end{bmatrix}$$

By choosing A as

$$
\begin{array}{c}
\begin{array}{cccccc} a & b & c & d & e & f \end{array} \\
\begin{array}{c} 1 \\ 2 \\ 3 \end{array}
\begin{bmatrix}
-1 & 1 & 0 & 0 & 0 & 0 \\
0 & -1 & -1 & -1 & 0 & 0 \\
0 & 0 & 0 & 1 & -1 & -1
\end{bmatrix}
\end{array}
$$

we have

$$V_e = A^t N(V) = \begin{bmatrix} -1 & 0 & 0 \\ 1 & -1 & 0 \\ 0 & -1 & 0 \\ 0 & -1 & 1 \\ 0 & 0 & -1 \\ 0 & 0 & -1 \end{bmatrix} \begin{bmatrix} v_1 \\ v_2 \\ v_3 \end{bmatrix} = \begin{bmatrix} v_a \\ v_b \\ v_c \\ v_d \\ v_e \\ v_f \end{bmatrix} \begin{bmatrix} -v_1 \\ v_1 - v_2 \\ -v_2 \\ -v_2 + v_3 \\ -v_3 \\ -v_3 \end{bmatrix}$$

If we use a cut-set matrix Q instead of incidence matrix A, Eq. 7-3-6 becomes

$$V_e = \begin{bmatrix} Q_{11}{}^t (Q_{12}{}^t)^{-1} \\ U \end{bmatrix} V_t = \begin{bmatrix} Q_{11}{}^t \\ Q_{12}{}^t \end{bmatrix} (Q_{12}{}^{-1})^t V_t = Q^t N(V) \quad (7\text{-}3\text{-}13)$$

where

$$N(V) = (Q_{12}{}^{-1})^t V_t \quad (7\text{-}3\text{-}14)$$

This is called a generalized node transformation where when Q is a fundamental cut-set matrix with respect to a forest (tree) t, the entries of $N(V)$

indicate all branch voltages of branches in t. For example, if we choose a tree in a linear graph in Fig. 7-3-1 as (a, c, d), then the fundamental cut-set matrix Q_f with respect to (a, c, d) is

$$Q_f = \begin{array}{c} \begin{array}{cccccc} b & e & f & a & c & d \end{array} \\ \begin{bmatrix} -1 & 0 & 0 & 1 & 0 & 0 \\ 1 & 1 & 1 & 0 & 1 & 0 \\ 0 & -1 & -1 & 0 & 0 & 1 \end{bmatrix} \end{array}$$

Then $\underline{N}(V)$ will be

$$\underline{N}(V) = \begin{bmatrix} v_a \\ v_c \\ v_d \end{bmatrix}$$

and by Eq. 7-3-13, we have

$$\begin{bmatrix} v_b \\ v_e \\ v_f \\ v_a \\ v_c \\ v_d \end{bmatrix} = \begin{bmatrix} -1 & 1 & 0 \\ 0 & 1 & -1 \\ 0 & 1 & -1 \\ 1 & 0 & 0 \\ 0 & 1 & 0 \\ 0 & 0 & 1 \end{bmatrix} \begin{bmatrix} v_a \\ v_c \\ v_d \end{bmatrix} = \begin{bmatrix} -v_a + v_c \\ v_c - v_d \\ v_c - v_d \\ v_a \\ v_c \\ v_d \end{bmatrix}$$

Note that when we choose $n_v - \rho$ linearly independent voltages corresponding to the rows of cut-set matrix Q (or incidence matrix) and express each branch voltage as a function of these chosen voltages by Eq. 7-3-13 (Eq. 7-3-8), then a system automatically satisfies Kirchhoff's voltage law.

7-4 DETERMINANT OF ADMITTANCE MATRIX

Suppose we use Kirchhoff's current law to write a system equation of an electrical network as

$$AI_e = 0 \tag{7-4-1}$$

(known as a node equation) where

$$I_e = f(J_1 \cdots J_p v_1 \cdots v_{n_e}) \tag{7-4-2}$$

$J_1, \ldots,$ and J_p are current generators and $v_1, \ldots,$ and v_{n_e} are branch voltages including voltage generators. Note that each branch current is a function of J's and v's by Eq. 7-4-2. Equation 7-4-1 does not represent an electrical network because an electrical network must satisfy both Kirchhoff's current and voltage laws. However, Eq. 7-4-1 forces a system to satisfy Kirchhoff's current law but not Kirchhoff's voltage law. For example, consider the electrical network in Fig. 7-4-1a whose corresponding linear graph is shown

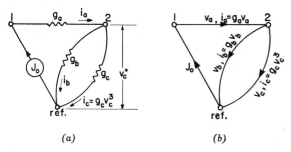

(a)   (b)

Fig. 7-4-1.   (a) An electrical network; (b) the corresponding linear graph.

in Fig. 7-4-1b where g_c indicates a nonlinear resistor characterized by $i_c = g_c v_c{}^3$. An incidence matrix A will be

$$A = \begin{array}{c} \\ 1 \\ 2 \end{array} \begin{array}{c} J_0 \quad g_a \; g_b \; g_c \\ \begin{bmatrix} -1 & 1 & 0 & 0 \\ 0 & -1 & 1 & 1 \end{bmatrix} \end{array}$$

and current vector I_e will be

$$I_e = \begin{bmatrix} J_0 \\ i_a \\ i_b \\ i_c \end{bmatrix} = \begin{bmatrix} J_0 \\ g_a v_a \\ g_b v_b \\ g_c v_c{}^3 \end{bmatrix}$$

where v_a, v_b, and v_c are the branch voltages of g_a, g_b, and g_c, respectively. Then

$$\begin{bmatrix} -1 & 1 & 0 & 0 \\ 0 & -1 & 1 & 1 \end{bmatrix} \begin{bmatrix} J_0 \\ i_a \\ i_b \\ i_c \end{bmatrix} = \begin{bmatrix} -J_0 + i_a \\ -i_a + i_b + i_c \end{bmatrix} = \begin{bmatrix} -J_0 + g_a v_a \\ -g_a v_a + g_b v_b + g_c v_c{}^3 \end{bmatrix} = 0$$

which means the system satisfies Kirchhoff's current law but not Kirchhoff's voltage law.

In order that a system satisfy Kirchhoff's voltage law, we need to use the node transformation given in the previous section. For the preceding example, if we take $\underline{A}$ where

$$\underline{A} = \begin{array}{c} \\ 1 \\ 2 \end{array} \overset{\begin{array}{ccc} g_a & g_b & g_c \end{array}}{\left[\begin{array}{ccc} 1 & 0 & 0 \\ -1 & 1 & 1 \end{array} \right]}$$

which is an incidence matrix of a network without current source J_0, the node transformation will be

$$\begin{bmatrix} v_a \\ v_b \\ v_c \end{bmatrix} = \underline{A}^t \begin{bmatrix} v_1 \\ v_2 \end{bmatrix} = \begin{bmatrix} v_1 - v_2 \\ v_2 \\ v_2 \end{bmatrix}$$

Substituting this into $AI_e = 0$, we have

$$AI_e = \begin{bmatrix} -J_0 + g_a(v_1 - v_2) \\ -g_a(v_1 - v_2) + g_b v_2 + g_v v_2{}^3 \end{bmatrix} = 0$$

which is the system equation for the network in Fig. 7-4-1.

This procedure also holds when we use a generalized Kirchhoff's current law, which is

$$QI_e = 0 \qquad\qquad (7\text{-}4\text{-}3)$$

where Q is a cut-set matrix. Similarly, we can use Kirchhoff's voltage law to write a system equation

$$BV_e = 0 \qquad\qquad (7\text{-}4\text{-}4)$$

where

$$V_e = f(E_1 \cdots E_p, i_1 \cdots i_{n_e}) \qquad\qquad (7\text{-}4\text{-}5)$$

$E_1, \ldots, E_p$ are voltage generators, and $i_1, \ldots, i_{n_e}$ are branch currents including current generators. Again, this equation is not enough to express an electrical network unless we replace $i_1, \ldots, i_{n_e}$ according to Kirchhoff's current law. A simple way is to employ the mesh transformation of the previous section.

When each component y_k of an electrical network can be expressed as

$$y_k v_k = i_k \qquad\qquad (7\text{-}4\text{-}6)$$

where i_k and v_k are the branch current and the branch voltage of component y_k, then the analysis becomes rather simple. In this section, we develop the so-called topological analysis of electrical networks in which each component satisfies Eq. 7-4-6.

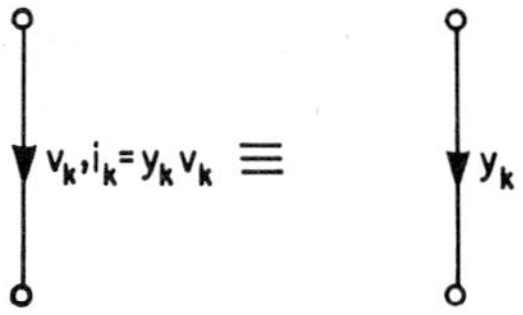

Fig. 7-4-2. Weights of an edge.

For convenience, when the weights i_k and v_k of an edge as shown in Fig. 7-4-2 satisfy Eq. 7-4-6, then we use y_k as the weight of edge k; that is, specifying weight y_k for edge k is equivalent to specifying two weights i_k and v_k which satisfy Eq. 7-4-6 as shown in Fig. 7-4-2. The relationship given by Eq. 7-4-6 is valid to represent a linear, lumped, time invariant component with its initial condition being zero. A component can be a resistance, an inductance, a capacitance, or a two-terminal electrical network as long as its initial condition is zero.

Consider an electrical network corresponding to a weighted linear graph in which each edge has weight y_k (admittance). Then we have one equation for each edge as follows:

$$
\begin{aligned}
y_1 v_1 &= i_1 \\
y_2 v_2 &= i_2 \\
&\vdots \\
y_{n_e} v_{n_e} &= i_{n_e}
\end{aligned}
\tag{7-4-7}
$$

This can be expressed as

$$
\begin{bmatrix}
y_1 & & & \\
 & y_2 & & 0 \\
 & & y_3 & \\
0 & & & \ddots \\
 & & & & y_{n_e}
\end{bmatrix}
\begin{bmatrix}
v_1 \\ v_2 \\ v_3 \\ \vdots \\ v_{n_e}
\end{bmatrix}
=
\begin{bmatrix}
i_1 \\ i_2 \\ i_3 \\ \vdots \\ i_{n_e}
\end{bmatrix}
\tag{7-4-8}
$$

For convenience, we use the symbols Y_e, V_e, and I_e to represent matrices in Eq. 7-4-8 so that the equation can be written as

$$
Y_e V_e = I_e
\tag{7-4-9}
$$

Note that Y_e is a diagonal matrix. Let A be an incidence matrix of a given linear graph. Then from Eq. 7-1-2 (in order to satisfy Kirchhoff's current law), we have

$$
A I_e = A Y_e V_e = 0
\tag{7-4-10}
$$

For convenience, suppose we use symbol V_n rather than $N(V)$ to represent a column matrix of node voltages in Eq. 7-3-8. Then by using a node transformation given by Eq. 7-3-8, we change Eq. 7-4-10 to

$$A Y_e V_e = A Y_e A^t V_n = 0 \qquad (7\text{-}4\text{-}11)$$

The matrix $A Y_e A^t$ is known as a *node admittance matrix of an electrical network*. In order to relate the determinant of a node admittance matrix and the subgraphs of a given linear graph, it is convenient to know the following Binet-Cauchy theorem.

Suppose

$$\begin{bmatrix} a_{11} + b_{11} & a_{12} & a_{13} & \cdots & a_{1n} \\ a_{21} + b_{21} & a_{22} & a_{23} & \cdots & a_{2n} \\ & & \cdots & & \\ a_{n1} + b_{n1} & a_{n2} & a_{n3} & \cdots & a_{nn} \end{bmatrix}$$

Then from the properties of determinants, we know that it is equal to

$$\begin{vmatrix} a_{11} + b_{11} & a_{22} & \cdots & a_{1n} \\ a_{21} + b_{21} & a_{22} & \cdots & a_{2n} \\ & \cdots & & \\ a_{n1} + b_{n1} & a_{n2} & \cdots & a_{nn} \end{vmatrix} = \begin{vmatrix} a_{11} & a_{12} & \cdots & a_{1n} \\ a_{21} & a_{22} & \cdots & a_{2n} \\ & \cdots & & \\ a_{n1} & a_{n2} & \cdots & a_{nn} \end{vmatrix} + \begin{vmatrix} b_{11} & a_{12} & \cdots & a_{1n} \\ b_{21} & a_{22} & \cdots & a_{2n} \\ & \cdots & & \\ b_{n1} & a_{n2} & \cdots & a_{nn} \end{vmatrix}$$

$$(7\text{-}4\text{-}12)$$

In general, if we have

$$R = [(r_{11} + r_{21})(r_{12} + r_{22}) \cdots (r_{1n} + r_{2n})] \qquad (7\text{-}4\text{-}13)$$

where

$$r_{1k} = \begin{bmatrix} a_{1k} \\ a_{2k} \\ \vdots \\ a_{nk} \end{bmatrix} \qquad \text{and} \qquad r_{2k} = \begin{bmatrix} b_{1k} \\ b_{2k} \\ \vdots \\ b_{nk} \end{bmatrix} \qquad (7\text{-}4\text{-}14)$$

then

$$|R| = \sum |r_{j_1 1} r_{j_2 2} \cdots r_{j_n n}| \qquad (7\text{-}4\text{-}15)$$

where $\sum$ means the summation of all possible determinants of all possible matrices formed by taking either r_{1k} or r_{2k} as the kth column for all $k = 1, 2, \ldots, n$.

Now consider two matrices A and B where $A = [a_{ij}]_{nm}$ and $B = [b_{ij}]_{mn}$. The product of A and B is

$$AB = \begin{bmatrix} \sum_{j_1=1}^{m} a_{1j_1}b_{j_11} & \sum_{j_2=1}^{m} a_{1j_2}b_{j_22} & \cdots & \sum_{j_n=1}^{m} a_{1j_n}b_{j_nn} \\ \sum_{j_1=1}^{m} a_{2j_1}b_{j_11} & \sum_{j_2=1}^{m} a_{2j_2}b_{j_22} & \cdots & \sum_{j_n=1}^{m} a_{2j_n}b_{j_nn} \\ & & \cdots & \\ \sum_{j_1=1}^{m} a_{nj_1}b_{j_11} & \sum_{j_2=1}^{m} a_{nj_2}b_{j_22} & \cdots & \sum_{j_n=1}^{m} a_{nj_n}b_{j_nn} \end{bmatrix} \quad (7\text{-}4\text{-}16)$$

Now from Eq. 7-4-15 we can see that

$$|AB| = \sum_{\substack{j_p=1 \\ \text{for all } p}}^{m} \begin{vmatrix} a_{1j_1}b_{j_11} & a_{1j_2}b_{j_22} & \cdots & a_{1j_n}b_{j_nn} \\ a_{2j_1}b_{j_11} & a_{2j_2}b_{j_22} & \cdots & a_{2j_n}b_{j_nn} \\ & & \cdots & \\ a_{nj_1}b_{j_11} & a_{nj_2}b_{j_22} & \cdots & a_{nj_n}b_{j_nn} \end{vmatrix} \quad (7\text{-}4\text{-}17)$$

or

$$|AB| = \sum_{\substack{j_p=1 \\ \text{for all } p}}^{m} \begin{vmatrix} a_{1j_1} & a_{1j_2} & \cdots & a_{1j_n} \\ a_{2j_1} & a_{2j_2} & \cdots & a_{2j_n} \\ & & \cdots & \\ a_{nj_1} & a_{nj_2} & \cdots & a_{nj_n} \end{vmatrix} (b_{j_11}b_{j_22}\cdots b_{j_nn}) \quad (7\text{-}4\text{-}18)$$

If there are two j's that take the same number, then there will be two columns in the right-hand side of Eq. 7-4-18 which are identical. Thus we only need to make the summation of the determinant of matrices that have different j's. Hence

$$|AB| = \sum_{\substack{j_p=1 \\ j_1 \neq j_2 \neq \ldots \neq j_n}}^{m} \begin{vmatrix} a_{1j_1} & a_{1j_2} & \cdots & a_{1j_n} \\ a_{2j_1} & a_{2j_2} & \cdots & a_{2j_n} \\ & & \cdots & \\ a_{nj_1} & a_{nj_2} & \cdots & a_{nj_n} \end{vmatrix} (b_{j_11}b_{j_22}\cdots b_{j_nn}) \quad (7\text{-}4\text{-}19)$$

Note that when $n > m$, it is impossible to choose all distinct j's. Hence $|AB|$ is zero.

Consider the case when j's are chosen from the set of numbers $(1, 2, \ldots, n)$ so that all j's are distinct. One particular choice would be

$$j_p = p \quad \text{for} \quad p = 1, 2, \ldots, n \quad (7\text{-}4\text{-}20)$$

With this choice, the right-hand side of Eq. 7-4-19 becomes

$$\begin{vmatrix} a_{11} & a_{12} & \cdots & a_{1n} \\ a_{21} & a_{22} & \cdots & a_{2n} \\ & & \cdots & \\ a_{n1} & a_{n2} & \cdots & a_{nn} \end{vmatrix}(b_{11}b_{22}\cdots b_{nn})$$

The different choice of j's such as

$$j_p = \begin{cases} p+1 & \text{for } p = 1, 2, \ldots, n-1 \\ 1 & \text{for } p = n \end{cases} \tag{7-4-21}$$

will give

$$\begin{vmatrix} a_{12} & a_{13} & \cdots & a_{1n} & a_{11} \\ a_{22} & a_{23} & \cdots & a_{2n} & a_{21} \\ & & \cdots & & \\ a_{n2} & a_{n3} & \cdots & a_{nn} & a_{n1} \end{vmatrix}(b_{21}b_{32}\cdots b_{nn-1}b_{1n})$$

which can be rearranged so that the determinant of the submatrix of A will have the same arrangement as the previous one. The change of signs caused by such rearrangement of columns is clearly related to the permutation of j's. Hence by considering all possible choices of j to have numbers from 1 to n, we can see that

$$\sum_{j_1 \neq j_2 \neq \ldots \neq j_n = 1}^{n} \begin{vmatrix} a_{1j_1} & a_{1j_2} & \cdots & a_{1j_n} \\ a_{2j_1} & a_{2j_2} & \cdots & a_{2j_n} \\ & & \cdots & \\ a_{nj_1} & a_{nj_2} & \cdots & a_{nj_n} \end{vmatrix}(b_{j_11}b_{j_22}\cdots b_{j_nn})$$

$$= \begin{vmatrix} a_{11} & a_{12} & \cdots & a_{1n} \\ a_{21} & a_{22} & \cdots & a_{2n} \\ & & \cdots & \\ a_{n1} & a_{n2} & \cdots & a_{nn} \end{vmatrix} \sum_{j_1 \neq j_2 \neq \ldots \neq j_n}^{n} \epsilon_{j_1 j_2 \ldots j_n}(b_{j_11}b_{j_22}\cdots b_{j_nn})$$

$$= \begin{vmatrix} a_{11} & a_{12} & \cdots & a_{1n} \\ a_{21} & a_{22} & \cdots & a_{2n} \\ & & \cdots & \\ a_{n1} & a_{n2} & \cdots & a_{nn} \end{vmatrix} \begin{vmatrix} b_{11} & b_{12} & \cdots & b_{1n} \\ b_{21} & b_{22} & \cdots & b_{2n} \\ & & \cdots & \\ b_{n1} & b_{n2} & \cdots & b_{nn} \end{vmatrix} \tag{7-4-22}$$

where $\epsilon_{j_1 \ldots j_n}$ is 1 if the number of permutations necessary to change $j_1 \cdots j_n$

in the natural order is even; otherwise $\epsilon_{j_1 \ldots j_n}$ is -1. This is true for any distinct choice of j. Hence we have the *Binet-Cauchy theorem*, which is

$$|AB| = \sum \text{major determinant of } A \times \text{corresponding} $$
$$\text{major determinant of } B \qquad (7\text{-}4\text{-}23)$$

where $\sum$ means "for all possible." Note that a major determinant of A is the determinant of an n by n submatrix of A. If a major determinant of A is the determinant of a submatrix of A obtained by taking columns $j_1, j_2, \ldots, j_n$, then the corresponding major determinant of B is the determinant of a submatrix of B formed by rows $j_1, j_2, \ldots, j_n$.

For example, suppose A and B are given so that

$$AB = \begin{bmatrix} 1 & 0 & 1 \\ 0 & 2 & 2 \end{bmatrix} \begin{bmatrix} 1 & 0 \\ 2 & 1 \\ 3 & 0 \end{bmatrix}$$

Then the determinant of AB (by using Eq. 7-4-23) is

$$\left| \begin{bmatrix} 1 & 0 & 1 \\ 0 & 2 & 2 \end{bmatrix} \begin{bmatrix} 1 & 0 \\ 2 & 1 \\ 3 & 0 \end{bmatrix} \right| = \begin{vmatrix} 1 & 0 \\ 0 & 2 \end{vmatrix} \begin{vmatrix} 1 & 0 \\ 2 & 1 \end{vmatrix} + \begin{vmatrix} 1 & 1 \\ 0 & 2 \end{vmatrix} \begin{vmatrix} 1 & 0 \\ 3 & 0 \end{vmatrix}$$
$$+ \begin{vmatrix} 0 & 1 \\ 2 & 2 \end{vmatrix} \begin{vmatrix} 2 & 1 \\ 3 & 0 \end{vmatrix} = 8$$

Now we are ready to go back to discuss node admittance matrices. By expressing $A Y_e A^t$ as

$$A Y_e A^t = H A^t \qquad (7\text{-}4\text{-}24)$$

where

$$A Y_e = H \qquad (7\text{-}4\text{-}25)$$

we can use the Binet-Cauchy theorem:

$$|HA^t| = \sum \text{major determinant of } H \times \text{corresponding} $$
$$\text{major determinant of } A^t \qquad (7\text{-}4\text{-}26)$$

Since Y_e is a diagonal matrix,

$$H = [H_1 \quad H_2 \quad \cdots \quad H_{n_e}] = [A_1 \quad A_2 \quad \cdots \quad A_{n_e}] \begin{bmatrix} y_1 & & & \\ & y_2 & & 0 \\ & & \ddots & \\ 0 & & & y_{n_e} \end{bmatrix}$$
$$= [y_1 A_1 \quad y_2 A_2 \quad \cdots \quad y_{n_e} A_{n_e}] \qquad (7\text{-}4\text{-}27)$$

Hence the determinant of any major submatrix H_m of H can be expressed as

$$|H_m| = |[H_{j_1} \quad H_{j_2} \quad \cdots \quad H_{j_m}]| = |[y_{j_1}A_{j_1} \quad y_{j_2}A_{j_2} \quad \cdots \quad y_{j_m}A_{j_m}]|$$
$$= (y_{j_1}y_{j_2}\cdots y_{j_m})|[A_{j_1} \quad A_{j_2} \quad \cdots \quad A_{j_m}]| \qquad (7\text{-}4\text{-}28)$$

Thus by using the symbol $A(j_1 j_2 \cdots j_m)$ to indicate a submatrix of A consisting of columns $j_1, j_2, \ldots, j_m$, we can express Eq. 7-4-26 as

$$|A Y_e A^t| = \sum_{(j)} (y_{j_1}y_{j_2}\cdots y_{j_m})|A(j_1 j_2 \cdots j_m)| \, |A(j_1 j_2 \cdots j_m)^t| \quad (7\text{-}4\text{-}29)$$

Since the determinant of a square matrix and the determinant of transposed of the same matrix are the same, Eq. 7-3-29 is equal to

$$|A Y_e A^t| = \sum_{(j)} (y_{j_1}y_{j_2}\cdots y_{j_m})|A(j_1 j_2 \cdots j_m)|^2 \qquad (7\text{-}4\text{-}30)$$

Previously we saw that a major submatrix of an incidence matrix A is nonsingular if and only if the edges corresponding to the columns of the submatrix form a tree (a forest in the case of separated linear graphs). We also know that the determinant of such a nonsingular submatrix is either $+1$ or -1. Since, in Eq. 7-4-30, the determinant of a major submatrix of A is squared and corresponding admittances (weights of edges corresponding to the columns of a major submatrix) are in the form of product, we can state that

$$|A Y_e A^t| = \sum \text{tree admittance product} \qquad (7\text{-}4\text{-}31)$$

where $\sum$ means "for all possible."

Definition 7-4-1. A *tree admittance product* is the product of all admittances which are the weights of edges in a tree.

Theorem 7-4-1. The determinant of a node admittance matrix of an electrical network consisting of passivelike elements is equal to the sum of all possible tree admittance products, where a *passivelike* element is one that can be represented by one edge as shown in Fig. 7-4-1.

For convenience, *the symbol Δ' is employed to represent the determinant of node admittance matrix.* Let N be a network consisting of passivelike elements. Let G be the corresponding oriented weighted graph of N. Then it can be seen from Eq. 7-4-31 that changing the orientation of edges in G does not influence Δ'. We can find all possible trees directly from network N rather than from the corresponding oriented weighted graph G. Hence it is not necessary to draw G from a given network N to obtain Δ'. For example,

Δ' of the network in Fig. 7-4-3 can be obtained as follows. All possible trees are

$$(g_a, C_b s), \left(g_a, \frac{1}{L_c s}\right), (g_a, C_d s), \left(C_b s, \frac{1}{L_c s}\right), \quad \text{and} \quad (C_b s, C_d s)$$

Hence

$$\Delta' = C_b C_d s^2 + g_a(C_b + C_d)s + \frac{C_b}{L_c} + \frac{g_a}{L_c s}$$

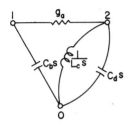

Fig. 7-4-3. An electrical network N.

In Eq. 7-4-31, if we set every admittance equal to 1, each tree admittance product is 1. Hence

$$\Delta'_{\text{with all } y\text{'s being 1}} = \text{number of trees} \qquad (7\text{-}4\text{-}32)$$

On the other hand, from Eq. 7-4-30,

$$\Delta'_{\text{with all } y\text{'s being 1}} = |A U A^t| = |A A^t| \qquad (7\text{-}4\text{-}33)$$

Let

$$A = \begin{bmatrix} A_1 \\ A_2 \\ \vdots \\ A_m \end{bmatrix}$$

Then

$$AA^t = \begin{bmatrix} A_1 \\ A_2 \\ \vdots \\ A_m \end{bmatrix} \begin{bmatrix} A_1^t & A_2^t & \cdots & A_m^t \end{bmatrix} = \begin{bmatrix} A_1 A_1^t & A_1 A_2^t & \cdots & A_1 A_m^t \\ A_2 A_1^t & A_2 A_2^t & \cdots & A_2 A_m^t \\ & & \cdots & \\ A_m A_1^t & A_m A_2^t & \cdots & A_m A_m^t \end{bmatrix}$$

$$(7\text{-}4\text{-}34)$$

Let $A_k = [a_{k1} \quad a_{k2} \quad \cdots \quad a_{kn_e}]$. Then

$$|A_k A_k^t| = \sum_{p=1}^{n_e} a_{kp}^2 \tag{7-4-35}$$

Since a_{kp} is nonzero only if edge p is incident at vertex k, by the definition of incidence matrices, $|A_k A_k^t|$ is equal to the number of edges incident at vertex k, or

$$|A_k A_k^t| = \text{number of edges incident at vertex } k \tag{7-4-36}$$

On the other hand,

$$|A_k A_q^t| = \sum_{p=1}^{n_e} a_{kp} a_{qp} \tag{7-4-37}$$

It can be seen that $a_{kp} a_{qp}$ is nonzero if and only if edge p is incident at both vertices k and q. Hence edge p must be connected between vertices k and q. Furthermore, if edge p is connected between vertices k and q, the sign of a_{kp} is opposite to that of a_{qp} (by the definition of incidence matrices). Thus

$$a_{kp} a_{qp} = \begin{cases} -1 & \text{if edge } i \text{ is connected between} \\ & \text{vertices } k \text{ and } q \\ 0 & \text{otherwise} \end{cases} \tag{7-4-38}$$

Hence

$$|A_k A_q^t| = -(\text{number of edges connected between vertices } k \text{ and } q) \tag{7-4-39}$$

Equations 7-4-32 and 7-4-39 give the following theorem.

Theorem 7-4-2. Let matrix T be

$$T = [t_{ij}] = AA^t \tag{7-4-40}$$

Then t_{ii} is the number of edges connected to vertex i, and t_{ij} is the negative of the number of edges connected between the vertices i and j. Furthermore,

$$|T| = \text{number of trees} \tag{7-4-41}$$

For example, the number of trees in a network in Fig. 7-4-3 can be obtained by

$$|T| = \begin{matrix} & \begin{matrix} 1 & \quad 2 \end{matrix} \\ \begin{matrix} 1 \\ 2 \end{matrix} & \left| \begin{bmatrix} 2 & -1 \\ -1 & 3 \end{bmatrix} \right| \end{matrix} = 5$$

Instead of using incidence matrices, if we use cut-set matrices as in Eqs.

7-1-5 and 7-3-13, Equation 7-4-31 will be changed slightly. To show this, recall the Kirchhoff's current law, which is

$$AI_e = 0 \qquad (7\text{-}4\text{-}42)$$

By multiplying a nonsingular matrix D, we have

$$DAI_e = Q_1 I_e = 0 \qquad (7\text{-}4\text{-}43)$$

Hence, we can use a cut-set matrix Q_1 to satisfy the Kirchhoff's current law. Thus Eq. 7-4-43, with Eq. 7-4-9, gives

$$Q_1 Y_e V_e = 0 \qquad (7\text{-}4\text{-}44)$$

The node transformation given by Eq. 7-3-13 with this equation gives

$$Q_1 Y_e Q^t N'(V) = 0 \qquad (7\text{-}4\text{-}45)$$

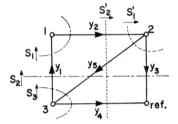

Fig. 7-4-4. A weighted oriented linear graph.

For example, consider the weighted oriented linear graph in Fig. 7-4-4. An incidence matrix A can be chosen as

$$
A = \begin{array}{c}
 \\
1 \\
2 \\
3
\end{array}
\begin{array}{ccccc}
y_1 & y_2 & y_3 & y_4 & y_5 \\
\left[\begin{array}{ccccc}
-1 & 1 & 0 & 0 & 0 \\
0 & -1 & 1 & 0 & 1 \\
1 & 0 & 0 & 1 & -1
\end{array}\right]
\end{array}
$$

Hence a node admittance matrix would be

$$
A Y_e A^t = \begin{bmatrix}
-1 & 1 & 0 & 0 & 0 \\
0 & -1 & 1 & 0 & 1 \\
1 & 0 & 0 & 1 & -1
\end{bmatrix}
\begin{bmatrix}
y_1 & & & & \\
& y_2 & & & 0 \\
& & y_3 & & \\
& 0 & & y_4 & \\
& & & & y_5
\end{bmatrix}
\begin{bmatrix}
-1 & 0 & 1 \\
1 & -1 & 0 \\
0 & 1 & 0 \\
0 & 0 & 1 \\
0 & 1 & -1
\end{bmatrix}
$$

$$
= \begin{bmatrix}
y_1 + y_2 & -y_2 & -y_1 \\
-y_2 & y_2 + y_3 + y_5 & -y_5 \\
-y_1 & -y_5 & y_1 + y_4 + y_5
\end{bmatrix}
$$

Choosing Q_1 and Q in Eq. 7-4-45 as

$$
Q_1 = \begin{array}{c} \\ S_1 \\ S_2 \\ S_3 \end{array}
\begin{array}{ccccc} y_1 & y_2 & y_3\ y_4 & y_5 \\ \left[\begin{array}{ccccc} 1 & -1 & 0 & 0 & 0 \\ 1 & 0 & -1 & 0 & -1 \\ 1 & 0 & 0 & 1 & -1 \end{array}\right] \end{array}
$$

and

$$
Q = \begin{array}{c} \\ S_1' \\ S_2' \\ S_3 \end{array}
\begin{array}{ccccc} y_1\ y_2 & y_3\ y_4 & y_5 \\ \left[\begin{array}{ccccc} 0 & 1 & -1 & 0 & -1 \\ 0 & 1 & 0 & 1 & -1 \\ 1 & 0 & 0 & 1 & -1 \end{array}\right] \end{array}
$$

we have an admittance matrix $Q_1 Y_e Q^t$:

$$
Q_1 Y_e Q^t = \left[\begin{array}{ccc} -y_2 & -y_2 & y_1 \\ y_3 + y_5 & y_5 & y_1 + y_5 \\ y_5 & y_4 + y_5 & y_1 + y_4 + y_5 \end{array}\right]
$$

This is not a node admittance matrix; however, this is a coefficient matrix of linearly independent equations which represents a given network.
    Let

$$
Q_1 = DA \tag{7-4-46}
$$

and

$$
Q = D'A \tag{7-4-47}
$$

Then we can see from Eqs. 7-4-29, 7-4-30, and 7-4-31 that

$$
Q_1 Y_e Q^t = |D|\,|D'| \sum \text{tree admittance product} \tag{7-4-48}
$$

Moreover, if Q_1 and Q are fundamental cut-set matrices, then $|D|$ and $|D'|$ are 1. Hence the determinant of $Q_1 Y_e Q^t$ is the same as that of $A Y_e A^t$.

7-5   OPEN-CIRCUIT NETWORK FUNCTIONS

In order to analyze electrical networks, consider the network shown in Fig. 7-5-1 in which there is a current generator J_k connected from the reference vertex (reference vertices when $\rho > 1$) to every vertex k for $k = 1, 2, \ldots, m$. Suppose all components in network N excluding these current

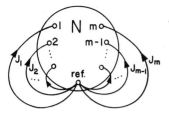

Fig. 7-5-1. A network N with current generators.

generators are passivelike elements (a passivelike element is an element that can be represented by one edge as shown in Fig. 7-4-2). Then we have

$$AI_e = J_n \tag{7-5-1}$$

where

$$J_n = \begin{bmatrix} J_1 \\ J_2 \\ \vdots \\ J_m \end{bmatrix} \tag{7-5-2}$$

and A is an incidence matrix of N itself with the above reference vertex. Note that $m = n_v - \rho$ where n_v is the number of vertices and ρ is the number of maximal connected subgraphs in the network.

Owing to the existence of current generators $J_1, \ldots, J_m$, we modify Eq. 7-4-11 to satisfy Kirchhoff's current law:

$$AY_eA^tV_n = J_n \tag{7-5-3}$$

Premultiplying both sides of Eq. 7-5-3 by $(AY_eA^t)^{-1}$, we have

$$V_n = (AY_eA^t)^{-1}J_n \tag{7-5-4}$$

Since each entry in $(AY_eA^t)^{-1}$ of this equation indicates an open-circuit network function and we have already learned the topological formula to obtain the determinant of AY_eA^t directly from a network, if we know how to obtain cofactors of matrix AY_eA^t directly from a network, we can obtain open-circuit network functions between any vertices topologically.

For convenience, Δ'_{pq} is employed to represent a cofactor of AY_eA^t which is $(-1)^{p+q}$ times the determinant of submatrix M_{pq} obtained by deleting the pth row and the qth column of AY_eA^t. In other words,

$$\Delta'_{pq} = (-1)^{p+q}|M_{pq}| \tag{7-5-5}$$

By defining

$$H = AY_e \tag{7-5-6}$$

we have

$$HA^t = \begin{bmatrix} H_1A_1{}^t & H_1A_2{}^t & \cdots & H_1A_m{}^t \\ H_2A_1{}^t & H_2A_2{}^t & \cdots & H_2A_m{}^t \\ & & \cdots & \\ H_mA_1{}^t & H_mA_2{}^t & \cdots & H_mA_m{}^t \end{bmatrix} \qquad (7\text{-}5\text{-}7)$$

where

$$H = \begin{bmatrix} H_1 \\ H_2 \\ \vdots \\ H_m \end{bmatrix} \qquad (7\text{-}5\text{-}8)$$

and

$$A = \begin{bmatrix} A_1 \\ A_2 \\ \vdots \\ A_m \end{bmatrix} \qquad (7\text{-}5\text{-}9)$$

The qth column of HA^t is

$$\begin{bmatrix} H_1A_q{}^t \\ H_2A_q{}^t \\ \vdots \\ H_mA_q{}^t \end{bmatrix} = \begin{bmatrix} H_1 \\ H_2 \\ \vdots \\ H_m \end{bmatrix} A_q{}^t \qquad (7\text{-}5\text{-}10)$$

and there is no other entry of HA^t that contains $A_q{}^t$. Thus the deletion of the qth column of HA^t can be achieved by deleting the qth row of A. Similarly, the pth row of HA^t is

$$[H_pA_1{}^t \quad H_pA_2{}^t \quad \cdots \quad H_pA_m{}^t] = H_p[A_1{}^t \quad A_2{}^t \quad \cdots \quad A_m{}^t] \qquad (7\text{-}5\text{-}11)$$

and there is no other entry in HA^t which contains H_p. Thus deleting the pth row of HA^t is equivalent to deleting the pth row of H. However,

$$H = \begin{bmatrix} A_1 \\ A_2 \\ \vdots \\ A_m \end{bmatrix} \begin{bmatrix} y_1 & & & \\ & y_2 & & 0 \\ & & \ddots & \\ 0 & & & y_{n_e} \end{bmatrix} \qquad (7\text{-}5\text{-}12)$$

Thus deleting the pth row of H is equivalent to deleting the pth row of A. Hence if we use the symbol A_{-r} to indicate a submatrix of A obtained by deleting the rth row, then the submatrix of $A Y_e A^t$ obtained by deleting the pth row and the qth column is equal to $A_{-p} Y_e (A_{-q})^t$. Now Eq. 1-5-5 can be written as

$$\Delta'_{pq} = (-1)^{p+q} |A_{-p} Y_e (A_{-q})^t| \qquad (7\text{-}5\text{-}13)$$

First we study a particular cofactor Δ'_{pp}. In this case, Eq. 7-5-13 becomes

$$\Delta'_{pp} = |A_{-p} Y_e (A_{-p})^t| \qquad (7\text{-}5\text{-}14)$$

Let vertex j be the reference vertex of weighted oriented graph G corresponding to a network (which means that in incidence matrix A there is no row corresponding to vertex j). In the case of separated linear graph, there are several reference vertices. However, as far as network functions are concerned, these vertices can be coincided without changing the network functions. Hence we may consider without the loss of generality that a weighted oriented graph is connected.

We have seen that A_{-p} is an incidence matrix of linear graph $G(p = j)$ obtained by coinciding vertices j and p in G. Thus by Eq. 7-4-31, Eq. 7-5-14 can be expressed as

$$|A_{-p} Y_e (A_{-p})^t| = \sum \text{tree admittance product of } G(p = j) \qquad (7\text{-}5\text{-}15)$$

Let a tree in $G(p = j)$ be $t(p = j)$. Since $G(p = j)$ consists of $n'_v = n_v - 1$ vertices, $t(p = j)$ consists of $n'_v - 1 = n_v - 2$ edges of $G(p = j)$. Hence subgraph g_s consisting of edges in $t(p = j)$ of G is not connected, but it will be connected when vertices p and j are coincided. Therefore we know g_s consists of two parts, each of which is connected, with one containing vertex p and the other vertex j. In other words, subgraph g_s becomes a tree when vertices p and j are coincided. This tree $t(p = j)$ of $G(p = j)$ is called a *2-tree* of G and is symbolized by $t_{2_{p,j}}$.

Definition 7-5-1. A *2-tree* $t_{2_{p,j}}$ of a connected linear graph G is a subgraph which satisfies the following conditions:

1. $t_{2_{p,j}}$ consists of $n_v - 2$ edges, contains all vertices of G, and contains no circuits.

2. $t_{2_{p,j}}$ consists of two connected parts including the case when a part consists of one isolated vertex.

3. One of two parts contains vertex p and the other contains vertex j.

With this definition, Eq. 7-5-15 can be written as

$$|A_p Y_e (A_{-p})^t| = \Delta'_{pp} = \sum 2\text{-tree } t_{2_{p,j}} \text{ admittance product} \qquad (7\text{-}5\text{-}16)$$

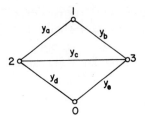

Fig. 7-5-2. A weighted linear graph G.

For example, all possible 2-trees $t_{2_{1,0}}$ of weighted linear graph G in Fig. 7-5-2 are

$(y_a y_b)$, $(y_a y_c)$, $(y_b y_c)$, $(y_a y_e)$, $(y_b y_d)$, $(y_c y_d)$, $(y_c y_e)$, and $(y_a y_e)$

Hence with reference vertex 0, Δ'_{11} is

$$\Delta'_{11} = y_a(y_b + y_c + y_e) + y_b(y_c + y_d) + y_c(y_d + y_e) + y_d y_e$$

We now can define a symbol we shall use later.

Definition 7-5-2. The symbol $t_{2_{a_1 a_2 \ldots a_p, b_1 b_2 \ldots b_q}}$ denotes a subgraph representing a 2-tree of a linear graph having the following properties:

1. One part of the two connected parts contains vertices $a_1, a_2, \ldots, a_p$ but not vertices $b_1, b_2, \ldots, b_q$.

2. The other part contains vertices $b_1, b_2, \ldots, b_q$ but not vertices $a_1, a_2, \ldots, a_p$.

Definition 7-5-3. The symbol $T_{2_{a_1 a_2 \ldots a_p, b_1 b_2 \ldots b_q}}$ represents the set of all possible $t_{2_{a_1 a_2 \ldots a_p, b_1 b_2 \ldots b_q}}$ in a given linear graph.

With these definitions, we have

$$T_{2_{i,j}} = T_{2_{ip,j}} \cup T_{2_{i,pj}} \tag{7-5-17}$$

where p is a vertex in linear graph G. Note that

$$T_{2_{ip,j}} \cap T_{2_{i,pj}} = \emptyset \tag{7-5-18}$$

When $j = p$,

$$T_{2_{ip,j}} = \emptyset \tag{7-5-19}$$

and

$$T_{2_{i,pj}} = T_{2_{i,j}} \tag{7-5-20}$$

Now we are ready to discuss cofactor Δ'_{pq}. From Eq. 7-5-13 and the Binet-Cauchy theorem, we know that

$$\begin{aligned}
\Delta'_{pq} &= (-1)^{p+q} |A_{-p} Y_e (A_{-q})^t| \\
&= (-1)^{p+q} \sum \text{major determinant of } A_{-p} Y_e \times \text{corresponding} \\
&\quad \text{major determinant of } A_{-q}{}^t \\
&= (-1)^{p+q} \sum (y_{k_1} y_{k_2} \cdots y_{k_{m'}}) |A_{-p}(k_1 k_2 \cdots k_{m'})| \, |A_{-q}{}^t(k_1 k_2 \cdots k_{m'})|
\end{aligned} \tag{7-5-21}$$

where $A_{-p}(k_1 k_2 \cdots k_{m'})$ is a submatrix consisting of columns $k_1, k_2, \ldots, k_{m'}$ of A_{-p} and $A_{-q}(k_1 k_2 \cdots k_{m'})$ is a submatrix consisting of columns $k_1, k_2, \ldots, k_{m'}$ of A_{-q}. We also know that a nonzero major determinant of A_{-p} corresponds to 2-tree $t_{2_{p,j}}$ where vertex j is the reference vertex. Hence if

$$A_{-p}(k_1 k_2 \cdots k_{m'})$$

is nonsingular, it corresponds to a $t_{2_{p,j}}$. Similarly, if and only if $A_{-q}(k_1 k_2 \cdots k_{m'})$ is nonsingular, it corresponds to a 2-tree $t_{2_{q,j}}$. Thus in order that

$$|A_{-p}(k_1 k_2 \cdots k_{m'})| \, |A_{-q}(k_1 k_2 \cdots k_{m'})|$$

be nonzero, a subgraph consisting of edges corresponding to columns $k_1, k_2, \ldots, k_{m'}$ must be not only a $t_{2_{p,j}}$ but also a $t_{2_{q,j}}$. This means that the subgraph must be a 2-tree in $T_{2_{p,j}}$ and also in $T_{2_{q,j}}$. From

$$T_{q_{p,j}} = T_{2_{pq,j}} \cup T_{2_{p,qj}} \tag{7-5-22}$$

and

$$T_{2_{q,j}} = T_{2_{pq,j}} \cup T_{2_{q,pj}} \tag{7-5-23}$$

we can see that only the 2-trees in $T_{2_{pq,j}}$ satisfy the foregoing conditions. Hence whenever the product $|A_{-p}(k_1 k_2 \cdots k_{m'})| \, |A_{-q}{}^t(k_1 k_2 \cdots k_{m'})|$ is nonzero, columns $k_1, k_2, \ldots, k_{m'}$ correspond to edges in a 2-tree $t_{2_{pq,j}}$. Thus Eq. 7-5-21 becomes

$$\Delta'_{pq} = (-1)^{p+q} \sum \epsilon \text{ 2-tree } t_{2_{pq,j}} \text{ admittance product} \tag{7-5-24}$$

where

$$\epsilon = |A_{-p}(k_1 k_2 \cdots k_{m'})| \, |A_{-q}{}^t(k_1 k_2 \cdots k_{m'})| \tag{7-5-25}$$

In order to find the value ϵ, we consider submatrix A_s of A consisting of columns $k_1, k_2, \ldots, k_{m'}$ as shown below:

$$A_s = \begin{bmatrix} a_{1k_1} & a_{1k_2} & \cdots & a_{1k_{m'}} \\ a_{2k_1} & a_{2k_2} & \cdots & a_{2k_{m'}} \\ & & \cdots & \\ a_{pk_1} & a_{pk_2} & \cdots & a_{pk_{m''}} \\ & & \cdots & \\ a_{qk_1} & a_{qk_2} & \cdots & a_{qk_{m'}} \\ & & \cdots & \\ a_{mk_1} & a_{mk_2} & \cdots & a_{mk_{m'}} \end{bmatrix} \tag{7-5-26}$$

Note that $m = n_v - 1$ (recall that we assume that a network is connected) and $m' = n_v - 2$ because the number of edges in a 2-tree is $n_v - 2$. More-over, A_s consists of the same columns as those in $A_{-p}(k_1 k_2 \cdots k_{m'})$ and $A_{-q}(k_1 k_2 \cdots k_{m'})$. Hence, if we delete the pth row from A_s, the resultant matrix is $A_{-p}(k_1 k_2 \cdots k_{m'})$. On the other hand, if we delete the qth row from A_s, the resultant matrix is $A_{-q}(k_1 k_2 \cdots k_{m'})$.

Since one of two connected parts of $t_{2_{pq,j}}$ contains both vertices p and q, there exists a path between vertices p and q in $t_{2_{pq,j}}$. Let this path be $(e_1, e_2, \cdots, e_r)$ as shown in Fig. 7-5-3.

Fig. 7-5-3. Path between vertices p and q in $t_{2_{pq,j}}$.

The columns of A_s corresponding to edges $e_1, e_2, \ldots, e_r$ of the path have the following properties. For convenience, let the columns corresponding to edges $e_1, e_2, \ldots, e_r$ be the e_1th, the e_2th, $\ldots$, and the e_rth columns of A_s, respectively. Let the rows corresponding to vertices $p, v_1, v_2, \ldots, v_{r-1}$, and q be the pth, the v_1th, the v_2th, $\ldots$, the v_{r-1}th, and qth rows of A_s, respectively. The e_1th column has a nonzero entry in the pth row, the e_1th and the e_2th columns have nonzeros in the v_2th row, $\ldots$, and the e_rth column has a nonzero in the qth row. Suppose the e_1th, the e_2th, $\ldots$, and the e_rth columns are the first r columns of A_s and the first $r + 1$ rows of A_s are the pth, the v_1th, $\ldots$, the v_{r-1}th, and the qth rows. This assumption is not valid in general. We will soon see that this assumption is not necessary for our purposes here. However, it will make the following discussion very clear. With this assumption,

$$
A_s = \begin{matrix} & \begin{matrix} e_1 & e_2 & e_3 & \cdots & e_r & \cdots \end{matrix} \\ \begin{matrix} p \\ v_1 \\ v_2 \\ \vdots \\ q \end{matrix} & \begin{bmatrix} \pm 1 & 0 & 0 & \cdots & 0 & \cdots \\ \mp 1 & \pm 1 & 0 & \cdots & 0 & \cdots \\ 0 & \mp 1 & \pm 1 & \cdots & 0 & \cdots \\ \vdots & \vdots & \vdots & \vdots & \vdots & \\ 0 & 0 & 0 & \cdots & \mp 1 & \cdots \end{bmatrix} \end{matrix} \qquad (7\text{-}5\text{-}27)
$$

We can add (± 1) times the e_2th column to the e_1th column such that the resultant e_1th column has two nonzeros, one in the pth row and the other in the v_2th row:

$$
A_s = \begin{array}{c} \\ p \\ v_1 \\ v_2 \\ \vdots \\ \vdots \\ q \end{array}
\begin{array}{ccccc}
e_1 & e_2 & e_3 & \cdots & e_r & \cdots \\
\left[\begin{array}{ccccc}
\pm 1 & 0 & 0 & \cdots & 0 & \cdots \\
0 & \pm 1 & 0 & \cdots & 0 & \cdots \\
\mp 1 & \mp 1 & \pm 1 & \cdots & 0 & \cdots \\
0 & 0 & & \cdots & & \cdots \\
\vdots & \vdots & \vdots & & \vdots & \\
0 & 0 & 0 & \cdots & \mp 1 & \cdots
\end{array}\right]
\end{array}
\tag{7-5-28}
$$

Note that in the e_1th column, this operation shifts a nonzero from the v_1th row to the v_2th row. Similarly, we can add (± 1) times the e_3th column to the e_1th column, so that in the e_1th column one of nonzeros shifts from the v_2th row to the v_3th row. By successive operations, we can shift one of the nonzeros in the e_1th column so that, finally, one nonzero is in the pth row and the other is in the qth row:

$$
A_s = \begin{array}{c} \\ p \\ v_1 \\ v_2 \\ \vdots \\ q \end{array}
\begin{array}{ccccc}
e_1 & e_2 & e_3 & \cdots & e_r & \cdots \\
\left[\begin{array}{ccccc}
\pm 1 & 0 & 0 & \cdots & 0 & \cdots \\
0 & \pm 1 & 0 & \cdots & 0 & \cdots \\
0 & \mp 1 & \pm 1 & \cdots & 0 & \cdots \\
\vdots & \vdots & \vdots & \vdots & \vdots & \\
\mp 1 & 0 & 0 & \cdots & \mp 1 & \cdots
\end{array}\right]
\end{array}
\tag{7-5-29}
$$

Note that we will have the same result regardless of the locations of the rows corresponding to vertices p, v_1, v_2, ..., v_{r-1}, and q and the columns corresponding to edges $e_1, e_2, \ldots, e_r$.

Without the loss of generality, let $q > p$. Then from Eq. 7-1-7, we can see that

$$
|A_{-q}(k_1 k_2 \cdots k_{m'})| = \pm(-1)^{p+e_1}|A_{-p-q}|]
\tag{7-5-30}
$$

and

$$
|A_{-p}(k_1 k_2 \cdots k_{m'})| = \mp(-1)^{q-1+e_1}|A_{-p-q}|
\tag{7-5-31}
$$

where A_{-p-q} is a submatrix of A_s obtained by deleting the pth and qth rows and the e_1th column. In Eq. 7-5-31, we have $q - 1 + e_1$ rather than $q + e_1$ because in A_{-p} the row corresponding to vertex q is not the qth row but the $(q - 1)$th row. With these results, we have

$$
\epsilon = |A_{-p}(k_1 k_2 \cdots k_{m'})|\,|A_{-q}(k_1 k_2 \cdots k_{m'})| = (-1)^{p+q}
\tag{7-5-32}
$$

Since A_{-p-q} is a nonsingular submatrix of an incidence matrix, $|A_{-p-q}|^2$ is 1. Hence ϵ is $(-1)^{p+q}$ for any $t_{2_{pq,j}}$. Thus by Eq. 7-5-24, we have the following theorem.

Theorem 7-5-1. A cofactor at the (p, q) position of a node admittance matrix of a passivelike network is

$$\Delta'_{pq} = \sum 2\text{-tree } t_{2_{pq,j}} \text{ admittance product} \qquad (7\text{-}5\text{-}33)$$

where j is the reference vertex.

For example, cofactor Δ'_{12} of the network in Fig. 7-5-4 can be obtained by

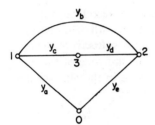

Fig. 7-5-4. A network.

finding all possible 2-trees $t_{2_{12,0}}$ (where 0 is the reference vertex) as follows: All possible 2-trees $t_{2_{12,0}}$ are $(y_b y_c)$, $(y_b y_d)$, and $(y_c y_d)$. Hence

$$\Delta'_{12} = y_b y_c + y_b y_d + y_c y_d$$

For convenience, we use the following symbols.

Definition 7-5-4. The symbols V and W represent $\sum$ tree admittance product and $\sum$ 2-tree admittance products as

$$V = \sum \text{tree admittance product} \qquad (7\text{-}5\text{-}34)$$

$$W_{a_1 a_2 \cdots a_r, b_1 b_2 \cdots b_s} = \sum \text{tree } t_{2_{a_1 a_2 \cdots a_r, b_1 b_2 \cdots b_s}} \text{ admittance product} \qquad (7\text{-}5\text{-}35)$$

By using Δ' and Δ'_{pq}, Eq. 7-5-4 can be written as

$$V_n = \frac{1}{\Delta'} \begin{bmatrix} \Delta'_{11} & \Delta'_{21} & \cdots & \Delta'_{m1} \\ \Delta'_{12} & \Delta'_{22} & \cdots & \Delta'_{m2} \\ & & \cdots & \\ \Delta'_{1m} & \Delta'_{2m} & \cdots & \Delta'_{mm} \end{bmatrix} J_n \qquad (7\text{-}5\text{-}36)$$

Consider the network in Fig. 7-5-5 where N consists of passivelike elements only. As far as the network functions between vertices p, q, and j (the reference vertex) are concerned, Eq. 7-5-36 can be simplified to

$$\begin{bmatrix} v_p \\ v_q \end{bmatrix} = \frac{1}{\Delta'} \begin{bmatrix} \Delta'_{pp} & \Delta'_{qp} \\ \Delta'_{pq} & \Delta'_{qq} \end{bmatrix} \begin{bmatrix} J_p \\ J_q \end{bmatrix}$$

(7-5-37)

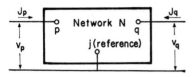

Fig. 7-5-5. Three-terminal network functions.

Therefore the open-circuit network functions are

$$\left. \frac{J_p}{v_p} \right|_{J_q = 0} = \frac{V}{W_{p,j}}$$

(7-5-38)

$$\left. \frac{J_p}{v_q} \right|_{J_q = 0} = \frac{V}{W_{pq,j}}$$

(7-5-39)

and

$$\left. \frac{v_q}{v_p} \right|_{J_q = 0} = \left. \frac{v_q/J_p}{v_p/J_p} \right|_{J_q = 0} = \frac{W_{pq,j}/V}{W_{p,j}/V} = \frac{W_{pq,j}}{W_{p,j}}$$

(7-5-40)

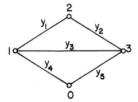

Fig. 7-5-6. A network.

For example, open-circuit network functions of the network in Fig. 7-5-6 can be obtained as follows:

$$V = y_1 y_2 y_4 + y_1 y_2 y_5 + y_1 y_3 y_4 + y_1 y_3 y_5 + y_1 y_4 y_5 + y_2 y_3 y_4 \\ + y_2 y_3 y_5 + y_2 y_4 y_5$$

and

$$W_{1,0} = y_1 y_2 + y_1 y_3 + y_1 y_5 + y_2 y_3 + y_2 y_5$$

Hence

$$\frac{J_1}{V_1} = \frac{V}{W_{1,0}}$$

$$= \frac{y_1[y_2(y_4 + y_5) + y_3(y_4 + y_5) + y_4 y_5] + y_2[y_3(y_4 + y_5) + y_4 y_5]}{y_1[y_2 + (y_3 + y_5)] + y_2(y_3 + y_5)}$$

Furthermore,

$$W_{2,0} = y_1 y_2 + y_1 y_3 + y_1 y_5 + y_2 y_4 + y_3 y_4 + y_4 y_5 + y_2 y_3 + y_3 y_5$$

Thus

$$\frac{J_2}{V_2} = \frac{V}{(y_1 + y_4)(y_2 + y_3 + y_5) + y_3(y_2 + y_5)}$$

and

$$W_{12,0} = y_1 y_2 + y_1 y_3 + y_1 y_5 + y_2 y_3$$

We now have

$$\left.\frac{J_1}{V_2}\right|_{J_2=0} = \frac{V}{[y_1(y_2 + y_3 + y_5) + y_2 y_3]}$$

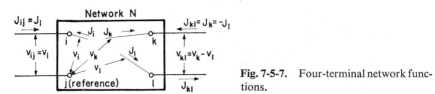

Fig. 7-5-7.   Four-terminal network functions.

Consider the network in Fig. 7-5-7 in which N is assumed to be of passive-like elements only. From Eq. 7-5-36, we have

$$\begin{bmatrix} v_i \\ v_k \\ v_l \end{bmatrix} = \frac{1}{\Delta'} \begin{bmatrix} \Delta'_{ii} & \Delta'_{ki} & \Delta'_{li} \\ \Delta'_{ik} & \Delta'_{kk} & \Delta'_{lk} \\ \Delta'_{il} & \Delta'_{kl} & \Delta'_{ll} \end{bmatrix} \begin{bmatrix} J_i \\ J_k \\ J_l \end{bmatrix} \qquad (7\text{-}5\text{-}41)$$

Since

$$v_{ij} = v_i \qquad\qquad (7\text{-}5\text{-}42)$$

$$J_{ij} = J_i \qquad\qquad (7\text{-}5\text{-}43)$$

$$v_{kl} = v_k - v_l \qquad\qquad (7\text{-}5\text{-}44)$$

and

$$J_{kl} = J_k = -J_l \qquad\qquad (7\text{-}5\text{-}45)$$

where v_p is the voltage from vertex p to the reference vertex j and J_p is the current from the reference vertex j to vertex p for $p = i, k, l$, Eq. 7-5-41 can be rewritten as

$$\begin{bmatrix} v_{ij} \\ v_{kl} \end{bmatrix} = \frac{1}{\Delta'} \begin{bmatrix} \Delta'_{ii} & \Delta'_{ki} - \Delta'_{li} \\ \Delta'_{ik} - \Delta'_{il} & \Delta'_{kk} + \Delta'_{ll} - \Delta'_{kl} - \Delta'_{lk} \end{bmatrix} \begin{bmatrix} J_{ij} \\ J_{kl} \end{bmatrix} \qquad (7\text{-}5\text{-}46)$$

Since

$$W_{pq,j} = \sum_{t_{2pq,j} \in T_{2pq,j}} 2\text{-tree } t_{2pq,j} \text{ admittance product} \qquad (7\text{-}5\text{-}47)$$

and

$$T_{2pq,j} = T_{2pqk,j} \cup T_{2pq,kj} \qquad (7\text{-}5\text{-}48)$$

we can express Δ'_{ki} as

$$\Delta'_{ki} = W_{ki,j} = W_{kil,j} + W_{ki,lj} \qquad (7\text{-}5\text{-}49)$$

Similarly,

$$\Delta'_{li} = W_{li,j} = W_{kil,j} + W_{li,kj} \qquad (7\text{-}5\text{-}50)$$

Hence

$$\Delta'_{ki} - \Delta'_{li} = W_{ki,lj} - W_{li,kj} \qquad (7\text{-}5\text{-}51)$$

Likewise

$$\Delta'_{ki} - \Delta'_{li} = W_{ki,lj} - W_{li,kj} \qquad (7\text{-}5\text{-}52)$$

Also, from

$$\Delta'_{kk} = W_{kj} = W_{kl,j} + W_{k,lj} \qquad (7\text{-}5\text{-}53)$$

$$\Delta'_{ll} = W_{lj} = W_{kl,j} + W_{l,kj} \qquad (7\text{-}5\text{-}54)$$

$$\Delta'_{kl} = W_{kl,j} \qquad (7\text{-}5\text{-}55)$$

and

$$\Delta'_{lk} = W_{kl,j} \qquad (7\text{-}5\text{-}56)$$

we have

$$\Delta'_{kk} + \Delta'_{ll} - \Delta'_{kl} - \Delta'_{lk} = W_{k,lj} + W_{kj,l} = W_{k,l} \qquad (7\text{-}5\text{-}57)$$

Thus Eq. 7-5-46 can be expressed as

$$\begin{bmatrix} v_{ij} \\ v_{kl} \end{bmatrix} = \frac{1}{V} \begin{bmatrix} W_{i,j} & W_{ki,lj} - W_{li,kj} \\ W_{ki,lj} - W_{li,kj} & W_{k,l} \end{bmatrix} \begin{bmatrix} J_{ij} \\ J_{kl} \end{bmatrix} \qquad (7\text{-}5\text{-}58)$$

In other words, open-circuit four-terminal network functions are

$$\frac{J_{ij}}{v_{ij}}\bigg|_{J_{kl}=0} = \frac{V}{W_{i,j}} \tag{7-5-59}$$

$$\frac{J_{ij}}{v_{kl}}\bigg|_{J_{kl}=0} = \frac{V}{W_{ki,lj} - W_{li,kj}} \tag{7-5-60}$$

and

$$\frac{v_{kl}}{v_{ij}}\bigg|_{J_{kl}=0} = \frac{W_{ki,lj} - W_{li,kj}}{W_{i,j}} \tag{7-5-61}$$

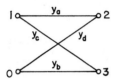

Fig. 7-5-8.

For example, open-circuit voltage ratio v_{23}/v_{10} of the network in Fig. 7-5-8 can be obtained by forming $W_{1,0}$ and $W_{12,30} - W_{13,20}$:

$$W_{1,0} = y_a y_b + y_a y_c + y_b y_d + y_c y_d$$

and

$$W_{12,30} - W_{13,20} = y_a y_b - y_c y_d$$

Hence

$$\frac{v_{23}}{v_{10}} = \frac{y_a y_b - y_c y_d}{y_a y_b + y_a y_c + y_b y_d + y_c y_d}$$

By defining

$$W_{ki,lj} = W_{ki,j} \qquad \text{for} \quad j = 1 \tag{7-5-62}$$

and

$$W_{li,kj} = 0 \qquad \text{for} \quad j = 1 \tag{7-5-63}$$

we can see that Eqs. 7-5-60 and 7-5-61 become Eqs. 7-5-39 and 7-5-40 when $j = 1$.

7-6 TOPOLOGICAL FORMULAS FOR SHORT-CIRCUIT NETWORK FUNCTIONS

From Eq. 7-5-46 we obtain

$$
\begin{bmatrix} J_{ij} \\ J_{kl} \end{bmatrix} = \frac{\Delta'}{\Delta'_{ii}(\Delta'_{kk} + \Delta'_{ll} - \Delta'_{kl} - \Delta'_{lk}) - (\Delta'_{ki} - \Delta'_{li})(\Delta'_{ik} - \Delta'_{il})}
$$
$$
\times \begin{bmatrix} \Delta'_{kk} + \Delta'_{ll} - \Delta'_{kl} - \Delta'_{lk} & -(\Delta'_{ki} - \Delta'_{li}) \\ -(\Delta'_{ik} - \Delta'_{il}) & \Delta'_{ii} \end{bmatrix} \begin{bmatrix} v_{ij} \\ v_{kl} \end{bmatrix} \tag{7-6-1}
$$

To simplify this equation, we must find a relationship among cofactors. Consider a square matrix $A = [a_{ij}]$ of order n. Let M be the adjugate (classical adjoint) of A. That is,

$$
M = \begin{bmatrix} \Delta_{11} & \Delta_{21} & \cdots & \Delta_{n1} \\ \Delta_{12} & \Delta_{22} & \cdots & \Delta_{n2} \\ & & \cdots & \\ \Delta_{1n} & \Delta_{2n} & \cdots & \Delta_{nn} \end{bmatrix} \tag{7-6-2}
$$

where Δ_{ij} is the cofactor of a_{ij} of A. Then we know that

$$
AM = \begin{bmatrix} D & & & 0 \\ & D & & \\ & & \ddots & \\ 0 & & & D \end{bmatrix} \tag{7-6-3}
$$

where D is the determinant of A. Suppose we rearrange rows and columns of A so that rows $r_1, r_2, \ldots, r_k$ are the first k rows and columns $j_1, j_2, \ldots, j_k$ are the first k columns. Let the resultant matrix be $\underline{A}$. Also let $\underline{M}$ be the adjugate of $\underline{A}$ which can be obtained by rearranging rows and columns of M.

Now we modify $\underline{M}$, by replacing all columns other than the first k columns:

$$
\underline{M}_k = \begin{bmatrix} \Delta_{r_1 j_1} & \Delta_{r_2 j_1} & \cdots & \Delta_{r_k j_1} & & & \\ \Delta_{r_1 j_2} & \Delta_{r_2 j_2} & \cdots & \Delta_{r_k j_2} & & 0 & \\ & & \cdots & & & & \\ \Delta_{r_1 j_k} & \Delta_{r_2 j_k} & \cdots & \Delta_{r_k j_k} & & & \\ \Delta_{r_1 1} & \Delta_{r_2 1} & \cdots & \Delta_{r_k 1} & 1 & & \\ \Delta_{r_1 2} & \Delta_{r_2 2} & \cdots & \Delta_{r_k 2} & & 1 & 0 \\ & & \cdots & & & & \ddots \\ \Delta_{r_1 n} & \Delta_{r_2 n} & \cdots & \Delta_{r_k n} & & 0 & & 1 \end{bmatrix} \tag{7-6-4}
$$

where $r_1 < r_2 < \cdots < r_k$ and $j_1 < j_2 < \cdots < j_k$. The product of $\underline{A}$ and $\underline{M}_k$ is

$$
\underline{A} \cdot \underline{M}_k =
\begin{bmatrix}
a_{r_1 j_1} & \cdots & a_{r_1 j_k} & a_{r_1 1} & \cdots & a_{r_1 n} \\
a_{r_2 j_1} & \cdots & a_{r_2 j_k} & a_{r_2 1} & \cdots & a_{r_2 n} \\
& \cdots & & & \cdots & \\
a_{r_k j_1} & \cdots & a_{r_k j_k} & a_{r_k 1} & \cdots & a_{r_k n} \\
a_{1 j_1} & \cdots & a_{1 j_k} & a_{11} & \cdots & a_{1n} \\
a_{2 j_1} & \cdots & a_{2 j_k} & a_{21} & \cdots & a_{2n} \\
& \cdots & & & \cdots & \\
a_{n j_1} & \cdots & a_{n j_k} & a_{n1} & \cdots & a_{nn}
\end{bmatrix}
$$

$$
\times
\begin{bmatrix}
\Delta_{r_1 j_1} & \cdots & \Delta_{r_k j_1} & & & \\
\Delta_{r_1 j_2} & \cdots & \Delta_{r_k j_2} & & 0 & \\
& \cdots & & & & \\
\Delta_{r_1 j_k} & \cdots & \Delta_{r_k j_k} & & & \\
\Delta_{r_1 1} & \cdots & \Delta_{r_k 1} & 1 & & \\
\Delta_{r_1 2} & \cdots & \Delta_{r_k 2} & 1 & & 0 \\
& \cdots & & & \cdots & 1 \\
\Delta_{r_1 n} & \cdots & \Delta_{r_k n} & 0 & & 1
\end{bmatrix}
$$

$$
=
\begin{bmatrix}
D & & & & a_{r_1 1} & \cdots & a_{r_1 n} \\
& D & & & a_{r_2 1} & \cdots & a_{r_2 n} \\
& & \ddots & & & \cdots & \\
& & & D & a_{r_k 1} & \cdots & a_{r_k n} \\
& & & & a_{11} & \cdots & a_{1n} \\
& 0 & & & a_{21} & \cdots & a_{2n} \\
& & & & & \cdots & \\
& & & & a_{n1} & \cdots & a_{nn}
\end{bmatrix}
\qquad (7\text{-}6\text{-}5)
$$

Hence, from the left-hand side of Eq. 7-6-5, the determinant of $\underline{A} \cdot \underline{M}_k$ is

$$
|\underline{A} \cdot \underline{M}_k| = |\underline{A}|\,|\underline{M}_k| = \mathscr{E} D
\begin{vmatrix}
\Delta_{r_1 j_1} & \cdots & \Delta_{r_k j_1} \\
\Delta_{r_1 j_2} & \cdots & \Delta_{r_k j_2} \\
& \cdots & \\
\Delta_{r_1 j_k} & \cdots & \Delta_{r_k j_k}
\end{vmatrix}
\qquad (7\text{-}6\text{-}6)
$$

where $\mathscr{E} = (-1)^{r_1 + r_2 + \cdots + r_k + j_1 + \cdots + j_k}$.

On the other hand, the determinant of $\underline{A} \cdot \underline{M}_k$ obtained from the right-hand side of Eq. 7-6-5 is

$$|\underline{A} \cdot \underline{M}_k| = D^k \left| A\left(\frac{r_1 r_2 \cdots r_k}{j_1 j_2 \cdots j_k}\right) \right| \tag{7-6-7}$$

where $A(\overline{r_1 r_2 \cdots r_k / j_1 j_2 \cdots j_k})$ is obtained from A by deleting rows $r_1, r_2, \ldots, r_k$ and columns $j_1, j_2, \ldots, j_k$. Thus

$$\begin{vmatrix} \Delta_{r_1 j_1} & \cdots & \Delta_{r_k j_1} \\ \Delta_{r_1 j_2} & \cdots & \Delta_{r_k j_2} \\ & \cdots & \\ \Delta_{r_1 j_k} & \cdots & \Delta_{r_k j_k} \end{vmatrix} = D^{k-1} \tilde{D}\left(\frac{r_1 r_2 \cdots r_k}{j_1 j_2 \cdots j_k}\right) \tag{7-6-8}$$

where

$$\tilde{D}\left(\frac{r_1 r_2 \cdots r_k}{j_1 j_2 \cdots j_k}\right) = (-1)^{r_1 + r_2 + \cdots + r_k + j_1 + j_2 + \cdots + j_k} \left| A\left(\frac{r_1 r_2 \cdots r_k}{j_1 j_2 \cdots j_k}\right) \right| \tag{7-6-9}$$

When the right-hand side of Eq. 7-6-8 consists of Δ_{ij}, Δ_{pj}, Δ_{iq}, and Δ_{pq}, then

$$\begin{vmatrix} \Delta_{ij} & \Delta_{pj} \\ \Delta_{iq} & \Delta_{pq} \end{vmatrix} = D\tilde{D}\left(\frac{\overline{ip}}{jq}\right) \tag{7-6-10}$$

or

$$\Delta_{ij}\Delta_{pq} - \Delta_{pj}\Delta_{iq} = D\tilde{D}\left(\frac{\overline{ip}}{jq}\right) \tag{7-6-11}$$

We have been using Δ for D. Usually $\tilde{D}(\overline{ip}/\overline{jq})$ is represented by Δ_{ijpq}. With these symbols, Eq. 7-6-11 can be expressed as

$$\Delta_{ij}\,\Delta_{pq} - \Delta_{pj}\,\Delta_{iq} = \Delta\,\Delta_{ijpq} \tag{7-6-12}$$

which is known as *Jacobi's relationship*. With this relationship, the denominator of the first term on the right-hand side of Eq. 7-6-1 becomes

$$\begin{aligned} \Delta_{ii}'(\Delta_{kk}' &+ \Delta_{ll}' - \Delta_{kl}' - \Delta_{lk}') - (\Delta_{ki}' - \Delta_{li}')(\Delta_{ik}' - \Delta_{il}') \\ &= (\Delta_{ii}'\Delta_{kk}' - \Delta_{ik}'\Delta_{ki}') + (\Delta_{ii}'\Delta_{ll}' - \Delta_{il}'\Delta_{li}') \\ &\quad - (\Delta_{ii}'\Delta_{kl}' - \Delta_{ki}'\Delta_{il}') - (\Delta_{ii}'\Delta_{lk}' - \Delta_{li}'\Delta_{ik}') \\ &= \Delta'(\Delta_{iikk}' + \Delta_{iill}' - \Delta_{iikl}' - \Delta_{iilk}') \end{aligned} \tag{7-6-13}$$

Thus Eq. 7-6-1 becomes

$$\begin{bmatrix} J_{ij} \\ J_{kl} \end{bmatrix} = \frac{1}{\Delta_{iikk}' + \Delta_{iill}' - \Delta_{iikl}' - \Delta_{iilk}'} \\ \times \begin{bmatrix} \Delta_{kk}' + \Delta_{ll}' - \Delta_{kl}' - \Delta_{lk}' & -(\Delta_{ki}' - \Delta_{li}') \\ -(\Delta_{ik}' - \Delta_{il}') & \Delta_{ii}' \end{bmatrix} \begin{bmatrix} v_{ij} \\ v_{kl} \end{bmatrix} \tag{7-6-14}$$

We can see that Δ'_{iipq} can be expressed as

$$\Delta'_{iipq} = (-1)^{p+q} |A_{-i-p} Y (A_{-i-p})^t| \tag{7-6-15}$$

where A_{-i-p} is a submatrix of incidence matrix A obtained by deleting rows i and p. Likewise A_{-i-q} is obtained from A by deleting rows i and q. Since Y is a diagonal matrix, the Binet-Cauchy theorem gives

$$\Delta'_{iipq} = (-1)^{p+q} \sum_{(k)} y_{k_1} y_{k_2} \cdots y_{k_{m''}} |A_{-i-p}(k_1 k_2 \cdots k_{m''})| \, |A^t_{-i-q}(k_1 k_2 \cdots k_{m''})| \tag{7-6-16}$$

where $A_{-i-p}(k_1 k_2 \cdots k_{m''})$ consists of columns $k_1, k_2, \ldots, k_{m''}$ of A_{-i-p} and $A_{-i-q}(k_1 k_2 \cdots k_{m''})$ consists of columns $k_1, k_2, \ldots, k_{m''}$ of A_{-i-q}.

Consider a linear graph $G(i = j)$ which is obtained from a given linear graph G by coinciding vertices i and j (with j being the reference vertex). We can see that a node admittance matrix of $G(i = j)$ can be obtained from a node admittance matrix of G by deleting the row and the column corresponding to vertex i. In other words,

$$A_{-i} Y (A_{-i})^t \quad \text{of} \quad G = A Y A^t \quad \text{of} \quad G(i = j) \tag{7-6-17}$$

Hence

$$A_{-i-p} Y (A_{-i-q})^t \quad \text{of} \quad G = A_{-p} Y (A_{-q})^t \quad \text{of} \quad G(i = j) \tag{7-6-18}$$

Thus from Eq. 7-5-33,

$$|A_{-i-p} Y (A_{-i-q})^t| \quad \text{of} \quad G = \sum 2\text{-tree } t_{2_{pq,j}} \text{ admittance product of } G(i = j) \tag{7-6-19}$$

Let g_s be a subgraph of G which becomes $t_{2_{pq,j}}$ of $G(i = j)$ when vertices i and j are coincided. Then g_s must consist of three connected parts, one containing vertices p and q, another containing vertex i, and the third containing vertex j (the reference vertex). Such a subgraph is called a 3-tree and is indicated by the symbol $t_{3_{i,pq,j}}$.

Definition 7-6-1 A 3-tree $t_{3_{i_1 i_2 \ldots i_p, j_1 j_2 \ldots j_u, k_1 k_2 \ldots k_r}}$ is a subgraph of a linear graph G (of n_v vertices) satisfying the following conditions:

1. $t_{3_{i_1 i_2 \ldots i_p, j_1 j_2 \ldots j_u, k_1 k_2 \ldots k_r}}$ consists of $n_v - 3$ edges, n_v vertices and no circuits.

2. $t_{3_{i_1 i_2 \ldots i_t, j_1 j_2 \ldots j_u, k_1 k_2 \ldots k_r}}$ consists of three parts (three maximal connected subgraphs) each of which is connected.

3. One of three parts contains vertices $i_1, i_2, \ldots, i_t$, another contains vertices $j_1, j_2, \ldots, j_u$, and the other contains vertices $k_1, k_2, \ldots, k_v$ as shown in Fig. 7-6-1. A part can be an isolated vertex if the number vertices which must be in the part is one.

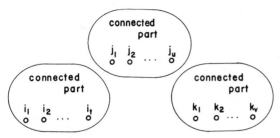

Fig. 7-6-1. A 3-tree $t_{3i_1i_2...i_t,j_1j_2...j_u,k_1k_2...k_v}$.

With this definition and Eq. 7-6-19, we have Theorem 7-6-1.

Theorem 7-6-1. A cofactor Δ'_{iipq} of a node admittance matrix of a passivelike network is

$$\Delta'_{iipq} = \sum 3\text{-tree } t_{3_i,pq,j} \text{ admittance product} \qquad (7\text{-}6\text{-}20)$$

where j is the reference vertex.

For convenience, we define the symbol $U_{i,pq,j}$ as follows.

Definition 7-6-2.

$$U_{i,pq,j} = \sum 3\text{-tree } t_{3_i,pq,j} \text{ admittance product} \qquad (7\text{-}6\text{-}21)$$

Now cofactor Δ'_{iikk} can be expressed as

$$\Delta'_{iikk} = U_{i,k,j} \qquad (7\text{-}6\text{-}22)$$

Since vertex l can be in any one of three parts, Eq. 7-6-22 can be written as

$$U_{i,k,j} = U_{i,kl,j} + U_{il,k,j} + U_{i,k,lj} \qquad (7\text{-}6\text{-}23)$$

Similarly, cofactors Δ'_{iill} and Δ'_{iikl} can be expressed as

$$\Delta'_{iill} = U_{i,l,j} = U_{i,kl,j} + U_{ik,l,j} + U_{i,l,kj} \qquad (7\text{-}6\text{-}24)$$

and

$$\Delta'_{iikl} = U_{i,kl,j} \qquad (7\text{-}6\text{-}25)$$

Hence

$$\Delta'_{iikk} + \Delta'_{iill} - \Delta'_{iikl} - \Delta'_{iilk} = U_{ik,l,j} + U_{i,k,lj} + U_{il,k,j} + U_{i,l,kj} \qquad (7\text{-}6\text{-}26)$$

Definition 7-6-3. The symbol $\sum U$ is defined as

$$\sum U = U_{ik,l,j} + U_{i,k,lj} + U_{il,k,j} + U_{i,l,kj} \qquad (7\text{-}6\text{-}27)$$

By using Definition 7-6-3 in the right-hand side of Eq. 7-6-26, Eq. 7-6-14 can be expressed as

$$\begin{bmatrix} J_{ij} \\ J_{kl} \end{bmatrix} = \frac{1}{\sum U} \begin{bmatrix} W_{k,l} & -(W_{ik,lj} - W_{il,kj}) \\ -(W_{ik,lj} - W_{il,kj}) & W_{i,j} \end{bmatrix} \begin{bmatrix} v_{ij} \\ v_{kl} \end{bmatrix} \qquad (7\text{-}6\text{-}28)$$

Note that when $j = l$, $\sum U$ in Eq. 7-6-27 becomes simply

$$\sum U = U_{i,k,j} \qquad (7\text{-}6\text{-}27)$$

Thus from Eq. 7-6-28, we obtain

$$\begin{bmatrix} J_p \\ J_q \end{bmatrix} = \frac{1}{U_{i,k,j}} \begin{bmatrix} W_{q,j} & -W_{qp,j} \\ -W_{pq,j} & W_{p,j} \end{bmatrix} \begin{bmatrix} v_p \\ v_q \end{bmatrix} \qquad (7\text{-}6\text{-}30)$$

Equations 7-6-28 and 7-6-30 give topological formulas for short-circuit network functions for four- and three-terminal networks. We now can obtain current ratio J_{kl}/J_{ij} as follows:

$$\left. \frac{J_{kl}}{J_{ij}} \right|_{v_{kl}=0} = \frac{J_{kl}/v_{ij}}{J_{ij}/v_{ij}} = \frac{-(W_{ik,lj} - W_{il,kj})}{W_{k,l}} \qquad (7\text{-}6\text{-}31)$$

For example, consider the network in Fig. 7-5-8. By obtaining

$$U = U_{13,2,0} + U_{1,3,20} + U_{12,3,0} + U_{1,2,30} = y_c + y_d + y_a + y_b$$

we can find the short-circuit transfer function $J_{10}/v_{23}|_{v_{10}=0}$:

$$\left. \frac{J_{10}}{v_{23}} \right|_{v_{10}=0} = \frac{-(W_{12,30} - W_{13,20})}{U} = \frac{y_c y_d - y_a y_b}{y_a + y_b + y_c + y_d}$$

When initial conditions of components are nonzero or there are voltage and (or) current generators in a network, the formulas developed in this chapter must be modified.

PROBLEMS

1. Obtain the open circuit driving point V_{10}/I_{10} and the transfer V_{20}/I_{10} functions of all three networks in Fig. P-7-1.

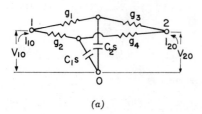

(a)

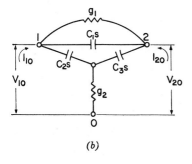

(b)

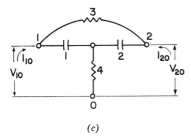

(c)   **Fig. P-7-1.**

2. Obtain the voltage ratio V_{23}/V_{10} of the two networks in Fig. P-7-2.

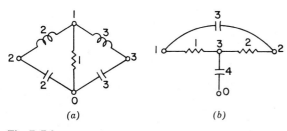

(a)   (b)

Fig. P-7-2.

3. Find the short-circuit transfer function $I_{20}/V_{10}|_{v_{20}=0}$ of the networks in Problem 1.

4. Find the open-circuit driving point admittance function of a network which is a complete graph consisting of n_v vertices with each edge being a 1-mho conductor.

5. Prove by using topological formulas that an open-circuit driving point function LC network containing no mutual couplings is the ratio of an even and an odd function of s.

6. Let Y_n be the node admittance matrix of a common terminal passive-like network with all admittances being nonnegative. Let

$$\Delta_{11} = \text{cofactor } (1, 1) \text{ of } Y_n = \frac{1}{s^{m-2}} [a_n s^n + a_{n-1} s^{n-1} + \cdots + a_0]$$

and

$$\Delta_{12} = \text{cofactor } (1, 2) \text{ of } Y_n = \frac{1}{s^{m-2}} [b_n s^n + b_{n-1} s^{n-1} + \cdots + b_0]$$

Then prove that $0 \le b_k \le a_k$ for $0 \le k \le n$.

7. Prove that

$$\left| \begin{array}{c} A \\ B_f \end{array} \right| = \pm (\text{number of trees})$$

where B_f is a fundamental circuit matrix.

8. Prove that BYB^t is a mesh impedance matrix where B is a circuit matrix.

9. Prove that $|BB^t|$ is the number of trees in a linear graph.

CHAPTER

8

TOPOLOGICAL FORMULAS FOR ACTIVE NETWORK, UNISTOR NETWORK, AND EQUIVALENT TRANSFORMATION

8-1 CURRENT AND VOLTAGE GRAPHS

There are several ways of analyzing networks by linear graphs; each has advantages and disadvantages. A node admittance matrix of the form $A Y A^t$ discussed previously has the property that Y is a diagonal matrix, and because of this property the determinant of such a matrix becomes a sum of tree admittance products. We will see that a node admittance matrix of an active network can also be expressed as $A_1 Y A_2{}^t$ such that Y is a diagonal matrix. However, incidence matrices A_1 and A_2 in $A_1 Y A_2{}^t$ are different. Hence the determinant of $A_1 Y A_2{}^t$ is not just a sum of tree admittance products.

Definition 8-1-1. An admittance y defined by

$$y v_{r,s} = i_{t,u} \tag{8-1-1}$$

is represented by two edges as shown in Fig. 8-1-1 where $v_{r,s}$ is the voltage from vertex r to vertex s and $i_{t,u}$ is the corresponding current directed from vertex t to vertex u. The edge representing $i_{t,u}$ is called a *current edge* and the

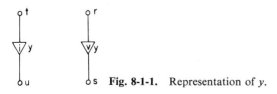

Fig. 8-1-1. Representation of y.

edge representing $v_{r,s}$ is called a *voltage edge* of admittance y. As shown in Fig. 8-1-1, the symbol i inside of the arrow indicates that the edge is a current edge and the symbol v inside of the arrow indicates that the edge is a voltage edge. This representation is known as a *topological representation of an admittance y*.

For example, a three-terminal transconductance g can be defined by the equation

$$gv_{G,k} = i_{p,k} \tag{8-1-2}$$

which is represented by two edges as shown in Fig. 8-1-2. When $r = t$ and $s = u$, the topological representation of admittance y in Fig. 8-1-1 is that of Fig. 8-1-3a. It is convenient to represent such an admittance by one edge as

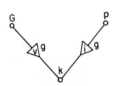

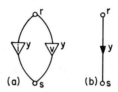

Fig. 8-1-2. Three-terminal transconductance g.

Fig. 8-1-3. Passivelike element.

shown in Fig. 8-1-3b because in this case y becomes a passivelike element (see Eq. 7-4-6 and Fig. 7-4-1). Similarly, when $r = u$ and $s = t$, we have

$$yv_{r,s} = i_{s,r} \tag{8-1-3}$$

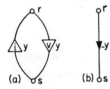

Fig. 8-1-4. Negative admittance.

The representation of such an admittance y is shown in Fig. 8-1-4a. Since

$$i_{s,r} = -i_{r,s} \tag{8-1-4}$$

Eq. 8-1-5 can be written as

$$(-y)v_{r,s} = i_{r,s} \tag{8-1-5}$$

Thus we can represent such an admittance by one edge with $-y$ as the weight (Fig. 8-1-4b). Note that because

$$v_{r,s} = -v_{s,r} \tag{8-1-6}$$

and by Eq. 8-1-4, the orientation of passivelike elements is arbitrary. It must be understood that one edge representing a passivelike element in Fig. 8-1-3b (also Fig. 8-1-4b) indicates both the voltage edge and the current edge of admittance y, that is, the representation in Fig. 8-1-3b is a shorthand notation of the representation of Fig. 8-1-3a.

To familiarize ourselves with voltage and current edges, consider a node basis equation

$$\begin{bmatrix} y_{11} & y_{12} & y_{13} \\ y_{21} & y_{22} & y_{23} \\ y_{31} & y_{32} & y_{33} \end{bmatrix} \begin{bmatrix} v_1 \\ v_2 \\ v_3 \end{bmatrix} = \begin{bmatrix} J_1 \\ J_2 \\ J_3 \end{bmatrix} \tag{8-1-7}$$

where voltage v_p is from vertex p to the reference vertex and current J_p is from the reference vertex to vertex p for $p = 1, 2, 3$. Then we can draw the linear graph corresponding to the network whose node admittance matrix is the same as that of Eq. 8-1-7 (Fig. 8-1-5). Hence the representation given in Fig. 8-1-1 is a very useful one. Note that in Fig. 8-1-5, every voltage edge and every current edge are connected to the reference vertex.

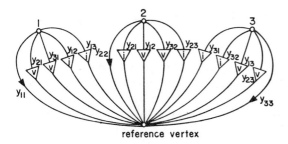

reference vertex

Fig. 8-1-5. A topological representation of a network.

It would be interesting to know where admittance y will be located in a node admittance matrix when the voltage and current edges of y are arbitrary. Consider an admittance y which satisfies Eq. 8-1-1. Note that under this assumption, the voltage edge of y is connected from vertex r to vertex s and the current edge of y is connected from vertex t to vertex u. Let node voltages of vertices r and s be v_r and v_s. Then Eq. 8-1-1 can be written as

$$y(v_r - v_s) = i_{t,u} \tag{8-1-8}$$

Let

$$
\begin{bmatrix}
y_{11} & y_{12} & \cdots & y_{1n} \\
y_{21} & y_{22} & \cdots & y_{2n} \\
y_{31} & y_{32} & \cdots & y_{3n} \\
& & \cdots & \\
y_{n1} & y_{n2} & \cdots & y_{nn}
\end{bmatrix}
\begin{bmatrix}
v_1 \\ v_2 \\ v_3 \\ \vdots \\ v_n
\end{bmatrix}
=
\begin{bmatrix}
J_1 \\ J_2 \\ J_3 \\ \vdots \\ J_n
\end{bmatrix}
\tag{8-1-9}
$$

be a node basis equation of a network which contains admittance y. Then the tth row of the node admittance matrix multiplied by the voltage vector in Eq. 8-1-9 is the sum of currents at vertex t, which must be equal to the current J_t (an external current). Since $i_{t,u}$ of Eq. 8-1-8 is from vertex t to vertex u, y must be at the (t, r) entry and $-y$ must be at the (t, s) entry of the node admittance matrix. Similarly, $-y$ must be at (u, r) entry and y must be at (u, s) entry of the node admittance matrix. These entries, where admittance y must be, are shown in Fig. 8-1-6, in which the two circled entries indicate

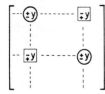

Fig. 8-1-6. Locations of admittance y in node admittance matrix.

that the y in these entries have the same sign. Similarly, y in the two entries which have squares must have the same sign. Furthermore, y in the entry indicated by a circle and y in the entry indicated by a square must be of opposite sign. For example, if $r < s < t < u$ in the node admittance matrix, y will appear as in Fig. 8-1-7. Note that when $t = r$ and $u = s$, the entries

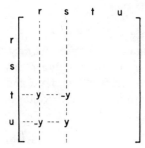

Fig. 8-1-7. Location of y when $r < s < t < u$.

indicated by circles in Fig. 8-1-6 are diagonal entries. Furthermore, we know that under these conditions, admittance y is a passivelike element. This is also true when $t = s$ and $u = r$ if we use Eq. 8-1-5. Hence if and only if admittance y is a passivelike element, locations of y in the node admittance matrix are symmetric with respect to the diagonal entries.

Suppose either vertex t or vertex u is the reference vertex. Then the column corresponding to the vertex will not be in the node admittance matrix. Hence only two entries in one column in Fig. 8-1-6 have admittance y. In other words, one of these two columns which have a circle and a square in Fig. 8-1-6 will be absent. Similarly, if either vertex r or vertex s is the reference vertex, then one of two rows in Fig. 8-1-6 which have a circle and a square will be absent. For example, if $t < u$ and s is the reference vertex, admittance y of Eq. 8-1-1 will appear in the node admittance matrix in Eq. 8-1-9 as shown in Fig. 8-1-8.

Suppose $u = s$, and s is the reference vertex. Then we can see that the uth column and the sth row will be absent. Thus only one entry, (t, r), will have admittance (Fig. 8-1-9).

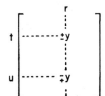

Fig. 8-1-8. Location of y when $t < u$ and s is the reference vertex.

Fig. 8-1-9. Location of y when $u = s$ and s is the reference vertex.

It is easily seen that a given node admittance matrix can be expressed as a sum of matrices each of which contains exactly one admittance. For example, we can express the following node admittance matrix as a sum of matrices:

$$
\begin{array}{c} \\ 1 \\ 2 \\ 3 \end{array}
\begin{array}{ccc} 1 & 2 & 3 \end{array}
\begin{bmatrix}
y_1 + y_2 & -y_2 - y_3 & y_3 \\
-y_2 + y_4 & y_2 + y_3 + y_5 & -y_3 \\
0 & -y_5 & y_5 + y_6
\end{bmatrix}
$$

$$
= \begin{bmatrix}
y_1 & 0 & 0 \\
0 & 0 & 0 \\
0 & 0 & 0
\end{bmatrix}
+ \begin{bmatrix}
y_2 & -y_2 & 0 \\
-y_2 & y_2 & 0 \\
0 & 0 & 0
\end{bmatrix}
+ \begin{bmatrix}
0 & -y_3 & y_3 \\
0 & y_3 & -y_3 \\
0 & 0 & 0
\end{bmatrix}
$$

$$+ \begin{bmatrix} 0 & 0 & 0 \\ y_4 & 0 & 0 \\ 0 & 0 & 0 \end{bmatrix} + \begin{bmatrix} 0 & 0 & 0 \\ 0 & y_5 & -y_5 \\ 0 & -y_5 & y_5 \end{bmatrix} + \begin{bmatrix} 0 & 0 & 0 \\ 0 & 0 & 0 \\ 0 & 0 & y_6 \end{bmatrix}$$

where each matrix contains only one admittance. From these matrices we can see which one is a passivelike element and which is an activelike element *where an activelike element is one which is defined by Eq. 8-1-1 but cannot be represented by one edge.* Furthermore, the locations of voltage and current edges of each admittance can easily be obtained. In this example, the first, second, fifth, and sixth matrices show that y_1, y_2, y_5, and y_6 are passivelike admittances and that y_3 and y_4 are activelike admittances. From these matrices, we can obtain a linear graph of a network whose admittance matrix is the same as the one given as shown in Fig. 8-1-10. It is not necessary to change a given admittance matrix into a sum of such matrices to obtain those informations, obviously.

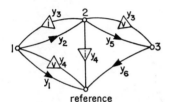

Fig. 8-1-10. A network.

Consider a transformer given by

$$\begin{bmatrix} \dfrac{L_{22}}{s\Delta_m} & \dfrac{-M_{12}}{s\Delta_m} \\ \dfrac{-M_{21}}{s\Delta_m} & \dfrac{L_{11}}{s\Delta_m} \end{bmatrix} \begin{bmatrix} v_{11'} \\ v_{22'} \end{bmatrix} = \begin{bmatrix} i_{11'} \\ i_{22'} \end{bmatrix} \tag{8-1-10}$$

where

$$\Delta_m = \begin{bmatrix} L_{11} & M_{12} \\ M_{21} & L_{22} \end{bmatrix} \tag{8-1-11}$$

The topological representation of such a transformer is given in Fig. 8-1-11.

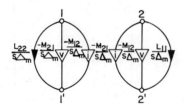

Fig. 8-1-11. Topological representation of a transformer.

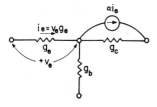

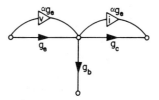

Fig. 8-1-12. An equivalent circuit for a transistor.

Fig. 8-1-13. Topological representation of a transistor.

The topological representation of an equivalent circuit for the transistor in Fig. 8-1-12 is shown in Fig. 8-1-13. Another example is the linear graph of the network in Fig. 8-1-14, shown in Fig. 8-1-15, where $g_1' = \alpha g_1$, $y_1 = L_2/s\Delta_m$, $y_2 = y_3 = -M_{12}/s\Delta_m$, and $y_4 = L_1/s\Delta_m$. Note that we use y_2 and y_3 to distinguish different admittances even if the values of these admittances are the same. Also note that we give arbitrary orientations to all passivelike elements in the figure.

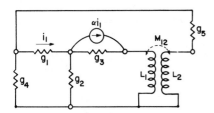

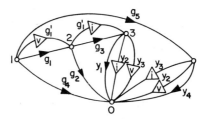

Fig. 8-1-14. A network containing a transistor.

Fig. 8-1-15. A linear graph corresponding to the network in Fig. 8-1-14.

Consider a linear graph corresponding to a network such as that in Fig. 8-1-15. Each admittance is represented by two edges, one a voltage edge and the other a current edge. For a passivelike admittance, one edge can represent both voltage and current edges by definition. Hence we can form two linear graphs G_v and G_i such that G_v consists of all voltage edges and all vertices of G and G_i consists of all current edges and all vertices of G. Note that if there is an admittance y in G_v, there is y in G_i and vice versa.

Definition 8-1-2. Let G be a linear graph representing a network. Linear graph G_v, which consists of all voltage edges of G, is called a *voltage graph*, and linear graph G_i, consisting of all current edges in G, is called a *current graph*.

Since there are no current edges in voltage graph G_v, it is not necessary to use the symbol v inside of the arrow of each edge. Thus each edge in G_v can be drawn as one used for a passivelike element. Similarly, each edge in current graph G_i can be drawn without using the symbol i. For example, the voltage and the current graphs of a linear graph in Fig. 8-1-15 are shown in Fig. 8-1-16. Note that when a network consists only of passivelike admittances, the voltage and the current graphs are identical.

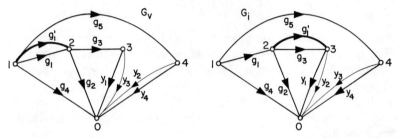

Fig. 8-1-16. Voltage and current graphs G_v and G_i of the linear graph in Fig. 8-1-15.

Each admittance in a network satisfies Eq. 8-1-1. Hence

$$\begin{aligned} y_1 v_1' &= i_1' \\ y_2 v_2' &= u_2' \\ &\cdots \\ y_e v_e' &= i_e' \end{aligned} \tag{8-1-12}$$

or

$$YV_e = I_e \tag{8-1-13}$$

where

$$V_e = \begin{bmatrix} v_1' \\ v_2' \\ \vdots \\ v_e' \end{bmatrix} \tag{8-1-14}$$

$$I_e = \begin{bmatrix} i_1' \\ i_2' \\ \vdots \\ i_e' \end{bmatrix} \tag{8-1-15}$$

and

$$Y = \begin{bmatrix} y_1 & & & 0 \\ & y_2 & & \\ & & \ddots & \\ 0 & & & y_e \end{bmatrix} \qquad (8\text{-}1\text{-}16)$$

Since Kirchhoff's current law is a condition on currents but not on voltages in a network, the incidence matrix (cut-set matrix) in Eq. 7-1-2 must be an incidence matrix of the current graph of a network. Let A_i be an incidence matrix of current graph G_i. Then Eq. 7-1-2 becomes

$$A_i I_e = 0 \qquad (8\text{-}1\text{-}17)$$

With Eq. 8-1-13, we have

$$A_i Y V_e = 0 \qquad (8\text{-}1\text{-}18)$$

Since the node transformation given by Eq. 7-3-8 is a relationship among voltages but not currents, the incidence matrix (cut-set matrix) in Eq. 7-3-8 must be an incidence matrix (cut-set matrix) of a voltage graph. Let A_v be an incidence matrix of voltage graph G_v. Then Eq. 7-3-8 can be written as

$$V_e = A_v^t V_n \qquad (8\text{-}1\text{-}19)$$

where each element in V_n is a node voltage. This, with Eq. 8-1-18, gives

$$A_i Y A_v^t V_n = 0 \qquad (8\text{-}1\text{-}20)$$

where $A_i Y A_v^t$ is called a node admittance matrix.

By the use of the Binet-Cauchy theorem, the determinant of node admittance matrix $A_i Y A_v^t$ can be expressed as

$$|A_i Y A_v^t| = \sum \text{major determinant of } A_i Y$$
$$\times \text{ corresponding major determinant of } A_v^t \qquad (8\text{-}1\text{-}21)$$

Since Y is a diagonal matrix, Eq. 8-1-21 can be written as

$$|A_i Y A_v^t| = \sum y_{k_1} y_{k_2} \cdots y_{k_m} |A_i(k_1 k_2 \cdots k_m)| \, |A_v^t(k_1 k_2 \cdots k_m)| \qquad (8\text{-}1\text{-}22)$$

where $A_i(k_1 k_2 \cdots k_m)$ is a submatrix of A_i consisting of columns $k_1, k_2, \ldots, k_m$ and $A_v(k_1 k_2 \cdots k_m)$ is a submatrix of A_v consisting of columns $k_1, k_2, \ldots, k_m$ where m is the number of rows in A_i (and A_v) which is $n_v - 1$ for a connected linear graph with n_v vertices.

We have seen that $A_i(k_1 k_2 \cdots k_m)$ is nonsingular if and only if the edges corresponding to columns $k_1, k_2, \ldots, k_m$ is a tree of G_i (in the case of a separated linear graph, it must be a forest). Similarly, $A_v(k_1 k_2 \cdots k_m)$ is nonsingular if and only if the edges corresponding to columns $k_1, k_2, \ldots, k_m$

form a tree of G_v. Thus in order that both $A_i(k_1k_2\cdots k_m)$ and $A_v(k_1k_2\cdots k_m)$ be nonsingular, edges whose weights are $y_{k_1}, y_{k_2}, \ldots, y_{k_m}$ must be tree in both G_i and G_v. Such a set of edges is called a complete tree (another name for the set is a common tree).

Definition 8-1-3.   A *complete tree* is a set of edges which is a tree of both voltage and current graphs.

With this definition, Eq. 8-1-22 can be expressed as

$$|A_i\,YA_v{}^t| = \sum_{(k)} \epsilon_k \text{ complete tree admittance product} \qquad (8\text{-}1\text{-}23)$$

where

$$\epsilon_k = |A_i(K_1k_2\cdots k_m)|\,|A_v(k_1k_2\cdots k_m)| = \pm 1 \qquad (8\text{-}1\text{-}24)$$

In the case of networks (consisting of passivelike elements), we can see that complete trees in Eq. 8-1-23 become just trees and ϵ_k is always $+1$. However, for networks containing activelike elements, we will find that ϵ_k for a complete tree can be either $+1$ or -1. Hence we must find some way of calculating ϵ_k for each complete tree.

8-2   SIGN-PERMUTATION

We will now see that the sign-permutation of a complete tree is the sign ϵ_k in Eq. 8-1-24. To study a sign-permutation, we must define a principal vertex as follows.

Definition 8-2-1.   Let G_t be a tree of a connected oriented graph G. Let u and v be the endpoints of edge y in G_t. Then vertex u is called the *principal vertex* of edge y if the path between vertex u and the reference vertex contains edge y.

Since a tree is a connected graph and contains no circuits, there exists exactly one path between any two vertices in the tree. Hence if the path from vertex u to the reference vertex contains edge y, the path from vertex v to the reference vertex does not contain edge y. Thus exactly one of two endpoints of an edge is the principal vertex of the edge. For example, the principal vertex of each edge in G_t in Fig. 8-2-1 is shown in Table 8-2-1. Note that the

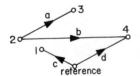

Fig. 8-2-1.   A tree G_t.

reference vertex cannot be the principal vertex of any edge. Moreover, since G_t is a tree, any vertex is the principal vertex of exactly one edge.

Table 8-2-1   Principal Vertices

Edge	Principal Vertex
a	3
b	2
c	1
d	4

Suppose G_t and G_u are two linear graphs, each representing a tree and both consisting of the same admittances as shown in Fig. 8-2-2. Then for

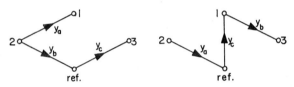

Fig. 8-2-2.   Linear graphs G_t and G_u.

each edge, there exist two principal vertices, one from G_t and the other from G_u. For example, each pair of principal vertices of edges in G_t and G_u in Fig. 8-2-2 is shown in Table 8-2-2. We can now define the sign-permutation,

Table 8-2-2   Pair of Principal Vertices

Edge	Principal Vertices	
	G_t	G_u
y_a	1	2
y_b	2	3
y_c	3	1

corresponding to two trees G_t and G_u, both of which consist of the same admittances.

Definition 8-2-2.   Let G_1 and G_2 be connected oriented graphs that consist of the same edges and the same vertices and contain no circuits. Then a *sign-*

permutation of G_1 and G_2 is an alley of two rows with the following characteristics:

1. Row p represents G_p for $p = 1, 2, \ldots$.
2. Each column represents a principal vertex.
3. The (p, q) entry for $p = 1, 2$ is

y if vertex q is the principal vertex of y and the orientation of y is
away from q in G_p.

and

y^- if vertex q is the principal vertex of y and the orientation of y is
toward q in G_p.

For example, the sign-permutation (SP) of the set of edges corresponding
to G_t and G_u in Fig. 8-2-2 is as follows:

$$\text{SP} = \begin{array}{c} \\ G_t \\ G_u \end{array} \begin{array}{ccc} 1 & 2 & 3 \\ \left(\begin{matrix} y_a^- & y_b & y_c^- \\ y_c^- & y_a & y_b^- \end{matrix}\right) \end{array}$$

Definition 8-2-3.   A sign-permutation is said to be even if the sum of the
following quantities is even; otherwise it is said to be odd:

1. The number of negative superscripts in the sign-permutation.
2. The number of transpositions necessary to rearrange the second row to
make it identical to the first row.

Accordingly, the sign-permutation of the preceding example is even.

Definition 8-2-4.   The value of a sign-permutation is $+1$ if the sign-
permutation is even and -1 otherwise.

For example, since the sign-permutation of the foregoing example is even,
the value of the sign-permutation, symbolized by $V[\text{SP}]$ is

$$V\left[\left(\begin{matrix} y_a^- & y_b & y_c^- \\ y_c^- & y_a & y_b^- \end{matrix}\right)\right] = 1$$

8-3   PRINCIPAL TREE

To show that the value of the sign-permutation of a complete tree is ϵ_k in
Eq. 8-1-23, we begin by studying the sign-permutation of a pair of particular
trees known as principal trees. Then we show that the value of a sign-permutation is invariant when a complete tree is transformed into a pair of principal
trees.

The structure of a tree shown in Fig. 8-3-1 is called a *principal tree*. In other words, if all edges $y_1, y_2, \ldots, y_m$ representing a tree are connected to the reference vertex with no restrictions on the orientations of each edge, the result is called a principal tree.

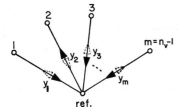

Fig. 8-3-1. A principal tree.

Consider an incidence matrix A_p of a principal tree. We know that A_p is of order $n_v - 1$ and is nonsingular where n_v is the number of vertices in the tree. Since one of two endpoints of every edge is the reference vertex, each column of A_p has only one nonzero, which is either $+1$ or -1. Furthermore, from the definition of the principal vertex of an edge, we can say that the location of the nonzero element in A_p corresponding to edge y is the intersection of the column representing y and the row representing the principal vertex of y. Finally, since A_p is an incidence matrix, it is known that the determinant of A_p is either $+1$ or -1. The incidence matrix A_p of a principal tree in Fig. 8-3-1 is as follows:

$$
A_p = \begin{array}{c} \\ 1 \\ 2 \\ 3 \\ \\ m \end{array}
\begin{array}{c} \begin{array}{ccccc} y_1 & y_2 & y_3 & \cdots & y_m \end{array} \\
\left[\begin{array}{ccccc}
1 & 0 & 0 & \cdots & 0 \\
0 & -1 & 0 & \cdots & 0 \\
0 & 0 & 1 & \cdots & 0 \\
& & \cdots & & \\
0 & 0 & 0 & \cdots & 1
\end{array} \right] \end{array}
\qquad (8\text{-}3\text{-}1)
$$

By interchanging two rows of this matrix, the sign of the determinant of A_p changes. Suppose the principal vertex of edge y_k is v and the principal vertex of edge y_p is u. Then there are ± 1 at the (v, k) entry and ± 1 at the (u, p) entry of A_p. Interchanging row v and row u changes ± 1 at the (v, k) entry to the (u, k) entry and ± 1 at the (u, p) entry to the (v, p) entry, which means that the principal vertex of edge y_k is changed to vertex u and the principal vertex of edge y_p is changed to vertex v. Hence this interchange is equivalent to the interchange of the locations of edges y_k and y_p. For example, interchanging the first and the second rows of the preceding matrix gives the following matrix:

$$
A_p' = \begin{array}{c} \\ 1 \\ 2 \\ 3 \\ \\ m \end{array}
\begin{array}{ccccc}
y_1 & y_2 & y_3 & \cdots & y_m \\
\left[\begin{array}{ccccc}
0 & -1 & 0 & \cdots & 0 \\
1 & 0 & 0 & \cdots & 0 \\
0 & 0 & 1 & \cdots & 0 \\
& & \cdots & & \\
0 & 0 & 0 & \cdots & 1
\end{array}\right]
\end{array}
\qquad (8\text{-}3\text{-}2)
$$

This new matrix A_p' is an incidence matrix of the principal tree shown in Fig. 8-3-1 with the locations of y_1 and y_2 interchanged. The number N of interchanges of locations of edges in a principal tree therefore affects the determinant of A_p by $(-1)^N$.

The multiplication of a column of A_p by -1 is equivalent to changing the orientation of the edge corresponding to the column. For example, multiplying column y_2 by -1 gives

$$
A_p'' = \begin{array}{c} \\ 1 \\ 2 \\ 3 \\ \\ m \end{array}
\begin{array}{ccccc}
y_1 & y_2 & y_3 & \cdots & y_m \\
\left[\begin{array}{ccccc}
1 & 0 & 0 & \cdots & 0 \\
0 & 1 & 0 & \cdots & 0 \\
0 & 0 & 1 & \cdots & 0 \\
& & \cdots & & \\
0 & 0 & 0 & \cdots & 1
\end{array}\right]
\end{array}
\qquad (8\text{-}3\text{-}3)
$$

This new matrix A_p'' is an incidence matrix of the principal tree in Fig. 8-3-1 except that the orientation of y_2 is opposite.

Suppose there are M edges in a principal tree whose orientations are away from the reference vertex. Then by changing the orientations of all these edges by multiplying corresponding columns by -1, all nonzero entries of incidence matrix A_p of the principal tree become positive. Let the resultant matrix of A_p'''. Then the determinant of A_p is equal to $(-1)^M$ times the determinant of A_p'''. For example, consider incidence matrix A_p of the principal tree shown in Fig. 8-3-2:

$$
A_p = \begin{array}{c} \\ 1 \\ 2 \\ 3 \\ 0 \end{array}
\begin{array}{cccc}
y_1 & y_2 & y_3 & y_4 \\
\left[\begin{array}{cccc}
0 & 0 & -1 & 0 \\
0 & 1 & 0 & 0 \\
-1 & 0 & 0 & 0 \\
0 & 0 & 0 & -1
\end{array}\right]
\end{array}
$$

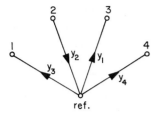

ref. **Fig. 8-3-2.** A principal tree.

By multiplying columns y_1, y_3, and y_4 by -1, which is equivalent to reversing the orientations of edges y_1, y_3, and y_4, we have a new incidence matrix A_p''' where

$$
\begin{array}{c}
\quad\;\; y_1 \;\; y_2 \;\; y_3 \;\; y_4 \\
\begin{array}{c}1\\2\\3\\4\end{array}
\begin{bmatrix}
0 & 0 & 1 & 0 \\
0 & 1 & 0 & 0 \\
1 & 0 & 0 & 0 \\
0 & 0 & 0 & 1
\end{bmatrix}
\end{array}
$$

Hence

$$|A_p| = (-1)^3 |A_p'''|$$

Consider two principal trees t_i and t_v shown in Fig. 8-3-3 where t_i and t_v consist of the same admittances (weights of edges). Let A_{t_i} and A_{t_v} be the

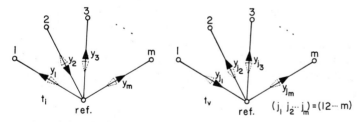

Fig. 8-3-3. Pair of principal tree.

incidence matrices of t_i and t_v, respectively. Let M_i and M_v be the number of times we multiply by -1 columns of A_{t_i} and A_{t_v}, respectively, in order to make every nonzero entry positive, and A_{t_i}'' and A_{t_v}'' be the resultant incidence matrices. Then

$$|A_{t_i}|\,|A_{t_v}| = (-1)^{M_i + M_v}|A_{t_i}''|\,|A_{t_v}''| \tag{8-3-4}$$

Let N be the number of interchanges of the locations of edges in t_v necessary to make the location of every edge in t_v the same as that of the corresponding edge (the edge that has the same weight) in tree t_i. Then the determinant of A''_{t_v} is equal to $(-1)^N$ times the determinant of A''_{t_i}, or

$$|A''_{t_i}| = (-1)^N |A''_{t_v}| \tag{8-3-5}$$

Thus, with Eq. 8-3-4, we have

$$|A_{t_i}| \, |A_{t_v}| = (-1)^{M_i + M_v + N} |A''_{t_v}|^2 \tag{8-3-6}$$

Since A''_{t_v} is an incidence matrix of an oriented linear graph and is non-singular, the determinant of A''_{t_v} is either $+1$ or -1. Thus Eq. 8-3-6 becomes

$$|A_{t_i}| \, |A_{t_v}| = (-1)^{M_i + M_v + N} \tag{8-3-7}$$

It can be seen that $M_i + M_v$ is equal to the number of negative superscripts in the sign-permutation of t_i and t_v and N is equal to the number of interchanges of edges in one row of the sign-permutation in order to make two rows identical. Hence $(-1)^{M_i + M_v + N}$ is equal to the value of the sign-permutation of t_i and t_v, or

$$(-1)^{M_i + M_v + N} = V[\text{SP} \quad \text{of } t_i \text{ and } t_v] \tag{8-3-8}$$

where t_i and t_v are principal trees.

Consider a set of admittances $(y_1, y_2, \ldots, y_m)$ corresponding to a complete tree, that is, edges $y_1, y_2, \ldots, y_m$ form a tree t_i in current graph G_i and at the same time they form a tree t_v in voltage graph G_v. Let $A_i(12 \cdots m)$ and $A_v(12 \cdots m)$ be incidence matrices of t_i and t_v, respectively. Thus $A_i(12 \cdots m)$ is a submatrix of incidence matrix A_i of current graph G_i and $A_v(12 \cdots m)$ is a submatrix of incidence matrix A_v of voltage graph G_v. We are going to show that the determinant of $A_i(12 \cdots m) \cdot A_v(12 \cdots m)$ is equal to the value of the sign-permutation of t_i and t_v even if t_i and t_v are not principal trees. To show this, we first show that we can change any tree to a principal tree by elementary operations on incidence matrix A_t of a tree so that the determinant is not affected.

Let $G_t(n_v - 1)$ be a linear graph representing a tree, where $G_t(n_v - 1)$ contains n_v vertices. Hence $G_t(n_v - 1)$ contains $n_v - 1$ edges and no circuits.

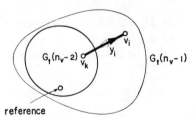

reference   **Fig. 8-3-4.**   A tree $G_t(n_v - 1)$.

Since $G_t(n_v - 1)$ is a tree, there are at least two vertices of degree 1, which have been studied before. Let one of these two vertices which is not the reference vertex be v_i and edge y_i be connected to vertex v_i. Note that because v_i is of degree 1, only edge y_i is connected to v_i in $G_v(n_v - 1)$.

Let v_k be the other endpoint of y_i as shown in Fig. 8-3-4. Consider incidence matrix A_t of $G_t(n_v - 1)$. Since y_i is only the edge connected to vertex v_i and v_i is not the reference vertex, the row in A_t corresponding to v_i has only one nonzero entry at the column corresponding to y_i as shown below:

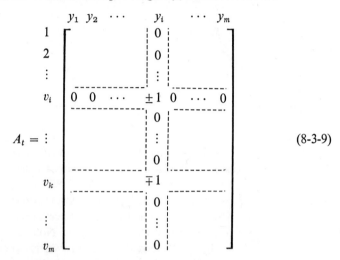

$$(8\text{-}3\text{-}9)$$

There is ∓ 1 at the (v_k, y_i) entry because y_i is connected between vertices v_i and v_k unless v_k is the reference vertex.

By adding row v_i to row v_k, we can remove ∓ 1 at (v_k, y_i) entry:

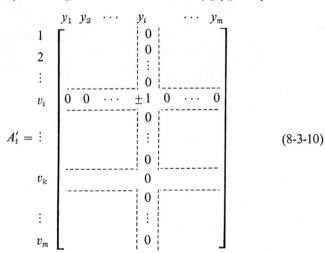

$$(8\text{-}3\text{-}10)$$

Note that the determinant of A_t is not changed by this operation. On the other hand, the resultant matrix A'_t is an incidence matrix of a tree which is the same as $G_t(n_v - 1)$ except that edge y_i is connected between vertex v_i and the reference vertex, not vertex v_k, as shown in Fig. 8-3-5. Note that in $G_t(n_v - 1)$, vertex v_i is the principal vertex of y_i. If vertex v_k is the reference vertex, the preceding operation is unnecessary.

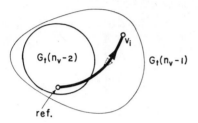

Fig. 8-3-5.   A modified tree of $G_t(n_v - 1)$.

Subgraph $G_t(n_v - 2)$ formed by taking edge y_i and vertex v_i out of $G_t(n_v - 1)$ has $n_v - 1$ vertices, $n_v - 2$ edges, and no circuits. Hence $G_t(n_v - 2)$ can be considered a tree of a linear graph. Thus there exist at least two vertices of degree 1 in $G_t(n_v - 2)$. Let v_j be one of these two vertices which is not the reference vertex. Also let edge y_j be the one connected to vertex v_j. Suppose edge y_j is connected between vertices v_m and v_j. Note that vertex v_j is the principal vertex of edge y_j. If v_m is the reference vertex, then the column y_j has only one nonzero. If this vertex is not the reference vertex, then by the operation mentioned previously, edge y_j can be shifted to the reference vertex without changing the determinant of the incidence matrix of this tree.

Continuing the process, we can change a tree to a principal tree without changing the determinant of the incidence matrix of the tree. Furthermore, each edge in a tree is shifted so that the edge is connected between the principal vertex of the edge and the reference vertex. Suppose we call the principal tree of a given tree resulting from this process the *corresponding principal tree* of a given tree. Then we can see that the determinant of an incidence matrix of a tree is equal to the determinant of an incidence matrix of the corresponding principal tree.

Theorem 8-3-1.   Let $y_{k_1}, y_{k_2}, \ldots, y_{k_m}$ be edges in a complete tree, where $t_i = (y_{k_1} y_{k_2} \cdots y_{k_m})$ is a tree in current graph G_i and $t_v = (y_{k_1} y_{k_2} \cdots y_{k_m})$ is a tree in voltage graph G_v. Also let $A_i(k_1 k_2 \cdots k_m)$ be an incidence matrix of t_i and $A_v(k_1 k_2 \cdots k_m)$ be an incidence matrix of t_v both of which define the same reference vertex. Then

$$\epsilon_k = |A_i(k_1 k_2 \cdots k_m)| \, |A_v(k_1 k_2 \cdots k_m)| = V[\text{SP of } t_i \text{ and } t_v] \quad (8\text{-}3\text{-}11)$$

Since we can obtain the principal vertex of each edge directly from t_i and t_v, we can obtain ϵ_k directly from the sign-permutation of t_i and t_v without changing them to principal trees.

It is clear that if a complete tree consists of passivelike elements only, then $A_i(k_1 k_2 \cdots k_m)$ and $A_v(k_1 k_2 \cdots k_m)$ are identical.

Rule If a complete tree consists of passivelike elements only, then ϵ_k of the complete tree is $+1$.

Example 8-3-1. Consider linear graph G shown in Fig. 8-3-6. We can

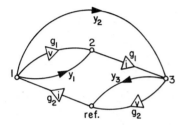

Fig. 8-3-6. A linear graph G.

obtain current graph G_i and voltage graph G_v as shown in Fig. 8-3-7. From these, all complete trees and the sign of each complete tree can be obtained

Table 8-3-1   Complete Trees and Their Signs

Tree of G_i	Is Tree on the Left Side a Tree of G_v?	Sign	
$y_1 y_2 y_3$	Yes	$+1$	Only passivelike elements
$y_1 y_2 g_2$	Yes	-1	$V\begin{bmatrix} t_i \\ t_v \end{bmatrix}\begin{pmatrix} \overset{1}{g_2^-} & \overset{2}{y_1^-} & \overset{3}{y_2^-} \\ y_2 & y_1^- & g_2 \end{pmatrix} = -1$
$y_1 g_1 y_3$	No		
$y_1 g_1 g_2$	No		$V\begin{bmatrix} t_i \\ t_v \end{bmatrix}\begin{pmatrix} \overset{1}{y_2} & \overset{2}{g_1} & \overset{3}{y_3} \\ y_2 & g_1^- & y_3 \end{pmatrix} = -1$
$y_2 y_3 g_1$	Yes	-1	
$y_2 y_3 g_2$	No		
$y_2 g_1 g_2$	Yes	$+1$	$V\begin{bmatrix} t_i \\ t_v \end{bmatrix}\begin{pmatrix} \overset{1}{g_2^-} & \overset{2}{g_1} & \overset{3}{y_2^-} \\ y_2 & g_1^- & g_2 \end{pmatrix} = 1$
$y_3 g_1 g_2$	No		

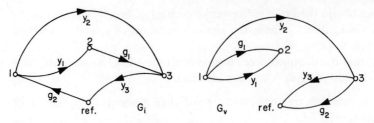

Fig. 8-3-7.   Current and voltage graphs.

shown in Table 8-3 Hence

$$|A_i Y (A_v)^t| = y_1 y_2 y_3 - y_1 y_2 g_2 - y_2 y_3 g_1 + y_2 g_1 g_2$$

8-4   OPEN-CIRCUIT DRIVING POINT FUNCTION

Since a topological formula for the determinant of a node admittance matrix has been studied in the previous section, we need study only a cofactor Δ_{ii} of a node admittance matrix to obtain a topological formula for an open-circuit driving point admittance function.

Let Δ' be the determinant of a node admittance matrix of the network shown in Fig. 8-4-1 which consists of a given network N and an admittance y connected between the input vertices i and o of N.

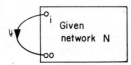

Fig. 8-4-1.   A network N with y.

From Eq. 8-1-23, Δ' can be written as

$$\Delta' = \sum_{(j)} \epsilon_j \text{ complete tree } t_j \text{ admittance product of network} \qquad (8\text{-}4\text{-}1)$$
$$N \text{ with } y$$

Some of these complete trees contain y and other do not. The complete trees that do not contain y are those of given network N. Hence Eq. 8-4-1 can be expressed as

$$\Delta' = \sum_{(j)} \epsilon_j \text{ complete tree } t_j \text{ admittance product of } N + \sum_{(j)} \epsilon_i$$
$$\text{complete tree } t_i \text{ admittance product of complete tree } t_i \qquad (8\text{-}4\text{-}2)$$
$$\text{containing } y$$

Or

$$\Delta' = V + yW_{(i,o;i,o)} \tag{8-4-3}$$

where V and $W_{(i,o;i,o)}$ are defined as follows.

Definition 8-4-1.

$$V = \sum_{(j)} \epsilon_j \text{ complete tree } t_j \text{ admittance product of a given} \tag{8-4-4}$$
network N

and

$$W_{(i,o;i,o)} = \frac{1}{y} \sum_{(j)} \epsilon_r \text{ complete tree } t_r(y) \text{ admittance product}$$
where $t_r(y)$ is a complete tree containing y \tag{8-4-5}

We can now define the symbol $W_{(j,k;i,o)}$ in general.

Definition 8-4-2.

$$W_{(j,k;i,o)} = \frac{1}{y} \sum_{(k)} \epsilon_k \text{ complete tree } t_k(y) \text{ admittance product} \tag{8-4-6}$$

where $t_k(y)$ is a complete tree containing y. The location of y is indicated by the subscripts of $W_{(j,k;i,o)}$, that is, the current edge of y is from j to k and the voltage edge of y is from i to o.

Since there are no external sources in the network in Fig. 8-4-1, we have

$$A_i Y(A_v)^t V_n = 0 \tag{8-4-7}$$

In order to have nontrivial solutions,

$$|A_i Y(A_v)^t| = \Delta' = 0 \tag{8-4-8}$$

Hence, by Eq. 8-4-3,

$$V + yW_{(i,o;i,o)} = 0 \tag{8-4-9}$$

or

$$-y = \frac{V}{W_{(i,o;i,o)}} \tag{8-4-10}$$

The topological representation of $-y$ is shown in Fig. 8-4-2. It can be seen that, by definition,

$$-y = \frac{J_i}{v_i} \tag{8-4-11}$$

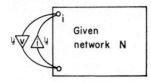

Fig. 8-4-2. Network N with y.

where J_i is the current from the reference vertex o to vertex i and v_i is the voltage from vertex i to the reference vertex o. Hence $-y$ is the open-circuit driving point admittance $Y_{o_{i,o}}$ of network N between vertices i and o. Thus the right-hand side of Eq. 8-4-10 gives the topological formula for the driving point admittance function between vertices i and o:

$$Y_{o_{i,o}} = \frac{V}{W_{(i,o;i,o)}} \qquad (8\text{-}4\text{-}12)$$

From Eq. 8-4-6, we can see that $W_{(i,o;i,o)}$ can be obtained by the following process:

1. Place y between vertices i and o.
2. Find all complete trees containing y.
3. Determine the sign of each complete tree containing y.
4. Sum all these complete tree admittance products without y and with proper signs found by step 3. The result is $W_{(i,o;i,o)}$.

For example, $W_{(1,3;1,3)}$ of G in Fig. 8-3-6 where 3 is chosen as the reference vertex can be found by the following process:

1. Place y between vertices 1 and 3 of G as shown in Fig. 8-4-3.

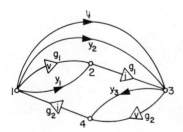

Fig. 8-4-3. Linear graph G with y between vertices 1 and 3.

2. Obtain current graph G'_i and voltage graph G'_v of the resultant graph as shown in Fig. 8-4-4.

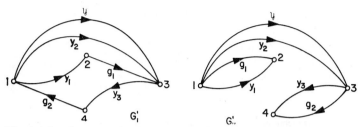

Fig. 8-4-4. Current and voltage graphs of the linear graph of Fig. 8-3-7.

3. Find all possible trees containing y of current graph G_i':

$$yy_1y_3, \quad yy_1g_2, \quad yy_3g_1, \quad yy_3g_2, \quad yg_1g_2$$

4. Select the trees obtained by Step 3 which are also trees of voltage graph G_v':

$$yy_1y_3, \quad yy_1g_2, \quad yy_3g_1, \quad yg_1g_2$$

5. Determine the sign of each complete tree obtained by Step 4. Since y and y_2 are in parallel, the signs of these complete trees are the same as those in Table 8-3-1 by replacing y_2 by y.

6. Form $W_{(1,3;1,3)}$:

$$W_{(1,3;1,3)} = y_1y_3 - y_1g_2 - y_3g_1 + g_1g_2$$

We can obtain the same result differently. It is obvious that the open-circuit driving point functions can be obtained by knowing the determinant and cofactor Δ_{pp}' of a node admittance matrix $A_i Y(A_v)^t$. Since we have already studied the determinant of $A_i Y(A_v)^t$, we need only consider cofactor Δ_{pp}' of $A_i Y(A_v)^t$. It is also clear that

$$\Delta_{pp}' = |A_{i-p} Y(A_{v-p})^t| \tag{8-4-13}$$

where A_{i-p} is obtained from A_i by deleting row p and A_{v-p} is obtained from A_v by deleting row p. It is known that A_{i-p} is an incidence matrix of linear graph $G_i(p = o)$ obtained from G_i by coinciding vertex p and the reference vertex o. Moreover, A_{v-p} is an incidence matrix of linear graph $G_v(p = o)$ obtained from G_v by coinciding vertex p and the reference vertex o. Thus

$$\Delta_{pp}' = \sum_{(j)} \epsilon_j \text{ complete tree admittance product of } G_i(p = o)$$

$$\text{and } G_v(p = o) \tag{8-4-14}$$

Let a set of edges $(y_{k_1} y_{k_2} \cdots y_{k_m})$ be a complete tree of $G_i(p = o)$ and $G_v(p = o)$. Then the subgraph consisting of edges $y_{k_1}, y_{k_2}, \ldots, y_{k_m}$ is called a complete 2-tree, symbolized by $t_{2_{p,o}}$. Since $t_{2_{p,o}}$ becomes a complete tree if vertex p and the reference vertex o are coincided, $t_{2_{p,o}}$ consists of two parts, one containing vertex p and the other containing vertex o. Furthermore, these two parts contain all vertices in G and contain no circuits.

With this complete 2-tree, Eq. 8-4-14 can be expressed as

$$\Delta'_{pp} = \sum_{(j)} \epsilon_j \text{ complete 2-tree } t_{2_{p,o}} \text{ admittance product} \qquad (8\text{-}4\text{-}15)$$

where ϵ_j is the sign-permutation of complete 2-tree $t_{2_{p,o}}$ by considering vertices p and o the same (and vertex o being the reference vertex). It is clear that if we add edge y between vertex p and the reference vertex, y with all edges in the complete 2-tree $t_{2_{p,o}}$ become a tree in the resulting graph. Thus Eq. 8-4-5 is true and is equal to Δ'_{pp}. For example, coinciding vertices 1 and 3 of the current and the voltage graphs in Fig. 8-3-7 gives $G_i(1 = 3)$ and $G_v(1 = 3)$ as shown in Fig. 8-4-5, from which we can obtain $W_{(1,3;1,3)}$ easily.

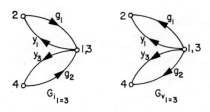

Fig. 8-4-5. $G_i(1 = 3)$ and $G_v(1 = 3)$ of G_i and G_v in Fig. 8-3-7.

8-5 OPEN-CIRCUIT TRANSFER FUNCTIONS

In Fig. 8-4-1, we connect admittance y between input vertices i and o of network N to obtain a cofactor Δ_{ii}. To calculate a cofactor for an open-circuit transfer function topologically, admittance y satisfying

$$y i_{i,o} = v_{j,k} \qquad (8\text{-}5\text{-}1)$$

is used as shown in Fig. 8-5-1 where i and o are the input vertices and j and k are the output vertices of network N.

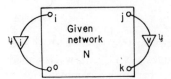

Fig. 8-5-1. Network with activelike y

Now the determinant of a node admittance matrix Δ' of this network is

$$\Delta' = \sum_{(j)} \epsilon_j \text{ complete tree admittance product}$$

$$= \sum_{(j)} \epsilon_j \text{ complete tree admittance product of complete}$$

trees which do not contain y

$$+ \sum_{(j')} \epsilon_{j'} \text{ complete tree admittance product of complete}$$

trees which contain y (8-5-2)

Or

$$\Delta' = V + y W_{(i,o;j,k)} \tag{8-5-3}$$

where

$$V = \sum_{(j)} \epsilon_j \text{ complete tree admittance product of network } N \tag{8-5-4}$$

and

$$W_{(i,o;j,k)} = \frac{1}{y} \sum_{(j')} \epsilon_{j'} \text{ complete tree admittance product}$$

of complete trees which contain y (8-5-5)

With the same argument as that given previously, Δ' must be zero or

$$-y = \frac{V}{W_{(i,o;j,k)}} \tag{8-5-6}$$

Since $-y$ is equivalent to reversing the orientation of the current edge of y, as shown in Fig. 8-5-2, we can see that $-y$ is equal to the open-circuit transfer admittance function $Y_{o_{i,o;j,k}}$. Hence

$$Y_{o_{i,o;j,k}} = \frac{V}{W_{(i,o;j,k)}} \tag{8-5-7}$$

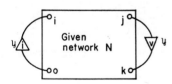

Fig. 8-5-2. Network N with y.

The procedure of finding $W_{(i,o;j,k)}$ is the same as that of finding $W_{(i,o;i,o)}$ except the current edge of y is connected from vertex i to vertex o and the voltage edge of y is connected from vertex j to vertex k. Note that the orientation of the current edge of y is opposite to the direction of current $J_{i,o}$ of the definition of network functions (which is from vertex o to vertex i). For example, $W_{(1,3;2,4)}$ of G in Fig. 8-3-6 can be obtained by the following procedure:

1. Place the current edge of y from vertex 1 to vertex 3 and the voltage edge of y from vertex 2 to vertex 4 (Fig. 8-5-3).

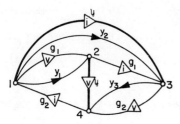

Fig. 8-5-3.   Linear graph G with y, $G(y)$.

2. Draw the current and the voltage graphs of the resulting linear graph (Fig. 8-5-4).

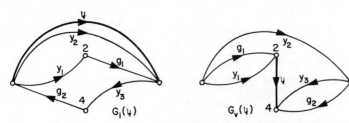

Fig. 8-5-4.   The current graph $G_i(y)$ and the voltage graph $G_v(y)$ of G of Fig. 8-5-3.

3. Find all possible trees containing y of the current graph:

$$yy_1y_3, \quad yy_1g_2, \quad yy_3g_1, \quad yg_1g_2$$

4. Select the trees obtained by Step 3 which are also the trees in the voltage graph:

$$yy_1y_3, \quad yy_1g_2, \quad yy_3g_1, \quad yg_1g_2$$

5. Determine the sign of each complete tree obtained by Step 4:

$$V\begin{bmatrix} \overset{1}{t_i}\begin{pmatrix} y \\ t_v\end{pmatrix}\begin{pmatrix} \overset{2}{y_1^-} \\ y_1 \end{pmatrix} \begin{pmatrix} \overset{3}{y_3} \\ y_3 \end{pmatrix} \end{bmatrix} = 1, \qquad V\begin{bmatrix} \overset{1}{t_i}\begin{pmatrix} g_2^- \\ t_r\end{pmatrix}\begin{pmatrix} y_1 \end{pmatrix}\begin{pmatrix} \overset{2}{y_1^-} \\ y \end{pmatrix} \begin{pmatrix} \overset{3}{y^-} \\ g_2 \end{pmatrix} \end{bmatrix} = -1$$

$$V\left[t_i\left(\begin{matrix} 1 & 2 & 3 \\ y & g_1 & y_3 \\ g_1 & y & y_3 \end{matrix}\right)\right] = -1, \qquad V\left[t_i\left(\begin{matrix} 1 & 2 & 3 \\ g_2^- & g_1 & y^- \\ g_1 & y & g_2 \end{matrix}\right)\right] = +1$$

6. Form $W_{(1r3;2,4)}$:

$$W_{(1,3;2,4)} = y_1 y_3 - y_1 g_2 - y_3 g_1 + g_1 g_2$$

From a node basis equation with the reference vertex o,

$$\begin{bmatrix} v_{i,o} \\ v_{j,k} \end{bmatrix} = \frac{1}{\Delta'} \begin{bmatrix} \Delta'_{ii} & \Delta'_{ji} - \Delta'_{ki} \\ \Delta'_{ij} - \Delta'_{ik} & \Delta'_{ji} + \Delta'_{kk} - \Delta'_{jk} - \Delta'_{kj} \end{bmatrix} \begin{bmatrix} J_{i,o} \\ J_{j,k} \end{bmatrix} \qquad (8\text{-}5\text{-}8)$$

Using Eqs. 8-4-12 and 8-5-7, we can write the following equation:

$$\begin{bmatrix} v_{i,o} \\ v_{j,k} \end{bmatrix} = \frac{1}{V} \begin{bmatrix} W_{(i,o;i,o)} & W_{(j,k;i,o)} \\ W_{(i,o;j,k)} & W_{(j,k;j,k)} \end{bmatrix} \begin{bmatrix} J_{i,o} \\ J_{j,k} \end{bmatrix}$$

which shows the topological formulas for open-circuit network functions.

8-6 SHORT-CIRCUIT NETWORK FUNCTIONS

From Eq. 8-5-8, we can obtain

$$\begin{bmatrix} J_{i,o} \\ J_{j,k} \end{bmatrix} = \frac{1}{\Delta'_{iijj} + \Delta'_{iikk} - \Delta'_{iijk} - \Delta'_{iikj}}$$

$$\times \begin{bmatrix} \Delta'_{jj} + \Delta'_{kk} - \Delta'_{jk} - \Delta'_{kj} & -(\Delta'_{ji} - \Delta'_{ki}) \\ -(\Delta'_{ij} - \Delta'_{ik}) & \Delta'_{ii} \end{bmatrix} \begin{bmatrix} v_{i,o} \\ v_{j,k} \end{bmatrix} \qquad (8\text{-}6\text{-}1)$$

The entries on the right-hand side of this equation indicate the short-circuit network functions. By comparing this with Eq. 8-5-8, we can see that all but $(\Delta'_{iijj} + \Delta'_{iikk} - \Delta'_{iijk} - \Delta'_{iikj})$ have been studied previously. Hence we need only find a topological formula for $(\Delta'_{iijj} + \Delta'_{iikk} - \Delta'_{iijk} - \Delta'_{iikj})$ to obtain topological formulas for short-circuit network functions.

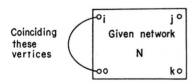

Coinciding these vertices

Fig. 8-6-1. A network $N(i = 0)$.

Consider a network $N(i = o)$ obtained from a network N by coinciding vertices i and o as shown in Fig. 8-6-1. From a node basis equation of $N(i = o)$, we have

$$\begin{bmatrix} v_{p,o} \\ v_{j,k} \end{bmatrix} = \frac{1}{\underline{\underline{\Delta}}} \begin{bmatrix} \underline{\underline{\Delta}}_{pp} & \underline{\underline{\Delta}}_{jp} - \underline{\underline{\Delta}}_{kp} \\ \underline{\underline{\Delta}}_{pj} - \underline{\underline{\Delta}}_{pk} & \underline{\underline{\Delta}}_{jj} + \underline{\underline{\Delta}}_{kk} - \underline{\underline{\Delta}}_{jk} - \underline{\underline{\Delta}}_{kj} \end{bmatrix} \begin{bmatrix} J_{p,o} \\ J_{j,k} \end{bmatrix} \tag{8-6-2}$$

Since $\underline{\underline{\Delta}}$ is the determinant of a node admittance matrix of network $N(i = o)$ which is obtained from N by coinciding vertices i and o, $\underline{\underline{\Delta}}$ is equal to cofactor Δ'_{ii} of a node admittance matrix $A_i Y(A_v)^t$ of given network N. Furthermore, cofactor $\underline{\underline{\Delta}}_{pp}$ in Eq. 8-6-2 is the determinant of matrix obtained from node admittance matrix $A_i Y(A_v)^t$ of N by deleting the ith and the pth rows and columns corresponding to vertices i and p. From Eq. 8-5-9, $W_{(j,k;j,k)}$ of network $N(i = o)$ is equal to

$$W_{(j,k;j,k)} = \underline{\underline{\Delta}}_{jj} + \underline{\underline{\Delta}}_{kk} - \underline{\underline{\Delta}}_{jk} - \underline{\underline{\Delta}}_{kj} \tag{8-6-3}$$

Hence

$$W_{(j,k;j,k)} \quad \text{of} \quad N(i = o) = \Delta'_{iijj} + \Delta'_{iikk} - \Delta'_{iijk} - \Delta'_{iikj}$$
$$\text{of} \quad A_i Y(A_v)^t \quad \text{of} \quad N \tag{8-6-4}$$

From Eqs. 8-4-15, 8-5-8, and 8-5-9,

$$W_{(j,k;j,k)} \quad \text{of} \quad N(i = o) = \sum_{(j)} \epsilon_j \text{ complete 2-tree } t_{2_{j,k}} \text{ admittance}$$
$$\text{product of } N(i = o) \tag{8-6-5}$$

It can be seen that each complete 2-tree $t_{2_{j,k}}$ of $N(i = o)$ is a complete tree of $N(i = o, j = k)$ obtained from $N(i = o)$ by coinciding vertices j and k. Hence

$$W_{(j,k;j,k)} \text{ of } N(i = o) = \sum_{(j)} \epsilon_j \text{ complete tree admittance product}$$
$$\text{of } N(i = o, j = k) \tag{8-6-6}$$

Let $U_{(i,o;j,k)}$ be defined as follows:

$$U_{(i,o;j,k)} = \sum_{(j)} \epsilon_j \text{ complete tree admittance product of } N(i = o, j = k)$$
$$\tag{8-6-7}$$

With Eqs. 8-6-7, 8-5-8, and 8-5-9, Eq. 8-6-1 can be expressed as

$$\begin{bmatrix} J_{i,o} \\ J_{j,k} \end{bmatrix} = \frac{1}{U_{(i,o;j,k)}} \begin{bmatrix} W_{(j,k;j,k)} & -W_{(j,k;i,o)} \\ -W_{(i,o;j,k)} & W_{(i,o;i,o)} \end{bmatrix} \begin{bmatrix} v_{i,o} \\ v_{j,k} \end{bmatrix} \tag{8-6-8}$$

which gives the topological formulas for short-circuit network functions.

Instead of coinciding vertices i and o and vertices j and k, we can obtain $U_{(i,o;j,k)}$ by the following process:

1. Place a passivelike admittance y_1 between vertices i and o and a passivelike admittance y_2 between vertices j and k.

2. Form V with complete trees all of which contain both y_1 and y_2. In other words, form

$$V(y_1 y_2) = \sum_{(j)} \epsilon_j \text{ complete tree admittance product of complete}$$

trees which contain both y_1 and y_2 \hfill (8-6-9)

3. $U_{(i,o;j,k)}$ is

$$U_{(i,o;j,k)} = \frac{1}{y_1 y_2} V(y_1 y_2) \hfill (8\text{-}6\text{-}10)$$

It is generally easier to obtain $U_{(i,o;j,k)}$ by obtaining complete trees of $N(i = o, j = k)$. For example, $U_{(1,3;2,4)}$ of G in Fig. 8-3-6 can be obtained by

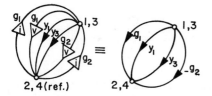

Fig. 8-6-2. $G(1 = 3, 2 = 4)$ of G in Fig. 8-3-6.

constructing $G(1 = 3, 2 = 4)$ as shown in Fig. 8-6-2. Then obtaining all possible complete trees of $G(1 = 3, 2 = 4)$, we obtain

$$U_{(1,3;2,4)} = y_1 + y_3 - g_1 - g_2$$

8-7 UNISTOR NETWORK

The analysis of networks by topological formulas given previously have the advantage of having no cancellation term if the admittances in a network are considered to be arbitrary. In other words, if edges $y_{j_1}, y_{j_2}, \ldots, y_{j_m}$ are a tree (or a complete tree in the case of activelike networks), term $y_{j_1} y_{j_2} \cdots y_{j_m}$ will be in the determinant obtained by the topological formulas. Since there can be no other tree that consists of $y_{j_1}, y_{j_2}, \ldots, y_{j_m}$, the term $y_{j_1} y_{j_2} \cdots y_{j_m}$ will not be cancelled. This is also true for cofactors. Hence, as long as we are using topological formulas, we do not need to worry about cancellation. On the other hand, in the case of activelike networks, the sign of each complete tree must be enumerated, which may be considered a disadvantage of this technique. We can avoid this sign problem if we allow some cancelling terms by restricting the locations of the voltage edge and the current edge of each admittance y. One way of achieving this is by using only unistors to represent networks.

Definition 8-7-1. A *unistor y* is defined as

$$yv_{po} = i_{pk} \tag{8-7-1}$$

where v_{po} is the voltage from vertex p to the reference vertex o, and i_{pk} is the current from vertex p to vertex k. The topological representation of the unistor specified by Eq. 8-7-1 is defined to be the one shown in Fig. 8-7-1.

<table>
<tr><td>o o (reference vertex)</td><td>o (reference)</td></tr>
<tr><td>**Fig. 8-7-1.** Representation of unistor.</td><td>**Fig. 8-7-2.** A conductor *g*.</td></tr>
</table>

Consider a conductor connected between vertices p and q as shown in Fig. 8-7-2. From

$$g(v_{po} - v_{qo}) = i_{pq} \tag{8-7-2}$$

we can see that if we define

$$gv_{po} = i'_{pq} \tag{8-7-3}$$

and

$$gv_{qo} = i''_{pq} \tag{8-7-4}$$

then

$$i_{pq} = i'_{pq} - i''_{pq} \tag{8-7-5}$$

and we can represent conductor g by two unistors satisfying Eqs. 8-7-3 and 8-7-4 as shown in Fig. 8-7-3.

In general, suppose we have an admittance y as shown in Fig. 8-7-4a where

$$yv_{rs} = i_{tu} \tag{8-7-6}$$

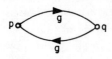

o
o(reference)   **Fig. 8-7-3.**   Unistors representing conductor *g*.

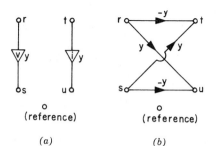

Fig. 8-7-4. Admittance y. (a) Voltage current edges of y; (b) unistor representation of y.

Consider the unistor network shown in Fig. 8-7-4b. We can see that the net current at vertex r is

$$(-y)v_{ro} + yv_{ro} = 0 \qquad (8\text{-}7\text{-}7)$$

Likewise, the net current at vertex $s = 0$. On the other hand, the net current at vertex t is

$$-[(-y)v_{ro} + yv_{so}] = y(v_{ro} - v_{so}) = i_{tu} \qquad (8\text{-}7\text{-}8)$$

where i_{tu} is the current given by Eq. 8-7-6. Furthermore, the net current at the vertex u is

$$-[yv_{ro} + (-y)v_{so}] = -y(v_{ro} - v_{so}) = -i_{tu} \qquad (8\text{-}7\text{-}9)$$

Thus the unistor network in Fig. 8-7-4b represents the admittance y in Fig. 8-7-4a.

When $s = u$, the unistor network in Fig. 8-7-4b is as shown in Fig. 8-7-5a. When $s = u$ and $r = t$, the unistor network is as shown in Fig. 8-7-5b, which is the representation of a passivelike admittance element.

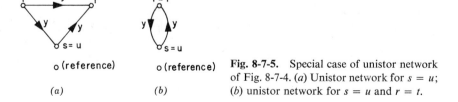

Fig. 8-7-5. Special case of unistor network of Fig. 8-7-4. (a) Unistor network for $s = u$; (b) unistor network for $s = u$ and $r = t$.

Note that *any unistor which is connected to the reference vertex and whose orientation is away from the reference vertex can be omitted* because of the definition of unistors. Now we know that any admittance can be represented by unistors. For example, the network in Fig. 8-7-6 can be represented by the

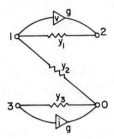

Fig. 8-7-6.   A network.

unistor network shown in Fig. 8-7-7a. By taking vertex o as the reference vertex, the unistor network in Fig. 8-7-7a can be simplified to that of Fig. 8-7-7b, which represents the network in Fig. 8-7-6. Note that we replace two unistors by one by adding their weights when these two unistors are connected between the same vertices (in parallel) and have the same orientation.

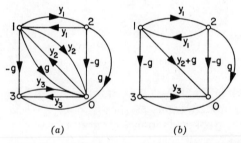

(a)   (b)

Fig. 8-7-7.   (a) Unistor network N; (b) simplified unistor network of N with 0 as the reference vertex.

Let N_u be a network consisting of unistors. Then we have the following equations for all unistors in N_u with o being the reference vertex:

$$
\begin{aligned}
y_1 v_{p_1 o} &= i_{p_1 k_1} \\
y_2 v_{p_2 o} &= i_{p_2 k_2} \\
&\;\vdots \\
y_n v_{p_n o} &= i_{p_n k_n}
\end{aligned}
\qquad (8\text{-}7\text{-}10)
$$

which can be written in a matrix form as

$$
YV_e = I_e \qquad (8\text{-}7\text{-}11)
$$

where

$$Y = \begin{bmatrix} y_1 & & & \\ & y_2 & & 0 \\ & & \ddots & \\ 0 & & & y_n \end{bmatrix} \qquad (8\text{-}7\text{-}12)$$

$$V_e = \begin{bmatrix} v_{p_1 0} \\ v_{p_2 0} \\ \vdots \\ v_{p_n 0} \end{bmatrix} \qquad (8\text{-}7\text{-}13)$$

and

$$I_e = \begin{bmatrix} i_{p_1 k_1} \\ i_{p_2 k_2} \\ \vdots \\ i_{p_n k_n} \end{bmatrix} \qquad (8\text{-}7\text{-}14)$$

Let G_i and G_v be the current and voltage graphs of N_u by considering that each unistor connected from vertex i to vertex j as shown in Fig. 8-7-8a is

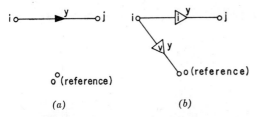

(a)   (b)

Fig. 8-7-8.   (a) Unistor y; (b) voltage and current edges of unistor y.

the shorthand notation of the current edge from i to j and the voltage edge from i to the reference vertex as shown in Fig. 8-7-8b. We can easily see that the current graph G_i of N_u is exactly the same as N_u. Hence the incidence matrix A_i of current graph G_i is the incidence matrix A of N_u. Thus by Kirchhoff's current law, we have

$$A I_e = A Y V_e = 0 \qquad (8\text{-}7\text{-}15)$$

In a node transformation, from Eq. 8-1-19,

$$V_e = A_v{}^t V_n \qquad (8\text{-}7\text{-}16)$$

where A_v is the incidence matrix of voltage graph G_v. Since every voltage edge is connected from the vertex, from which the corresponding current edge is connected, the location of $+1$ in A_v is exactly the same as that of $+1$ in A_i. On the other hand, every voltage edge is connected to the reference vertex, which may not be the case for current edges. Hence there is no -1 in A_v. Thus A_v can be obtained from A_i by changing all -1 to 0.

Definition 8-7-2. The symbol A^+ is defined as the matrix obtained from A by replacing all -1 by 0.

With this definition, we have

$$A_v = A^+ \quad \text{of} \quad N_u \qquad (8\text{-}7\text{-}17)$$

Thus the node transformation in Eq. 8-7-16 becomes

$$V_e = A^{+t}V_n \qquad (8\text{-}7\text{-}18)$$

This with Eq. 8-7-15 gives

$$A Y A^{+t}V_n = 0 \qquad (8\text{-}7\text{-}19)$$

which is the node basis equation of network N_u. When there are external currents connected from the reference vertex to vertices in N_u, it is obvious that Eq. 8-7-19 will be changed to

$$A Y A^{+t}V_n = J \qquad (8\text{-}7\text{-}20)$$

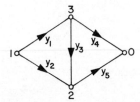

Fig. 8-7-9. Network N_u.

Example 8-7-1. Consider network N_u consisting of unistors as shown in Fig. 8-7-9. The incidence matrix A with o being the reference vertex is

$$
A = \begin{array}{c} \\ 1 \\ 2 \\ 3 \end{array}
\begin{array}{c} \begin{array}{ccccc} y_1 & y_2 & y_3 & y_4 & y_5 \end{array} \\
\left[\begin{array}{ccccc}
1 & 1 & 0 & 0 & 0 \\
0 & -1 & -1 & 0 & 1 \\
-1 & 0 & 1 & 1 & 0
\end{array} \right]
\end{array}
$$

Thus

$$
A^+ = \begin{array}{c} 1 \\ 2 \\ 3 \end{array} \begin{bmatrix} \overset{y_1}{1} & \overset{y_2}{1} & \overset{y_3}{0} & \overset{y_4}{0} & \overset{y_5}{0} \\ 0 & 0 & 0 & 0 & 1 \\ 0 & 0 & 1 & 1 & 0 \end{bmatrix}
$$

Then the node admittance matrix of N_u

$$
A Y A^{+t} = \begin{bmatrix} 1 & 1 & 0 & 0 & 0 \\ 0 & -1 & -1 & 0 & 1 \\ -1 & 0 & 1 & 1 & 0 \end{bmatrix} \begin{bmatrix} y_1 & & & & \\ & y_2 & & 0 & \\ & & y_3 & & \\ & 0 & & y_4 & \\ & & & & y_5 \end{bmatrix} \begin{bmatrix} 1 & 0 & 0 \\ 1 & 0 & 0 \\ 0 & 0 & 1 \\ 0 & 0 & 1 \\ 0 & 1 & 0 \end{bmatrix}
$$

$$
= \begin{bmatrix} y_1 + y_2 & 0 & 0 \\ -y_2 & y_5 & -y_3 \\ -y_1 & 0 & y_3 + y_4 \end{bmatrix}
$$

By the use of the Binet-Cauchy theorem, the determinant of node admittance matrix $A Y A^{+t}$ can be expressed as

$$
|A Y A^{+t}| = \Sigma \text{ major determinant of } A Y \text{ corresponding major} \\ \text{determinant of } A^{+t}
$$

$$
= \underset{(j)}{\Sigma} (y_{j_1} y_{j_2} \cdots y_{j_m}) |A(j_1 j_2 \cdots j_m)| \; |A^+(j_1 j_2 \cdots j_m)| \qquad (8\text{-}7\text{-}21)
$$

where $A(j_1 j_2 \cdots j_m)$ and $A^+(j_1 j_2 \cdots j_m)$ are the square matrices obtained by columns $j_1 j_2, \ldots, j_m$ of A and A^+, respectively. We know that $A(j_1 j_2 \cdots j_m)$ is nonzero if edges $y_{j_1}, y_{j_2}, \ldots, y_{j_m}$ form a tree in N_u. Moreover, $A^+(j_1 j_2 \cdots j_m)$ is nonzero if edges $y_{j_1}, y_{j_2}, \ldots, y_{j_m}$ form a tree in the voltage graph G_v of N_u.

It is now useful to define a directed tree as follows.

Definition 8-7-3. A *directed tree* is a tree in N_u in which there exists a directed path from any vertex to the reference vertex.

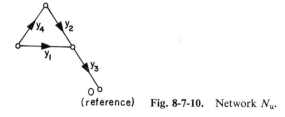

(reference)   **Fig. 8-7-10.** Network N_u.

For example, in Fig. 8-7-10, (y_1, y_2, y_3) is a directed tree but (y_1, y_3, y_4) is not a directed tree where G is the reference vertex.

With the definition of directed trees, Eq. 8-7-20 can be expressed as

$$|AYA^{+t}| = \sum \text{directed tree } t_o \text{ admittance product} \qquad (8\text{-}7\text{-}22)$$

where t_o is a directed tree with o being the reference vertex. The proof is left to the reader.

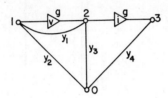

Fig. 8-7-11.   An electrical network N.

Example 8-7-2.   Consider the network in Fig. 8-7-11. The corresponding unistor network N_u consisting of unistors, shown in Fig. 8-7-12a, can be

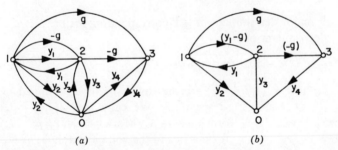

(a)   *(b)*

Fig. 8-7-12.   *(a)* Unistor network of N; *(b)* simplified unistor network N_u.

simplified as shown in Fig. 8-7-12b. From this network N_u, we can obtain the determinant Δ of a node admittance matrix as

$$\Delta = |ATA^{+t}| = y_1 y_2 y_4 + (y_1 - g)y_3 y_4 + y_3 y_4 y_5 + y_2 y_4(-g) + y_2 y_3 y_4$$

which can be simplified to

$$\Delta = y_1 y_2 y_4 + y_1 y_3 y_4 + y_2 y_3 y_4 - y_2 y_4 y_5$$

We can obtain topological formulas for cofactors, for example,

$$\Delta_{ii} = \sum \text{directed 2-tree } t_{2_{i,o}} \text{ admittance product} \qquad (8\text{-}7\text{-}23)$$

However, it may be simpler to insert an admittance y, as we previously did in the topological analysis of active network, to obtain desired solutions.

This topological analysis by unistor networks is not popular because of the existence of cancellation in the results.

There are several techniques that combine topological analysis of passive-like and activelike networks so that the determination of signs of each complete tree becomes simple. However, we do not discuss them here.

8-8 EQUIVALENT TRANSFORMATION FOR ELECTRICAL NETWORKS

There are many electrical networks that have the same network functions between fixed vertices. These are called equivalent networks under specified vertices. Here we consider an algorithm to construct an equivalent network from a given network under a set of vertices topologically, which is known as an equivalent transformation.

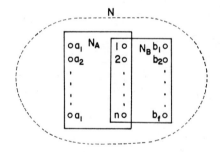

Fig. 8-8-1. A network N consisting of N_A and N_B.

Consider a network N consisting of two subnetworks N_A and N_B joined by vertices $1, 2, \ldots, n$ as shown in Fig. 8-8-1. Vertices $a_1, a_2, \ldots, a_l$ and vertices $1, 2, \ldots, n$ are in N_A, and vertices $b_1, b_2, \ldots, b_f$ and vertices $1, 2, \ldots, n$ are in N_B. Without the loss of generality, let vertex n be the reference vertex. Let V_A be the column vector in which each element v_{a_p} is the node voltage from vertex a_p to the reference vertex n for $p = 1, 2, \ldots, l$; let V_B be the column vector in which each element v_{b_q} is the node voltage from vertex b_q to the reference vertex n for $q = 1, 2, \ldots, f$, and let V_C be the column vector where each element v_c be the node voltage from vertex c to *the reference vertex n* for $c = 1, 2, \ldots, n - 1$. Similarly, let J_A be the column vector consisting of j_{a_q}, which is the current from the reference vertex n to vertex a_p for $p = 1, 2, \ldots, l$; let J_B be the column vector in which each j_{b_q} is the current from the reference vertex n to vertex b_q for $q = 1, 2, \ldots, f$; and let

J_C be the current vector where entry j_c is the current from the reference vertex n to vertex c for $c = 1, 2, \ldots, n-1$.

The node basis equations of network N can be expressed as

$$\begin{bmatrix} Y_{A_{11}} & Y_{A_{12}} & 0 \\ Y_{A_{21}} & Y_{A_{22}} & 0 \\ 0 & 0 & 0 \end{bmatrix}\begin{bmatrix} V_A \\ V_C \\ V_B \end{bmatrix} + \begin{bmatrix} 0 & 0 & 0 \\ 0 & Y_{B_{11}} & Y_{B_{12}} \\ 0 & Y_{B_{21}} & Y_{B_{22}} \end{bmatrix}\begin{bmatrix} V_A \\ V_C \\ V_B \end{bmatrix} = \begin{bmatrix} J_A \\ J_C \\ J_B \end{bmatrix} \quad (8\text{-}8\text{-}1)$$

where

$$\begin{bmatrix} Y_{A_{11}} & Y_{A_{12}} \\ Y_{A_{12}} & Y_{A_{22}} \end{bmatrix}$$

is from subnetwork N_A and

$$\begin{bmatrix} Y_{B_{11}} & Y_{B_{12}} \\ Y_{B_{21}} & Y_{B_{22}} \end{bmatrix}$$

is from subnetwork N_B. By rewriting Eq. 8-8-1, we have

$$\begin{bmatrix} Y_{A_{11}} & Y_{A_{12}} \\ Y_{A_{21}} & Y_{A_{22}} \end{bmatrix}\begin{bmatrix} V_A \\ V_C \end{bmatrix} + \begin{bmatrix} 0 & 0 \\ Y_{B_{11}} & Y_{B_{12}} \end{bmatrix}\begin{bmatrix} V_C \\ V_B \end{bmatrix} = \begin{bmatrix} J_A \\ J_C \end{bmatrix} \quad (8\text{-}8\text{-}2)$$

and

$$Y_{B_{21}}V_C + Y_{B_{22}}V_B = J_B \quad (8\text{-}8\text{-}3)$$

From Eq. 8-8-3, V_B can be expressed as

$$V_B = (Y_{B_{22}})^{-1}J_B - (Y_{B_{22}})^{-1}Y_{B_{21}}V_C \quad (8\text{-}8\text{-}4)$$

Substituting this into Eq. 8-8-2, we have

$$\begin{bmatrix} Y_{A_{11}} & Y_{A_{12}} \\ Y_{A_{21}} & Y_{A_{22}} \end{bmatrix}\begin{bmatrix} V_A \\ V_C \end{bmatrix} + \begin{bmatrix} & 0 \\ Y_{B_{11}} - Y_{B_{12}}(Y_{B_{22}})^{-1}Y_{B_{21}} \end{bmatrix}V_C$$
$$= \begin{bmatrix} J_A \\ J_C - Y_{B_{12}}(Y_{B_{22}})^{-1}J_B \end{bmatrix} \quad (8\text{-}8\text{-}5)$$

under the assumption that $Y_{B_{22}}$ is nonsingular. Obviously, Eqs. 8-8-1 and 8-8-5 give the same network functions between the vertices in subnetwork N_A. On the other hand, Eq. 8-8-5 shows that we can replace subnetwork N_B by new subnetwork N_C, whose node admittance matrix is equal to $Y_{B_{11}} - Y_{B_{12}}(Y_{B_{22}})^{-1}Y_{B_{21}}$ having current generators given by $-Y_{B_{12}}(Y_{B_{22}})^{-1}J_B$ as shown in Fig. 8-8-2.

When we consider subnetwork N_B alone, and calculate n-terminal short-circuit admittance functions between every pair of vertices in $1, 2, \ldots, n$, we

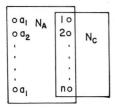

Fig. 8-8-2. Equivalent network of N.

can see that the solution is exactly equal to $Y_{B_{11}} - Y_{B_{12}}(Y_{B_{22}})^{-1} Y_{B_{21}}$. Furthermore, $-Y_{B_{12}}(Y_{B_{22}})^{-1} J_B$ indicates the equivalent n-terminal short-circuit current generators from reference vertex n to vertex c for $c = 1, 2, \ldots, n-1$ caused by J_B in N_B. Thus as long as $Y_{B_{22}}$ is nonsingular, which is necessary for the existence of n-terminal short-circuit admittance functions between a pair of vertices in $1, 2, \ldots, n$, we can obtain subnetwork N_B so that we can eliminate vertices $b_1, b_2, \ldots, b_f$ without changing the network functions of N between any pair of vertices in subnetwork N_A. The rules of obtaining subnetwork N_C are as follows.

Rule a. From the reference vertex n to every vertex u in $1, 2, \ldots, n-1$, we give a current generator J_u in N_C which is equal to the n-terminal short-circuit current i_u of N_B:

$$J_u \quad \text{in } N_C = i_u \quad \text{in } N_B|_{v_1 = v_2 = \cdots = v_{n-1} = 0} \tag{8-8-6}$$

where v_p is the voltage from vertex p to the reference vertex n for $p = 1, 2, \ldots, n-1$, and i_u is the current from vertex u to the reference vertex n.

Rule b. Between every pair (t, u) of vertices $1, 2, \ldots, n-1$, we give admittances y_{tu} and y_{ut} in N_C which satisfy

$$y_{t_u} v_u = i_t \tag{8-8-7}$$

and

$$y_{ut} v_t = i_u \tag{8-8-8}$$

where y_{tu} and y_{ut} are equal to the n-terminal short-circuit transfer admittance functions of N'_B (N_B without current generators), that is,

$$y_{tu} = \left. \frac{J_t}{v_u} \right|_{\substack{v_1 = v_2 = \cdots = v_{n-1} = 0 \\ v_u \neq 0}} \quad \text{in } N'_B \tag{8-8-9}$$

and

$$y_{ut} = \left. \frac{J_u}{v_t} \right|_{\substack{v_1 = v_2 = \cdots = v_{n-1} = 0 \\ v_t \neq 0}} \quad \text{in } N'_B \tag{8-8-10}$$

Rule c. From every vertex in u in $1, 2, \ldots, n-1$ to the reference vertex n, we give y_{un} in N_C defined by

$$y_{un} v_u = i_u \tag{8-8-11}$$

which is equal to the n-terminal short-circuit driving point admittance function between vertex u and the reference vertex n of N_B', that is,

$$y_{un} = \left. \frac{J_u}{v_u} \right|_{\substack{v_1 = v_2 = \cdots = v_{n-1} = 0 \\ v_u \neq 0}} \quad \text{in } N_B' \tag{8-8-12}$$

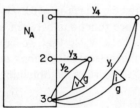

Fig. 8-8-3.   A given network.

Example 8-8-1.   Consider the network in Fig. 8-8-3. Note that the use of nonoriented edges for representing passivelike elements as in the figure is permissible because the orientations of passivelike elements are arbitrary. Moreover, it is convenient because we can assign suitable orientations whenever necessary. Suppose we only need to consider network functions between the vertices in N_A. Since there is no current generator in N_B, we start by using Rule b. Let vertex 3 be the reference vertex. Then

$$y_{12} = \left. \frac{J_1}{v_2} \right|_{v_1 = 0} = \frac{y_3 y_4 g}{(y_1 + y_4)(y_2 + y_3)}$$

$$y_{21} = \left. \frac{J_2}{v_1} \right|_{v_2 = 0} = 0$$

By Rule c,

$$y_{13} = \left. \frac{J_1}{v_1} \right|_{v_2 = 0} = \frac{y_1 y_4}{y_1 + y_4}$$

$$y_{23} = \left. \frac{J_2}{v_2} \right|_{v_1 = 0} = \frac{y_2 y_3}{y_2 + y_3}$$

The resulting network is shown in Fig. 8-8-4.

When subnetwork N_B contains no activelike elements (including mutual couplings), Rules b and c become one rule.

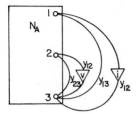

Fig. 8-8-4. Equivalent network.

Rule b'. Between every pair (t, u) of vertices in $1, 2, \ldots, n$, we give passivelike admittance $\tilde{y}_{tu}$ between t and u in N_C which is equal to the negative of the n-terminal short-circuit transfer admittance function between t and u, with the reference vertex being other than t and u, of N_B':

$$\tilde{y}_{tu} = \frac{J_t'}{v_u'}\bigg|_{\substack{v_1' = v_2' = \cdots = v_n' = 0 \\ v_u' \neq 0}} = \frac{J_u'}{v_t'}\bigg|_{\substack{v_1' = v_2' = \cdots = v_n' = 0 \\ v_t' \neq 0}} \quad \text{in } N_B' \quad (8\text{-}8\text{-}13)$$

where J_p' is the current from the arbitrary chosen reference vertex r, which is one of $1, 2, \ldots, n$ other than t and u, to vertex p, and v_p' is the voltage from vertex p to the chosen reference vertex r for $p = t, u$.

In order to give topological formulas for n-terminal short-circuit network functions, we use the following symbols.

Definition 8-8-1. The symbol $W_{n_0, n_1 | \overline{n_2 \cdots n_k}}$ with respect to a linear graph G is defined as

$$W_{n_0, n_1 | \overline{n_2 \cdots n_k}} = \sum \epsilon_j \text{ complete } k\text{-tree } t_{n_0, n_1 | \overline{n_2 \cdots n_k}} \text{ admittance product}$$

$$(8\text{-}8\text{-}14)$$

Here the symbol $t_{n_0, n_1 | \overline{n_2 \cdots n_k}}$ indicates a subgraph of G which becomes a complete tree with an additional edge y when vertices $n_2, \ldots, n_k$ are coincided where the current edge of y is connected from n_0 to n_k and the voltage edge of y is from n_1 to n_k.

Definition 8-8-2. The symbol $V_{\overline{n_1 n_2 \cdots n_k}}$ with respect to a linear graph G is defined as

$$V_{\overline{n_1 n_2 \cdots n_k}} = \sum \epsilon_j \text{ complete } k\text{-tree } t_{\overline{n_1 n_2 \cdots n_k}} \text{ admittance product} \quad (8\text{-}8\text{-}15)$$

where subgraph $t_{\overline{n_1 n_2 \cdots n_k}}$ becomes a complete tree when vertices $n_1, n_2, \ldots, n_k$ are coincided.

With these definitions, we can give topological formulas for n-terminal short-circuit network functions.

Consider driving point function

$$y_{un} = \frac{J_u}{v_u}\bigg|_{\substack{v_1 = v_2 = \cdots = v_{n-1} = 0 \\ v_u \neq 0}} \tag{8-8-16}$$

given by Eq. 8-8-12. By coinciding vertices $1, 2, \ldots, n$ except vertex u in N_B', we have a new network symbolized by $N_B'[p = n; p = 1, 2, \ldots, n - 1, p \neq u]$. Then Eq. 8-8-16 becomes simply an open-circuit driving point admittance function of this new network $N_B'[p = n; p = 1, 2, \ldots, n - 1, p \neq u]$. Thus

$$y_{un} = \frac{V}{W_{u,u|\bar{n}}} \quad \text{of} \quad N_B'[p = n; p = 1, 2, \ldots, n - 1, \quad p \neq u] \tag{8-8-17}$$

where

$$V = \sum \epsilon_j \text{ complete tree admittance product of}$$
$$N_B'[p = n; p = 1, 2, \ldots, n - 1, \quad p \neq u] \tag{8-8-18}$$

Since $N_B'[p = n; p = 1, 2, \ldots, n - 1, \quad p \neq u]$ is obtained from N_B' by coinciding vertices $1, 2, \ldots, n$ except u,

$$V \text{ of } N_B'[p = n; p = 1, 2, \ldots, n - 1, \quad p \neq u]$$
$$= V_{\overline{1 \ 2 \ \cdots \ u-1 \ u+1 \ \cdots \ n}} \quad \text{of } N_B' \tag{8-8-19}$$

and

$$W_{u,u|\bar{n}} \text{ of } N_B'[p = n; p = 1, 2, \ldots, n - 1, \quad p \neq u]$$
$$= V_{\overline{1 \ 2 \ \cdots \ n}} \quad \text{of } N_B' \tag{8-8-20}$$

Thus

$$y_{un} = \frac{V_{\overline{1 \ 2 \ \cdots \ u-1 \ u+1 \ \cdots \ n}}}{V_{\overline{1 \ 2 \ \cdots \ n}}} \quad \text{of } N_B' \tag{8-8-21}$$

Note that the subscript of $V_{\overline{1 \ 2 \ \cdots \ u-1 \ u+1 \ \cdots \ n}}$ in the numerator of this equation does not contain vertex u. This is the topological formula for an n-terminal short-circuit driving point admittance function.

Admittance y_{tu} in Eq. 8-8-9 (and Eq. 8-8-10) can be considered as

$$y_{tu} = \frac{J_t}{v_u}\bigg|_{v_t = 0} \quad \text{of } N_B'[p = n; p = 1, 2, \ldots, n - 1, \quad p \neq t, u] \tag{8-8-22}$$

where $N_B'[p = n; p = 1, 2, \ldots, n - 1, \quad p \neq t, u]$ is obtained from N_B' by coinciding vertices $1, 2, \ldots, n$ except vertices t and u. From Eq. 8-6-8 and Def. 8-8-2 we know that

$$y_{tu} = \frac{J_t}{v_u}\bigg|_{v_t = 0} = \frac{-W_{u,t|n}}{V_{\overline{u \ t \ n}}} \quad \text{of } N_B'[p = n; p = 1, 2, \ldots, n - 1, \quad p \neq t, u]$$
$$\tag{8-8-23}$$

we have

$$y_{tu} = \frac{-W_{u,t|\overline{1\ 2\ \cdots\ t-1\ z+1\ \cdots\ u-1\ u+1\ \cdots\ n}}}{V_{\overline{1\ 2\ \cdots\ n}}} \quad \text{of } N'_B \qquad (8\text{-}8\text{-}24)$$

Note that when N'_B contains no activelike element (including mutual couplings), this equation becomes

$$\tilde{y}_{tu} = \tilde{y}_{ut} = -y_{tu} = \frac{W_{u,t|\overline{1\ 2\ \cdots\ t-1\ t+1\ \cdots\ u-1\ u+1\ n}}}{V_{\overline{1\ 2\ \cdots\ n}}} \quad \text{of } N'_B \qquad (8\text{-}8\text{-}25)$$

Equivalent current generators can also be obtained by a topological formula. Let J_b be a current generator connected from the reference vertex n to vertex b in N_B where $b \neq 1, 2, \ldots, n-1$. Then the equivalent current generator J_t corresponding to J_b is given from the reference vertex n to vertex t for $t = 1, 2, \ldots, n$ by the equivalent transformation. Since

$$\left.\frac{J_t}{J_b}\right|_{v_1 = v_2 = \cdots = v_{n-1} = 0} \quad \text{of } N''_B$$

$$= \left.\frac{J_t}{J_b}\right|_{v_t = 0} \quad \text{of } N''_B[p = n; p = 1, 2, \ldots, n-1, \quad p \neq t] \qquad (8\text{-}8\text{-}26)$$

Also

$$\left.\frac{J_t}{J_b}\right|_{v_t = 0} \quad \text{of } N''_B[p = n; p = 1, 2, \ldots, n-1, \quad p \neq t]$$

$$= \left.\frac{J_t/v_b}{J_b/v_b}\right|_{v_t = 0} \quad \text{of } N'_B[p = n; p = 1, 2, \ldots, n-1, \quad p \neq t] \qquad (8\text{-}8\text{-}27)$$

where N''_B is obtained from N_B by deleting all current generators except J_b and $N''_B[p = n; p = 1, 2, \ldots, n-1, \quad p \neq t]$ is obtained from N''_B by coinciding vertices $1, 2, \ldots, n$ except vertex t. The right-hand side of Eq. 8-8-27 can be considered the ratio of two terminal short-circuit admittance functions. Hence from Eq. 8-6-8, we have

$$\left.\frac{J_t}{J_b}\right|_{v_t = 0} = \frac{W_{b,t|\overline{\tilde{n}}}}{W_{t,t|\overline{\tilde{n}}}} \quad \text{of } N'_B[p = n; p = 1, 2, \ldots, n-1, \quad p \neq t] \qquad (8\text{-}8\text{-}28)$$

or

$$\left.\frac{J_t}{J_b}\right|_{v_1 = v_2 = \cdots = v_{n-1} = 0} = \frac{W_{b,t|\overline{1\ 2\ \cdots\ t-1\ t+1\ \cdots\ n}}}{V_{\overline{1\ 2\ \cdots\ n}}} \quad \text{of } N'_B \qquad (8\text{-}8\text{-}29)$$

Note that if we use Eq. 8-8-29, we must find equivalent current generators $J_1, J_2, \ldots, J_{n-1}$ for each current generator in N_B.

Suppose subnetwork N_B not only consists of passivelike elements but also contains only one vertex b other than vertices $1, 2, \ldots, n$ (which are in

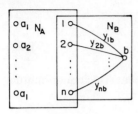

Fig. 8-8-5. A network with N_B having a special structure.

common with subnetwork N_A) as shown in Fig. 8-8-5. For convenience, y_{rb} is the admittance connected between vertices r and b for $r = 1, 2, \ldots, n$. We can see that the denominator of Eq. 8-8-25 is the sum of tree admittance products of the network obtained from N_B by coinciding vertices $1, 2, \ldots, n$.

Fig. 8-8-6. N_B with coinciding vertices $1, 2, \ldots, n$.

The network obtained by coinciding vertices $1, 2, \ldots, n$ is shown in Fig. 8-8-6. We can see readily that the sum of tree admittance products of this network is just the sum of each admittance. Hence

$$V_{\overline{1\,2\,\cdots\,n}} = \sum_{r=1}^{n} y_{rb} \qquad (8\text{-}8\text{-}30)$$

On the other hand, $W_{ut|\overline{1\,2\,\cdots\,t-1\,t+1\,\cdots\,u-1\,u+1\,\cdots\,n}}$ is the sum of 2-tree $t_{2_{ut,n}}$ admittance products of the network obtained from N_B by coinciding vertices $1, 2, \ldots, n$ except vertices t and u. The network obtained by coinciding vertices $1, 2, \ldots, n$ except t and u is shown in Fig. 8-8-7. Hence

$$W_{ut|\overline{1\,2\,\cdots\,t-1\,t+1\,\cdots\,u-1\,u+1\,\cdots\,n}} = y_{tb}y_{ub} \qquad (8\text{-}8\text{-}31)$$

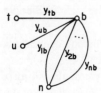

Fig. 8-8-7. N_B with coinciding vertices $1, 2, \ldots, n$ except t and u.

Thus

$$y_{tu} = \frac{y_{tb} y_{ub}}{(\sum\limits_{r=1}^{n} y_{rb})} \tag{8-8-32}$$

This is the equation for the mesh-star transformation. When $n = 3$, Eq. 8-8-32 is for the well-known t-π transformation.

PROBLEMS

1. Find the open-circuit driving point I_{10}/V_{10} and transfer I_{20}/V_{10} functions of the networks in Fig. P-8-1 by using the topological formulas given by Eqs. 8-1-23, 8-3-11, 8-4-1, and 8-5-5.

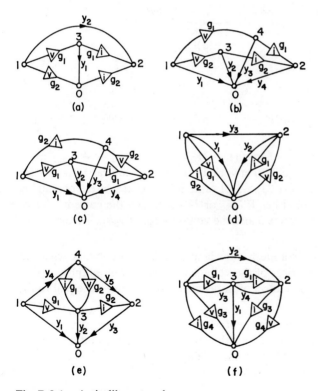

Fig. P-8-1. Activelike networks.

   2. Find the open-circuit transfer functions I_{23}/V_{10} of the networks in Fig. P-8-2 by using Eqs. 8-1-23, 8-3-11, and 8-5-5.

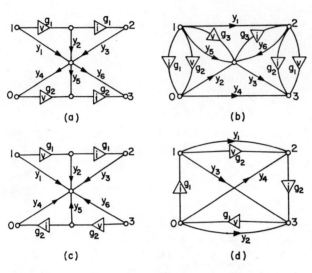

Fig. P-8-2. Activelike networks.

   3. Find the voltage ratio V_{23}/V_{10} of the networks in Fig. P-8-2 by using Eqs. 8-4-1 and 8-5-5.

   4. Prove the topological formula $I_{23}/V_{10} = V/(W_{12,30} - W_{13,20})$ given by Eq. 7-5-60 for a passivelike network by using topological formulas for activelike networks (Eqs. 8-1-23, 8-3-11, and 8-5-5). *Hint:* A tree containing edge $y_{23;10}$ participates in either $W_{12,30}$ or $W_{13,20}$ where the current edge of $y_{23,30}$ is from vertex 2 to vertex 3 and the voltage edge of $y_{23;10}$ is from vertex 1 to vertex 0.

   5. Show that $B_v Z B_i^t$ is a mesh impedance matrix of a network where B_v and B_i are the circuit matrices of the voltage and the current graphs, respectively, and Z is a diagonal matrix of impedances (impedance j is equal to the inverse of admittance y).

   6. Prove or disprove the statement: To determine the sign of a complete tree, we short all passivelike elements, then we form a sign-permutation of the resultant graphs. In other words, the sign of a complete tree depends only on the activelike elements in the complete tree.

   7. Find the open-circuit driving point I_{10}/V_{10} and transfer I_{20}/V_{10}

functions of the networks in Fig. P-8-1 by changing them to unistor networks then using Eq. 8-7-22.

8. Develop a method of obtaining a voltage ratio of a unistor network. Test your method by obtaining the voltage ratio V_{23}/V_{10} of the unistor networks in Fig. P-8-8.

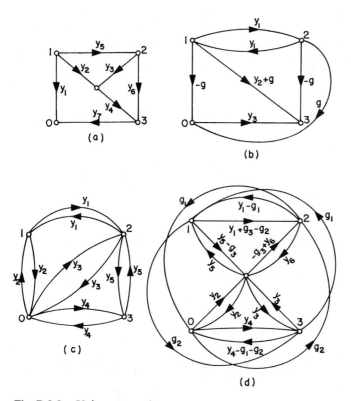

Fig. P-8-8. Unistor networks.

9. Find conditions when unistor networks have no cancelling terms in
$$\Delta = \Sigma \text{ directed trees } t_0 \text{ admittance product}$$

10. Develop a formula which gives the number of directed trees of a unistor network with a fixed reference vertex.

11. Develop a formula to obtain the determination of a mesh impedance matrix of a unistor network.

12. Prove Eq. 8-7-23.

13. Develop the topological formula for Δ_{ij} similar to that given by Eq. 8-7-23.

14. Develop the formulas for the short-circuit network functions for unistor networks.

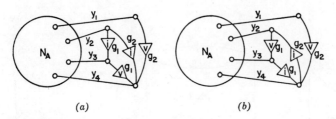

(a)   (b)

Fig. P-8-15.

15. Obtain equivalent networks of the networks in Fig. P-8-15 by using Eqs. 8-8-9, 8-8-10, and 8-8-12.

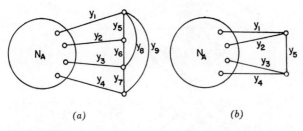

(a)   (b)

Fig. P-8-16.

16. Obtain equivalent networks of the networks in Fig. P-8-16 by using Eqs. 8-8-12 and 8-8-13.

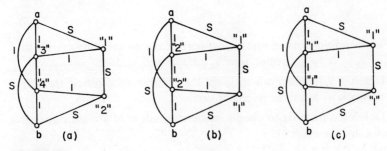

(a)   (b)   (c)

Fig. P-8-17.

17. Find open-circuit driving point V_{ab}/I_{ab} functions of the networks in Fig. P-8-17 by using Eqs. 8-8-21 and 8-8-25 or (and) 8-8-32 successively so that vertices indicated by "1" are removed at once, then all vertices indicated by "2" are removed by one step, next all vertices indicated by "3" are removed, and finally the vertex indicated by "4" is removed.

CHAPTER
9
GENERATION OF TREES

9-1 NECESSITY OF GENERATING TREES

Consider the matrix

$$M = \begin{bmatrix} m_{11} & m_{12} & \cdots & m_{1n} \\ m_{21} & m_{22} & \cdots & m_{2n} \\ & & \cdots & \\ m_{n1} & m_{n2} & \cdots & m_{nn} \end{bmatrix} \qquad (9\text{-}1\text{-}1)$$

The determinant of M can be expressed as

$$|M| = \sum_{(j)} \delta_{j_1 j_2 \cdots j_n} m_{1 j_1} m_{2 j_2} \cdots m_{n j_n} \qquad (9\text{-}1\text{-}2)$$

We can see that as long as all entries of M are distinct, any two terms $m_{1 j_1} m_{2 j_2} \cdots m_{n j_n}$ and $m_{1 j_1'} m_{2 j_2'} \cdots m_{n j_n'}$ are different. However, if M is a node admittance matrix (or any other admittance matrix $Q_1 Y Q_2{}^t$ where Q_1 and Q_2 are cut-set matrices) admittance y can appear in more than one entry. Hence when we expand to obtain the determinant of such a matrix, generally

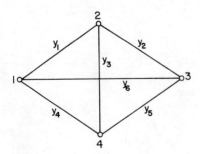

Fig. 9-1-1. A network.

there are many terms that can cancel each other. For example, a node admittance matrix of the network in Fig. 9-1-1 is

$$A\,YA^t = \begin{bmatrix} y_1 + y_4 + y_6 & -y_1 & -y_6 \\ -y_1 & y_1 + y_2 + y_3 & -y_2 \\ -y_6 & -y_2 & y_2 + y_5 + y_6 \end{bmatrix} \quad (9\text{-}1\text{-}3)$$

The determinant of this matrix is

$$
\begin{aligned}
|A\,YA^t| &= (y_1 + y_4 + y_6)[(y_1 + y_2 + y_3)(y_2 + y_5 + y_6) - y_2^2] \\
&\quad + y_1[(-y_1)(y_2 + y_5 + y_6) - y_2 y_6] \\
&\quad - y_6[y_1 y_2 + y_6(y_1 + y_2 + y_3)] \\
&= y_1^2 y_2 + y_1^2 y_5 + y_1^2 y_6 + y_1 y_2^2 + y_1 y_2 y_5 + y_1 y_2 y_6 + y_1 y_2 y_3 \\
&\quad + y_1 y_3 y_5 + y_1 y_3 y_6 + y_1 y_2 y_4 + y_1 y_4 y_5 + y_1 y_4 y_6 + y_2^2 y_4 \\
&\quad + y_2 y_4 y_5 + y_2 y_4 y_6 + y_2 y_3 y_4 + y_3 y_4 y_5 + y_3 y_4 y_6 + y_1 y_2 y_6 \\
&\quad + y_1 y_5 y_6 + y_1 y_6^2 + y_2^2 y_6 + y_2 y_5 y_6 + y_2 y_6^2 + y_2 y_3 y_6 \\
&\quad + y_3 y_5 y_6 + y_3 y_6^2 - y_1 y_2^2 - y_2^2 y_4 - y_2^2 y_6 - y_1^2 y_2 - y_1^2 y_5 \\
&\quad - y_1^2 y_6 - y_1 y_2 y_6 - y_1 y_2 y_6 - y_1 y_6^2 - y_2 y_6^2 - y_3 y_6^2 \\
&= y_1 y_2 y_5 + y_1 y_2 y_3 + y_1 y_3 y_5 + y_1 y_3 y_6 + y_1 y_2 y_4 + y_1 y_4 y_5 \\
&\quad + y_1 y_4 y_6 + y_2 y_4 y_5 + y_2 y_4 y_6 + y_2 y_3 y_4 + y_3 y_4 y_5 + y_3 y_4 y_6 \\
&\quad + y_1 y_5 y_6 + y_2 y_5 y_6 + y_2 y_3 y_6 + y_3 y_5 y_6
\end{aligned}
$$

which of course shows many cancellations. We can see from this that one of the advantages of a topological analysis of networks given previously (for all networks except unistors) is that there will be no such cancellations as long as admittances in a given network are all distinct.

The topological analysis of networks which we have studied requires that we know all possible trees of networks. Suppose we have a network consisting of n_v vertices and n_e edges. In order to obtain all possible trees, suppose we check all possible combinations of $n_v - 1$ edges. Then $\binom{n_e}{n_v - 1}$ sets of edges must be tested. For example, for a complete graph of 5 vertices and 10 edges, there are $\binom{10}{4} = 210$ sets of 4 edges. However, there are $5^3 = 125$ trees. For a complete graph of 6 vertices, there are 3003 sets of 5 edges. However, there are $6^4 = 1296$ trees. There are 54,264 sets of 6 edges but only 16,807 trees in a complete graph of 7 vertices.

These facts indicate that the testing of all possible sets of $n_v - 1$ edges is not an efficient method of obtaining all possible trees. A desirable method would be to generate trees systematically so that we do not worry about whether or not we have all possible trees. Moreover, a systematic method should insure that the same tree will not be generated more than once, thus eliminating the necessity of testing for duplications. Two methods of generating trees are considered here. Both methods produce no duplications and are rather simple procedures. Also both require about the same amount of computation to obtain all possible trees.

9-2 GENERATION OF TREES BY ELEMENTARY TREE TRANSFORMATIONS

Let t_1 and t_2 be trees in a linear graph G. We have found that if there are k edges in $t_1 - t_2$, then the distance between t_1 and t_2 is k. It is clear that if there are k edges in $t_1 - t_2$, then there are k edges in $t_2 - t_1$. Suppose

$$t_1 - t_2 = (e) \tag{9-2-1}$$

and

$$t_2 - t_1 = (e') \tag{9-2-2}$$

(the distance between t_1 and t_2 is 1), then we can write

$$t_2 = t_1 \oplus (e, e') \tag{9-2-3}$$

This is the elementary tree transformation from t_1 to t_2.

Consider a fundamental cut-set S with respect to tree t_1. Suppose S contains edge e which is in t_1. Then we know that no other edge in t_1 will be in S. *This cut-set is represented by the symbol $S_e(t_1)$ where subscript e indicates the edge in $S \cap t_1$ and t_1 in the parentheses indicates that cut-set $S_e(t_1)$ is a fundamental cut-set with respect to tree t_1.* Since there is only one fundamental cut-set with respect to tree t_1 which contains edge e (in t_1), $S_e(t_1)$ is the fundamental cut-set with respect to t_1 that contains edge e. Let $(e_1, e_2, \ldots, e_m)$ be t_1. Then the set of fundamental cut-sets with respect to t_1 consists of $S_{e_1}(t_1), S_{e_2}(t_1), \ldots, S_{e_m}(t_1)$. Note that we use m rather than $n_v - 1$ for simplicity only.

Suppose $S_e(t_1)$ consists of edges $e, a_1, a_2, \ldots, a_k$. Also suppose tree t_2 is of distance 1 from tree t_1 and does not contain edge e. Let

$$t_2 - t_1 = (b) \tag{9-2-4}$$

Then by inserting edge b to tree t_1, we have a circuit containing edge e. Thus, by the definition of fundamental cut-sets, edge b must be in $S_e(t_1)$.

Theorem 9-2-1. Let t_1 and t_2 be trees and be of distance one. Also let

$$t_1 - t_2 = (e) \tag{9-2-5}$$

and

$$t_2 - t_1 = (b) \tag{9-2-6}$$

Then edge b is in $S_e(t_1)$. Conversely, if edge b is in $S_e(t_1)$, then $t_1 \oplus (e, b)$ is a tree.

We may now extend Theorem 9-2-1.

Theorem 9-2-2. For a tree t_0, let $t_1, t_2, \ldots, t_p$ be all trees such that

$$t_0 - t_r = (e) \qquad \text{for all } r = 1, 2, \ldots, p \qquad (9\text{-}2\text{-}7)$$

Let

$$t_r - t_0 = (b_r) \qquad \text{for all } r = 1, 2, \ldots, p \qquad (9\text{-}2\text{-}8)$$

Then

$$S_e(t_0) = (e, b_1, b_2, \ldots, b_p) \qquad (9\text{-}2\text{-}9)$$

Conversely, suppose $S_e(t_0)$ is given by Eq. 9-2-9. Then $t_0 \oplus (e, b_r)$ for $r = 1, 2, \ldots, p$ are all the trees satisfying Eq. 9-2-7.

Example 9-2-1. Consider the linear graph in Fig. 9-1-1. Let t_0 be (y_1, y_2, y_3). Then, trees

$$t_1 = (y_2, y_3, y_4)$$

and

$$t_2 = (y_2, y_3, y_6)$$

are all the trees satisfying the condition in Theorem 9-2-2 with y_1, that is,

$$t_0 - t_1 = (y_1)$$

and

$$t_0 - t_2 = (y_1)$$

Hence fundamental cut-set $S_{y_1}(t_0)$ is

$$S_{y_1}(t_0) = (y_1, y_4, y_6)$$

By using Theorem 9-2-2, we can obtain all trees of distance one from a tree t_0 by knowing the set of fundamental cut-sets with respect to tree t_0, which is shown in the next example.

Example 9-2-2. Consider a linear graph in Fig. 9-1-1. The set of fundamental cut-sets with respect to tree $t_0 = (y_1, y_2, y_3)$ consists of

$$S_{y_1}(t_0) = (y_1, y_4, y_6)$$
$$S_{y_2}(t_0) = (y_2, y_5, y_6)$$

and

$$S_{y_3}(t_0) = (y_3, y_4, y_5)$$

Thus all trees which are the distance one from t_0 are as follows: by $S_{y_1}(t_0)$,

$$t_1 = t_0 \oplus (y_1, y_4) = (y_2, y_3, y_4)$$
$$t_2 = t_0 \oplus (y_1, y_6) = (y_2, y_3, y_6)$$

by $S_{y_2}(t_0)$,

$$t_3 = t_0 \oplus (y_2, y_5) = (y_1, y_3, y_5)$$
$$t_4 = t_0 \oplus (y_2, y_6) = (y_1, y_3, y_6)$$

and by $S_{y_3}(t_0)$,

$$t_5 = t_0 \oplus (y_3, y_4) = (y_1, y_2, y_4)$$

and

$$t_6 = t_0 \oplus (y_3, y_5) = (y_1, y_2, y_5)$$

We now define a set T^e of trees.

Definition 9-2-1.

$$T^e = \{t_0 \oplus (eb); \quad b \in S_e(t_0), \quad b \neq e\} \qquad (9\text{-}2\text{-}10)$$

With this symbol, we have Lemma 9-2-1.

Lemma 9-2-1. Every tree of distance one from tree t_0 is in exactly one of the following sets where $t_0 = (e_1, e_2, \ldots, e_m)$:

$$T^{e_1}, \quad T^{e_2}, \quad \ldots, \quad T^{e_m}$$

The proof is easily carried out using Theorem 9-2-2. Note that because each tree t in T^{e_r} can be obtained by replacing edge e_r in t_0, tree t cannot have edge e_r. On the other hand, because tree t in T^{e_r} is of distance one from t_0, all edges in t_0 except edge e_r will be in t. Hence tree t cannot be in T^{e_s} if $s \neq r$. Thus all trees in T^{e_k} for $k = 1, 2, \ldots, m$ and t_0 are different.

In general, trees in $T^{e_1}, T^{e_2}, \ldots, T^{e_m}$ plus t_0 are not all possible trees in a linear graph. If we take each tree t in one of these sets, say T^{e_1}, and obtain all trees of distance one from t, we may obtain more trees.

To investigate this, let $\underset{\sim}{T}^{e_r e_s}$ be the set of trees defined by

$$\underset{\sim}{T}^{e_r e_s} = \{t' \oplus (e_s b); \quad b \in S_{e_s}(t'), \ t' \in T^{e_r}, \quad b \neq e_s\} \qquad (9\text{-}2\text{-}11)$$

For example, from $T^{y_1} = \{(y_3, y_3, y_4), (y_2, y_3, y_6)\}$ in Example 9-2-2 where $t_0 = (y_1, y_2, y_3)$, we can obtain $\underset{\sim}{T}^{y_1 y_2}$ as follows. Cut-set $S_{y_2}(t_1)$, where $t_1 = (y_2, y_3, y_4)$ (from the linear graph of Fig. 9-1-1), is

$$S_{y_2}(t_1) = (y_2, y_5, y_6)$$

and cut-set $S_{y_2}(t_2)$, where $t_2 = (y_2, y_3, y_6)$ is

$$S_{y_2}(t_2) = (y_1, y_2, y_4, y_5)$$

Note that $S_{y_2}(t_1)$ and $S_{y_2}(t_2)$ are fundamental cut-sets with respect to tree t_1 and tree t_2, respectively. Furthermore, edge y_2 must be in these fundamental cut-sets. Hence set $\underset{\sim}{T}^{y_1 y_2}$ consists of trees given by

$$
\begin{aligned}
t_1' &= t_1 \oplus (y_2, y_5) = (y_3, y_4, y_5) \\
t_2' &= t_1 \oplus (y_2, y_6) = (y_3, y_4, y_6) \qquad \text{by } S_{y_2}(t_1) \\
t_3' &= t_2 \oplus (y_2, y_1) = (y_1, y_3, y_6) \\
t_4' &= t_2 \oplus (y_2, y_4) = (y_3, y_4, y_6) \\
t_5' &= t_2 \oplus (y_2, y_5) = (y_3, y_5, y_6) \qquad \text{by } S_{y_2}(t_2)
\end{aligned}
$$

There are two troubles with this set $\underset{\sim}{T}^{y_1 y_2}$. One is tree t_3', which is of distance one from t_0. Thus t_3' is in one of the sets T^{y_1}, T^{y_2}, and T^{y_3}. The second problem is that tree t_2' obtained from t_1 and tree t_4' obtained from t_2 are the same. Hence we must find more restrictions for such a set to make sure we will have only trees which are distance two from t_0.

Definition 9-2-2.

$$
T^{e_1 e_2} = \{t' \oplus (e_2 b); \quad b \in S_{e_2}(t') \cap S_{e_2}(t_0), t' \in T^{e_1}, \quad b \neq e_2\} \quad (9\text{-}2\text{-}12)
$$

For example, since $S_{y_2}(t_1)$ and $S_{y_2}(t_0)$ in Examples 9-2-1 and 9-2-2 give

$$
S_{y_2}(t_1) \cap S_{y_2}(t_0) = (y_2, y_5, y_6) \cap (y_2, y_5, y_6) = (y_2, y_5, y_6)
$$

t_1' and t_2' are in $T^{y_1 y_2}(t_0)$. On the other hand,

$$
S_{y_2}(t_2) \cap S_{y_2}(t_0) = (y_1, y_2, y_4, y_5) \cap (y_2, y_5, y_6) = (y_2, y_5)
$$

Hence only t_5' among t_3', t_4', and t_5' is in $T^{y_1 y_2}$, or

$$
T^{y_1 y_2} = \{(y_3, y_4, y_5), (y_3, y_4, y_6), (y_3, y_5, y_6)\}
$$

Since $S_{e_2}(t_0)$ is a fundamental cut-set with respect to t_0 and contains edge e_2 and edges e_1 and e_2 are in t_0, cut-set $S_{e_2}(t_0)$ does not contain edge e_1. Thus $S_{e_2}(t') \cap S_{e_2}(t_0)$ does not contain edge e_1. Hence it is impossible to replace edge e_2 by edge e_1 to obtain a tree in $T^{e_1 e_2}$. This means that every tree in $T^{e_1 e_2}$ contains all edges in t_0 except edges e_1 and e_2. Hence every tree in $T^{e_1 e_2}$ is of distance two from t_0.

In general, we define set $T^{e_1 e_2 \cdots e_k}$ as follows.

Definition 9-2-3.

$$
\begin{aligned}
T^{e_1 e_2 \cdots e_{k-1} e_k} &\\
&= \{t' \oplus (e_k b); \quad b \in S_{e_k}(t') \cap S_{e_k}(t_0), t' \in T^{e_1 e_2 \cdots e_{k-1}}, \quad b \neq e_k\} \quad (9\text{-}2\text{-}13)
\end{aligned}
$$

for $k \leq m$ where $t_0 = (e_1, e_2, \ldots, e_m)$.

We will show that by Eq. 9-2-13 with t_0, which is called a reference tree, all trees can be generated. In other words, the tree generation procedure using

Eq. 9-2-13 is the following. We generate a first tree $t_0 = (e_1, e_2, \cdots, e_m)$, the reference tree. We replace each branch e_i of t_0 by a chord in the fundamental cut-set $S_{e_i}(t_0)$, to generate the classes T^{e_i} for $i = 1, 2, \ldots, m$. We take each of these classes T^{e_i} and replace a branch e_j, $1 < j \le m$, by a chord $e_k \in S_{e_j}(t) \cap S_{e_j}(t_0)$. Then we start again from $T^{e_i e_j}$ to produce $T^{e_i e_j e_k}$ where $i < j < k \le m$, each time requiring the chord to be in the fundamental cut-sets with respect to the current (parent) tree and the reference tree t_0. Now we can show two facts. First, we can generate all possible trees by this process, and second, no duplications arise. To prove these facts we use the following definitions.

Definition 9-2-4. Let t and t' be two trees and t_0 be the reference tree. Then the ordered pair (e_i, e_j) of edges is said to possess the *forward property* with respect to t' and t_0 if

1. $e_j \notin t'$, $e_i \in t' \cap t_0$ (obviously $e_j \ne e_i$).
2. $e_j \in S_{e_i}(t') \cap S_{e_i}(t_0)$.

The ordered pair (e_i, e_j) of edges is said to possess the *backward property* with respect to t and t_0 if

1. $e_j \in t$, $e_i \in t_0$, $e_i \ne e_j$.
2. $e_i \in S_{e_j}(t)$, $e_j \in S_{e_i}(t_0)$.

Using Eq. 9-2-13, we can see that if (e_i, e_j) possesses the forward property with respect to t' and t_0, then we can go forward to a new tree t'' by our procedure, replacing e_i by e_j. If (e_i, e_j) possesses the backward property with respect to t and t_0, by replacing e_j by e_i in t a new tree t''' is produced. From this new tree t''', t can be generated by our procedure. That is, we can go backward through the procedure to a tree t''', which has one more branch in common with the reference tree t_0. To prove that this is true, we need only observe that if we generate t''' from t by replacing e_j by e_i, the fundamental cut-set involved is invariant. That is,

$$S_{e_j}(t) = S_{e_i}(t''') \tag{9-2-14}$$

Now we show that if t_0 is the reference tree and t is any tree, we can find a tree t' such that t' and t_0 have one more branch in common than t and t_0, and that t could be generated from t' by our procedure. Then by induction we can generate any tree from the reference tree by our procedure.

Theorem 9-2-3. Let t_0 be the reference tree and t be any tree ($t \ne t_0$). For every edge $e_i \in t_0 - t$, there exists an edge $e_j \in t - t_0$ such that (e_i, e_j) possesses the backward property with respect to t and t_0. Hence there exists a tree t' such that (e_i, e_j) possesses the forward property with respect to t' and t_0.

Proof. Let $C_{e_i}(t)$ be the fundamental circuit produced by chord e_i to tree t, that is, $C_{e_i}(t)$ is a circuit formed by inserting edge e_i to tree t. Since $S_{e_i}(t_0)$ is a cut-set, $S_{e_i}(t_0) \cap C_{e_i}(t)$ contains an even number of edges and is not empty. Hence there is another edge e_j in the intersection. Since $e_j \in C_{e_i}(t)$, $e_j \in t$ and $e_i \in S_{e_j}(t)$. Since $e_j \in S_{e_i}(t_0)$, $e_j \notin t_0$. Hence (e_i, e_j) possesses the backward property with respect to t and t_0 and the rest follows. QED

Thus every tree of a linear graph can be generated by this process. However, if $S_{e_i}(t_0) \cap C_{e_i}(t)$ contains more than two edges, it is possible to generate t several times. Unless we take some precaution, duplications cannot be avoided, as shown by the following example.

Example 9-2-13. Consider the linear graph in Fig. 9-1-1. Let t_0 be (y_1, y_3, y_5). Then

$$T^{y_1} = \{(y_3, y_4, y_5), (y_3, y_5, y_6)\}$$

and

$$T^{y_1 y_5} = \{(y_2, y_3, y_4), (y_3, y_4, y_6), (y_2, y_3, y_6)\}$$

Since

$$S_{y_3}(t_0) = (y_2, y_3, y_4, y_6) \quad \text{and} \quad S_{y_3}(y_2, y_3, y_4) = (y_1, y_3, y_5, y_6)$$

(by Eq. 9-2-13), $(y_2, y_3, y_4) \oplus (y_3, b)$ for all b which are in $S_{y_3}(y_2, y_3, y_4) \cap S_{y_3}(t_0)$ except y_3 will be (y_2, y_4, y_6). Since $S_{y_3}(y_3, y_4, y_6) = (y_1, y_2, y_3)$, $(y_3, y_4, y_6) \oplus (y_3, b)$ for all b which are in $S_{y_3}(y_3, y_4, y_6) \cap S_{y_3}(t_0)$ except y_3, which again is (y_2, y_4, y_6). Thus we can see that there is duplication in $T^{y_1 y_5 y_3}$ obtained by Eq. 9-2-13.

Suppose we use the sequence (y_1, y_3, y_5), that is, we form T^{y_1}, then we form $T^{y_1 y_3}$ and $T^{y_1 y_3 y_5}$ to obtain tree (y_2, y_4, y_6). Then

$$T^{y_1 y_3} = \{(y_2, y_5, y_6), (y_2, y_4, y_5)\}$$

Now $(y_2, y_5, y_6) \oplus (y_5, b)$ for all b which are in $S_{y_5}(y_2, y_5, y_6) \cap S_{y_5}(t_0)$ except y_3 gives no tree because there is no edge b in this case. Now (y_2, y_4, y_5) $\oplus (y_5, b)$ for all b which are in $S_{y_5}(y_2, y_4, y_5) \cap S_{y_3}(t_0)$ except y_5 gives (y_2, y_4, y_6). Thus there is no duplication in $T^{y_1 y_3 y_5}$ obtained from $T^{y_1 y_3}$ by Eq. 9-2-13.

The technique for tree generation which avoids duplication is based on the following elementary observation.

Lemma 9-2-2. Let e_i be an edge in t_0 which is connected to a vertex of degree 1 in t_0. Also let t be any tree such that $e_i \notin t$. Then there exists exactly one edge $e_j \in t$ such that (e_i, e_j) possesses the backward property with respect to t and t_0.

Proof. As is clear from the proof of Theorem 9-2-3, we need show only that

$$S_{e_i}(t_0) \cap C_{e_i}(t)$$

contains only two edges e_i and e_j. Since e_i is an edge of t_0 and is connected to a vertex of degree 1 in t_0, the fundamental cut-set of e_i with respect to t_0 consists of the edges incident at the vertex (the vertex of degree 1 in t_0 to which e_i is connected). Clearly, only one of them can be in $C_{e_i}(t)$ since every vertex of $C_{e_i}(t)$ is of degree 2 and e_i is one of two edges at that vertex. Thus only (e_i, e_j) has the backward property with respect to t and t_0.   QED

This lemma guarantees that for any tree t in $T^{e_1 e_2 \cdots e_{k-1} e_k}$ if edge e_k is connected to a vertex of degree 1 in a reference tree t_0, then there is only one tree in $T^{e_1 e_2 \cdots e_{k-1}}$ from which tree t can be obtained by Eq. 9-2-13.

We may also note that any duplications that arise are within class $T^{e_{i_1} e_{i_2} \cdots e_{i_r}}$. In other words, if

$$(i_1, i_2, \ldots, i_r) \neq (j_1, j_2, \ldots, j_p) \qquad (9\text{-}2\text{-}15)$$

$$T^{e_{i_1} e_{i_2} \cdots e_{i_r}} \cap T^{e_{j_1} e_{j_2} \cdots e_{j_p}} = \emptyset \qquad (9\text{-}2\text{-}16)$$

This result follows immediately since if t and t' are trees from the two classes,

$$t_0 - t = (e_{i_1}, e_{i_2}, \ldots, e_{i_r}) \qquad (9\text{-}2\text{-}17)$$

and

$$t_0 - t' = (e_{j_1}, e_{j_2}, \ldots, e_{j_p}) \qquad (9\text{-}2\text{-}18)$$

which are different by hypothesis. Thus it is sufficient to find a procedure that avoids duplication within a class.

Definition 9-2-5. The ordered set $(e_{i_1} e_{i_2} \cdots e_{i_k})$ of edges constitutes an *M-sequence* if every subset $(e_{i_1} e_{i_2} \cdots e_{i_r})$ for $r = 1, 2, \ldots, k$ consists of edges of a connected subgraph.

It is well known that the edges in a tree can always be ordered as an M-sequence. In fact, if the starting tree t_0 is generated by a computer, it is easier to obtain edges in t_0 as an M-sequence. With this definition of a M-sequence, we have Theorem 9-2-4.

Theorem 9-2-4. Let $t_0 = (e_1, e_2, \ldots, e_m)$ be the reference tree. If $(e_1 e_2 \cdots e_m)$ is an M-sequence, the trees in $T^{e_{i_1} e_{i_2} \cdots e_{i_p}}$ where $1 \leq i_1 < i_2 < \cdots < i_p \leq m$, generated from $T^{e_{i_1} e_{i_2} \cdots e_{i_{p-1}}}$ according to Eq. 9-2-13 are all different.

Proof. Since the result is obvious for the first step where we go from t_0 to $T^{e_{i_1}}$, we may assume for induction that the tree in $T^{e_{i_1} e_{i_2} \cdots e_{i_{p-1}}}$ are all different. Hence, if two trees in this class generate the same t in $T^{e_{i_1} e_{i_2} \cdots e_{i_p}}$ by replacement of e_p, different edges e_i and e_j must replace e_{i_p} in the two cases. To prove the required result, it is sufficient to show that for any tree t in the class $T^{e_{i_1} e_{i_2} \cdots e_{i_p}}$ there is only one edge e_j such that the ordered pair (e_{i_p}, e_j) has the backward property with respect to t and t_0. For then t has only one predecessor in $T^{e_{i_1}e_{i_2}\cdots e_{i_p-1}}$. We prove this result by transforming the linear graph into one in which t and t_0' become trees t' and t_0' and e_{i_p} becomes an end edge. Then Theorem 9-2-4 follows from Lemma 9-2-2.

We note that the edges which follow e_{i_p} in M-sequence $(e_1 e_2 \cdots e_m)$ are common to t and t_0 (they have not been replaced yet). Let us short these edges of given linear graph G to obtain a new linear graph G'. Let t' and t_0' be the remaining edges of t and t_0, respectively, after shorting all edges in $t \cap t_0$. Hence t' and t_0' are trees of G'. Then the fundamental cut-set of e_{i_p} with respect to either tree remains unaffected. That is,

$$S_{i_{i_p}}(t') = S_{e_{i_p}}(t) \tag{9-2-19}$$

and

$$S_{e_{i_p}}(t_0') = S_{e_{i_p}}(t_0) \tag{9-2-20}$$

Similarly, the trees in $T^{e_{i_1}e_{i_2}\cdots e_{i_p-1}}$ without these edges in $t \cap t_0$ become trees in G' and $S_{e_i}(t')$ for any t' in this class will be unaffected. Hence as many new trees are produced in G' as in G, by replacing e_{i_p}. However, since $(e_1 e_2 \cdots e_m)$ is an M-sequence, $(e_1, e_2, \ldots, e_{i_p-1})$ is connected and so is $(e_1, e_2, \ldots, e_{i_p})$. Neither contains a circuit. Thus e_{i_p} is an end edge in t_0'. Therefore, by Lemma 9-2-2, there is only one edge $e_j \in t'$ such that (e_{i_p}, e_j) possesses the backward property with respect to t' and t_0' and hence there is only one tree in $T^{e_{i_1}e_{i_2}\cdots e_{i_p-1}}$ without edges in $t \cap t_0$ from which t' can be generated. This means that there exists one tree in $T^{e_{i_1}e_{i_2}\cdots e_{i_p-1}}$ from which t can be generated.   QED

We illustrate the complete procedure by the following example.

Example 9-2-4. The following steps will generate all possible trees in the linear graph in Fig. 9-2-1 without duplication.

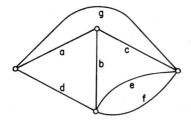

Fig. 9-2-1. A linear graph G.

STEP 1.   Take a tree t_0 to be the reference tree. Let

$$t_0 = (a, b, g)$$

Choose one M-sequence of edges in t_0. Let (abg) be the chosen M-sequence.

STEP 2.   The fundamental cut-sets with respect to t_0 are

$$S_a(t_0) = (a, c, d, e, f), \quad S_b(t_0) = (b, d, e, f), \quad \text{and} \quad S_g(t_0) = (c, e, f, g)$$

STEP 3.   Form T^a by Eq. 9-2-10:

$$\begin{aligned} T^a &= \{t_0 \oplus (a, e_i); \quad e_i \in S_a(t_0), \quad e_i \neq a\} \\ &= \{(b, c, g), (b, d, g), (b, e, g), (b, f, g)\} \end{aligned}$$

STEP 4.   Form T^{ab}:

$$\begin{aligned} T^{ab} &= \{t' \oplus (be_i); \quad e_i \in S_b(t') \cap S_b(t_0), \, t' \in T^a, \quad e_i \neq b\} \\ &= \{(c, d, g), (c, e, g), (c, f, g)\} \end{aligned}$$

where

$$\begin{aligned} S_b(b, c, g) \cap S_b(t_0) &= (b, d, e, f) \cap (b, d, e, f) = (b, d, e, f) \\ S_b(b, d, g) \cap S_b(t_0) &= (a, b, c) \cap (b, d, e, f) = (b) \\ S_b(b, e, g) \cap S_b(t_0) &= (b) \end{aligned}$$

and

$$S_b(b, f, g) \cap S_b(t_0) = (b)$$

STEP 5.   Form T^{abg}:

$$\begin{aligned} T^{abg} &= \{t' \oplus (ge_i); \quad e_i \in S_g(t') \cap S_g(t_0), \, t' \in T^{ab}, \quad e_i \neq g\} \\ &= \{(c, d, e), (c, d, f)\} \end{aligned}$$

where

$$\begin{aligned} S_g(c, d, g) \cap S_g(t_0) &= (a, b, e, f, g) \cap (c, e, f, g) = (e, f, g) \\ S_g(c, e, g) \cap S_g(t_0) &= (a, d, g) \cap (c, e, f, g) = (g) \end{aligned}$$

and

$$S_g(c, f, g) \cap S_g(t_0) = (g)$$

STEP 6.   Form T^{ag}:

$$\begin{aligned} T^{ag} &= \{t' \oplus (ge_i); \quad e_i \in S_g(t') \cap S_g(t_0), \, t' \in T^a, \quad e_i \neq g\} \\ &= \{(b, c, d), (b, d, e), (b, d, f)\} \end{aligned}$$

where

$$\begin{aligned} S_g(b, c, g) \cap S_g(t_0) &= (a, d, g) \cap (c, e, f, g) = (g) \\ S_g(b, d, g) \cap S_g(t_0) &= (c, e, f, g) \cap (c, e, f, g) = (c, e, f, g) \\ S_g(b, e, g) \cap S_g(t_0) &= (a, d, g) \cap (c, e, f, g) = (g) \end{aligned}$$

and

$$S_g(b, f, g) \cap S_g(t_0) = (g)$$

STEP 7.   Form T^b:

$$T^b = \{t_0 \oplus (b, e_i); \quad e_i \in S_b(t_0), \quad e_i \neq b\}$$
$$= \{(a, d, g), (a, e, g), (a, f, g)\}$$

STEP 8.   Form T^{bg}:

$$T^{bg} = \{(a, c, d), (a, d, e), (a, d, f), (a, c, e), (a, c, f)\}$$

where

$$S_g(a, d, g) \cap S_g(t_0) = (c, e, f, g) \cap (c, e, f, g) = (c, e, f, g)$$
$$S_g(a, e, g) \cap S_g(t_0) = (b, c, d, g) \cap (c, e, f, g) = (c, g)$$

and

$$S_g(a, f, g) \cap S_g(t_0) = (b, c, d, g) \cap (c, e, f, g) = (c, g)$$

STEP 9.   Form T^g:

$$T^g = \{t_0 \oplus (ge_i); \quad e_i \in S_g(t_0), \quad e_i \neq g\}$$
$$= \{(a, b, c), (a, b, e), (a, b, f)\}$$

All possible trees in G are those in the classes T^a, T^{ab}, T^{abg}, T^{ag}, T^b, T^{bg}, and T^g plus the reference tree t_0.

As given, the procedure seems to imply that we must examine at most 2^{m-1} replacement sets where $m = n_v - 1$. However, we note that if any replacement yields any empty set, the following replacements need not be examined. Thus the number examined will be less than $2^m - 1$.

In generating complete trees (for active networks), we may use an M-sequence of a current graph for both current and voltage graphs, allowing the output lists to be similarly ordered. In this case, there will be duplications in the sets of trees of the voltage graph, which, however, do not complicate matters. A disadvantage of using this generation of trees for active networks is that we may generate more trees than necessary. This is because the number of complete trees generally is much smaller than the number of trees in either the current graph or the voltage graph.

9-3   GENERATION OF COMPLETE TREES

Here we develop a procedure of generating all possible complete trees by modifying cut-set matrices. Since complete trees become trees when the current and voltage graphs are the same, the procedure is applicable for generating all trees in a linear graph.

Let Q_i and Q_v be cut-set (incidence) matrices of the current graph G_i and voltage graph G_v of network G, respectively. Also let the ith columns of both Q_i and Q_v represent admittance y_i of G. With matrices Q_i and Q_v, the following operations are called elementary operations:

1. Adding and subtracting one row from another.
2. Multiplying a row by (-1).
3. Interchanging rows.
4. Interchanging columns.

Associated with operations 2 and 3, we define a number N called an *M-number* which indicates the number of times we use these two operations in order to make a given matrix a desired form. Particularly, M-numbers N_i and N_v, which are necessary in order to change cut-set matrices (or incidence matrices) Q_i and Q_v into fundamental forms as $[Q_{i_{11}} \quad U]$ and $[Q_{v_{11}} \quad U]$, respectively, are called *fundamental M-numbers* with respect to complete tree t_0 where t_0 corresponds to the unit matrices of $[Q_{i_{11}} \quad U]$ and $[Q_{v_{11}} \quad U]$.

Consider a complete tree t. Let $A_i(t)$ and $A_v(t)$ be incidence matrices of current and voltage graphs of t. It is clear that we can change these matrices into the unit matrix by elementary operations even if we keep the correspondence between the columns of $A_i(t)$ and $A_v(t)$ [i.e., if we interchange the ith and the jth columns of $A_i(t)$, we interchange the ith and the jth columns of $A_v(t)$, and conversely]. Operation 1 does not change the determinant. Hence the determinant of $A_i(t) \cdot A_v(t)$ is equal to

$$|A_i(t)| \, |A_v(t)| = (-1)^{N_i + N_v} \tag{9-3-1}$$

Consider incidence matrices A_i and A_v where

$$A_i = [A_{i_1} A_i(t)] \tag{9-3-2}$$

and

$$A_v = [A_{v_1} A_v(t)] \tag{9-3-3}$$

We can change these matrices into fundamental forms as $[Q_{i_{11}} \quad U]$ and $[Q_{v_{11}} \quad U]$ by elementary operations. Furthermore, we use operations 2 and 3 N_i times for A_i and N_v times for A_v in order that $A_i(t)$ and $A_v(t)$ become unit matrices in the fundamental forms. Hence the sign of complete tree t given in Eq. 8-1-23 is equal to $(-1)^{N_i + N_v}$.

Example 9-3-1. Let incidence matrices A_i and A_v of a given network be

$$A_i = \begin{array}{c} 1 \\ 2 \\ 3 \end{array} \begin{bmatrix} \begin{array}{ccccc} y_1 & y_2 & y_3 & y_4 & y_5 \\ 1 & 0 & 1 & 0 & 0 \\ 0 & 1 & -1 & 1 & 0 \\ 0 & -1 & 0 & 0 & -1 \end{array} \end{bmatrix} \qquad A_v = \begin{array}{c} 1 \\ 2 \\ 3 \end{array} \begin{bmatrix} \begin{array}{ccccc} y_1 & y_2 & y_3 & y_4 & y_5 \\ 1 & -1 & -1 & 0 & 0 \\ 0 & 0 & 1 & 0 & -1 \\ -1 & 0 & 0 & -1 & 0 \end{array} \end{bmatrix}$$

By using elementary operations, we can change these matrices into fundamental forms (fundamental cut-set matrices):

$$Q_i' = \begin{matrix} & \begin{matrix} y_1 & y_2 & y_3 & y_4 & y_5 \end{matrix} \\ \begin{matrix} 1 \\ 2 \\ 3 \end{matrix} & \begin{bmatrix} 1 & 0 & 1 & 0 & 0 \\ 1 & 1 & 0 & 1 & 0 \\ 0 & 1 & 0 & 0 & 1 \end{bmatrix} \end{matrix} \qquad Q_v' = \begin{matrix} & \begin{matrix} y_1 & y_2 & y_3 & y_4 & y_5 \end{matrix} \\ \begin{matrix} 1 \\ 3 \\ 2 \end{matrix} & \begin{bmatrix} -1 & 1 & 1 & 0 & 0 \\ 1 & 0 & 0 & 1 & 0 \\ -1 & 1 & 0 & 0 & 1 \end{bmatrix} \end{matrix}$$

with M-numbers $N_i = 1$ and $N_v = 4$. Thus the sign of complete tree
$t_0 = (y_3, y_4, y_5)$ is $(-1)^{N_i + N_v} = -1$.

*For convenience, the sign of a complete tree t is given by the positive or
negative superscript.* For example, since the sign of tree t_0 in the preceding
example is -1, we express it as $(y_3, y_4, y_5)^-$. It is clear that all complete trees
of distance one from a complete tree t_0 are in the set

$$\bigcup_{a \in t_0} T^a = \bigcup_{a \in t_0} \{t_0 \oplus (ae); \quad e \in S_a{}^i(t_0) \cap S_a{}^v(t_0), \quad e \neq a\} \qquad (9\text{-}3\text{-}4)$$

where $S_a{}^i(t_0)$ is the fundamental cut-set with respect to t_0 that contains edge a
(a branch of t_0) in the current graph and $S_a{}^v(t_0)$ is the fundamental cut-set
with respect to t_0 that contains branch a in the voltage graph of a given
network. To obtain these complete trees in the set given by Eq. 9-3-4 directly
from matrices Q_i and Q_v, we use the following definition.

Definition 9-3-1. An operation $\odot$ of two m by n matrices $P = [p_{ij}]$ and
$R = [r_{ij}]$ is the product of corresponding entries of P and R, that is,

$$P \odot R = [p_{ij} r_{ij}] \qquad (9\text{-}3\text{-}5)$$

For example, if P and R are

$$P = \begin{bmatrix} 1 & -1 & 0 \\ 0 & 1 & -1 \end{bmatrix} \qquad \text{and} \qquad R = \begin{bmatrix} 0 & 1 & -1 \\ 1 & -1 & -1 \end{bmatrix}$$

then

$$P \odot R = \begin{bmatrix} 0 & -1 & 0 \\ 0 & -1 & 1 \end{bmatrix}$$

With this definition, all complete trees of distance one from t_0 can be
obtained from matrices $[Q_{i_{11}} \quad U]$ and $[Q_{v_{11}} \quad U]$ with t_0 corresponding to U
as follows:

STEP 1. Form $[Q_{i_{11}} \quad U] \odot [Q_{v_{11}} \quad U]$. Note that each row j of

$$[Q_{i_{11}} \quad U] \odot [Q_{v_{11}} \quad U]$$

represents

$$S_{a_j}{}^i(t_0) \cap S_{a_j}{}^v(t_0)$$

where row j of $[Q_{i_{11}} \quad U]$ and $[Q_{v_{11}} \quad U]$ represent $S_{a_j}{}^i(t_0)$ and $S_{a_j}{}^v(t_0)$, respectively.

STEP 2.   Form all complete trees in T^{a_j} (where $a_j \in t_0$) by replacing a_j by edge e whose corresponding column in $[Q_{i_{11}} \quad U] \odot [Q_{v_{11}} \quad U]$ has a non-zero at row j (where row j is the row in which the unit matrix U has 1 at the column corresponding to edge a_j).   It can be seen that this replacement is to obtain another fundamental cut-set matrix with respect to another tree. If this nonzero is $+1$, the sign of the newly obtained complete tree is the same as that of t_0. Otherwise, the sign of the new complete tree is opposite from that of t_0. The following example will illustrate these steps.

Example 9-3-2.   All complete trees of distance 1 from $t_0^- = (y_3, y_4, y_5)$ can be obtained from Q_i' and Q_v' in Example 9-3-1 by first forming $Q_i' \odot Q_v'$:

$$
Q_i' \odot Q_v' = \begin{array}{c} \\ 1 \\ 2 \\ 3 \end{array}
\begin{array}{c} y_1 \ y_2 \ y_3 \ y_4 \ y_5 \\ \left[\begin{array}{ccc|ccc} -1 & 0 & 1 & 0 & 0 \\ 1 & 0 & 0 & 1 & 0 \\ 0 & 1 & 0 & 0 & 1 \end{array} \right] \end{array}
$$

Now T^{y_3} is obtained from t_0 by replacing edge y_3 by edge y_1 because only column y_1 has a nonzero entry at the first row representing $S_{y_3}{}^i(t_0) \cap S_{y_3}{}^v(t_0)$. Or

$$ T^{y_3} = (y_1, y_4, y_5)^+ $$

The sign of (y_1, y_4, y_5) is the opposite of that of t_0 because the nonzero entry is -1. Similarly,

$$ T^{y_4} = (y_1, y_3, y_5)^- $$

and

$$ T^{y_5} = (y_2, y_3, y_4)^- $$

In order to obtain all trees whose distances are more than one away from t_0, consider cut-set matrix Q which can be partitioned as

$$
Q = [Q_{11} \quad U] = \begin{array}{c} b_1 \cdots b_r \ \ b_1' \cdots b_k' \ \quad e_1 \cdots e_k \ \ e_{k+1} \cdots e_m \\ \left[\begin{array}{cc|c|cc} Q_a & Q_b & U_1 & 0 \\ \hline Q_c & Q_d & 0 & U_2 \end{array} \right] \end{array} \qquad (9\text{-}3\text{-}6)
$$

where $(e_1, \ldots, e_k, e_{k+1}, \ldots, e_m)$ is tree t_0. Suppose we interchange all columns $e_1, \ldots, e_k$ by columns $b_1', \ldots, b_k'$ as

$$
\begin{bmatrix} Q_a & U_1 & Q_b & 0 \\ Q_c & 0 & Q_d & U_2 \end{bmatrix}
$$

It can be seen that matrix

$$\begin{bmatrix} Q_b & 0 \\ Q_d & U_2 \end{bmatrix}$$

is nonsingular if and only if Q_b is nonsingular. Thus if and only if Q_b can be changed to a unit matrix by elementary operations, edges $b'_1, \ldots, b'_k, e_{k+1}, \ldots, e_m$ form a tree. Therefore, to obtain all possible trees by changing edges $e_1, e_2, \ldots, e_k$ of t_0, we need to know all possible nonsingular major submatrices of $[Q_a \quad Q_b]$. In other words, if G is a linear graph corresponding to cut-set matrix Q, then we can see that $[Q_a \quad Q_b]$ is a cut-set matrix of linear graph $G(\overline{e_{k+1}\cdots e_m};\ \overline{e_1\cdots e_k})$ obtained from G by shorting edges $e_{k+1}, \ldots, e_m$ and deleting edges $e_1, \ldots, e_k$. A nonsingular major submatrix of $[Q_a \quad Q_b]$ represents a tree in $G(\overline{e_{k+1}\cdots e_m};\ \overline{e_1\cdots e_k})$, and the edges in this tree plus edges $e_{k+1}, \ldots, e_m$ form a tree in the original linear graph. By knowing all nonsingular submatrices of $[Q_a \quad Q_b]$, we can obtain all trees in $T^{e_1\cdots e_k}$.

Consider Q_i and Q_v which are partitioned as

$$Q_i = \begin{array}{c} \begin{array}{ccccc} b_1\cdots b_r & b'_1\cdots b'_k & e_1\cdots e_k & e_{k+1}\cdots e_m \end{array} \\ \left[\begin{array}{c:c:c:c} Q_{i_a} & Q_{i_b} & U_1 & 0 \\ \hdashline Q_{i_c} & Q_{i_d} & 0 & U_2 \end{array} \right] \end{array} \qquad (9\text{-}3\text{-}7)$$

and

$$Q_v = \begin{array}{c} \begin{array}{ccccc} b_1\cdots b_r & b'_1\cdots b'_k & e_1\cdots e_k & e_{k+1}\cdots e_m \end{array} \\ \left[\begin{array}{c:c:c:c} Q_{v_a} & Q_{v_b} & U_1 & 0 \\ \hdashline Q_{v_c} & Q_{v_d} & 0 & U_2 \end{array} \right] \end{array} \qquad (9\text{-}3\text{-}8)$$

Then if and only if

$$\begin{bmatrix} Q_{i_b} & 0 \\ Q_{i_d} & U_2 \end{bmatrix} \quad \text{and} \quad \begin{bmatrix} Q_{v_b} & 0 \\ Q_{v_d} & U_2 \end{bmatrix}$$

are nonsingular, edges $b'_1, \ldots, b'_k, e_{k+1}, \ldots, e_m$ form a complete tree. Hence if we know all possible complete trees of current and voltage graphs $G_i(\overline{e_{k+1}\cdots e_m};\ \overline{e_1\cdots e_k})$ and $G_v(\overline{e_{k+1}\cdots e_m};\ \overline{e_1\cdots e_k})$ whose cut-set matrices are $[Q_{i_a} \quad Q_{i_b}]$ and $[Q_{v_a} \quad Q_{v_b}]$, respectively, we can obtain all possible complete trees that can be formed from t_0 by replacing edges $e_1, \ldots, e_k$ (i.e., all complete trees in $T^{e_1\cdots e_k}$). In other words, to obtain all complete trees in $T^{e_1\cdots e_k}$, we consider $[Q_{i_a} \quad Q_{i_b}]$ and $[Q_{v_a} \quad Q_{v_b}]$ as the cut-set matrices of the current and voltage graphs of a given network and obtain all possible complete trees from them.

Suppose t'_0 is a complete tree in $G_i(\overline{e_{k+1}\cdots e_m};\ \overline{e_1\cdots e_k})$ and $G_v(\overline{e_{k+1}\cdots e_m};\ \overline{e_1\cdots e_k})$ whose sign is ϵ'_0. Also suppose the sign of complete tree t_0 of a given network is ϵ_0. Then we can see that the sign ϵ_1 of complete tree t which is obtained from t_0 by replacing edges $e_1, \ldots, e_k$ by all edges in t_0' is

$$\epsilon_1 = \epsilon_0 \cdot \epsilon_0' \qquad (9\text{-}3\text{-}9)$$

This can be seen readily if we know that $\epsilon_0 = (-1)^{N_i + N_v}$ and $\epsilon_0' = (-1)^{N_i' + N_v'}$.

The process of obtaining all complete trees from incidence matrices A_i and A_v consists of the following three steps:

STEP 1. Change matrices A_i and A_v to fundamental forms $[Q_{i_{11}} \quad U]$ and $[Q_{v_{11}} \quad U]$. Then we know a complete tree t_0 has its sign corresponding to unit matrix U. Let t_0 be $(e_1, e_2, \ldots, e_m)$.

STEP 2. Compute $[Q_{i_{11}} \quad U] \odot [Q_{v_{11}} \quad U]$. Then obtain all complete trees of distance 1 from t_0 and their signs.

STEP 3. For every combination of k edges in t_0 for $k > 1$, we form class $T^{e_{i_1} \cdots e_{i_k}}$ of complete trees with respect to t_0 where $1 \le i_1 < \cdots < i_k \le m$. To obtain all complete trees in each $T^{e_{i_1} \cdots e_{i_k}}$, we form submatrices $Q_i(e_{i_1} \cdots e_{i_k})$ and $Q_v(e_{i_1} \cdots e_{i_k})$ of $[Q_{i_{11}} \quad U]$ and $[Q_{v_{11}} \quad U]$, respectively, where $Q_i(e_{i_1} \cdots e_{i_k})$ consists of columns other than those of unit matrix U in $[Q_{i_{11}} \quad U]$ and rows corresponding to $S_e(t_0)$ (i.e., rows which have 1 at column e_{i_p}) for $p = 1$, $2, \ldots, k$. Then we consider $Q_i(e_{i_1} \cdots e_{i_k})$ and $Q_v(e_{i_1} \cdots e_{i_k})$ as a pair of given cut-set matrices and obtain all possible complete trees from $Q_i(e_{i_1} \cdots e_{i_k})$ and $Q_v(e_{i_1} \cdots e_{i_k})$, which can be accomplished by repeated (if necessary) execution of these three steps by considering these matrices as A_i and A_v in Step 1. To each of these complete trees obtained from $Q_i(e_{i_1} \cdots e_{i_k})$ and $Q_v(e_{i_1} \cdots e_{i_k})$, we insert all edges in $t_0 - (e_{i_1}, \ldots, e_{i_k})$ to form a complete tree in $T^{e_{i_1} \cdots e_{i_k}}$. Note that to obtain all complete trees, we start with incidence matrices A_i and A_v. Otherwise the sign of the first complete tree t_0 must be found by some other technique such as a sign-permutation.

The following example shows the complete procedure for finding all possible complete trees.

Example 9-3-3. Let incidence matrices of a network be

$$A_i = \begin{array}{c} \\ 1 \\ 2 \\ 3 \end{array} \begin{array}{cccccccc} y_0 & y_1 & y_2 & y_3 & y_4 & y_5 & y_6 \\ \left[\begin{array}{ccccccc} 1 & 0 & 1 & 0 & 1 & 0 & 0 \\ -1 & 1 & 0 & 1 & -1 & 1 & 0 \\ 0 & -1 & -1 & 0 & 0 & 0 & -1 \end{array}\right] \end{array}$$

and

$$A_v = \begin{array}{c} \\ 1 \\ 2 \\ 3 \end{array} \begin{array}{cccccccc} y_0 & y_1 & y_2 & y_3 & y_4 & y_5 & y_6 \\ \left[\begin{array}{ccccccc} 1 & 0 & 1 & -1 & -1 & 0 & 0 \\ 0 & 1 & 0 & 0 & 1 & 0 & -1 \\ 0 & -1 & -1 & 0 & 0 & -1 & 0 \end{array}\right] \end{array}$$

then the following steps will produce complete trees.

STEP 1.   Change A_i and A_v into fundamental forms (fundamental cut-set matrices):

$$Q_i' = \begin{array}{c} \\ 1' \\ 2' \\ 3' \end{array} \begin{array}{cccccccc} y_0 & y_1 & y_2 & y_3 & y_4 & y_5 & y_6 \\ \left[\begin{array}{cccc|ccc} 1 & 0 & 1 & 0 & 1 & 0 & 0 \\ 0 & 1 & 1 & 1 & 0 & 1 & 0 \\ 0 & 1 & 1 & 0 & 0 & 0 & 1 \end{array}\right] \end{array}$$

$$Q_v' = \begin{array}{c} \\ 1' \\ 3' \\ 2' \end{array} \begin{array}{ccccccc} y_0 & y_1 & \;\; y_2 & y_3 & y_4 & y_5 & y_6 \\ \left[\begin{array}{ccccc|ccc} -1 & 0 & -1 & 1 & 1 & 0 & 0 \\ 0 & 1 & 1 & 0 & 0 & 1 & 0 \\ -1 & -1 & -1 & 1 & 0 & 0 & 1 \end{array}\right] \end{array}$$

Thus $t_0 = (y_4, y_5, y_6)^-$. Note that $N_i = 1$ and $N_v = 4$.

STEP 2.   Form $Q_i' \odot Q_v'$:

$$Q_i' \odot Q_v' = \begin{array}{ccccccc} y_0 & y_1 & \;\; y_2 & y_3 & y_4 & y_5 & y_6 \\ \left[\begin{array}{ccccc|ccc} -1 & 0 & -1 & 0 & 1 & 0 & 0 \\ 0 & 1 & 1 & 0 & 0 & 1 & 0 \\ 0 & -1 & -1 & 0 & 0 & 0 & 1 \end{array}\right] \end{array}$$

Thus all complete trees of distance 1 from t_0 are

$$(y_0, y_5, y_6)^+, \quad (y_2, y_5, y_6)^+, \quad (y_1, y_4, y_6)^-, \quad (y_2, y_4, y_6)^-,$$
$$(y_1, y_4, y_5)^+, \quad \text{and} \quad (y_2, y_4, y_5)^+$$

STEP 3.   To obtain $T^{y_4 y_5}$, we form submatrices of Q_i' and Q_v':

$$Q_i(\underline{y_4 y_5}) = \begin{array}{c} \\ 1' \\ 2' \end{array} \begin{array}{cccc} y_0 & y_1 & y_2 & y_3 \\ \left[\begin{array}{cccc} 1 & 0 & 1 & 0 \\ 0 & 1 & 1 & 1 \end{array}\right] \end{array} \qquad Q_v(\underline{y_4 y_5}) = \begin{array}{c} \\ 1' \\ 3' \end{array} \begin{array}{cccc} y_0 & y_1 & y_2 & y_3 \\ \left[\begin{array}{cccc} -1 & 0 & -1 & 1 \\ 0 & 1 & 1 & 0 \end{array}\right] \end{array}$$

Note that $Q_i(\underline{y_4 y_5})$ is obtained by taking rows $1'$ and $2'$ which are the rows having 1 at columns y_4 and y_5, and all columns other than those of a unit matrix of Q_i'. Similarly, $Q_v(\underline{y_4 y_5})$ is obtained by taking rows $1'$ and $3'$ and all columns other than those of a unit matrix of Q_v'.

STEP 4.   We execute Step 1 with $Q_i(\underline{y_4 y_5})$ and $Q_v(\underline{y_4 y_5})$:

$$Q_i'(\underline{y_4 y_5}) = \begin{array}{c} \\ 1' \\ 2'' \end{array} \begin{array}{cccc} y_0 & y_1 & y_2 & y_3 \\ \left[\begin{array}{cc|cc} 1 & 0 & 1 & 0 \\ -1 & 1 & 0 & 1 \end{array}\right] \end{array} \qquad Q_v'(\underline{y_4 y_5}) = \begin{array}{c} \\ 3' \\ 1'' \end{array} \begin{array}{cccc} y_0 & y_1 & y_2 & y_3 \\ \left[\begin{array}{cc|cc} 0 & 1 & 1 & 0 \\ -1 & 1 & 0 & 1 \end{array}\right] \end{array}$$

From these fundamental forms, we have

$$t_{01} = (y_2 y_3)^-$$

STEP 5.   From

$$Q_i'(\underline{y_4 y_5}) \odot Q_v'(\underline{y_4 y_5}) = \begin{matrix} y_0 & y_1 & y_2 & y_3 \\ \begin{bmatrix} 0 & 0 & 1 & 0 \\ 1 & 1 & 0 & 1 \end{bmatrix} \end{matrix}$$

we know that there are two complete trees of distance 1 from t_{01},

$$(y_0, y_2)^- \quad \text{and} \quad (y_1, y_2)^-$$

STEP 6.   By Step 3 with $Q_i'(\underline{y_4 y_5})$ and $Q_v'(\underline{y_4 y_5})$, we can obtain complete trees of distance 2 from t_{01} by considering submatrices of $Q_i'(\underline{y_4 y_5})$ and $Q_v'(\underline{y_4 y_5})$ which are

$$Q_i(\underline{y_4 y_5}:\underline{y_2 y_3}) = \begin{matrix} y_0 & y_1 \\ \begin{bmatrix} 1 & 0 \\ -1 & 1 \end{bmatrix} \end{matrix} \qquad Q_v(\underline{y_4 y_5}:\underline{y_2 y_3}) = \begin{matrix} y_0 & y_1 \\ \begin{bmatrix} 0 & 1 \\ -1 & 1 \end{bmatrix} \end{matrix}$$

STEP 7.   By Step 1, we change these matrices into fundamental forms:

$$Q_i'(\underline{y_4 y_5}:\underline{y_2 y_3}) = \begin{matrix} y_0 & y_1 \\ \begin{bmatrix} 1 & 0 \\ 0 & 1 \end{bmatrix} \end{matrix} \qquad Q_v'(\underline{y_4 y_5}:\underline{y_2 y_3}) = \begin{matrix} y_0 & y_1 \\ \begin{bmatrix} 1 & 0 \\ 0 & 1 \end{bmatrix} \end{matrix}$$

Hence $t_{02} = (y_0, y_1)^-$ and there are no complete trees of distance 1 from t_{02}. Thus $(y_0, y_1)^-$ is the only complete tree of distance 2 from t_{01}. All complete trees of linear graphs $G_i(\overline{y_0}; \overline{y_4 y_5})$ and $G_v(\overline{y_0}; \overline{y_4 y_5})$ (which are obtained from the given current and voltage graphs by shorting y_0 and deleting y_4 and y_5) are

$$t_{01} = (y_2, y_3)^-, \quad (y_0, y_2)^-, \quad (y_1, y_2)^-, \quad \text{and} \quad (y_0, y_1)^-$$

With these, we can obtain complete trees in $T^{y_4 y_5}$ of the given current and the voltage graphs by replacing y_4 and y_5 of t_0 and changing their signs:

$$T^{y_4 y_5} = [(y_2, y_3, y_6)^+, (y_0, y_2, y_6)^+, (y_1, y_2, y_6)^+, (y_0, y_1, y_6)^+]$$

STEP 8.   For $T^{y_4 y_6}$, we use submatrices of Q_i' and Q_v' which are

$$Q_i(\underline{y_4 y_6}) = \begin{matrix} & y_0 & y_1 & y_2 & y_3 \\ 1' \\ 3' \end{matrix}\begin{bmatrix} 1 & 0 & 1 & 0 \\ 0 & 1 & 1 & 0 \end{bmatrix} \qquad Q_v(\underline{y_4 y_6}) = \begin{matrix} & y_0 & y_1 & y_2 & y_3 \\ 1' \\ 2' \end{matrix}\begin{bmatrix} -1 & 0 & -1 & 1 \\ -1 & -1 & -1 & 1 \end{bmatrix}$$

STEP 9.   Change these submatrices into fundamental forms:

$$Q_i'(\underline{y_4 y_6}) = \begin{matrix} y_3 & y_0 & y_1 & y_2 \\ \begin{bmatrix} 0 & -1 & 1 & 0 \\ 0 & 1 & 0 & 1 \end{bmatrix} \end{matrix} \qquad Q_v'(\underline{y_4 y_6}) = \begin{matrix} y_3 & y_0 & y_1 & y_2 \\ \begin{bmatrix} 0 & 0 & 1 & 0 \\ -1 & 1 & 0 & 1 \end{bmatrix} \end{matrix}$$

Thus $t_{03} = (y_1, y_2)$.

STEP 10. From

$$Q_i'(\underline{y_4 y_6}) \odot Q_v'(\underline{y_4 y_6}) = \begin{array}{cccc} y_3 & y_0 & y_1 & y_2 \end{array} \\ \begin{bmatrix} 0 & 0 & 1 & 0 \\ 0 & 1 & 0 & 1 \end{bmatrix}$$

we can see that $(y_0 y_1)$ is the only complete tree of distance 1 from t_{03}.

STEP 11. To obtain complete trees of distance 2 from t_{03}, we use

$$Q_i(\underline{y_4 y_6} : \underline{y_1 y_2}) = \begin{array}{cc} y_3 & y_0 \end{array} \\ \begin{bmatrix} 0 & -1 \\ 0 & 1 \end{bmatrix} \qquad Q_v(\underline{y_4 y_6} : \underline{y_1 y_2}) = \begin{array}{cc} y_3 & y_0 \end{array} \\ \begin{bmatrix} 0 & 0 \\ -1 & 1 \end{bmatrix}$$

Since these are singular matrices, there are no complete trees of distance 2 from t_{03} in $G_i(\bar{y}_5 : y_4 y_6)$ and $G_v(\bar{y}_5 : y_4 y_6)$. Thus

$$T^{y_4 y_6} = \{(y_1, y_2, y_5)^- \quad (y_0, y_1, y_5)^-\}.$$

STEP 12. For $T^{y_5 y_6}$, we consider

$$Q_i(\underline{y_5 y_6}) = \begin{array}{c} 2' \\ 3' \end{array} \begin{array}{cccc} y_0 & y_1 & y_2 & y_3 \end{array} \\ \begin{bmatrix} 0 & 1 & 1 & 1 \\ 0 & 1 & 1 & 0 \end{bmatrix} \qquad Q_v(\underline{y_5 y_6}) = \begin{array}{c} 3' \\ 2' \end{array} \begin{array}{cccc} y_0 & y_1 & y_2 & y_3 \end{array} \\ \begin{bmatrix} 0 & 1 & 1 & 0 \\ -1 & -1 & -1 & 1 \end{bmatrix}$$

STEP 13. Change these submatrices into fundamental forms:

$$Q_i'(\underline{y_5 y_6}) = \begin{array}{cccc} y_0 & y_1 & y_2 & y_3 \end{array} \\ \begin{bmatrix} 0 & 1 & 1 & 0 \\ 0 & 0 & 0 & 1 \end{bmatrix} \qquad Q_v'(\underline{y_5 y_6}) = \begin{array}{cccc} y_0 & y_1 & y_2 & y_3 \end{array} \\ \begin{bmatrix} 0 & 1 & 1 & 0 \\ -1 & 0 & 0 & 1 \end{bmatrix}$$

Then $t_{04} = (y_2 y_3)$.

STEP 14. Form $Q_i'(\underline{y_5 y_6}) \odot Q_v'(\underline{y_5 y_6})$:

$$Q_i'(\underline{y_5 y_6}) \odot Q_v'(\underline{y_5 y_6}) = \begin{array}{cccc} y_0 & y_1 & y_2 & y_3 \end{array} \\ \begin{bmatrix} 0 & 1 & 1 & 0 \\ 0 & 0 & 0 & 1 \end{bmatrix}$$

Hence (y_1, y_3) is the only complete tree of distance 1 from t_{04}.

STEP 15. To find complete trees of distance 2 from t_{04} in $G_i(\overline{\bar{y}_5} : \overline{y_5 y_6})$ and $G_v(\overline{y_4} : \overline{y_5 y_6})$, we consider

$$Q_i(\underline{y_5 y_6} : \underline{y_2 y_3}) = \begin{array}{cc} y_0 & y_1 \end{array} \\ \begin{bmatrix} 0 & 1 \\ 0 & 0 \end{bmatrix} \qquad Q_v(\underline{y_5 y_6} : \underline{y_2 y_3}) = \begin{array}{cc} y_0 & y_1 \end{array} \\ \begin{bmatrix} 0 & 1 \\ -1 & 0 \end{bmatrix}$$

Since $Q_i(\underline{y_5 y_6} : \underline{y_2 y_3})$ is singular, there is no complete trees of distance 2 from t_{04}. Thus

$$T^{y_5 y_6} = [(y_2, y_3, y_4)^- \quad (y_1, y_3, y_4)^-]$$

STEP 16.   To obtain $T^{y_4 y_5 y_6}$, we use

$$
Q_i(\underline{y_4 y_5 y_6}) = \begin{array}{c} \\ 1' \\ 2' \\ 3' \end{array}
\begin{array}{cccc} y_0 & y_1 & y_2 & y_3 \end{array}
\begin{bmatrix} 1 & 0 & 1 & 0 \\ 0 & 1 & 1 & 1 \\ 0 & 1 & 1 & 0 \end{bmatrix}
\qquad
Q_v(\underline{y_4 y_5 y_6}) = \begin{array}{c} \\ 1' \\ 3' \\ 2' \end{array}
\begin{array}{cccc} y_0 & y_1 & y_2 & y_3 \end{array}
\begin{bmatrix} -1 & 0 & -1 & 1 \\ 0 & 1 & 1 & 0 \\ -1 & -1 & -1 & 1 \end{bmatrix}
$$

STEP 17.   Change these matrices into fundamental forms:

$$
Q_i'(\underline{y_4 y_5 y_6}) = \begin{array}{cccc} y_0 & y_2 & y_1 & y_3 \end{array}
\begin{bmatrix} 1 & 1 & 0 & 0 \\ -1 & 0 & 1 & 0 \\ 0 & 0 & 0 & 1 \end{bmatrix}
\qquad
Q_v'(\underline{y_4 y_5 y_6}) = \begin{array}{cccc} y_0 & y_2 & y_1 & y_3 \end{array}
\begin{bmatrix} 0 & 1 & 0 & 0 \\ 0 & 0 & 1 & 0 \\ -1 & 0 & 0 & 1 \end{bmatrix}
$$

Hence $t_{05} = (y_1, y_2, y_3)$.

STEP 18.   Form $Q_i'(\underline{y_4 y_5 y_6}) \odot Q_v'(\underline{y_4 y_5 y_6})$:

$$
Q_i'(\underline{y_4 y_5 y_6}) \odot Q_v'(\underline{y_4 y_5 y_6}) = \begin{array}{cccc} y_0 & y_2 & y_1 & y_3 \end{array}
\begin{bmatrix} 0 & 1 & 0 & 0 \\ 0 & 0 & 1 & 0 \\ 0 & 0 & 0 & 1 \end{bmatrix}
$$

Hence there are no complete trees of distance 1 or more from t_{05} and

$$T^{y_4 y_5 y_6} = (y_1, y_2, y_3)^-$$

Thus all possible complete trees have been found.

It is clear that trees of a linear graph G become complete trees by taking G as the voltage graph and the current graph. Hence we can use the foregoing method to obtain all trees of G. However, since $A_i = A_v$ for this case, $Q_i' = Q_v'$ and it is not necessary to compute $Q_i' \odot Q_v'$. Furthermore, the sign of each complete tree becomes positive.

PROBLEMS

1.   Generate all possible trees of the linear graphs in Fig. P-9-1 by both methods. (One is given in Section 9-2 and the other in Section 9-3.)

2.   Generate all possible complete trees of the pairs of linear graphs in Fig. P-9-2.

3.   Suppose G_i and G_v are 2-isomorphic to each other. Then what kind of trees in G_i are also trees in G_v?

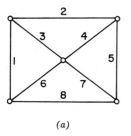

(a)

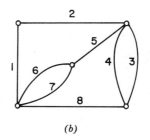

(b)

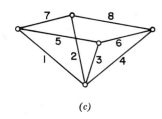

(c)

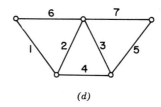

(d)

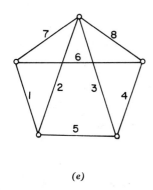

(e)

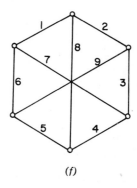

(f)

Fig. P-9-1.

4. Suppose G_i becomes identical to G_v by renaming the vertices as shown in Fig. P-9-2c. Then what kind of trees in G_i are also trees in G_v? What will be the signs of complete trees?

5. With the assumption that G_i and G_v are different, is there a condition on G_i and G_v such that all possible complete trees are all the possible trees of G_i?

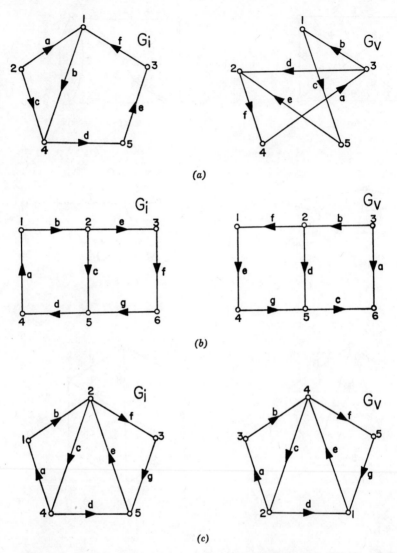

Fig. P-9-2. Pairs of linear graphs. (*a*) G_i and G_v; (*b*) G_i and G_v; (*c*) G_i and G_v.

CHAPTER
10
FLOW GRAPH AND SIGNAL FLOW GRAPH

10-1 FLOW GRAPH

In Chapter 7, we employed a weighted graph to represent an electrical network consisting of passivelike elements, and we found that a nonzero term of the determinant of an admittance matrix corresponds to a tree. In Chapter 8 a weighted graph for an activelike network was employed and we found that a nonzero term of the determinant of an admittance matrix is related to a (complete) tree. Then we studied a weighted graph for a unistor network and again observed that a nonzero term of the determinant of an admittance matrix corresponds to a (directed) tree. In this chapter we consider flow graphs and signal flow graphs, which are another form of weighted graphs. However, a nonzero term of a determinant associated with these weighted graphs does not correspond to a tree but to either a circuit or a vertex disjoint union of circuits.

Definition 10-1-1. For a square matrix $W = [w_{ij}]$ where $i, j = 1, 2, \ldots, n$, a *flow graph* G is a weighted oriented graph satisfying two conditions: (1) G consists of vertices $1, 2, \ldots, n$ and (2) the weight of each edge connected from vertex i to vertex j (whose orientation is from i to j) is w_{ji} for all $1 \leq i, j \leq n$.

For convenience, we use the symbol w_{ji} to indicate an edge as well as the weight of the edge. From the definition, we must note that edge w_{ji} is connected from i to j not from j to i. (The orientation of edge w_{ji} defined here is opposite from a conventional one. This makes the comparison between a flow graph and a signal flow graph, which will be studied in the next section, easier.)

As an example, suppose a matrix is

$$W = \begin{bmatrix} w_{11} & w_{12} & w_{13} \\ w_{21} & w_{22} & w_{23} \\ w_{31} & w_{32} & w_{33} \end{bmatrix}$$

Then a flow graph corresponding to this matrix consists of nine edges and three vertices as shown in Fig. 10-1-1. Note that in a flow graph, self-loops are allowed.

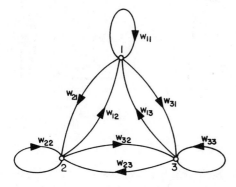

Fig. 10-1-1.   A flow graph.

An objective of employing flow graphs is to obtain the determinant of matrix W directly from the flow graph corresponding to W. Consider the determinant of matrix W which can be expressed as

$$|W| = \sum \delta_{j_1 j_2 \cdots j_n} w_{1j_1} w_{2j_2} \cdots w_{nj_n} \qquad (10\text{-}1\text{-}1)$$

Recall that (1) $\delta_{j_1 j_2 \cdots j_n}$ is zero if

$$\begin{pmatrix} 1 & 2 & \cdots & n \\ j_1 & j_2 & \cdots & j_n \end{pmatrix}$$

is not a permutation, (2) if

$$\begin{pmatrix} 1 & 2 & \cdots & n \\ j_1 & j_2 & \cdots & j_n \end{pmatrix}$$

is an even permutation (i.e., the even number of interchanges of j's in the second row is necessary in order that the second row become identical to the first row), then $\delta_{j_1 j_2 \cdots j_n} = 1$, and (3) if

$$\begin{pmatrix} 1 & 2 & \cdots & n \\ j_1 & j_2 & \cdots & j_n \end{pmatrix}$$

is an odd permutation, then $\delta_{j_1 j_2 \cdots jn} = -1$.

Let G be a flow graph corresponding to matrix W. Let g be a subgraph consisting of edges $w_{1j_1}, w_{2j_2}, \ldots, w_{nj_n}$. Note that the subscripts k and j_k of edge w_{kj_k} indicate that edge w_{kj_k} is connected from vertex j_k to vertex k. If

$$\begin{pmatrix} 1 & 2 & \cdots & n \\ j_1 & j_2 & \cdots & j_n \end{pmatrix}$$

corresponding to edges $w_{1j_1}, w_{2j_2}, \ldots, w_{nj_n}$ is a permutation, each vertex in g appears only once in the first row and only once in the second row of

$$\begin{pmatrix} 1 & 2 & \cdots & n \\ j_1 & j_2 & \cdots & j_n \end{pmatrix}$$

Hence there are exactly two edges connected to every vertex in g. Furthermore, $d^+(p) = d^-(p) = 1$ for every vertex p; in other words, one of the two edges connected to a vertex has its orientation away from the vertex and the other has the orientation toward the vertex. If subgraph g is connected, then g is a directed circuit.

Definition 10-1-2.   A *directed circuit* is a circuit whose orientation agrees with the orientation of every edge in the circuit.

Suppose subgraph g consists of k maximal connected subgraphs. Then each maximal connected subgraph must be a directed circuit, and g must be a vertex disjoint union of directed circuits.

Definition 10-1-3.   A *vertex disjoint union of directed circuits* is a set of directed circuits such that no two have vertices in common.

For convenience, we define a P-set cycle as follows.

Definition 10-1-4.   A *P-set cycle* is a subgraph which is either a directed circuit or a vertex disjoint union of directed circuits and which contains all vertices in a given oriented graph.

Since subgraph g contains all vertices in flow graph G, we can conclude that if

$$\begin{pmatrix} 1 & 2 & \cdots & n \\ j_1 & j_2 & \cdots & j_n \end{pmatrix}$$

is a permutation, subgraph g consisting of edges $w_{1j_1}, w_{2j_2}, \ldots, w_{nj_n}$ is a P-set cycle.

To show that the converse is also true, we assume that a subgraph g is a P-set cycle. Note that g must contain n edges. Let the edges in g be $w_{i_1j_1}$, $w_{i_2j_2}, \ldots, w_{i_nj_n}$. Then the sequence $(i_1 i_2 \cdots i_n)$ of subscripts must be a permutation of $(1\ 2\cdots n)$. Similarly, the sequence $(j_1 j_2 \cdots j_n)$ of subscripts must be a permutation of $(1\ 2\cdots n)$. Hence

$$\begin{pmatrix} i_1 & i_2 & \cdots & i_n \\ j_1 & j_2 & \cdots & j_n \end{pmatrix}$$

is a permutation and product $w_{i_1 j_1} w_{i_2 j_2} \cdots w_{i_n j_n}$ is a term in Eq. 10-1-1. In other words,

$$|W| = \sum \epsilon_j \ P\text{-set cycle } D_j \text{ product} \qquad (10\text{-}1\text{-}2)$$

where ϵ_j is either $+1$ or -1 depending on $\delta_{j_1 j_2 \cdots j_n}$.

To find the sign of ϵ_j, we assume that a P-set cycle D_j consisting of edges $w_{i_1 j_1}, w_{i_2 j_2}, \ldots, w_{i_n j_n}$ is a vertex disjoint union of k directed circuits. When $k = 1$, D_j is a directed circuit. Without the loss of generality, let

$$(w_{i_{p_1} j_{p_1}} w_{i_{p_1+1} j_{p_1+1}} \cdots w_{i_{p_1 + r_1} j_{p_1 + r_1}}),$$
$$(w_{i_{p_2} j_{p_2}} \cdots w_{i_{p_2+r_2} j_{p_2+r_2}}), \quad \ldots, \quad (w_{i_{p_k} j_{p_k}} \cdots w_{i_{p_k+r_k} j_{p_k+r_k}})$$

be these k directed circuits. Then

$$\begin{pmatrix} i_{p_s} & \cdots & i_{p_s+r_s} \\ j_{p_s} & \cdots & j_{p_s+r_s} \end{pmatrix}$$

must be a permutation for $s = 1, 2, \ldots, k$ because $i_{p_s}, i_{p_s+1}, \ldots, i_{p_s+r_s}$ represent all vertices in a directed circuit and so do $j_{p_s}, j_{p_s+1}, \ldots, j_{p_s+r_s}$.

Since we can always arrange the edges in a directed circuit such that sequence $(w_{i_{p_s} j_{p_s}} \cdots w_{i_{p_s+r_s} j_{p_s+r_s}})$ is a closed edge train in the direction equal to the orientation of the directed circuit, we have

$$i_{p_s+u-1} = j_{p_s+u} \qquad \text{for} \quad u = 1, 2, \ldots, r_s \qquad (10\text{-}1\text{-}3)$$

and

$$i_{p_s+r_s} = j_{p_s} \qquad (10\text{-}1\text{-}4)$$

Hence

$$\begin{pmatrix} i_{p_s} & \cdots & i_{p_s+r_s} \\ j_{p_s} & \cdots & j_{p_s+r_s} \end{pmatrix} = \begin{pmatrix} i_{p_s} & i_{p_s+1} & \cdots & i_{p_s+r_s-1} & i_{p_s+r_s} \\ i_{p_s+r_s} & i_{p_s} & \cdots & i_{p_s+r_s-2} & i_{p_s+r_s-1} \end{pmatrix} \qquad (10\text{-}1\text{-}5)$$

Thus $r_s - 1$ interchanges of entries of the second row of this permutation are sufficient to make the second row identical to the first row.

By assumption

$$\begin{pmatrix} i_1 & i_2 & \cdots & i_n \\ j_1 & j_2 & \cdots & j_n \end{pmatrix} = \begin{pmatrix} i_{p_1} & \cdots & i_{p_1+r_1} \\ j_{p_1} & \cdots & j_{p_1+r_1} \end{pmatrix} \begin{pmatrix} i_{p_2} & \cdots & i_{p_2+r_2} \\ j_{p_2} & \cdots & j_{p_2+r_2} \end{pmatrix} \cdots$$

$$\begin{pmatrix} i_{p_k} & \cdots & i_{p_k+r_k} \\ j_{p_k} & \cdots & j_{p_k+r_k} \end{pmatrix} \qquad (10\text{-}1\text{-}6)$$

and we have found that since $r_s - 1$ interchanges of entries of the second row of

$$\begin{pmatrix} i_{p_s} & \cdots & i_{p_s + r_s} \\ j_{p_s} & \cdots & j_{p_s + r_s} \end{pmatrix}$$

are sufficient to make the second row identical with the first row, $(r_1 - 1) + (r_2 - 1) + \cdots + (r_k - 1)$ interchanges of entries of the second row of

$$\begin{pmatrix} i_1 & i_2 & \cdots & i_n \\ j_1 & j_2 & \cdots & j_n \end{pmatrix}$$

are sufficient to make these two rows equal. Note that r_s is the number of edges in a directed circuit corresponding to

$$\begin{pmatrix} i_{p_s} & \cdots & i_{p_s + r_s} \\ j_{p_s} & \cdots & j_{p_s + r_s} \end{pmatrix}$$

Hence

$$(r_1 - 1) + (r_2 - 1) + \cdots + (r_k - 1) = r_1 + r_2 + \cdots + r_k - k$$
$$= n - k \qquad (10\text{-}1\text{-}7)$$

or $n - k$ interchanges are sufficient to make the second row of

$$\begin{pmatrix} i_1 & i_2 & \cdots & i_n \\ j_1 & j_2 & \cdots & j_n \end{pmatrix}$$

identical to the first row. Thus the sign of ϵ_j in Eq. 10-1-2 is

$$\epsilon_j = (-1)^{n+k} \qquad (10\text{-}1\text{-}8)$$

or

$$|W| = (-1)^n \sum_{(j)} (-1)^{k_j} \ P\text{-set cycle } D_j \text{ product} \qquad (10\text{-}1\text{-}9)$$

where k_j is the number of directed circuits in P-set cycle D_j.

Example 10-1-1. Consider the flow graph in Fig. 10-1-1 whose corresponding matrix is

$$W = \begin{bmatrix} w_{11} & w_{12} & w_{13} \\ w_{21} & w_{22} & w_{23} \\ w_{31} & w_{32} & w_{33} \end{bmatrix}$$

By Eq. 10-1-9, the determinant of W is

$$|W| = (-1)^3[(-1)^3 w_{11} w_{22} w_{33} + (-1)^2 w_{12} w_{21} w_{33} + (-1)^2 w_{13} w_{31} w_{22}$$
$$+ (-1)^2 w_{23} w_{32} w_{11} + (-1) w_{12} w_{23} w_{31} + (-1) w_{13} w_{32} w_{21}]$$

The corresponding subgraphs are shown in Fig. 10-1-2.

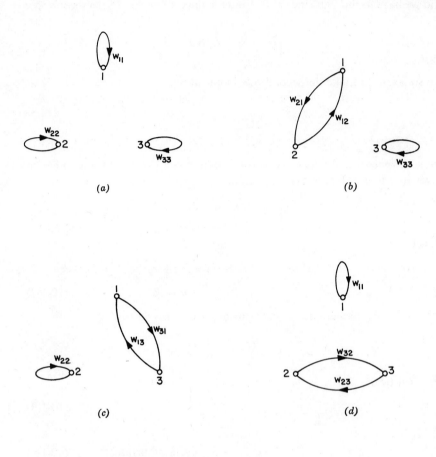

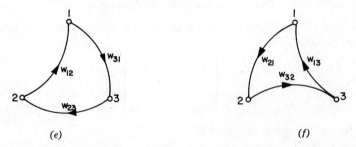

Fig. 10-1-2. *P*-set cycles. (*a*) $D_1 = (w_{11}, w_{22}, w_{33})$; (*b*) $D_2 = (w_{12}, w_{21}, w_{33})$; (*c*) $D_3 = (w_{13}, w_{31}, w_{22})$; (*d*) $D_4 = (w_{23}, w_{32}, w_{11})$; (*e*) $D_5 = (w_{12}, w_{23}, w_{31})$; (*f*) $D_6 = (w_{13}, w_{32}, w_{21})$.

370

10-2 SIGNAL FLOW GRAPH

A signal flow graph is a weighted oriented graph in which there are weights on both edges and vertices. For convenience, the symbol (i, j) is used for indicating an oriented edge which is connected from vertex i to vertex j.

Definition 10-2-1.   An equation

$$x_r = \sum_{p=0}^{n} m_{rp} x_p \qquad (0 \le r \le n) \tag{10-2-1}$$

is represented by an oriented graph G in which
1. The weight of vertex x_p is x_p.
2. For $m_{ru} \ne 0$ $(0 \le u \le n)$, there exists an edge (x_u, x_r) whose weight is m_{ru}.
3. Edges connected to vertex r whose orientations are toward vertex r are those having weight m_{ru}.

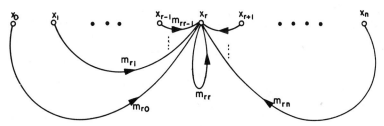

Fig. 10-2-1.   Signal flow graph.

Such an oriented graph G (Fig. 10-2-1) is called a *signal flow graph representing Eq. 10-2-1*. Equation 10-2-1 is called an equation at vertex x_r of signal flow graph G. Note that when $m_{ru} = 0$, either there is an edge (x_u, x_r) with weight 0 or there is no edge whose weight is m_{ru}.

Now consider the following simultaneous equations:

$$\left.\begin{aligned}
x_1 &= m_{11}x_1 + m_{12}x_2 + \cdots + m_{1n}x_n \\
x_2 &= m_{21}x_1 + m_{22}x_2 + \cdots + m_{2n}x_n \\
&\qquad\qquad\vdots \\
x_{n-1} &= m_{n-11}x_1 + m_{n-12}x_2 + \cdots + m_{n-1n}x_n \\
x_0 &= m_{01}x_1 + m_{02}x_2 + \cdots + m_{0n}x_n
\end{aligned}\right\} \tag{10-2-2}$$

The corresponding signal flow graph G consists of $n + 1$ vertices as shown in

Fig. 10-2-2. Note that x_n does not appear on the left-hand side of any equation and x_0 does not appear on the right-hand side of any equation in Eq. 10-2-2. Vertex x_n is called a *source* and vertex x_0 is called a *sink*.

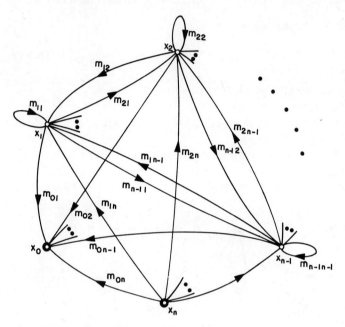

Fig. 10-2-2. A signal flow graph.

Three familiar rules which are used for simplifying a signal flow graph can be obtained as follows:

1. Suppose a signal flow graph G has parallel edges as shown in Fig. 10-2-3a. The equation at vertex x_q is

$$x_q = \sum_{\substack{r=0 \\ r \neq p}}^{n} m_{qr}x_r + m_{qp}x_p + m'_{qp}x_p = \sum_{\substack{r=0 \\ r \neq p}}^{n} m_{qr}x_r + (m_{qp} + m'_{qp})x_p \quad (10\text{-}2\text{-}3)$$

(a)   (b)

Fig. 10-2-3. (*a*) Signal flow graph G; (*b*) signal flow graph G'.

Signal flow graph G' in Fig. 10-2-3b is obtained from G by replacing these parallel edges by one edge whose weight is $m_{qp} + m'_{qp}$. It is clear that the equations at every vertex in G and G' are the same. Thus parallel edges whose orientation are the same can be replaced by one edge whose weight is the sum of the weights of the parallel edges.

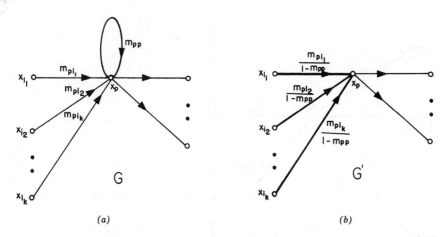

Fig. 10-2-4. (a) Signal flow graph G; (b) signal flow graph G'.

2. Consider signal flow graph G in Fig. 10-2-4a in which there is a self-loop at vertex x_p. The equation at vertex x_p is

$$x_p = m_{pp}x_p + \sum_{\substack{r=1 \\ i_r \neq p}}^{k} m_{pi_r}x_{i_r} \tag{10-2-4}$$

This can be changed to

$$(1 - m_{pp})x_p = \sum_{\substack{r=1 \\ i_r \neq p}}^{k} m_{pi_r}x_{i_r} \tag{10-2-5}$$

or

$$x_p = \sum_{\substack{r=1 \\ i_r \neq p}}^{k} \frac{m_{pi_r}}{1 - m_{pp}} x_{i_r} \tag{10-2-6}$$

under the condition that $m_{pp} \neq 1$.

Let G' in Fig. 10-2-4b be obtained from G by deleting self-loop (x_p, x_p) and changing the weight of edge (x_{i_r}, x_p) from m_{pi_r} to $m_{pi_r}/(1 - m_{pp})$ for $r = 1,$ $2, \ldots, k$. Then Eq. 10-2-6 is the equation at vertex x_p in G'. Furthermore, the equations at any other vertex in G' and G are the same. Thus we can state that a self-loop at a vertex x_p can be removed by dividing the weight of every edge which is connected at vertex x_p with its orientation toward x_p, by $(1 - m_{pp})$ if $m_{pp} \neq 1$.

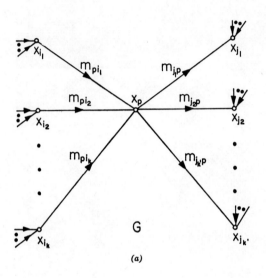

(a)

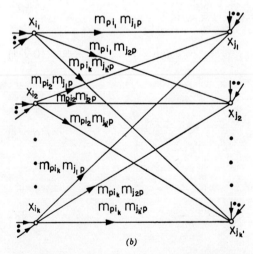

(b)

Fig. 10-2-5. Signal flow graph for elimination of a vertex. (a) Signal flow graph G; (b) signal flow graph G' obtained by eliminating vertex x_p.

3. Consider signal flow graph G in Fig. 10-2-5a where there are no self-loops at vertex x_p. The equation at vertex x_p is

$$x_p = \sum_{\substack{r=1 \\ i_r \neq p}}^{k} m_{pi_r} x_{i_r} \qquad (10\text{-}2\text{-}7)$$

The equation at vertex x_{j_u} in G can be written as

$$x_{j_u} = m_{j_u p} x_p + \sum_{\substack{s=0 \\ i_s \neq p}}^{n} m_{j_u i_s} x_{i_s} \qquad (10\text{-}2\text{-}8)$$

for $u = 1, 2, \ldots, k'$. Substituting Eq. 10-2-7 into Eq. 10-2-8, we have

$$x_{j_u} = \sum_{\substack{r=1 \\ i_r \neq p}}^{k} (m_{j_u p} m_{p i_r}) x_{i_r} + \sum_{\substack{s=0 \\ i_s \neq p}}^{n} m_{j_u i_s} x_{i_s} \qquad (10\text{-}2\text{-}9)$$

This equation indicates that we can eliminate vertex x_p by replacing every directed path, which consists of two edges m_{pi_r} and $m_{j_s p}$ ($1 \leq r \leq k$, $1 \leq s \leq k'$) by one edge from x_{i_r} to x_{j_s} whose weight is equal to $(m_{pi_r} m_{j_s p})$ as shown in Fig. 10-2-5b without changing the equations at all other vertices. Vertices x_{i_r} and x_{j_s} can be the same. However, if they are the same, m_{pi_r} and $m_{j_s p}$ will be a directed circuit rather than a directed path (just for the sake of terminology).

One objective of employing a signal flow graph is to find the relation between the source x_n and the sink x_0. By the successive application of the foregoing rules, if we obtain a signal flow graph that contains only two vertices x_n and x_0 and one edge from x_n to x_0, the weight of the edge gives the desired relation. We can see that the successive application of these rules to obtain a desired relation is nothing but substitution and simplification of simultaneous equations. An important advantage in using a signal flow graph is that a desired relation may be obtained topologically, which we discuss next.

Equation 10-2-2 can be expressed as

$$\begin{bmatrix} x_1 \\ x_2 \\ \vdots \\ x_{n-1} \\ x_0 \end{bmatrix} = \begin{bmatrix} m_{11} & m_{12} & \cdots & m_{1n} \\ m_{21} & m_{22} & \cdots & m_{2n} \\ & & \cdots & \\ m_{n-11} & & \cdots & m_{n-1n} \\ m_{01} & & \cdots & m_{0n} \end{bmatrix} \begin{bmatrix} x_1 \\ x_2 \\ \vdots \\ x_{n-1} \\ x_n \end{bmatrix} \qquad (10\text{-}2\text{-}10)$$

With

$$x_n = K x_0 \qquad (10\text{-}2\text{-}11)$$

Eq. 10-2-10 can be written as

$$
\begin{bmatrix} x_1 \\ x_2 \\ \vdots \\ x_{n-1} \\ x_n \\ x_0 \end{bmatrix} = \left[\begin{array}{ccccc|c} m_{11} & m_{12} & & m_{1n} & & 0 \\ m_{21} & m_{22} & & m_{2n} & & 0 \\ \vdots & \vdots & & & & \vdots \\ m_{n-11} & m_{n-12} & & m_{n-1n} & & 0 \\ \hline 0 & 0 & 0 & \cdots & 0 & 0 & K \\ \hline m_{01} & m_{02} & & m_{0n} & & 0 \end{array} \right] \begin{bmatrix} x_1 \\ x_2 \\ \vdots \\ x_{n-1} \\ x_n \\ x_0 \end{bmatrix} \qquad (10\text{-}2\text{-}12)
$$

or

$$
\begin{bmatrix} (m_{11} - 1) & m_{12} & \cdots & m_{1n} & 0 \\ m_{21} & (m_{22} - 1) & \cdots & m_{2n} & 0 \\ \vdots & \vdots & \cdots & \vdots & \vdots \\ m_{n-11} & m_{n-12} & \cdots & m_{n-1n} & 0 \\ 0 & 0 & 0 \cdots & 0 & -1 & K \\ m_{01} & m_{02} & \cdots & m_{0n} & -1 \end{bmatrix} \begin{bmatrix} x_1 \\ x_2 \\ \vdots \\ x_{n-1} \\ x_n \\ x_0 \end{bmatrix} = 0 \quad (10\text{-}2\text{-}13)
$$

For simplicity, let Eq. 10-2-13 be expressed as

$$(M - U)X = 0 \qquad (10\text{-}2\text{-}14)$$

Now compare the flow graph of matrix $(M - U)$ as shown in Fig. 10-2-6 and the signal flow graph in Fig. 10-2-2. We can see that these linear graphs are almost identical. In fact, the signal flow graph in Fig. 10-2-2 is a subgraph of the flow graph in Fig. 10-2-6. In other words, we can obtain the flow graph in Fig. 10-2-6 from the signal flow graph in Fig. 10-2-2 by inserting the self-loop of (-1) to every vertex and one edge K from vertex x_0 to vertex x_n. For convenience, the signal flow graph in Fig. 10-2-2 is called the *corresponding signal flow graph* of the flow graph in Fig. 10-2-6.

By Eq. 10-1-9, the determinant of $(M - U)$ is

$$|M - U| = (-1)^{n+1} \sum (-1)^{k_j} P\text{-set cycle } D_j \text{ product} \quad (10\text{-}2\text{-}15)$$

of the flow graph in Fig. 10-2-6.

Let $D_j(\bar{K})$ be a P-set cycle which does not contain edge K, and let $D_q(K)$ be a P-set cycle which contains edge K. Then Eq. 10-2-15 becomes

$$|M - U| = (-1)^{n+1}[C + KP] \qquad (10\text{-}2\text{-}16)$$

where

$$C = \sum (-1)^{k_j} P\text{-set cycle } D_j(\bar{K}) \text{ product} \qquad (10\text{-}2\text{-}17)$$

and

$$KP = \sum (-1)^{k_q} P\text{-set cycle } D_q(K) \text{ product} \qquad (10\text{-}2\text{-}18)$$

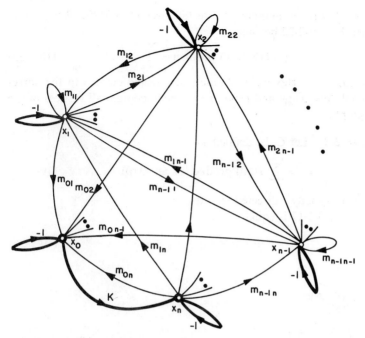

Fig. 10-2-6. Flow graph of $(M - U)$.

In order to have a nontrivial solution of Eq. 10-2-14, the determinant of $(M - U)$ must be zero, or

$$(-1)^{n+1}[C + KP] = 0 \qquad (10\text{-}2\text{-}19)$$

Hence by Eqs. 10-2-11 and 10-2-19,

$$\frac{x_0}{x_n} = \frac{1}{K} = \frac{-P}{C} \qquad (10\text{-}2\text{-}20)$$

Since C is $\sum (-1)^{k_j}$ P-set cycle $D_j(\overline{K})$ product, each P-set cycle $D_j(\overline{K})$ can be considered a union of two sets $D_j(0)$ and $D_j(-1)$ of directed circuits where $D_j(0)$ is the set of directed circuits consisting of edges in the corresponding signal flow graph and $D_j(-1)$ is the set of self-loops of (-1) which *are not* in the corresponding signal flow graph. Suppose there are u self-loops in $D_j(-1)$. Then

$$D_j(-1) \text{ product} = (-1)^u \qquad (10\text{-}2\text{-}21)$$

and

$$\begin{aligned}
(-1)^{k_j} D_j(\overline{K}) \text{ product} &= (-1)^{k_j}[D_j(0) \text{ product}][D_j(-1) \text{ product}] \\
&= (-1)^{k_j - u}(-1)^u[(D_j(0) \text{ product}](-1)^u \\
&= (-1)^{k_j} D_j(0) \text{ product} \qquad (10\text{-}2\text{-}22)
\end{aligned}$$

where $k'_j = k_j - u$ is the number of directed circuits in $D_j(0)$. When $D_j(0)$ is empty, then Eq. 10-2-22 becomes

$$(-1)^{k_j} D_j(\bar{K}) \text{ product} = 1 \qquad (10\text{-}2\text{-}23)$$

Note that $D_j(0)$ is a vertex disjoint union of directed circuits in the corresponding signal flow graph and *is not necessary* to contain all vertices of the signal flow graph.

Definition 10-2-2.   Let C_r be defined as

$$C_r = \sum D_j(0) \text{ product of set } D_j(0) \qquad (10\text{-}2\text{-}24)$$

which consists of exactly r directed circuits.

Then C in Eq. 10-2-17 is equal to

$$C = 1 - C_1 + C_2 - C_3 + \cdots \qquad (10\text{-}2\text{-}25)$$

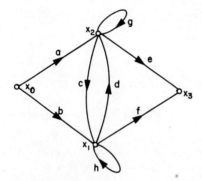

Fig. 10-2-7.   A signal flow graph.

For example, C of the signal flow graph in Fig. 10-2-7 can be obtained by getting $C_1, C_2, \ldots$ topologically as

$$C_1 = g + h + cd$$
$$C_2 = gh$$

and

$$C_u = 0 \qquad \text{for} \quad u \geq 3$$

Hence

$$C = 1 - g - h - cd + gh$$

Recall that a directed path from vertex v_0 to v_n is a path from vertex v_0 to vertex v_n such that if we give an orientation from v_0 to v_n to the path, the orientation of every edge in the path will agree with that of the path (Definition 6-6-7). For example, (a, e) in Fig. 10-2-7 is a directed path from vertex x_0 to vertex x_3.

Since KP of flow graph G_F is $\sum (-1)^{k_q}$ P-set cycle $D_q(K)$ product and each P-set cycle $D_q(K)$ must contain edge K, deleting edge K from $D_q(K)$ is a vertex disjoint union of one directed path P (from vertex x_n to vertex x_0) and directed circuits. It is now useful to define the symbols P_i and $G_F(\overline{\Omega(P_i)K})$.

Definition 10-2-3. The symbol P_i indicates a directed path from vertex x_0 to vertex x_n in a flow graph G_F (in Fig. 10-2-6), and the symbol $G_F(\overline{\Omega(P_i)K})$ is a subgraph of G_F obtained by deleting all vertices in P_i, all edges connected to these vertices, and edge K.

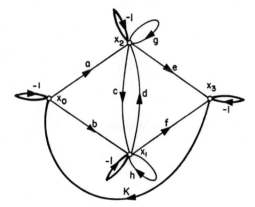

Fig. 10-2-8. A flow graph G_F.

An example of $G_F(\overline{\Omega(P_i)K})$ is shown in Fig. 10-2-9 when G_F is as shown in Fig. 10-2-8 and $P = (a, e)$.

Fig. 10-2-9. Subgraph $G_F(\overline{\Omega(P_i)K})$ of G_F where $P_i = (a, e)$.

With this definition, $D_q(K)$ without edge K which contains path P_i can be considered the set consisting of P_i and a vertex disjoint union of directed circuits in $G_F(\overline{\Omega(P_i)}\bar{K})$. If we take all $D_q(K)$ of flow graph G_F all of which contain directed path P_i, then

$$\sum_{\{D_q(K);P_i \in D_q(K)\}} (-1)^{k_q} D_q(K) \text{ product} = -K \sum_{(i)} (P_i \text{ product}) \times$$

$$[\sum(-1)^{k_q-1} P\text{-set cycle product of } G_F(\overline{\Omega(P_i)}\bar{K})] \quad (10\text{-}2\text{-}26)$$

where $\displaystyle\sum_{\{D_q(K);P_i \in D_q(K)\}}$ means "for all $D_q(K)$ which contain P_i."

Since $k_q - 1$ is the number of directed circuits in P-set cycle $D_q(K)$ without counting one which contains edge K, $k_q - 1$ is the number of directed circuits in the P-set cycle D_q' in $G_F(\overline{\Omega(P_i)}\bar{K})$. Thus by Eq. 10-2-25,

$$\sum(-1)^{k_q-1} P\text{-set cycle } D_q' \text{ product of } G_F(\overline{\Omega(P_i)}\bar{K})$$
$$= 1 - C_1 + C_2 - C_3 + \cdots \text{ of } G(\overline{\Omega(P_i)}) \quad (10\text{-}2\text{-}27)$$

where $G(\overline{\Omega(P_i)})$ is a subgraph of the corresponding signal flow graph G obtained by deleting all vertices in directed path P_i and deleting all edges connected to these deleted vertices. To avoid confusion, we use the symbol $C(\bar{P}_i)$ to indicate

$$C(\bar{P}_i) = 1 - C_1 + C_2 - C_3 + \cdots \text{ of } G(\overline{\Omega(P_i)}) \quad (10\text{-}2\text{-}28)$$

Then by Eq. 10-2-26, we have

$$\sum_{\{D_q(K);P_i \in D_q(K)\}} (-1)^{k_q} D_q(K) \text{ product} = -K(P_i \text{ product}) \cdot C(\bar{P}_i) \quad (10\text{-}2\text{-}29)$$

Hence the numerator of the right-hand side of Eq. 10-2-20 is

$$-P = \sum_{(i)} (P_i \text{ product}) \cdot C(\bar{P}_i) \quad (10\text{-}2\text{-}30)$$

Thus by Eqs. 10-2-11 and 10-2-20, the ratio of the source and the sink is equal to

$$\frac{x_0}{x_n} = \frac{\displaystyle\sum_{(i)}^{(i)} (P_i \text{ product}) \cdot C(\bar{P}_i)}{1 - C_1 + C_2 - C_3 + \cdots} \quad (10\text{-}2\text{-}31)$$

This equation is known as the Mason's formula. As an example, consider the signal flow graph in Fig. 10-2-7. We can obtain $\displaystyle\sum_{(i)} (P_j \text{ product}) \cdot C(\bar{P}_i)$ from x_0 to x_3 as

$$\sum_{(i)} (P_i \text{ product}) \cdot C(\bar{P}_i) = ae(1 - h) + bf(1 - g) + acf + bde$$

Hence

$$\frac{x_0}{x_3} = \frac{ae(1 - h) + bf(1 - g) + acf + bde}{1 - g - h - cd + gh}$$

10-3 EQUIVALENT TRANSFORMATION OF SIGNAL FLOW GRAPH

Instead of using the rules discussed at the beginning of the last section to change a signal flow graph into simpler structures, we can simplify a portion of a signal flow graph by one step. To study this procedure, consider a signal flow graph G consisting of two subgraphs G_M and G_R which are joined by

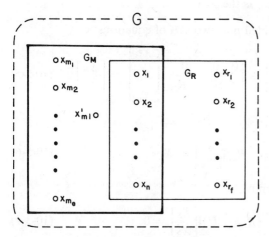

Fig. 10-3-1. Signal flow graph G consisting of subgraphs G_M and G_R.

vertices $x_1, x_2, \ldots, x_n$ as shown in Fig. 10-3-1. The equations corresponding to this signal flow graph are

$$\begin{bmatrix} M_{11} & M_{12} & 0 \\ M_{21} & M_{22} & 0 \\ 0 & 0 & 0 \end{bmatrix}\begin{bmatrix} X_M \\ X_I \\ X_R \end{bmatrix} + \begin{bmatrix} 0 & 0 & 0 \\ 0 & R_{11} & R_{12} \\ 0 & R_{21} & R_{22} \end{bmatrix}\begin{bmatrix} X_M \\ X_I \\ X_R \end{bmatrix} = \begin{bmatrix} \dot{X}_M \\ X_I \\ S_R \end{bmatrix} \quad (10\text{-}3\text{-}1)$$

where the matrix

$$M = \begin{bmatrix} M_{11} & M_{12} \\ M_{21} & M_{22} \end{bmatrix} \quad (10\text{-}3\text{-}2)$$

is from subgraph G_M and the matrix

$$R = \begin{bmatrix} R_{11} & R_{12} \\ R_{21} & R_{22} \end{bmatrix} \quad (10\text{-}3\text{-}3)$$

is from subgraph G_R. The vectors X_M, $\dot{X}_M$, X_I, and X_R are

$$X_M = \begin{bmatrix} x_{m_1} \\ x_{m_2} \\ \vdots \\ x_{m_e} \end{bmatrix} \qquad \underset{\sim}{X}_M = \begin{bmatrix} x'_{m_1} \\ x_{m_2} \\ \vdots \\ x_{m_e} \end{bmatrix}, \qquad X_I = \begin{bmatrix} x_1 \\ x_2 \\ \vdots \\ x_n \end{bmatrix}, \qquad \text{and} \qquad X_R = \begin{bmatrix} x_{r_1} \\ x_{r_2} \\ \vdots \\ x_{r_f} \end{bmatrix}$$

(10-3-4)

where x_{m_1} is the source and x'_{m_1} is the sink.

Equation 10-3-1 can be divided into two sets of equations:

$$\begin{bmatrix} M_{11} & M_{12} \\ M_{21} & M_{22} \end{bmatrix}\begin{bmatrix} X_M \\ X_I \end{bmatrix} + \begin{bmatrix} 0 & 0 & 0 \\ 0 & R_{11} & R_{12} \end{bmatrix}\begin{bmatrix} X_M \\ X_I \\ X_R \end{bmatrix} = \begin{bmatrix} \underset{\sim}{X}_M \\ X_I \end{bmatrix}$$

(10-3-5)

and

$$[R_{21} \quad R_{22}]\begin{bmatrix} X_I \\ X_R \end{bmatrix} = X_R$$

(10-3-6)

Equation 10-3-6 can be written as

$$[R_{21} \quad (R_{22} - U)]\begin{bmatrix} X_I \\ X_R \end{bmatrix} = 0$$

(10-3-7)

which can be solved for X_R under the assumption that $(R_{22} - U)$ is non-singular as

$$X_R = -[R_{22} - U]^{-1}R_{21}X_I$$

(10-3-8)

Substitution of this into Eq. 10-3-5 gives

$$\begin{bmatrix} M_{11} & M_{12} \\ M_{21} & M_{22} \end{bmatrix}\begin{bmatrix} X_M \\ X_I \end{bmatrix} + \begin{bmatrix} 0 & 0 \\ 0 & R_{11} - R_{12}[R_{22} - U]^{-1}R_{21} \end{bmatrix}\begin{bmatrix} X_M \\ X_I \end{bmatrix} = \begin{bmatrix} \underset{\sim}{X}_M \\ X_I \end{bmatrix}$$

(10-3-9)

The ratio between the source and the sink (x'_{m_1}/x_{m_1}) obtained from Eq. 10-3-1 is exactly the same as that obtained from Eq. 10-3-9. Hence the signal flow graph after eliminating all vertices $x_{r_1}, x_{r_2}, \ldots, x_{r_f}$ in subgraph G in Fig. 10-3-1 must be the one which is represented by Eq. 10-3-9. In other words, we can replace subgraph G_R by a new subgraph G_I consisting of vertices $x_1, x_2, \ldots, x_n$ which is given by

$$[R_{11} - R_{12}(R_{22} - U)^{-1}R_{21}]X_I = X_I$$

(10-3-10)

Thus to obtain a new subgraph G_I from given subgraph G_R we need only each entry of $[R_{11} - R_{12}(R_{22} - U)^{-1}R_{21}]$.

First, we define the following symbols for convenience.

Definition 10-3-1. For a given matrix R, let the symbol

$$R\left(\overline{\frac{t_1 t_2 \cdots t_v}{u_1 u_2 \cdots u_v}}\right) \quad \text{or} \quad R\left(\frac{\bar{t}_q:}{\bar{u}_q:} q = 1, 2, \ldots, v\right)$$

be the submatrix obtained from R by deleting rows $t_1, t_2, \ldots, t_v$ and columns $u_1, u_2, \ldots, u_v$. The symbol $G_R(\overline{u_1 u_2 \cdots u_v})$ or $G_R(\bar{u}_q: q = 1, 2, \ldots, v)$ is used for a linear graph obtained from linear graph G_R by deleting vertices $u_1, u_2, \ldots, u_v$ and all edges connected to these vertices.

Let

$$[R_{22} - U]^{-1} = \frac{1}{\Delta}\begin{bmatrix} \Delta_{h_1 h_1} & \Delta_{h_2 h_1} & \cdots & \Delta_{h_f h_1} \\ \Delta_{h_1 h_2} & \Delta_{h_2 h_2} & \cdots & \Delta_{h_f h_2} \\ & & \cdots & \\ \Delta_{h_1 h_f} & \Delta_{h_2 h_f} & \cdots & \Delta_{h_f h_f} \end{bmatrix} \qquad (10\text{-}3\text{-}11)$$

and

$$R_{11} - R_{12}[R_{22} - U]^{-1}R_{21} = [C_{tu}] \qquad (10\text{-}3\text{-}12)$$

Also let the rows and the columns of R be $1, 2, \ldots, n, h_1, h_2, \ldots, h_f$ in the order. Then from Eq. 10-3-12,

$$C_{tu} = r_{tu} - \frac{1}{\Delta}[r_{th_1} \quad r_{th_2} \quad \cdots \quad r_{th_f}]\begin{bmatrix} \Delta_{h_1 h_1} & \cdots & \Delta_{h_f h_1} \\ \Delta_{h_1 h_2} & \cdots & \Delta_{h_f h_2} \\ & \cdots & \\ \Delta_{h_1 h_f} & \cdots & \Delta_{h_f h_f} \end{bmatrix}\begin{bmatrix} r_{h_1 u} \\ r_{h_2 u} \\ \vdots \\ r_{h_f u} \end{bmatrix} \qquad (10\text{-}3\text{-}13)$$

or

$$C_{tu} = \frac{1}{\Delta}\left(r_{tu}\Delta - \sum_{p,q=1}^{f} r_{th_p}\Delta_{h_p h_q}r_{h_q u}\right) \qquad (10\text{-}3\text{-}14)$$

The numerator of the right-hand side of Eq. 10-3-14 is exactly the same as the row-column expansion of the matrix

$$\begin{bmatrix} r_{tu} & r_{th_1} & \cdots & r_{th_f} \\ \hline r_{h_1 u} & & & \\ r_{h_2 u} & & R_{22} - U & \\ \vdots & & & \\ r_{h_f u} & & & \end{bmatrix} = [R - U]\left(\frac{\bar{q}: q = 1, 2, \ldots, n, \quad q \neq t}{\bar{q}': q' = 1, 2, \ldots, n, \quad q' \neq u}\right)$$

$$(10\text{-}3\text{-}15)$$

Since

$$[R_{22} - U] = [R - U]\left(\begin{matrix}\bar{q}: \\ \bar{q}:\end{matrix} q = 1, 2, \ldots, n\right) \qquad (10\text{-}3\text{-}16)$$

we have

$$C_{tu} = \frac{\left|[R - U]\left(\begin{matrix}\bar{q}: q = 1, 2, \ldots, n, & q \neq t \\ \bar{q}: q' = 1, 2, \ldots, n, & q' \neq u\end{matrix}\right)\right|}{\left|[R - U]\left(\begin{matrix}\bar{w}: \\ \bar{w}:\end{matrix} w = 1, 2, \ldots, n\right)\right|} \qquad (10\text{-}3\text{-}17)$$

The signal flow graph corresponding to

$$R\left(\begin{matrix}\bar{w}: \\ \bar{w}:\end{matrix} w = 1, 2, \ldots, n\right)$$

is $G_R(\bar{x}_w: w = 1, 2, \ldots, n)$ because the deletion of row i_w and column i_w is equivalent to deleting vertex x_{i_w} and all edges connected to the vertex in G_R. Thus, by Eq. 10-1-9,

$$\left|[R - U]\left(\begin{matrix}\bar{w}: \\ \bar{w}:\end{matrix} w = 1, 2, \ldots, n\right)\right|$$
$$= (-1)^\alpha \sum (-1)^{k_j} P\text{-set cycle } D_j \text{ product of } \bar{G}_R(\bar{x}_w: w = 1, 2, \ldots, n) \qquad (10\text{-}3\text{-}18)$$

where $\bar{G}_R(\bar{x}_w: w = 1, 2, \ldots, n)$ is the flow graph obtained by inserting a self-loop with (-1) to every vertex in $G_R(\bar{x}_w: w = 1, 2, \ldots, n)$ and k_j is the number of directed circuits in P-set cycle D_j. Thus, by Eq. 10-2-26,

$$\left|[R - U]\left(\begin{matrix}w: \\ w:\end{matrix} w = 1, 2, \ldots, n\right)\right|$$
$$= (-1)^\alpha(1 - C_1 + C_2 - C_3 + \cdots) \text{ of signal flow graph}$$
$$G_R(\bar{x}_w: w = 1, 2, \ldots, n) \qquad (10\text{-}3\text{-}19)$$

Note that α is the number of vertices in $G_R(\bar{x}_w: w = 1, 2, \ldots, n)$.

For

$$\left|[R - U]\left(\begin{matrix}\bar{q}: q = 1, 2, \ldots, n, & q \neq t \\ \bar{q}': q' = 1, 2, \ldots, n, & q' \neq u\end{matrix}\right)\right|$$

we consider a flow graph $\bar{G}_F(K, \bar{w}_w: w = 1, 2, \ldots, n, \ w \neq t, u)$ which is obtained from signal flow graph $G_R(\bar{x}_w: w = 1, 2, \ldots, n, \ w \neq t, u)$ by inserting a self-loop with (-1) to every vertex and inserting edge K from x_t to x_u. We consider matrix F whose row u and column t are zeros except the

(u, t) entry which is K and which has the property that by deleting row u and column t, the resultant matrix is

$$[R - U]\begin{pmatrix}\bar{q}: & q = 1, 2, \ldots, n & q \neq t \\ \bar{q}': & q' = 1, 2, \ldots, n, & q' \neq u\end{pmatrix}$$

as

$$F = \begin{matrix} & & t & u & & r_1 & & & & r_f \\ t & \begin{bmatrix} \\ u \\ r_1 \\ \vdots \\ r_f \end{bmatrix} \end{matrix}$$

$$F = \begin{array}{c} t \\ u \\ r_1 \\ \vdots \\ r_f \end{array} \begin{bmatrix} 0 & & & & & & \\ K & 0 & & & & 0 \\ \hline 0 & & & & & & \\ \vdots & & [R - U]\begin{pmatrix}\bar{q}: & q = 1, 2, \ldots, n, & q \neq t \\ \bar{q}': & q' = 1, 2, \ldots, n, & q' \neq u\end{pmatrix} \\ 0 & & & & & & \end{bmatrix} \quad \text{(10-3-20)}$$

Now the determinant of F is

$$|F| = (-1)^{t+u} K \left| [R - U]\begin{pmatrix}\bar{q}: & q = 1, 2, \ldots, n, & q \neq t \\ \bar{q}': & q' = 1, 2, \ldots, n, & q' \neq u\end{pmatrix}\right| \quad \text{(10-3-21)}$$

Since F is a square matrix such that both column r and row r represent the same vertex x_r for all columns and rows, there is a flow graph G'_F corresponding to F such that

$$|F| = (-1)^{\alpha'} \sum (-1)^{k_r} \text{ P-set cycle } D_r \text{ product of } G'_F \quad \text{(10-3-22)}$$

We can also see that flow graph G'_F is the same as $\bar{G}_R(K, \bar{x}_w: w = 1, 2, \ldots, n,$ $w \neq t, u)$ except edges connected from x_t to x_u are absent. Thus

$$(-1)^{\alpha'} \sum (-1)^{k_r} \text{ P-set cycle } D_r \text{ product of } G'_F$$
$$= (-1)^{\alpha'} \sum (-1)^{k_r} \text{ P-set cycle } D_r(K) \text{ product}$$
$$\text{of } \bar{G}_F(K, \bar{x}_w: w = 1, 2, \ldots, n, \quad w \neq t, u) \quad \text{(10-3-23)}$$

where $D_r(K)$ indicates a P-set cycle which contains edge K. Thus, by Eq. 10-3-21, we have

$$\left| [R - U]\begin{pmatrix}\bar{q}: & q = 1, 2, \ldots, n, & q \neq t \\ \bar{q}': & q' = 1, 2, \ldots, n, & q' \neq u\end{pmatrix}\right|$$
$$= \frac{-1}{K} (-1)^{\alpha'} \sum (-1)^{k_r} \text{ P-set cycle } D_r(K) \text{ product}$$
$$\text{of } \bar{G}_F(K, \bar{x}_w: w = 1, 2, \ldots, n, \quad w \neq t, u) \quad \text{(10-3-24)}$$

Let a directed circuit in $\bar{G}_F(K, \bar{x}_w: w = 1, 2, \ldots, n, \quad w \neq t, u)$ be (K, P_i) where P_i is a directed path from x_u to x_t. Then

$$\sum_{\{D_j(K); P_i \in D_j(K)\}} (-1)^{k_j} D_j(K) \text{ product of } \bar{G}_F(K, \bar{x}_w: w = 1, 2, \ldots, n, \quad w \neq t, u)$$

is exactly the same as the left-hand side of Eq. 10-2-26. Thus we have

$$\left| [R - U] \begin{pmatrix} \bar{q}: & q = 1, 2, \ldots, n, & q \neq t \\ \bar{q}': & q' = 1, 2, \ldots, n, & q' \neq u \end{pmatrix} \right|$$
$$= (-1)^{\alpha'} \sum_{(i)} (P_i \text{ product}) \cdot C(\bar{P}_i) \text{ of } G_R(\bar{x}_w: w = 1, 2, \ldots, n, \quad w \neq t, u)$$

$$(10\text{-}3\text{-}25)$$

Note that $G_R(\bar{x}_w: w = 1, 2, \ldots, n, \quad w \neq t, u)$ is the signal flow graph obtained from G_R by deleting vertices x_w for $w = 1, 2, \ldots, n$ except vertices x_t and x_u. Since $\alpha' = \alpha + 2$, by using Eqs. 10-3-19 and 10-3-25, c_{tu} in Eq. 10-3-17 becomes

$$C_{tu} = \frac{\sum\limits_{(i)} (P_i \text{ product}) \cdot C(\bar{P}_i) \text{ of } G_R(\bar{x}_w: w = 1, 2, \ldots, n, \quad w \neq t, u)}{1 - C_1 + C_2 - C_3 + \cdots \text{ of } G_R(\bar{x}_w: w = 1, 2, \cdots, n)} \quad (10\text{-}3\text{-}26)$$

This is the same as the Mason's formula in Eq. 10-2-31 if we consider x_t as the source and x_u as the sink of $G_R(\bar{x}_w: w = 1, 2, \ldots, n, \quad w \neq t, u)$.

When $t = u$, Eq. 10-3-14 becomes

$$C_{tt} = \frac{1}{\Delta} \left(r_{tt} \Delta - \sum_{p,q=1} r_{th_p} \Delta_{h_p h_q} r_{h_q t} \right) \quad (10\text{-}3\text{-}27)$$

Thus

$$C_{tt} = \frac{\begin{vmatrix} r_{tt} & r_{th_1} & r_{th_2} & \cdots & r_{th_f} \\ r_{h_1 t} & & & & \\ r_{h_2 t} & & R_{22} - U & & \\ \vdots & & & & \\ r_{h_f t} & & & & \end{vmatrix}}{|R_{22} - U|} \quad (10\text{-}3\text{-}28)$$

By modifying this equation as

$$C_{tt} = C_{tt} - 1 + 1 \quad (10\text{-}3\text{-}29)$$

we find

$$C_{tt} = \frac{\begin{vmatrix} r_{tt} & r_{th_1} & \cdots & r_{th_f} \\ r_{h_1 t} & & & \\ r_{h_2 t} & & R_{22} - U & & -|R_{22} - U| \\ \vdots & & & \\ r_{h_f t} & & & \end{vmatrix}}{|R_{22} - U|} + 1$$

$$= \frac{\begin{vmatrix} r_{tt} - 1 & r_{th_1} & \cdots & r_{th_f} \\ r_{h_1 t} & & & \\ r_{h_2 t} & & R_{22} - U & \\ \vdots & & & \\ r_{h_f t} & & & \end{vmatrix}}{|R_{22} - U|} + 1$$

$$= \frac{\left| [R - U]\!\left(\begin{matrix} \bar{q}: \\ \bar{q}: \end{matrix} q = 1, 2, \ldots, n, \quad q \neq t\right) \right|}{\left| [R - U]\!\left(\begin{matrix} \bar{w}: \\ \bar{w}: \end{matrix} w = 1, 2, \ldots, n\right) \right|} + 1 \qquad (10\text{-}3\text{-}30)$$

Since the denominator of this equation is the same as that of Eq. 10-3-17, we need study only the numerator. Note that

$$\left| [R - U]\!\left(\begin{matrix} \bar{w}: \\ \bar{w}: \end{matrix} w = 1, 2, \ldots, n, \quad w \neq t\right) \right|$$

will have the same expression as that in Eq. 10-3-18 except that the flow graph under consideration is $\bar{G}_R(\bar{x}_w: w = 1, 2, \ldots, n, \quad w \neq t)$, or

$$\left| [R - U]\!\left(\begin{matrix} \bar{w}: \\ \bar{w}: \end{matrix} w = 1, 2, \ldots, n, \quad w \neq t\right) \right|$$
$$= (-1)^{\alpha + 1} \sum_{(j)} (-1)^{k_j} \; P\text{-set cycle } D_j \text{ product}$$
$$\text{of } \bar{G}_R(\bar{x}_w: w = 1, 2, \ldots, n, \quad w \neq t) \quad (10\text{-}3\text{-}31)$$

Suppose we divide the right-hand side of Eq. 10-3-31 into two parts:

$$\left| [R - U]\!\left(\begin{matrix} \bar{w}: \\ \bar{w}: \end{matrix} w = 1, 2, \ldots, n, \quad w \neq t\right) \right| = (-1)^{\alpha + 1}(F_1 + F_2) \quad (10\text{-}3\text{-}32)$$

where F_1 is the collection of terms corresponding to all possible P-set cycle $D_r(x_t)$ which contains the self-loop (-1) at vertex x_t, and F_2 is the collection of remaining terms. Then F_1 can be expressed as

$$F_1 = (-1) \sum_{(s)} (-1)^{k_s} \; P\text{-set cycle } D_s \text{ product of } \bar{G}_R(\bar{x}_w: w = 1, 2, \ldots, n)$$
$$(10\text{-}3\text{-}33)$$

This is because each $D_j(x_t)$ contain the self-loop (-1) at vertex x_t. Without this self-loop, the remaining set is a P-set cycle D_s of $\bar{G}_R(\bar{x}: w = 1, 2, \ldots, n)$. Thus Eq. 10-3-32 becomes

$$\left| [R - U]\!\left(\begin{matrix} \bar{w}: \\ \bar{w}: \end{matrix} w = 1, 2, \ldots, n, \quad w \neq t\right) \right|$$
$$= (-1)^{\alpha + 2} \{(-1)[1 - C_1 + C_2 - C_3 + \cdots \text{ of } G_R(\bar{x}_w: w = 1, 2, \ldots, n)]$$
$$+ \sum_{(j)} [d_j(x_t) \text{ product}] \cdot C(\bar{d}_j)\} \text{ of } G_R(\bar{x}_w: w = 1, 2, \ldots, n, \quad w \neq t)$$
$$(10\text{-}3\text{-}34)$$

where $d_j(x_t)$ is a directed circuit in signal flow graph $G_R(\bar{x}_w : w = 1, 2, \ldots, n, \\ w \neq t)$ which contains vertex x_t. Moreover, $C(\bar{d}_j)$ is $(1 - C_1 + C_2 - C_3 + \cdots)$ of subgraph $G''(\bar{d}_j)$ obtained from $G_R(\bar{x}_w : w = 1, 2, \ldots, n, \quad w \neq t)$ by deleting all vertices in directed circuit d_j and deleting all edges connected to these deleted vertices.

With Eqs. 10-3-19 and 10-3-34, Eq. 10-3-30 becomes

$$C_{tt} = \frac{\sum\limits_{(j)} [d_j(x_t) \text{ product}] \cdot C(\bar{d}_j) \text{ of } G_R(\bar{x}_w : w = 1, 2, \ldots, n, \quad w \neq t)}{1 - C_1 + C_2 - C_3 + \cdots \text{ of } G_R(\bar{x}_w : w = 1, 2, \ldots, n)}$$

(10-3-35)

Note that the two equations for the equivalent transformation (Eqs. 10-3-26 and 10-3-35) are very similar, that is, the only difference between these two equations is that one has $[d_j(x_t)$ product$]$. Since P_i is a directed path from vertex x_u to vertex x_t, coinciding vertices x_u and x_t, this directed path becomes a directed circuit $d_j(x_t)$ containing vertex x_t. Note that Eqs. 10-3-26 and 10-3-35 are obtained under the assumption that matrix $[R_{22} - U]$ is nonsingular. However we can obtain these equations without this assumption which will be left to the reader.

Example 10-3-1. Suppose we have the signal flow graph shown in Fig. 10-3-2 in which the source and the sink are in G_M. Then we can eliminate

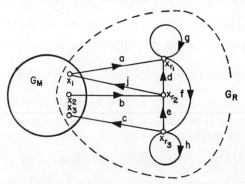

Fig. 10-3-2. A signal flow graph G consisting of two subgraphs G_M and G_R.

subgraph G_R as follows. Since the denominator D of Eqs. 10-3-26 and 10-3-35 are the same,

$$D = 1 - C_1 + C_2 - C_3 + \cdots \text{ of } G_R(\bar{x}_w : w = 1, 2, 3) \quad (10\text{-}3\text{-}36)$$

where $G_R(\bar{x}_w : w = 1, 2, 3)$ is given in Fig. 10-3-3, D is simply

$$D = 1 - g - h - def + gh$$

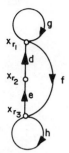

Fig. 10-3-3. Subgraph $G_R(\bar{x}_w: w = 1, 2, 3)$.

For k_{11}, the numerator N_{11} in Eq. 10-3-35 is

$$N_{11} = \sum_{(j)} [d_j(x_1) \text{ product}] \cdot C(d_j) \text{ of } G_R(\bar{x}_w: w = 2, 3)$$

where $G_R(\bar{x}_w: w = 2, 3)$ is shown in Fig. 10-3-4. Hence

$$N_{11} = aefj$$

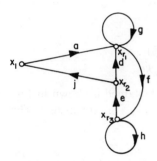

Fig. 10-3-4. Subgraph $G_R(\bar{x}_w: w = 2, 3)$.

Thus

$$K_{11} = \frac{aefj}{1 - g - h - def + gh}$$

For k_{31}, numerator N_{31} in Eq. 10-3-26 is

$$N_{31} = \sum_{(i)} (P_i \text{ product}) \cdot C(\bar{P}_i) \text{ of } G_R(\bar{x}_2)$$

where $G_R(\bar{x}_2)$ is shown in Fig. 10-3-5. Hence

$$N_{31} = acf$$

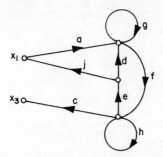

Fig. 10-3-5. $G_R(\bar{x}_2)$.

Now

$$k_{31} = \frac{acf}{1 - g - h - def + gh}$$

Similarly,

$$k_{32} = \frac{bcdf}{1 - g - h - def + gh}$$

and

$$k_{12} = \frac{bj(1 - g - h + gh)}{1 - g - h - def + gh}$$

where $G_R(\bar{x}_3)$ from which the numerator of k_{12} is obtained is shown in Fig. 10-3-6. It can easily be seen that others such as $k_{21}, k_{22}, \ldots$, are all zero. The

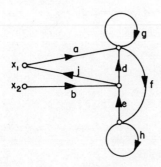

Fig. 10-3-6. $G_R(\bar{x}_3)$.

resultant signal flow graph after deleting vertices x_{r_1}, x_{r_2}, and x_{r_3} is shown in Fig. 10-3-7.

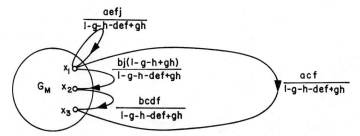

Fig. 10-3-7. Resultant signal flow graph after eliminating G_R.

PROBLEMS

1. Obtain the determinant of a matrix M of the flow graphs in Fig. P-10-1.

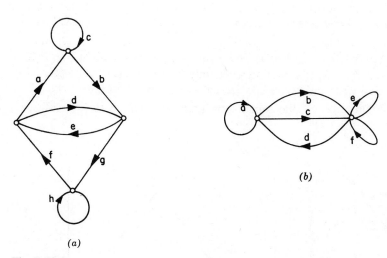

(a)

(b)

Fig. P-10-1.

2. Find the determinant of matrix M by first drawing the corresponding flow graph and then using Eq. 10-1-9.

$$(a) \ M = \begin{bmatrix} 1 & 2 & 3 \\ 1 & 3 & 4 \\ 4 & 1 & 1 \end{bmatrix} \qquad (b) \ M = \begin{bmatrix} 0 & 1 & 2 \\ 3 & 1 & 0 \\ 1 & 0 & 4 \end{bmatrix}$$

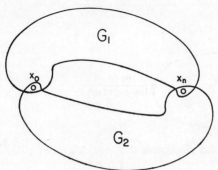

Fig. P-10-6.

3. Show that the determinant of a matrix M and that of M^t (transpose of M) are the same by using a flow graph.

4. Let G be a flow graph of matrix M. Obtain a topological formula for the determinant of matrix $[M - U]$ directly from G where U is a unit matrix.

5. Let G be a flow graph of matrix M. Find a topological formula which gives the characteristic polynomial of matrix M (i.e., the determinant of matrix $[M - \lambda U]$) directly from G.

6. Suppose a signal flow graph consists of two subgraphs G_1 and G_2 which have only two vertices x_0 and x_n in common as shown in Fig. P-10-6. Suppose $x_0/x_n = R_1$ is a solution when only G_1 is considered and $x_0/x_n = R_2$ is a solution when only G_2 is considered. What is x_0/x_n of G?

7. Obtain x_4/x_1 of the signal flow graphs in Fig. P-10-7 by Mason's formula (Eq. 10-2-31).

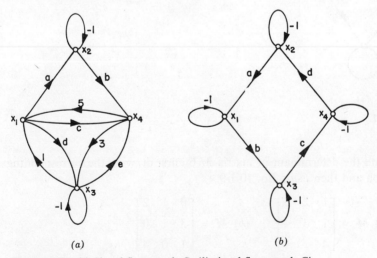

(a)   (b)

Fig. P-10-7.   (a) Signal flow graph G; (b) signal flow graph G'.

8. Obtain an equivalent signal flow graph which contains only vertices x_1, x_2, and x_3 from the graphs in Fig. P-10-8.

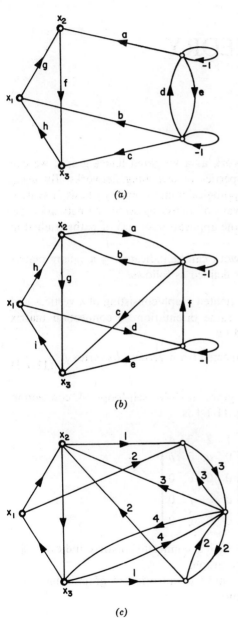

(a)

(b)

(c)

Fig. P-10-8. (a) Signal flow graph G_1; (b) signal flow graph G_2; (c) signal flow graph G_3.

CHAPTER
11
SWITCHING THEORY

11-1 CONNECTION MATRIX

By considering a switching network as a weighted linear graph, we can obtain interesting topological properties of switching networks. By using these properties, we can develop topological analysis and synthesis of switching networks. First we study two ways of analyzing switching networks, the so-called connection matrices and the application of sets of paths studied in Chapter 1.

The structure of a switching network can be indicated by a square matrix called a connection matrix which is defined as follows.

Definition 11-1-1. Let G be an oriented graph consisting of n vertices and having no parallel edges with the same orientation. A connection matrix $C = [c_{ij}]$ of order $n \times n$ is defined by

$$c_{ij} = \begin{cases} e & \text{if edge } e \text{ is connected from vertex } i \text{ to vertex } j \\ 0 & \text{otherwise} \end{cases} \qquad (11\text{-}1\text{-}1)$$

Note that we are allowing this graph to have self-loops. A connection matrix C of the linear graph in Fig. 11-1-1 is

$$C = \begin{array}{c} \\ 1 \\ 2 \\ 3 \\ 4 \end{array} \begin{array}{c} \begin{array}{cccc} 1 & 2 & 3 & 4 \end{array} \\ \begin{bmatrix} 0 & a & 0 & b \\ 0 & 0 & c & 0 \\ d & e & f & 0 \\ 0 & 0 & g & 0 \end{bmatrix} \end{array}$$

It can be seen from this example that a connection matrix indicates the location of every edge in an oriented graph.

In order to write a connection matrix of an oriented graph which has parallel edges, we modify the definition.

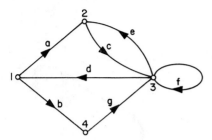

Fig. 11-1-1. An oriented graph.

Definition 11-1-2. Let $e_1, e_2, \ldots, e_k$ be all edges which are connected between vertices i and j and whose orientations are from i to j. Then entry c_{ij} of a connection matrix is $\sum_{r=1}^{k} e_r$.

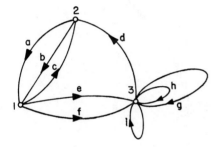

Fig. 11-1-2. An oriented graph.

For example, a connection matrix C of the oriented graph in Fig. 11-1-2 is

$$
C = \begin{array}{c} \\ 1 \\ 2 \\ 3 \end{array}
\begin{array}{c}
\begin{array}{ccc} 1 & 2 & 3 \end{array} \\
\left[\begin{array}{ccc}
0 & c & e+f \\
a+b & 0 & 0 \\
0 & d & g+h+i
\end{array} \right]
\end{array}
$$

We will show that a connection matrix and a matrix W corresponding to a flow graph are very similar. Recall that in a flow graph for a given matrix $W = [w_{ij}]$, w_{ij} is a weight given to an edge which is connected *from j to i* (Chapter 10). Let G be a flow graph of a matrix $W = [w_{ij}]$. Let W' be obtained from W by replacing w_{pq} by $\sum_{r=1}^{k} w_{r_{pq}}$. Let G' be obtained from G (in Fig. 11-1-3) by replacing edge w_{pq} by parallel edges $w_{r_{pq}}$ for $r = 1, 2, \ldots, k$

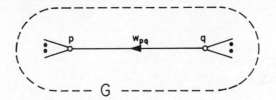

Fig. 11-1-3. A flow graph G.

whose orientations are identical with that of edge w_{pq} as shown in Fig. 11-1-4.

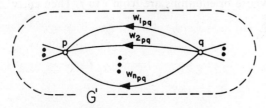

Fig. 11-1-4. Flow graph G'.

Then, the determinant of matrix W' can be expressed as

$$|W'| = \sum_{j_p \neq q} \delta_{j_1 j_2 \cdots j_n} w_{1j_1} w_{2j_2} \cdots w_{nj_n} + \sum_{j_p = q} \delta_{j_1 j_2 \cdots q \cdots j_n} w_{1j_1} w_{2j_2} \cdots w_{pq} \cdots w_{nj_n}$$

$$= \sum_{j_p \neq q} \delta_{j_1 j_2 \cdots j_n} w_{1j_1} w_{2j_2} \cdots w_{nj_n}$$

$$+ \sum_{j_p = q} \delta_{j_1 j_2 \cdots q \cdots j_n} w_{1j_1} w_{2j_2} \cdots \left(\sum_{r=1}^{k} w_{r_{pq}} \right) \cdots w_{nj_n} \qquad (11\text{-}1\text{-}2)$$

Since if $\delta_{j_1 j_2 \cdots q \cdots j_n}$ is nonzero, each $w_{1j_1} w_{2j_2} \cdots w_{r_{pq}} \cdots w_{nj_n}$ is a P-set cycle. Thus

$$|W'| = (-1)^n \sum (-1)^{k_j} \ P\text{-set cycle } D'_j \text{ product of } G' \qquad (11\text{-}1\text{-}3)$$

Obviously, this equation holds when there are several parallel edges. This shows that we can modify the definition of a flow graph so that parallel edges can exist. In other words, we can obtain a matrix W corresponding to a flow graph G by defining

$w_{ij} = $ the sum of weights of edges connected from vertex j to vertex i, when there are edges connected from j to i

and

$w_{ij} = 0$, when there are no edges connected from j to i

On the other hand, an entry c_{ji} of a connection matrix $C = [c_{ij}]$ of a flow graph G (by considering a flow graph as a weighted oriented graph) is

$c_{ji} =$ the sum of the weights of edges connected from vertex j to vertex i, when there are such edges

and

$c_{ji} = 0$, when there are no edges connected from vertex j to vertex i

Thus we can easily see that

$$C = W^t \qquad\qquad (11\text{-}1\text{-}4)$$

In other words, if C is a connection matrix of an oriented graph G, then G can be considered as a flow graph of matrix C^t. Hence, by Eq. 10-1-9, we have

$$|C| = (-1)^n \sum (-1)^{k_j} \ P\text{-set cycle } D_j \text{ product of } G \qquad (11\text{-}1\text{-}5)$$

Unfortunately, many oriented graphs have no P-set cycles. For example, the oriented graph in Fig. 11-1-1 has no P-set cycles. Thus $|C| = 0$, which is not an exciting result.

By modifying a connection matrix, we find the following important result.

Theorem 11-1-1. Let weights of all edges in an oriented graph G be Boolean variables. Let C be a connection matrix of G. Then Δ_{sr} is a switching function from vertex r to vertex s where Δ_{sr} is a cofactor at the (s, r) position of matrix $[C + U]$ and operations are Boolean sum and product.

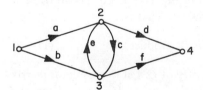

Fig. 11-1-5. An oriented graph G.

Example 11-1-1. Consider oriented graph G in Fig. 11-1-5. A connection matrix C is

$$C = \begin{array}{c c} & \begin{array}{cccc} 1 & 2 & 3 & 4 \end{array} \\ \begin{array}{c} 1 \\ 2 \\ 3 \\ 4 \end{array} & \begin{bmatrix} 0 & a & b & 0 \\ 0 & 0 & c & d \\ 0 & e & 0 & f \\ 0 & 0 & 0 & 0 \end{bmatrix} \end{array}$$

Then $[C + U]$ is

$$[C + U] = \begin{array}{c} \\ 1 \\ 2 \\ 3 \\ 4 \end{array} \begin{array}{cccc} 1 & 2 & 3 & 4 \\ \begin{bmatrix} 1 & a & b & 0 \\ 0 & 1 & c & d \\ 0 & e & 1 & f \\ 0 & 0 & 0 & 1 \end{bmatrix} \end{array}$$

A cofactor Δ_{41} of $[C + U]$ is

$$\Delta_{41} = \begin{vmatrix} a & b & 0 \\ 1 & c & d \\ e & 1 & f \end{vmatrix} = acf + ad + bf + bed$$

which is a switching function from vertex 1 to vertex 4.

Proof of Theorem 11-1-1. Recall that in Section 10-1 we found that

$$|W| = (-1)^n \sum (-1)^{k_j} \, P\text{-set cycle } D_j \text{ product} \qquad (11\text{-}1\text{-}6)$$

from the definition of a determinant

$$|W| = \sum \delta_{j_1 j_2 \cdots j_n} w_{1j_1} w_{2j_2} \cdots w_{nj_n} \qquad (11\text{-}1\text{-}7)$$

because each term $\delta_{j_1 j_2 \cdots j_n} w_{1j_1} w_{2j_2} \cdots w_{nj_n}$ is a P-set cycle product when $\delta_{j_1 j_2 \cdots j_n}$ is nonzero. Since there is $1:1$ correspondence between a P-set cycle product in Eq. 11-1-6 and a nonzero term in Eq. 11-1-7, Eq. 11-1-6 without signs is valid for a matrix of Boolean variables. That is,

$$|C'| = \sum P\text{-set cycle } D_j \text{ product} \qquad (11\text{-}1\text{-}8)$$

Let a variable Y be added to the (s, r) entry of C. Let the resultant matrix be $C(Y)$. From Eq. 11-1-8, the determinant of matrix $[C(Y) + U]$ can be written as

$$|C(Y) + U| = \sum P\text{-set cycle } D_j \text{ product of } G'_Y \qquad (11\text{-}1\text{-}9)$$

where G'_Y is obtained from given graph G by inserting edge Y from vertex s to vertex r and inserting a self-loop with weight 1 to every vertex in G.

There is a P-set cycle D_j which does not contain edge Y. Let this be denoted by $D_j(0)$. Also let a P-set cycle which contains Y be expressed as $YP_j(0)$. Then Eq. 11-1-9 can be expressed as

$$|C(Y) + U| = \sum D_j(0) \text{ product} + \sum YP_j(0) \text{ product of } G'_Y \quad (11\text{-}1\text{-}10)$$

It is clear that

$$|C + U| = \sum D_j(0) \text{ product} \tag{11-1-11}$$

and

$$\Delta_{sr} = \sum P_j(0) \text{ product} \tag{11-1-12}$$

Since $YP_j(0)$ is a P-set cycle in G'_Y, $P_j(0)$ must contain a directed path from vertex r to vertex s so that it becomes a directed circuit with edge Y. Let this directed path be P_i. Let $G'_Y(\overline{\Omega(P_i)})$ be an oriented graph obtained from G'_Y by deleting all vertices in P_i and all edges connected to vertices in P_i. Then any P-set cycle in $G'_Y(\overline{\Omega(P_i)})$ with directed circuit $(Y) \cup P_i$ is a P-set cycle in G'_Y. Furthermore, any P-set cycle in G'_Y which contains directed circuit $(Y) \cup P_i$ is a P-set cycle in $G'_Y(\overline{\Omega(P_i)})$ when we remove $(y) \cup P_i$ from it. Thus

$$\sum P_j(0) \text{ product} = \sum_{(i)} P_i \text{ product} \left(\sum P\text{-set cycle of } D'_j \text{ product of } G'_Y(\overline{\Omega(P_i)}) \right) \tag{11-1-13}$$

where $\sum_{(i)}$ means to take all possible directed paths from vertex r to vertex s in G'_Y.

For a nonempty $G'_Y(\overline{\Omega(P_i)})$ there is a P-set cycle which consists of self-loops with weight 1 only. Hence,

$$P\text{-set cycle } D'_j \text{ product of } G'_Y(\overline{\Omega(P_i)}) = 1 + \cdots \tag{11-1-14}$$

which is equal to 1 because of Boolean algebra. When $G'_Y(\overline{\Omega(P_i)})$ is empty, a $P_j(0)$ product is a P_j product. Hence Eq. 11-1-13 becomes

$$\sum P_j(0) \text{ product of } G'_Y = \sum_{(i)} P_i \text{ product of } G'_Y \tag{11-1-15}$$

Since a directed path from vertex r to vertex s other than Y itself in G'_Y is also in G and vice versa, we have

$$\Delta_{sr} = \sum_{(i)} P_i \text{ product of } G \tag{11-1-16}$$

which is a switching function from vertex r to vertex s.   QED

To analyze a nonoriented switching network, we first note that the same weights can be assigned to many edges in an oriented graph for Theorem 11-1-1. If we replace one nonoriented edge by two parallel oriented edges whose orientations are opposite each other and whose weights are the same as that of the nonoriented edge shown in Fig. 11-1-6, we can obtain an

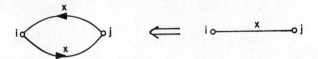

Fig. 11-1-6. Correspondence between oriented edges and nonoriented edges.

oriented graph G' corresponding to a nonoriented graph G which have the following properties:

1. For any path P between two vertices r and s in G, there are two directed paths P' and P'' in G' such that P' is from r to s and P'' is from s to r and

$$P' \text{ product } = P'' \text{ product } = P \text{ product}$$

2. For any directed path P' from r to s in G', there exists a directed path P between r and s in G such that

$$P' \text{ product } = P \text{ product}$$

Hence for switching functions, G and G' are the same. Thus Theorem 11-1-1 can be modified for a nonoriented graph.

Theorem 11-1-2. Let weights of all edges in a nonoriented graph G be Boolean variables. Let C be a connection matrix of G. Then cofactor Δ_{rs} $(= \Delta_{sr})$ of $[C + U]$ is a switching function between vertices r and s in G.

Example 11-1-2. A connection matrix C of the switching network in Fig. 11-1-7 is

$$C = \begin{array}{c} \\ 1 \\ 2 \\ 3 \\ 4 \end{array} \begin{array}{c} \begin{array}{cccc} 1 & 2 & 3 & 4 \end{array} \\ \begin{bmatrix} 0 & 0 & a & b \\ 0 & 0 & c & d \\ a & c & 0 & e \\ b & d & e & 0 \end{bmatrix} \end{array}$$

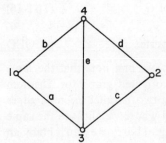

Fig. 11-1-7. Switching network G.

Then Δ_{12} of $[C + U]$ is

$$\begin{vmatrix} 0 & c & d \\ a & 1 & e \\ b & e & 1 \end{vmatrix} = ac + ade + bce + bd$$

which is the switching function between vertices 1 and 2.

11-2   ANALYSIS OF SWITCHING NETWORK

We have studied one way to obtain a switching function between two vertices by using Theorem 11-1-2 which uses a connection matrix. Here we are concerned with another way of analyzing switching networks, using the properties of sets of all possible paths in Section 1-2 and 1-6.

Let $\{E\}$ be a set of Euler graphs and $\{P_{ij}\}$ a set of all possible paths between vertices i and j of a linear graph G. Then we know from Theorem 1-6-12 that

$$\{P_{ij}\} = \min \{P \oplus E; \quad E \in \{E\}\} \tag{11-2-1}$$

where $P \in \{P_{ij}\}$. That is, if we know one path P between two vertices, then all possible paths between the same vertices can be obtained by the ring sum of P and Euler graphs of G. We also know from Theorem 1-5-3 that if linear graph G is nonseparable, then $\{E\}$ can be replaced by $\{C\}$ where $\{C\}$ is a set of all possible circuits in G. That is,

$$\{P_{ij}\} = \min \{P \oplus C; \quad C \in \{C\}\} \tag{11-2-2}$$

for a nonseparable graph G.

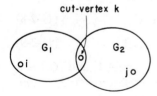

cut-vertex k

Fig. 11-2-1.   A separable graph G.

Suppose G is separable as shown in Fig. 11-2-1 where G_1 and G_2 are nonseparable. Then we can see easily that $\{P_{ij}\}$ can be expressed as

$$\{P_{ij}\} = \{P_{ik}\} \otimes \{P_{kj}\} \tag{11-2-3}$$

where $i \in G_1$, $j \in G_2$,

$$\{P_{ik}\} = \min \{P_1 \oplus C_1; \quad C_1 \in \{C_1\}\} \tag{11-2-4}$$

$$\{P_{kj}\} = \min \{P_2 \oplus C_2; \quad C_2 \in \{C_2\}\} \tag{11-2-5}$$

$P_1 \in \{P_{ik}\}$, $P_2 \in \{P_{kj}\}$, $\{C_1\}$ is a set of all possible circuits in G_1, and $\{C_2\}$ is a set of all possible circuits in G_2.

In general, we can state the following theorem.

Theorem 11-3-1. Suppose a separable linear graph G can be separated into k nonseparable subgraphs $G_1, G_2, \ldots, G_k$ by cutting cut-vertices $v_2, v_3, \ldots, v_{k'}$ ($k' \leq k$). (An example with $k = 6$ is given in Fig. 11-2-2.) Also suppose a

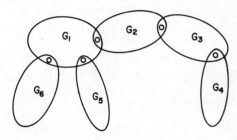

Fig. 11-2-2. A separable graph G.

path P_{ij} between i and j passes through cut-vertices $v_2, v_3, \ldots, v_m$ ($m \leq k'$). Let $i \in G_1$, G_1 and G_2 have cut-vertex v_2 in common, G_2 and G_3 have cut-vertex v_3 in common, $\ldots$, G_{m-1} and G_m have cut-vertex v_m in common, and $j \in G_m$. Then

$$\{P_{ij}\} = \{P_{iv_2}\} \otimes \{P_{v_2 v_3}\} \otimes \cdots \otimes \{P_{v_m j}\} \tag{11-2-6}$$

where

$$\{P_{iv_2}\} = \min \{P_1 \oplus C_1; \quad C_1 \in \{C_1\}\} \tag{11-2-7}$$

$$\{P_{v_r v_{r+1}}\} = \min \{P_r \oplus C_r; \quad C_r \in \{C_r\}\} \tag{11-2-8}$$

$$\{P_{v_m j}\} = \min \{P_m \oplus C_m; \quad C_m \in \{C_m\}\} \tag{11-2-9}$$

$P_1 \in \{P_{iv_2}\}$, $P_r \in \{P_{v_r v_{r+1}}\}$, $P_m \in \{P_{v_m j}\}$ for $r = 2, 3, \ldots, m - 1$, and $\{C_u\}$ is a set of all possible circuits in G_u and the empty set for $u = 1, 2, \ldots, m$. Furthermore, switching function F_{ij} can be expressed as

$$F_{ij} = F_{iv_1} \cdot F_{v_1 v_2} \cdot \ldots \cdot F_{v_m j} \tag{11-2-10}$$

where

$$F_{iv_1} = \sum_{P_1 \in \{P_{iv_1}\}} \text{path } P_1 \text{ product} \tag{11-2-11}$$

$$F_{v_r v_{r+1}} = \sum_{P_r \in \{P_{v_r v_{r+1}}\}} \text{path } P_r \text{ product for } r = 2, 3, \ldots, m \quad (11\text{-}2\text{-}12)$$

and

$$F_{v_m j} = \sum_{P_m \in \{P_{v_m j}\}} \text{path } P_m \text{ product} \quad (11\text{-}2\text{-}13)$$

Example 11-2-1. Consider the switching network in Fig. 11-2-3. Suppose we choose a path P_{ij} between i and j as $P_{ij} = (a_1, b_1, d_1, d_2)$. This path passes

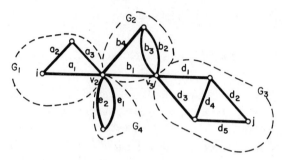

Fig. 11-2-3. A switching network.

through cut-vertices v_2 and v_3. It is clear that all we need to consider are subgraphs G_1, G_2, and G_3. From the sets $\{C_1\}$, $\{C_2\}$, and $\{C_3\}$ of all possible circuits and the empty set of G_1, G_2, and G_3 where

$$\{C_1\} = \{\emptyset, (a_1, a_2, a_3)\}$$
$$\{C_2\} = \{\emptyset, (b_1, b_2, b_4), (b_1, b_3, b_4), (b_2, b_3)\}$$

and

$$\{C_3\} = \{\emptyset, (d_1, d_3, d_4), (d_2, d_4, d_5), (d_1, d_2, d_3, d_5)\}$$

we can obtain

$$\{P_{iv_2}\} = \min\{(a_1) \oplus C_1; \quad C_1 \in \{C_1\}\} = \{(a_1), (a_2, a_3)\}$$
$$\{P_{v_2 v_3}\} = \min\{(b_1) \oplus C_2; \quad C_2 \in \{C_2\}\} = \{(b_1), (b_2, b_4), (b_3, b_4)\}$$

and

$$\{P_{v_3 j}\} = \min\{(d_1, d_2) \oplus C_3; \quad C_3 \in \{C_3\}\} = \{(d_1, d_2), (d_2, d_3, d_4), (d_1, d_4, d_5),$$
$$(d_3, d_5)\}$$

Thus switching function F_{ij} is

$$F_{ij} = F_{jv_2} \cdot F_{v_2 v_3} \cdot F_{v_3 j}$$
$$= (a_1 + a_2 a_3)(b_1 + b_2 b_4 + b_3 b_4)(d_1 d_2 + d_2 d_3 d_4 + d_1 d_4 d_5 + d_3 d_5)$$

The technique discussed here is suitable when a computer is used for analysis of switching network. Furthermore, if it is necessary to calculate more than one switching function, we may be able to use the relationship between sets of paths such as a triangular relationship given by Theorem 1-6-7 to simplify the calculations.

11-3 SYNTHESIS OF COMPLETELY SPECIFIED SWITCHING FUNCTION

A special switching network known as an SC-(single contact) network will be considered here.

Definition 11-3-1. An SC-network is a nonoriented graph in which the weight of each edge has a different Boolean variable.

Here, we consider only those switching (Boolean) functions that are completely specified (i.e., the corresponding truth table has either 1 or 0). The question which we are going to answer is whether there is an SC-network that satisfies a given completely specified switching function F. For example, a function

$$F = ab + acd + de$$

is not a switching function of an SC-network. On the other hand, a function

$$F = ab + acd$$

is clearly a switching function of an SC-network.

For convenience, we use the following definition.

Definition 11-3-2. $\{P_{rs}\}$ is said to be a set corresponding to completely specified switching function F_{rs} if

$$F_{rs} = \sum_{P_{rs} \in \{P_{rs}\}} P_{rs} \text{ product} \qquad (11\text{-}3\text{-}1)$$

where P_{rs} product is the product of all members in $\{P_{rs}\}$.

For example, if $F_{rs} = a(b + ce) + d(e + bc)$, then set $\{P_{rs}\}$ corresponding to F_{rs} is

$$\{P_{rs}\} = \{(a, b), (a, c, e), (d, e), (b, c, d)\}$$

Since the weight of each edge is different in an SC-network, we use the weight of an edge as the name of the edge, for convenience. By expressing a completely specified switching function F_{ij} as

$$F_{ij} = \sum_{P_{ij} \in \{P_{ij}\}} P_{ij} \text{ product} \qquad (11\text{-}3\text{-}2)$$

we can see that if a switching network corresponding to F_{ij} is an SC-network, set $\{P_{ij}\}$ must be a set of all possible paths between vertices i and j. From this reason, we have the following necessary condition for a switching function of an SC-network.

Theorem 11-3-1.   For any odd number of P_r's in $\{P_{ij}\}$, say $P_1, P_2, \ldots, P_{2k+1}$, where set $\{P_{ij}\}$ corresponds to F_{ij} (a completely specified switching function of an SC-network), there exists P' in $\{P_{ij}\}$ such that

$$P' \subset P_1 \oplus P_2 \oplus \cdots \oplus P_{2k+1} \tag{11-3-3}$$

For example, if $F_{ij} = a(b + ce) + bcd$, then $\{P_{ij}\} = \{(a, b)(a, c, e)(b, c, d)\}$. Since $(a, b) \oplus (a, c, e) \oplus (b, c, d) = (d, e)$ and there is no $P' \in \{P_{ij}\}$ satisfying Eq. 11-3-3, F_{ij} is not a completely specified switching function of an SC-network.

Proof of Theorem 11-3-1.   If $\{P_{ij}\}$ is a set of all possible paths between vertices i and j, then the ring sum of odd number of $P_r \in \{P_{ij}\}$ is an M-graph M_{ij} (Section 1-4), which is either a path P or an edge disjoint union of a path P and circuits where $P \in \{P_{ij}\}$. QED

It may be helpful to review some properties of sets of subgraphs which will be needed for synthesis of SC-networks. In Section 3-5, we learned to synthesize a linear graph from a fundamental cut-set matrix Q_f. From Eq. 3-4-24, we have a relationship between a fundamental cut-set matrix Q_f and a fundamental circuit matrix $B_f = [U \quad B_{f_{12}}]$:

$$Q_f = [B^t_{f_{12}} \quad U] \tag{11-3-4}$$

We know that a fundamental circuit matrix can easily be obtained from a set $\{C\}$ of all possible circuits. In Section 1-6, we saw that

$$\{C\} = \min \{E\} \tag{11-3-5}$$

where $\{E\}$ is a set of all possible Euler graphs of a linear graph G. Hence if a set $\{E\}$ is given, we can obtain a linear graph G which has $\{E\}$ as a set of all possible Euler graphs.

A switching function F_{ij} of an SC-network will give $\{P_{ij}\}$ (a set of all possible paths between i and j) but not $\{E\}$. However, from Sections 1-4 and 5-3, set $\{E, M_{ij}\}$ has $n_e - n_v + 2$ generators where n_e and n_v are the number of edges and the number of vertices in a connected linear graph G, respectively. Since we are trying to synthesize an SC-network G from a completely specified switching function F_{ij}, we can assume that G is connected. Furthermore, SC-network G synthesized from F_{ij} can have the property that every edge is in at least one path from i to j. Thus we can easily see that G is either nonseparable or consists of k nonseparable subgraphs $G_1, G_2, \ldots, G_k$ such that (1) no two of these subgraphs have edges in common, (2) any two of these subgraphs have at most one vertex in common, (3) there are $k - 1$ cut-vertices, (4) $i \in G_1$ and $j \in G_k$, and (5) every path from i to j passes all $k - 1$ cut-vertices as shown in Fig. 11-3-1. Thus it can be shown that

$$\{P_{ij}\} \otimes \{P_{ij}\} = \{C\} \tag{11-3-6}$$

where $\{C\}$ is a set of all circuits in G.

Fig. 11-3-1. SC-network G.

From set $\{C\}$, we can obtain a fundamental circuit matrix B_f. Hence a fundamental cut-set matrix can be obtained by Eq. 11-3-4. Now we can find a linear graph which has $\{C\}$ as a set of all possible circuits. However, G may not have $\{P_{ij}\}$ as a set of all possible paths from i to j. As an example, suppose $\{P_{ij}\} = \{(a, b), (c, d)\}$. Then

$$\{P_{ij}\} \otimes \{P_{ij}\} = \{\emptyset, (a, b, c, d)\}$$

Linear graph G in Fig. 11-3-2 has (a, b, c, d) as a circuit. Thus G has $\{C\}$ as a set of all possible circuits. However, G does not have $\{(a, b)(c, d)\}$ as a set of all possible paths between two vertices.

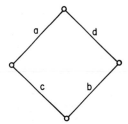

Fig. 11-3-2. A linear graph G.

In order to obtain a linear graph which has $\{P_{ij}\}$ as a set of all possible paths between i and j, we insert edge y whose endpoints are i and j. In other words, if there is an SC-network G which has $\{P_{ij}\}$ as a set of all possible paths between i and j, then by inserting edge y, a new linear graph $G \cup (y)$ will have the property that $(y) \cup P$ is a circuit for every $P \in \{P_{ij}\}$. Thus

$$\{(y) \cup P_{ij}\} = \{(y) \cup P; \quad P \in \{P_{ij}\}\} \tag{11-3-7}$$

is a subset of set $\{C_y\}$ of all possible circuits in $G \cup (y)$.

Since our SC-network has the property that

$$\{P_{ij}\} \otimes \{P_{ij}\} = \{C\} \qquad (11\text{-}3\text{-}8)$$

$\{(y) \cup P_{ij}\}$ will contain all linearly independent circuits which can generate all circuits of $G \cup (y)$. Thus we can obtain a circuit matrix B_y of $G \cup (y)$ by using circuits in $\{(y) \cup P_{ij}\}$. From this circuit matrix B_y, we can obtain a fundamental circuit matrix B_f and a fundamental cut-set matrix Q_f. By this matrix Q_f, we can construct a linear graph G'.

Now an important question is whether deleting edge y from G' gives an SC-network which has the given set $\{P_{ij}\}$ as a set of all possible paths between two vertices. The answer is given by the following theorem.

Theorem 11-3-2. Let $\{P_{ij}\}$ be a set corresponding to a completely specified switching function F_{ij} which satisfies Theorem 11-4. Let $P_1, P_2, \ldots, P_m$ be linearly independent paths in $\{P_{ij}\}$ such that any other path P' in $\{P_{ij}\}$ with these paths cannot be linearly independent. If and only if there exists a nonoriented linear graph G of nullity m whose linearly independent circuits are $(y) \cup P_1, (y) \cup P_2, \ldots, (y) \cup P_m$ where y is not in any path in $\{P_{ij}\}$, G without edge y is an SC-network satisfying F_{ij}.

Note that an SC-network satisfying F_{ij} means that there exist two vertices in the SC-network such that the switching function between these vertices is F_{ij}. The name *path* is used for a member in $\{P_{ij}\}$, although it may not be a precise terminology if there are no SC-networks satisfying F_{ij}.

Proof of Theorem 11-3-2. Suppose there is a nonoriented linear graph G of nullity m in which $(y) \cup P_1, (y) \cup P_2, \ldots, (y) \cup P_m$ are linearly independent circuits. Suppose there is a path P' between i and j in G which is not in $\{P_{ij}\}$. Since $(y) \cup P'$ is a circuit in G, $(y) \cup P'$ can be obtained by the ring sum of some of $(y) \cup P_1, (y) \cup P_2, \ldots, (y) \cup P_m$. Without the loss of generality, let

$$(y) \cup P' = [(y) \cup P_1] \oplus [(y) \cup P_2] \oplus \cdots \oplus [(y) \cup P_{2k+1}] \quad (11\text{-}3\text{-}9)$$

where $2k + 1 \le m$. Then

$$p' = P_1 \oplus P_2 \oplus \cdots \oplus P_{2k+1} \qquad (11\text{-}3\text{-}10)$$

which must be in $\{P_{ij}\}$ because F_{ij} satisfies Theorem 11-3-1. This contradicts the assumption that P' is not in $\{P_{ij}\}$.

Suppose G without edge y is an SC-network satisfying F_{ij}. Then it is obvious that G is of nullity m and $(y) \cup P_1, (y) \cup P_2, \ldots, (y) \cup P_m$ are linearly independent circuits.   QED

By use of this theorem, we can synthesize an SC-network from a completely specified switching function F_{ij} by the following procedure.

STEP 1.   Test whether a given function F_{ij} satisfied Theorem 11-3-1. If not, F_{ij} is not a switching function of an SC-network.

STEP 2.   From a set $\{P_{ij}\}$ corresponding to F_{ij}, take linearly independent paths $P_1, P_2, \ldots, P_m$. Form a circuit matrix B by circuits $(y) \cup P_1$, $(y) \cup P_2$, $\ldots, (y) \cup P_m$.

STEP 3.   By elementary row and column operations, change B to a fundamental circuit matrix $[U \quad B_{f_{12}}]$.

STEP 4.   Form a fundamental cut-set matrix Q_f:

$$Q_f = [B_{f_{12}}^t \quad U] \qquad\qquad (11\text{-}3\text{-}11)$$

STEP 5.   Use the method in Section 3-5 to obtain a linear graph G whose fundamental cut-set matrix Q_f is obtained by Step 4.

STEP 6.   Delete edge y. The resulting graph is an SC-network satisfying F_{ij} where i and j are the endpoints of edge y.

The following example illustrates this procedure.

Example 11-3-1. It can be seen that a completely specified switching function

$$F_{ij} = ab + acd + bce + de$$

satisfies Theorem 11-3-1. Thus we will go to Step 2.

STEP 2.   From

$$\{P_{ij}\} = \{(a, b), (a, c, d), (b, c, e), (d, e)\}$$

suppose we take $P_1 = (a, b)$, $P_2 = (a, c, d)$, and $P_3 = (b, c, e)$ as linearly independent paths. Note that $(d, e) = P_1 \oplus P_2 \oplus P_3$. A circuit matrix B is

$$
B =
\begin{array}{c}
(y) \cup P_1 \\
(y) \cup P_2 \\
(y) \cup P_3
\end{array}
\begin{array}{c}
\begin{array}{cccccc} a & b & c & d & e & y \end{array} \\
\left[\begin{array}{cccccc}
1 & 1 & 0 & 0 & 0 & 1 \\
1 & 0 & 1 & 1 & 0 & 1 \\
0 & 1 & 1 & 0 & 1 & 1
\end{array}\right]
\end{array}
$$

STEP 3.   We rearrange columns of B:

$$
B =
\begin{array}{c}
1 \\
2 \\
3
\end{array}
\begin{array}{c}
\begin{array}{cccccc} b & d & e & a & c & y \end{array} \\
\left[\begin{array}{cccccc}
1 & 0 & 0 & 1 & 0 & 1 \\
0 & 1 & 0 & 1 & 1 & 1 \\
1 & 0 & 1 & 0 & 1 & 1
\end{array}\right]
\end{array}
$$

Then adding row 1 to row 3, we have

$$
B_f = \begin{array}{c} \\ 1 \\ 2 \\ 3 \end{array}
\begin{array}{c} b \quad d \quad e \quad a \quad c \quad y \\
\left[\begin{array}{ccc:ccc}
1 & 0 & 0 & 1 & 0 & 1 \\
0 & 1 & 0 & 1 & 1 & 1 \\
0 & 0 & 1 & 1 & 1 & 0
\end{array}\right] = [U \quad B_{f_{12}}]
\end{array}
$$

STEP 4. We obtain Q_f:

$$
Q_f = \begin{array}{c} \\ 1 \\ 2 \\ 3 \end{array}
\begin{array}{c} b \quad d \quad e \quad a \quad c \quad y \\
\left[\begin{array}{ccc:ccc}
1 & 1 & 1 & 1 & 0 & 0 \\
0 & 1 & 1 & 0 & 1 & 0 \\
1 & 1 & 0 & 0 & 0 & 1
\end{array}\right]
\end{array}
$$

STEP 5. We first obtain an H-submatrix with respect to row 1:

$$
H_{(1)} = \begin{array}{c} \\ 2 \\ 3 \end{array}
\begin{array}{c} c \quad y \\
\left[\begin{array}{c:c}
1 & 0 \\ \hdashline
0 & 1
\end{array}\right]
\end{array}
$$

With this H-submatrix, we can obtain a pair of M-submatrices $M_1(1)$ and $M_2(1)$:

$$
M_1(1) = \begin{array}{c} \\ 1 \\ 2 \end{array}
\begin{array}{c} b \quad d \quad e \quad a \quad c \\
\left[\begin{array}{ccccc}
1 & 1 & 1 & 1 & 0 \\
0 & 1 & 1 & 0 & 1
\end{array}\right]
\end{array}
$$

and

$$
M_2(1) = \begin{array}{c} \\ 1 \\ 3 \end{array}
\begin{array}{c} b \quad d \quad e \quad a \quad y \\
\left[\begin{array}{ccccc}
1 & 1 & 1 & 1 & 0 \\
1 & 1 & 0 & 0 & 1
\end{array}\right]
\end{array}
$$

Two linear graphs G_1 and G_2 corresponding to $M_1(1)$ and $M_2(1)$ are shown in Fig. 11-3-3. Combining G_1 and G_2 to remove vertex 1, we have a linear graph G as shown in Fig. 11-3-4a.

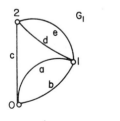 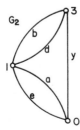

Fig. 11-3-3. Linear graphs for $M_1(1)$ and $M_2(1)$.

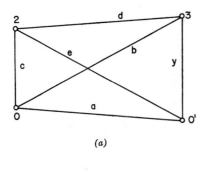

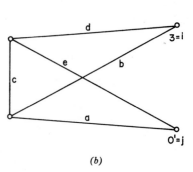

Fig. 11-3-4. (a) Linear graph G'; (b) an SC-network.

STEP 6. We delete edge y from G_y to obtain linear graph G (Fig. 11-3-4b), which is an SC-network satisfying a given F_{ij} between vertices 3 and 0' (between which edge y was connected).

11-4 SYNTHESIS OF INCOMPLETELY SPECIFIED SWITCHING FUNCTION BY SC-NETWORK

Consider the following truth table:

x_2	x_1	F
0	0	0
0	1	I
1	0	1
1	1	1

where the symbol I indicates that F is not specified (don't care). A switching function that has an incompletely specified situation such as one given by this truth table is called an incompletely specified switching function.

Definition 11-4-1. For an incompletely specified switching function F_{ij}, the symbols $\{R_{ij}\}$ and $\{I_{ij}\}$ indicate sets corresponding to F_{ij} which satisfy

1. $F_{ij}(1) = \displaystyle\sum_{R_r \in \{R_{ij}\}} R_r$ product   (11-4-1)

2. $F_{ij}(I) = \displaystyle\sum_{I_r \in \{I_{ij}\}} I_r$ product   (11-4-2)

where $F_{ij}(1)$ is obtained from F_{ij} by setting all I in the corresponding truth table to 0, and $F_{ij}(I)$ is obtained from F_{ij} by setting all 1 to 0 and all I to 1 in the corresponding truth table.

Note that

$$F_{ij} = F_{ij}(1) + F_{ij}(I) \qquad (11\text{-}4\text{-}3)$$

when we set all I in the corresponding truth table to 1. Also we can see that

$$\{R_{ij}\} \cap \{I_{ij}\} = \emptyset \qquad (11\text{-}4\text{-}4)$$

Instead of using a truth table, we can use $F_{ij}(1)$ and $F_{ij}(I)$ to specify an incompletely specified switching function. For example,

$$F_{ij}(1) = x_2$$
$$F_{ij}(I) = x_1 \bar{x}_2$$

gives an incompletely specified switching function whose truth table is the one given earlier. Sets $\{R_{ij}\}$ and $\{I_{ij}\}$ of this set of equations are

$$\{R_{ij}\} = (x_2)$$

and

$$\{I_{ij}\} = (x_1, \bar{x}_2)$$

When an incompletely specified switching function is given, then we can choose $\{P_{ij}\}$ such that

$$\{R_{ij}\} \subset \{P_{ij}\}$$

and any $I \in \{I_{ij}\}$ can be in $\{P_{ij}\}$. However, the choice of $\{P_{ij}\}$ must be made such that an SC-network can be synthesized by the procedure given in the previous section. The following example illustrates a process of choosing set $\{P_{ij}\}$.

Example 11-4-1. We have an incompletely specified switching function whose corresponding sets are given as

$$\{R_{ij}\} = \{(a, e, f), (b, d, f), (c, d, e), (a, b, c)\}$$

and

$$\{I_{ij}\} = \{(d, e, f), (a, b, f), (a, c, e), (b, c, d), (a, b, e)\}$$

Suppose we choose $\{P_{ij}\} = \{R_{ij}\}$. This choice will satisfy Theorem 11-3-1. By choosing $P_1 = (a, e, f)$, $P_2 = (b, d, f)$, and $P_3 = (c, d, e)$, we have

$$
B = \begin{array}{c} (y) \cup P_1 \\ (y) \cup P_2 \\ (y) \cup P_3 \end{array}
\begin{array}{cccccccc}
 & a & b & c & d & e & f & y \\
\left[\begin{array}{ccc|cccc}
1 & 0 & 0 & 0 & 1 & 1 & 1 \\
0 & 1 & 0 & 1 & 0 & 1 & 1 \\
0 & 0 & 1 & 1 & 1 & 0 & 1
\end{array}\right]
\end{array}
$$

Since this is a fundamental circuit matrix, we can obtain a fundamental cut-set matrix Q_f as

$$Q_f = \begin{array}{c} \\ 1 \\ 2 \\ 3 \\ \cdot 4 \end{array}\overset{\begin{array}{ccccccc} a & b & c & d & e & f & y \end{array}}{\left[\begin{array}{ccc|cccc} 0 & 1 & 1 & 1 & 0 & 0 & 0 \\ 1 & 0 & 1 & 0 & 1 & 0 & 0 \\ 1 & 1 & 0 & 0 & 0 & 1 & 0 \\ 1 & 1 & 1 & 0 & 0 & 0 & 1 \end{array}\right]}$$

This matrix is not realizable as a fundamental cut-set matrix. Hence $F_{ij}(1)$ is not a switching function of an SC-network.

Suppose we add (d, e, f) in $\{P_{ij}\}$. Then

$$(d, e, f) \oplus P_1 \oplus P_2 = (a, b, f) \in \{I_{ij}\}$$

By Theorem 11-3-1, (a, b, f) must be in $\{P_{ij}\}$. Similarly,

$$(d, e, f) \oplus P_1 \oplus P_3 = (a, c, e) \in \{I_{ij}\}$$

indicates that (a, c, e) must be chosen to be in $\{P_{ij}\}$. Finally,

$$(d, e, f) \oplus P_2 \oplus P_3 = (b, c, d) \in \{I_{ij}\}$$

shows that (b, c, d) must be in $\{P_{ij}\}$. Thus we have

$$\{P_{ij}\} = \{R_{ij}\} \cup \{(d, e, f), (a, b, f), (a, c, e), (b, c, d)\}$$

Obviously, this set satisfies Theorem 11-3-1. By choosing $P_4 = (d, e, f)$ as a member of linearly independent paths, we have

$$B = \begin{array}{c} \\ (y) \cup P_1 \\ (y) \cup P_2 \\ (y) \cup P_3 \\ (y) \cup P_4 \end{array}\overset{\begin{array}{ccccccc} a & b & c & d & e & f & y \end{array}}{\left[\begin{array}{ccccccc} 1 & 0 & 0 & 0 & 1 & 1 & 1 \\ 0 & 1 & 0 & 1 & 0 & 1 & 1 \\ 0 & 0 & 1 & 1 & 1 & 0 & 1 \\ 0 & 0 & 0 & 1 & 1 & 1 & 1 \end{array}\right]}$$

By adding the fourth row to the second and third rows, we have

$$B_f = \overset{\begin{array}{ccccccc} a & b & c & d & e & f & y \end{array}}{\left[\begin{array}{cccc|ccc} 1 & 0 & 0 & 0 & 1 & 1 & 1 \\ 0 & 1 & 0 & 0 & 1 & 0 & 0 \\ 0 & 0 & 1 & 0 & 0 & 1 & 0 \\ 0 & 0 & 0 & 1 & 1 & 1 & 1 \end{array}\right]}$$

From this, we obtain a fundamental cut-set matrix Q_f as

$$Q_f = \begin{array}{c} \begin{array}{ccccccc} a & b & c & d & e & f & y \end{array} \\ \left[\begin{array}{ccc|ccc} 1 & 1 & 0 & 1 & 1 & 0 & 0 \\ 1 & 0 & 1 & 1 & 0 & 1 & 0 \\ 1 & 0 & 0 & 1 & 0 & 0 & 1 \end{array}\right] \end{array}$$

By adding the third row to the second row, we have

$$Q = \begin{array}{c} \begin{array}{ccccccc} a & b & c & d & e & f & y \end{array} \\ \left[\begin{array}{ccccccc} 1 & 1 & 0 & 1 & 1 & 0 & 0 \\ 0 & 0 & 1 & 0 & 0 & 1 & 1 \\ 1 & 0 & 0 & 1 & 0 & 0 & 1 \end{array}\right] \end{array}$$

which has at most two nonzeros in each column. Thus we can obtain linear graph G shown in Fig. 11-4-1. Without edge y, we have an SC-network satisfying F_{ij}.

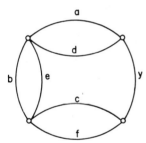

Fig. 11-4-1. Linear graph G.

Instead of choosing (d, e, f), suppose we choose (a, b, e) as a member of $\{P_{ij}\}$. That is,

$$\{P_{ij}\} \supset \{R_{ij}\} \cup (a, b, e)$$

Then

$$(a, b, e) \oplus P_1 \oplus P_2 = (d) \notin \{I_{ij}\}$$

Thus it is impossible for $\{P_{ij}\}$ to satisfy Theorem 11-3-1 by adding any more members in $\{I_{ij}\}$. Therefore the previous choice of $\{P_{ij}\}$ is the only choice that gives a desired SC-network.

It must be noted that a different choice of $\{P_{ij}\}$ will not change the number of edges in a resultant SC-network as long as the number of different Boolean variables appearing in $\{P_{ij}\}$ is the same. In other words, the number of edges in a resultant SC-network is determined by the number of different Boolean variables in $\{P_{ij}\}$.

11-5 MULTICONTACT SWITCHING NETWORKS

In an SC-network, the weight of each edge must be a different Boolean variable. In an SP-network, each Boolean variable can be the weights of two edges in a switching network. However, if two edges have the same Boolean variable x, one must be x and the other must be $\bar{x}$ (the complement of x).

Definition 11-5-1.   An *SP-network* is a nonoriented linear graph such that (1) the weight of each edge is a Boolean variable and (2) for each Boolean variable x, there is at most one edge whose weight is x and at most one edge whose weight is $\bar{x}$.

Consider a completely specified switching function F_{ij} of an SP-network G which is expressed by

$$F_{ij} = \sum_{P_r \in \{P_{ij}\}} P_r \text{ product} \qquad (11\text{-}5\text{-}1)$$

Set $\{P_{ij}\}$ may not be a set of all possible paths between i and j in G. The reason is that there may be a path that contains two edges whose weights complement each other, such as x and $\bar{x}$, which will not correspond to a nonzero term in F_{ij} because $x\bar{x} = 0$. For example, F_{ij} of the SP-network in Fig. 11-5-1 is

$$F_{ij} = x\bar{y} + yz + \bar{x}\bar{y}z$$

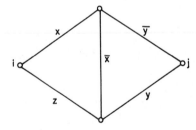

Fig. 11-5-1.   An SP-network.

Set $\{P_{ij}\}$ corresponding to F_{ij} is

$$\{P_{ij}\} = \{(x, \bar{y}), (y, z), (\bar{x}, \bar{y}, z)\}$$

which is obviously not a set of all possible paths between i and j in the SP-network.

Definition 11-5-2. When a path contains both a Boolean variable and its complement, the path is called a *0-path*.

It is clear now that $\{P_{ij}\}$ corresponding to a completely specified switching function F_{ij} of an SP-network does not, in general, satisfy Theorem 11-3-1. Thus we must insert 0-paths (which are paths between i and j) to $\{P_{ij}\}$ so that a resultant set $\{P'_{ij}\}$ will give an SP-network by the procedure in Section 11-3. This is illustrated by the following example.

Example 11-5-1. Suppose a completely specified switching function F_{ij} is

$$F_{ij} = ace + a\bar{c}\bar{d}e + \bar{a}bce + bd\bar{e} + \bar{a}b\bar{c}\bar{d}e$$

Then set $\{P_{ij}\}$ corresponding to F_{ij} is

$$\{P_{ij}\} = \{(a, c, e), (a, \bar{c}, \bar{d}, e), (\bar{a}, b, c, e), (b, d, \bar{e}), (\bar{a}, b, \bar{c}, \bar{d}, e)\}$$

which does not satisfy Theorem 11-3-1. This is because

$$(a, c, e) \oplus (a, \bar{c}, \bar{d}, e) \oplus (b, d, \bar{e}) = (b, d, \bar{d}, e) \notin \{P_{ij}\}$$
$$(a, c, e) \oplus (\bar{a}, b, c, e) \oplus (b, d, \bar{e}) = (a, \bar{a}, d, \bar{e}) \notin \{P_{ij}\}$$

and

$$(a, \bar{c}, \bar{d}, e) \oplus (\bar{a}, b, c, e) \oplus (b, d, \bar{e}) = (a, \bar{a}, d, \bar{d}, e) \notin \{P_{ij}\}$$

However, all of these are 0-paths. Hence we can insert these into $\{P_{ij}\}$ without changing F_{ij}. The resulting set $\{P'_{ij}\}$ is

$$\{P_{ij}\} = \{(a, c, e), (a, \bar{c}, \bar{d}, e), (\bar{a}, b, c, e), (b, d, \bar{e}), (\bar{a}, b, \bar{c}, \bar{d}, e), (b, d, \bar{d}, e),$$
$$(a, \bar{a}, d, \bar{e},), (a, \bar{a}, d, \bar{d}, e)\}$$

Let $P_1 = (a, c, e)$, $P_2 = (a, \bar{c}, \bar{d}, e)$, $P_3 = (\bar{a}, b, c, e)$, and $P_4 = (b, d, \bar{e})$ be the chosen linearly independent paths. Then we can obtain a circuit matrix B:

$$
B = \begin{array}{c}
\begin{array}{ccccccccc} a & \bar{d} & \bar{a} & d & b & c & e & \bar{e} & y \end{array} \\
\left[\begin{array}{ccccccccc}
1 & 0 & 0 & 0 & 0 & 1 & 1 & 0 & 1 \\
1 & 1 & 0 & 0 & 0 & 1 & 0 & 1 & 1 \\
0 & 0 & 1 & 0 & 1 & 1 & 1 & 0 & 1 \\
0 & 0 & 0 & 1 & 1 & 0 & 0 & 1 & 1
\end{array} \right]
\end{array}
$$

By adding the first row to the second row, we have

$$
B_f = \begin{array}{c}
\begin{array}{ccccccccc} a & \bar{d} & \bar{a} & d & b & c & e & \bar{e} & y \end{array} \\
\left[\begin{array}{cccc|ccccc}
1 & 0 & 0 & 0 & 0 & 1 & 1 & 0 & 1 \\
0 & 1 & 0 & 0 & 0 & 0 & 1 & 1 & 0 \\
0 & 0 & 1 & 0 & 1 & 1 & 1 & 0 & 1 \\
0 & 0 & 0 & 1 & 1 & 0 & 0 & 1 & 1
\end{array} \right]
\end{array}
$$

Hence the fundamental cut-set matrix Q_f is

$$
Q_f = \begin{array}{c}
\begin{array}{cccccccccc} a & \bar{d} & \bar{a} & d & b & c & e & \bar{e} & y \end{array} \\
\left[\begin{array}{cccc|ccccc}
0 & 0 & 1 & 1 & 1 & 0 & 0 & 0 & 0 \\
1 & 0 & 1 & 0 & 0 & 1 & 0 & 0 & 0 \\
1 & 1 & 1 & 0 & 0 & 0 & 1 & 0 & 0 \\
0 & 1 & 0 & 1 & 0 & 0 & 0 & 1 & 0 \\
1 & 0 & 1 & 1 & 0 & 0 & 0 & 0 & 1
\end{array}\right]
\end{array}
$$

If we add the first row to the fifth row and the second row to the third row, we obtain an incidence matrix:

$$
A = \begin{array}{c}
\begin{array}{cccccccccc} a & \bar{d} & \bar{a} & d & b & c & e & \bar{e} & y \end{array} \\
\left[\begin{array}{ccccccccc}
0 & 0 & 1 & 1 & 1 & 0 & 0 & 0 & 0 \\
1 & 0 & 1 & 0 & 0 & 1 & 0 & 0 & 0 \\
0 & 1 & 0 & 0 & 0 & 1 & 1 & 0 & 0 \\
0 & 1 & 0 & 1 & 0 & 0 & 0 & 1 & 0 \\
1 & 0 & 0 & 0 & 1 & 0 & 0 & 0 & 1
\end{array}\right]
\end{array}
$$

Linear graph G whose incidence matrix this is is shown in Fig. 11-5-2. Without edge y, the linear graph is an SP-network satisfying the given function F_{ij}.

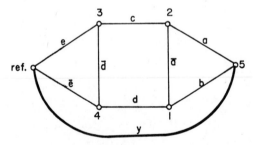

Fig. 11-5-2.   A linear graph G.

When a switching function of an SP-network is incompletely specified, we have sets $\{R_{ij}\}$ and $\{I_{ij}\}$. Then the procedure of synthesizing an SP-network is almost identical with that in Section 11-4 except that we can insert 0-paths into $\{P_{ij}\}$.

When a switching network can have multicontacts (i.e., a Boolean variable can be the weights of many edges in a nonoriented linear graph), synthesis of such a network by the procedure in Section 11-3 becomes very complicated. In this case each edge must have a distinct name in order to use the procedure and it is very difficult to find whether a Boolean variable in a given switching function F_{ij} is from one edge or not. For example, (a, b, c) in set $\{P_{ij}\}$ corresponding to a switching function F_{ij} may be from (a_1, a_2, b, c) where $a_1 = a_2$. In other words, suppose there is a path between i and j in a switching network containing two edges which have the same weight x. Then the path product of this path will show only one x in F_{ij}. When we use the procedure in Section 11-3, we must reconstruct this path with two edges having variable x, which is not an easy task. The following example will illustrate this fact.

Example 11-5-2. Consider a switching function F_{ij} where

$$F_{ij} = ab + ac + bc$$

Set $\{P_{ij}\}$ corresponding to F_{ij} is

$$\{P_{ij}\} = \{(a, b), (a, c), (b, c)\}$$

which does not satisfy Theorem 11-3-1 because

$$(a, b) \oplus (b, c) \oplus (b, c) = \emptyset$$

If we change

$$(a, b) \Rightarrow (a_1, b_1)$$
$$(a, c) \Rightarrow (a_2, c)$$

and

$$(b, c) \Rightarrow (b_2, c)$$

where $a_1 = a_2$ and $b_1 = b_2$, then

$$(a_1, b_1) \oplus (a_2, c) \oplus (b_2, c) = (a_1, a_2, b_1, b_2)$$

Thus set $\{P_{ij}\}$ where

$$\{P_{ij}\} = \{(a_1, b_1), (a_2, c), (b_2, c), (a_1, a_2, b_1, b_2)\}$$

satisfies Theorem 11-3-1 and the procedure in Section 11-3 gives a switching network satisfying F_{ij}.

PROBLEMS

1. Find a switching function from i to j of the networks in Fig. P-11-1 by the use of a connection matrix.

2. Find switching functions F_{12} and F_{13} of the networks in Fig. P-11-2 by the use of sets of paths in Section 11-2.

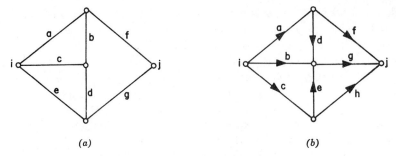

Fig. P-11.1. (a) SC-network G; (b) network G'.

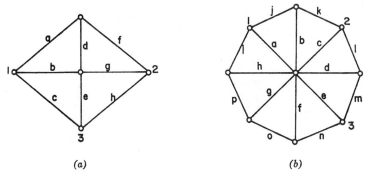

Fig. P-11-2. (a) SC-network G; (b) SC-network G'.

3. Synthesize an SC-network which satisfies a completely specified switching function

$$F = af + bdf + cdef + cg + beg + adeg$$

4. Synthesize an SC-network which satisfies an incompletely specified switching function whose corresponding sets $\{R_{ij}\}$ and $\{I_{ij}\}$ are

$$\{R_{ij}\} = \{(a, d, e), (b, d, f), (b, e), (c)\}$$

and

$$\{I_{ij}\} = \{(a, c, e), (a, f)\}$$

5. Is $F_{ij} = ab + ac + bc + ad$ a switching function of an SC-network?

6. Suppose F_1 is not a switching function of an SC-network, but F_2 is a switching function of an SC-network. Is $F = F_1 + F_2$ a switching function of an SC-network?

7. Synthesize an SP-network satisfying a completely specified switching function

$$F_{ij} = x\bar{y}z + \bar{x}yz + \overline{yu\bar{z}} + \bar{y}u\bar{z}$$

8. Suppose F_1 is a completely specified switching function of an SC-network. Is $F = xF_1 + \bar{x}F_1$ a completely specified switching function of an SP-network?

9. Let F_{ij} be a switching function between i and j, and F_{ip} be a switching function between i and p of SC-network G. Suppose we insert a new variable x between p and j in G. What is a switching function between i and j of the resulting SC-network?

10. Let BG be an SP-network. Also let i, j, and k be any three vertices in G. Is there a relationship among F_{ij}, F_{jk}, and F_{ik} where F_{pq} is a switching function between p and q for $p, q \in (i, j, k)$?

CHAPTER
12
COMMUNICATION NET THEORY—EDGE WEIGHTED CASE

12-1 SINGLE FLOW IN A NONORIENTED EWC NET

Consider a medium which transmits information from one station to another, such as a telephone line, a highway, a power line, or a utility line. There is some maximum quantity of information per unit of time that can be transmitted by such a medium. Each station may have the maximum quantity of information which can be handled by the station. A network of such media is called a *communication net*; it is represented by a linear graph where each vertex indicates a station and each edge indicates a medium (a link) by which information can be transmitted between two stations. When stations are not limiting the amount of information to be transmitted, the net is called an *EWC (Eage Weighted Communication) net*. When the maximum quantity of information that can be transmitted by each edge is large enough so that the transmission of information is limited only by vertices (stations), the net is called a *VWC (Vertex Weighted Communication) net*. We discuss EWC nets in this chapter and VWC nets in the next chapter.

Since only edges limit the transmission of information in EWC nets, only edges have capacities (weights) which indicate the maximum quantities of

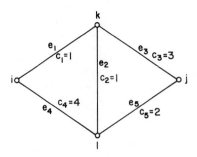

Fig. 12-1-1. A nonoriented EWC net.

information that can be transmitted by edges. These weights are called *edge capacities*. When all edges are nonoriented, a net is called a *nonoriented EWC net*. On the other hand, when all edges are oriented, we call it an *oriented EWC net*. For example, the linear graph in Fig. 12-1-1 is a nonoriented EWC net where the weight given to each edge indicates the edge capacity of the edge.

Instead of using the phrase "transmitting information from vertex i to vertex j," we use the phrase "assigning flow from i to j" for convenience. The following terminology is used here.

Definition 12-1-1. Let P_r be a path between vertices i and j in a nonoriented graph. A corresponding directed path from i to j is obtained by giving an orientation to every edge in P_r so that the path P_r becomes a directed path from i to j.

For example, the corresponding directed path from i to j of a path $p = (a, b, c)$ in the nonoriented linear graph in Fig. 12-1-2a is shown in Fig. 12-1-2b.

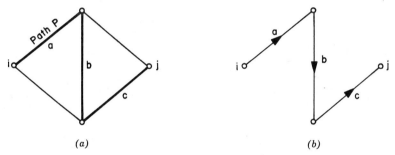

(a)   *(b)*

Fig. 12-1-2. *(a)* Linear graph G and path P; *(b)* the corresponding directed path from i to j of P.

Definition 12-1-2. The symbol $\psi_{r_{ij}}$ indicates a flow from a vertex i to vertex j. (Note that the subscripts indicate these vertices.)

To assigning a flow $\psi_{r_{ij}}$ to a path means transmitting $\psi_{r_{ij}}$ by the edges in the path.

Definition 12-1-3. Let P_r be a path between vertices i and j in a nonoriented graph. Assigning a flow $\psi_{r_{ij}}$ to path P_r means assigning $\psi_{r_{ij}}$ to each edge in P_r with the orientation which agrees with that of the corresponding directed path of P_r from i to j.

For example, to assign $\psi_{r_{ij}} = 1$ to path $P = (a, b, c)$, we give a flow of 1 unit

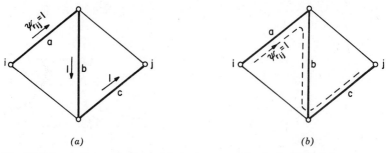

Fig. 12-1-3. Assignment of $\psi_{r_{ij}}$. (a) $\psi_{r_{ij}}$ to every edge in P; (b) line $\psi_{r_{ij}}$.

with a proper orientation to every edge in P as shown in Fig. 12-1-3a. For simplicity, a line normally is used for indicating $\psi_{r_{ij}}$ as shown in Fig. 12-1-3b.

When we say that a flow ψ_{ij} is assigned to a nonoriented EWC net G, we mean, in general, that

$$\psi_{ij} = \sum_{r=1}^{n} \psi_{r_{ij}} \qquad (12\text{-}1\text{-}1)$$

where each portion $\psi_{r_{ij}}$ of flow ψ_{ij} is assigned to a path P_r between i and j in G for $r = 1, 2, \ldots, n$, and $\sum_{r=1}^{n} \psi_{r_{ij}}$ means the summation of quantities of $\psi_{r_{ij}}$. For example, in nonoriented EWC net G in Fig. 12-1-4, there are three flows $\psi_{1_{ij}} = 1$, $\psi_{2_{ij}} = 2$, and $\psi_{3_{ij}} = 3$, assigned to paths P_1, P_2, and P_3, respectively. Hence we say that $\psi_{ij} = \sum_{r=1}^{3} \psi_{r_{ij}} = 6$ is assigned to G.

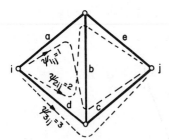

Fig. 12-1-4. Assignment of ψ_{ij}.

It must be noted from the definition of edge capacities, an assignment of flow $\psi_{r_{ij}}$ to path P_r ($r = 1, 2, \ldots, n$) should not produce a situation where the amount of flow assigned to an edge exceeds the capacity of the edge. For example, in the EWC net in Fig. 12-1-4, edge b must handle $\psi_{1_{ij}} + \psi_{2_{ij}} = 3$.

If the edge capacity of b is less than 3, the assignment indicated in Fig. 12-1-4 is not proper to handle flow $\psi_{ij} = 6$. In other words, there are certain restrictions for assigning flow $\psi_{r_{ij}}$ to a path P_r.

Suppose $\psi_{s_{ij}}$ $(s = 1, 2, \ldots, h)$ are flows assigned to paths $P_{s_{ij}}$ from i to j for all s. Then for any edge e' whose edge capacity is c', the equation

$$c' \geq \psi(e') \tag{12-1-2}$$

must satisfy where

$$\psi(e') = \sum_{s=1}^{h} \delta_s(e')\psi_{s_{ij}} \tag{12-1-3}$$

and

$$\delta_s(e') = \begin{cases} 1 & \text{if edge } e' \text{ is in path } P_s \\ 0 & \text{otherwise} \end{cases} \tag{12-1-4}$$

The symbol $\psi(e')$ in Eq. 12-1-2 indicates the total flow assigned to edge e' by the assignment of flows $\psi_{s_{ij}}$ to paths $P_{s_{ij}}$ for $s = 1, 2, \ldots, h$.

Suppose flows $\psi_{s_{ij}}$ have been assigned to paths $P_{s_{ij}}$ for $s = 1, 2, \ldots, h$. Then we know that Eq. 12-1-2 satisfies for every edge in EWC net G. In order to assign an additional flow $\psi_{t_{ij}}$ to a path $P_{t_{ij}}$, every edge e_u in path $P_{t_{ij}}$ must satisfy

$$\psi(e_u) + \psi_{t_{ij}} \leq c_u \tag{12-1-5}$$

where $\psi(e_u)$ is the total flow in edge e_u given by Eq. 12-1-3 and c_u is the capacity of edge e_u. Equation 12-1-5 can be rewritten as

$$\psi_{t_{ij}} \leq c_u - \psi(e_u) \tag{12-1-6}$$

which must be satisfied by every edge in $P_{t_{ij}}$. In other words,

$$\psi_{t_{ij}} \leq \min \{c_u - \psi(e_u); \quad e_u \in P_{t_{ij}}\} \tag{12-1-7}$$

where the right-hand side gives the upper bound of an additional flow which can be assigned to an EWC net by using a path $P_{t_{ij}}$.

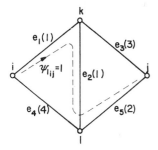

Fig. 12-1-5. Flow $\psi_{ij} = 1$.

Example 12-1-1. Suppose we give flow $\psi_{1_{ij}}$ to the EWC net in Fig. 12-1-1 as shown in Fig. 12-1-5. Then there is only one path $P_{ij} = (e_4, e_5)$ to which a non-zero flow from i to j can be assigned. Since min $\{4 - \psi(e_4); \ 2 - \psi(e_5)\}$ is 1, an additional flow $\psi_{2_{ij}}$ cannot be larger than 1.

Definition 12-1-4. The statement "flow ψ_{ij} can be assigned" means that there exists a set of path $P_{r_{ij}}$ (for $r = 1, 2, \ldots, h$), each of which is between i and j, and a set of flows $\psi_{r_{ij}}$ where $\psi_{ij} = \sum\limits_{r=1}^{h} \psi_{r_{ij}}$ such that $\psi_{r_{ij}}$ can be assigned to path $P_{r_{ij}}$ consecutively for $r = 1, 2, \ldots, h$.

In Example 12-1-1, we can see that $\psi_{ij} = 2$ can be assigned to the communication net in Fig. 12-1-1. However, if we do not assign $\psi_{1_{ij}} = 1$ as in Fig. 12-1-5, we can assign flow $\psi_{ij} = 4$ to the same net as shown in Fig. 12-1-6.

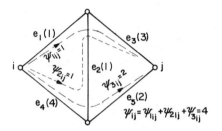

Fig. 12-1-6. Assigning $\psi_{ij} = 4$ to an EWC net.

This shows that the amount of flow ψ_{ij} that can be assigned depends on the assignment of flow. On the other hand, if an EWC net consists of a finite number of edges and each edge capacity is finite, we know that only a finite amount of flow can be assigned. Thus there must be the maximum flow which can be assigned from vertex i to vertex j. This maximum flow is called the *terminal capacity* from i to j and is symbolized by t_{ij}.

Definition 12-1-5. A terminal capacity t_{ij} of an EWC net G is the maximum flow from i to j which can be assigned to G.

Suppose a flow ψ_{ij} can be assigned to an EWC net G, and assume that ψ_{ij} is the maximum. That is, ψ_{ij} is equal to the terminal capacity t_{ij} from i to j. Then any flow $\psi'_{ij} \leq \psi_{ij}$ can be assigned to G because assigning $\psi'_{r_{ij}} = \alpha \psi_{r_{ij}}$ to path $P_{r_{ij}}$ for $r = 1, 2, \ldots, h$ satisfies Eq. 12-1-2 where

$$\alpha = \frac{\psi'_{ij}}{\psi_{ij}} \tag{12-1-8}$$

and

$$\psi_{ij} = \sum_{r=1}^{h} \psi_{r_{ij}} \tag{12-1-9}$$

Hence it is important to know a terminal capacity t_{ij} of an EWC net. To discuss terminal capacities, we need the following definition.

Definition 12-1-6.   An edge is said to be *saturated* (or called a *saturated edge*) under an assignment of a flow ψ_{ij}, if the flow through the edge is equal to the capacity of the edge.

For example, under the assignment of flow ψ_{ij} in Fig. 12-1-5, edges e_1 and e_2 are saturated. In Fig. 12-1-6, edges e_1, e_2, and e_5 are saturated.

Definition 12-1-7.   Under an assignment of flow ψ_{ij}, if a cut-set S consists of saturated edges, then S is called a *saturated cut-set*.

For example, in Fig. 12-1-6, edges e_1, e_2, and e_5 are saturated. Hence $S = (e_1, e_2, e_5)$ is a saturated cut-set under this assignment of flow ψ_{ij}. We choose the orientation of a saturated cut-set equal to the direction of a flow in an edge in the cut-set. It is possible that flows assigned to an edge may not be in the same direction.

Consider the EWC net in Fig. 12-1-7. In edge e_2, there are two flows of opposite directions. By taking one of these flows, say ψ_1, in e_2 to fix the orientation of cut-set S, we can say that the direction of flow in edge e_1 and the orientation of S do not coincide. Similarly, we can see that the direction of flow in edge e_5 is different from the orientation of S.

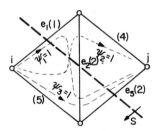

Fig. 12-1-7.   An EWC net with $\psi_{ij} = 3$ containing a saturated cut-set S.

Definition 12-1-8.   An edge is said to be a *basic saturated* edge if the edge is saturated and all nonzero flows in the edge are in the same direction (orientation).

For example, in Fig. 12-1-7, edge e_2 is saturated but not a basic saturated edge. On the other hand, edges e_1 and e_5 are basic saturated edges.

Definition 12-1-9.   When all edges in a saturated cut-set are basic saturated edges and the orientation of the cut-set and the direction of nonzero flows in every edge in the cut-set coincide, then S is called a *basic saturated cut-set*.

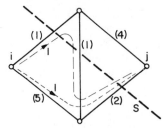

Fig. 12-1-8. Saturated cut-set S.

It is obvious that S in Fig. 12-1-7 is not a basic saturated cut-set. Similarly, saturated cut-set S in Fig. 12-1-8 is not a basic saturated cut-set. However, saturated cut-set (e_1, e_2, e_5) in Fig. 12-1-6 is a basic saturated cut-set.

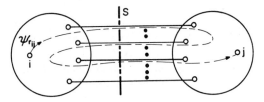

Fig. 12-1-9. Saturated cut-set and flow.

Consider the EWC net in Fig. 12-1-9 where S is a saturated cut-set but is not a basic saturated cut-set. Then it is easily seen that there is a path p_{ij} from i to j to which a nonzero flow $\psi_{r_{ij}}$ is assigned and which contains at least three edges of cut-set S as shown in the figure. On the other hand, if S is a basic saturated cut-set, then every path p_{ij} from i to j, to which a nonzero flow $\psi_{r_{ij}}$ is assigned, must contain exactly one edge of S as shown in Fig. 12-1-10.

Before relating the maximum flow and basic saturated cut-sets, we study the effect of the existence of a saturated cut-set on assigning additional flows.

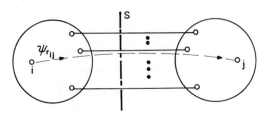

Fig. 12-1-10. A basic saturated cut-set and flow.

Lemma 12-1-1. By an assignment of flow ψ_{ij} to an EWC net, if there exists no saturated cut-set, then additional flow from i to j can be assigned.

Proof. Since there is no saturated cut-set, if we change the edge capacity c_r of each edge e_r in an EWC net by $c_r' = c_r - \psi(e_r)$, where $\psi(e_r)$ is the total flow given to edge e_r, and delete the edge whose new edge capacity c_r' is zero, then there must exist at least one path p_{ij} from i to j in the resultant net, because otherwise there must be a saturated cut-set. Since every edge capacity c_r' of every edge in path p_{ij} is nonzero, we can assign a nonzero flow ψ_{ij}' to the path where

$$\psi_{ij}' \le \min \{c_r': \quad e_r \in p_{ij}\} \tag{12-1-10}$$

$$\text{QED}$$

By Lemma 12-1-1, if we cannot assign any more flow ψ_{ij}' to an EWC net G, then we know that there exists at least one saturated cut-set which separate vertices i and j.

An important theorem about a maximum flow follows.

Theorem 12-1-1. By an assignment of flow ψ_{ij} to an EWC net G of a finite number of edges, if there exists a saturated cut-set but no basic saturated cut-set, then

$$\psi_{ij} < t_{ij} \tag{12-1-11}$$

Note that t_{ij} is the terminal capacity from i to j which is equal to the maximum flow from i to j.

Proof. Since the number of edges in G is finite, the number of cut-sets in G is also finite. Let $\{S_q(i; j)\}$ be the set of all cut-sets which separate i and j. Also let $\{\underline{S}_{q'}(i; j)\}$ consisting of $\underline{S}_1, \underline{S}_2, \ldots, \underline{S}_k$ be the set of all saturated cut-sets in $\{S_q(i; j)\}$ under an assignment of ψ_{ij}. Then we can see that there exists $\underline{S}_c$ in $\{\underline{S}_{q'}(i; j)\}$ which is the closest to vertex j. That is, if $\underline{S}_{q'} = \mathscr{E}(\Omega_{q'} \times \overline{\Omega}_{q'})^*$ for $q' = 1, 2, \ldots, c, \ldots, k$ where $j \in \overline{\Omega}_{q'}$, then $\overline{\Omega}_c \subset \overline{\Omega}_{q'}$ for all q'.

Consider EWC net G in Fig. 12-1-11 where $\underline{S}_c$ consists of edges $e_1, e_2, \ldots, e_r$ and is the closest to vertex j. Then none of the saturated cut-sets in $\{\underline{S}_{q'}(i; j)\}$ contains edges in g_{cj} where the vertices in g_{cj} are those in $\overline{\Omega}_c$. Note that g_{ci} and g_{cj} are the pair of subgraphs obtained by deleting all edges in $\underline{S}_c$.

Since $\underline{S}_c$ is a saturated cut-set but not a basic saturated cut-set, there exists a path p_{ij} which contains more than one edge of $\underline{S}_c$ and to which a nonzero flow $\psi_{r_{ij}}$ is assigned. Without the loss of generality, let edge e_1 be the first edge in $\underline{S}_c$ appearing in the edge train of p_{ij} from i to j as shown in Fig. 12-1-11. Let v_1 and v_1' be the vertices on which edge e_1 is connected and let

---

* See Definitions 2-7-4 and 5-1-1.

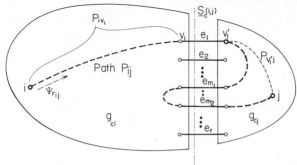

Fig. 12-1-11.   EWC net G with $\underline{S}_c(i;j)$.

v_1 belong to g_{ci}. Since there is no saturated cut-set in g_{cj} which separates vertices v_1' and j, we can find path p' from v_1' to j such that we can assign non-zero flow δ from v_1' to j.

Instead of assigning $\psi_{r_{ij}}$ to p_{ij}, we assign $\psi_{r_{ij}} - \delta$ to p_{ij} and assign δ to path $p_{ij}' = p_{iv_1} \cup (e_1) \cup p_{v_1'j}$ where p_{iv_1} is a path from i to v_1, which is a part of p_{ij} as shown in Fig. 12-1-11. By doing so, we have exactly the same amount of flow from i to j, that is, ψ_{ij}. However, because flows assigned to edges in $p_{ij} \cap \underline{S}_c$ except e_1 are reduced by δ, $\underline{S}_c$ is no longer a saturated cut-set. Also by choosing δ small enough, we will not produce any new saturated cut-sets. Thus this new assignment of flow ψ_{ij} gives at most $k - 1$ saturated cut-sets.

We can repeat this process until we have no saturated cut-sets under an assignment of ψ_{ij}. When there is no saturated cut-set, then we can assign additional flow by Lemma 12-1-1. Hence ψ_{ij} is not the maximum flow from i to j which can be assigned to the given EWC net.   QED

From Theorem 12-1-1, we can obtain the following theorem.

Theorem 12-1-2.   If and only if an assignment of ψ_{ij} produces at least one basic saturated cut-set which separates i and j,

$$\psi_{ij} = t_{ij} \tag{12-1-12}$$

Proof.   Consider any cut-set that separates i and j as shown in Fig. 12-1-10. Then we can see that flow ψ_{ij} must pass through edges in the cut-set. Thus the maximum amount that can be sent from vertex i to vertex j cannot exceed the sum of edge capacities of all edges in the cut-set. Since this is true for any cut-set separating i and j, this is true for a basic saturated cut-set which separates i and j. Moreover, the total flow through a basic saturated cut-set is equal to the sum of edge capacities of all edges in the cut-set. Thus Eq. 12-1-12 is true. This, with Theorem 12-1-1, gives the proof of this theorem.
                                        QED

Definition 12-1-10. The value of cut-set S, indicated by $V[S]$, is the sum of edge capacities of all edges in S.

Consider a set $\{S_q(i;j)\}$ of all cut-sets which separate vertices i and j. Note that for any cut-set S in $\{S_q(i;j)\}$, $V[S]$ cannot be smaller than the maximum flow ψ_{ij}. If an assignment of a flow ψ_{ij} produces a basic cut-set S' (which separates i and j), S' is in $\{S_q(i;j)\}$. The sum of flows assigned to edges in a basic saturated cut-set S' must be equal to ψ_{ij}. Hence we can conclude with the result of Theorem 12-1-2.

Theorem 12-1-3. (The Max Flow-Min Cut Theorem) For an EWC net,

$$t_{ij} = \min\{V[S]; \quad S \in \{S_q(i;j)\}\} \tag{12-1-13}$$

Example 12-1-2. Consider the EWC net in Fig. 12-1-12. The set of cut-sets which separates 1 and 2 is

$$\{(a, b), (a, c, e), (b, c, d), (d, e)\}$$

Hence

$$\begin{aligned} t_{12} &= \min\{V[a, c], V[a, c, e], V[b, c, d], V[d, e]\} \\ &= \min\{7, 6, 10, 7\} = 6 \end{aligned}$$

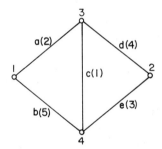

Fig. 12-1-12. An EWC net.

Suppose two assignments of ψ_{ij} give two different sets $\{S'_q(i;j)\}$ and $\{S''_q(i;j)\}$ of basic saturated cut-sets which separate vertices i and j. Let ψ'_{ij} be one such assignment and ψ''_{ij} be the other assignment. Let $\sum_{(r)} \psi'_{r_{ij}} = \psi'_{ij}$ where $\psi'_{r_{ij}}$ is assigned to path $P_{r_{ij}}$ and $\sum_{(r)} \psi''_{r_{ij}} = \psi''_{ij}$ where $\psi''_{r_{ij}}$ is assigned to path $P_{r_{ij}}$ for $r = 1, 2, \ldots, k$. Note that $\psi'_{r_{ij}}$ or $\psi''_{r_{ij}}$ can be zero. Instead of assigning $\psi'_{r_{ij}}$ to $P_{r_{ij}}$, we can assign $a\psi'_{r_{ij}} + (1 - a)\psi''_{r_{ij}}$ to path $P_{r_{ij}}$ for all $r = 1, 2, \ldots, k$, which gives the same amount of flow from i to j, where $0 < a < 1$. With this assignment, any cut-set S'_q in $\{S'_q(i;j)\}$ is not a basic saturated cut-set unless this cut-set is also a basic saturated cut-set in $\{S_r''(i;j)\}$. Hence by Theorem 12-1-2, $\{S_q'(i;j)\} \cap \{S_r''(i;j)\} \neq \emptyset$ in order that both ψ_{ij}' and ψ_{ij}'' give the maximum flow from i to j. By the same argument, we can show that all sets $\{S_p(i;j)\}$ have at least one cut-set in common where $\{S_p(i;j)\}$ is a set of basic saturated cut-sets separating i and j produced by an assignment of the maximum flow from i to j for $p = 1, 2, \ldots$. Such a cut-set is called a *cut-set corresponding to* t_{ij} or a *corresponding cut-set of* t_{ij}. For example, cut-set (a, c, e) in Fig. 12-1-12 is a corresponding cut-set of t_{ij}. The corresponding cut-sets of terminal capacities are very important for studying the relationship among terminal capacities, discussed next.

12-2 TERMINAL CAPACITY MATRIX OF A NONORIENTED EWC NET

One way of specifying an EWC net is to give terminal capacities between all possible pairs of vertices in the net. This can be done by a matrix $T = [t_{ij}]$ called a terminal capacity matrix.

Definition 12-2-1.   A *terminal capacity matrix* $T = [t_{ij}]$ of an EWC net G is a square matrix whose rth row and column represent vertex r and whose entry t_{ij} is

$$t_{ij} = \text{terminal capacity from } i \text{ to } j \text{ for } i \neq j$$

and

$$t_{ii} = d \text{ for all } i \tag{12-2-1}$$

For an EWC net, d in this definition can be anything unless we give some meaning to a terminal capacity from vertex i to itself. In some cases, it may be convenient to define d as zero. The symbol d is from the *diagonal* entry.

Example 12-2-1.   Consider EWC net G in Fig. 12-1-12. We have found that $t_{12} = 6$. Other terminal capacities are

$$t_{13} = \min \{V[a, b], \quad V[a, c, e], \quad V[a, c, d]\} = 6$$
$$t_{14} = \min \{V[a, b], \quad V[b, c, d], \quad V[b, c, e]\} = 7$$
$$t_{23} = \min \{V[d, e], \quad V[b, c, d], \quad V[a, c, d]\} = 7$$
$$t_{24} = 6$$

and

$$t_{34} = 6$$

Hence a terminal capacity matrix T of G is

$$
T = \begin{array}{c} \\ 1 \\ 2 \\ 3 \\ 4 \end{array} \begin{array}{cccc} 1 & 2 & 3 & 4 \\ \begin{bmatrix} d & 6 & 6 & 7 \\ 6 & d & 7 & 6 \\ 6 & 7 & d & 6 \\ 7 & 6 & 6 & d \end{bmatrix} \end{array}
$$

To discuss properties of a terminal capacity matrix of an EWC net, we define a principal partition of a matrix as follows.

Definition 12-2-2. Let M be a square matrix in which every diagonal entry is d and every off-diagonal entry is a real number. A partition of M as

$$
M = \left[\begin{array}{c:c} M_a & M_c \\ \hdashline M_d & M_b \end{array} \right] \tag{12-2-2}
$$

is called a *principal partition* if (1) every entry in M_c is identical and is the smallest number in M without considering d and (2) M_a and M_b are square submatrices with every diagonal entry being d. M_a and M_b are called the *resultant main submatrices* by a principal partition of M.

Since EWC nets discussed in this section are nonoriented, we can see that $t_{ij} = t_{ji}$. Thus a terminal capacity matrix T is symmetric. In order to study properties of terminal capacity matrix, we recall from the previous section that for every terminal capacity t_{ij}, there exists a corresponding cut-set S_1 which separates i and j and whose value is equal to t_{ij}. Let

$$
S_1 = \mathscr{E}(\Omega_1 \times \bar{\Omega}_1) \tag{12-2-3}
$$

Since S_1 is a corresponding cut-set of t_{ij}, $i \in \Omega_1$ and $j \in \bar{\Omega}_1$. Suppose t_{ij} is the smallest among all terminal capacities in an EWC net G. Then for any vertex p in Ω_1 and any vertex q in $\bar{\Omega}_1$, $t_{pq} = t_{ij}$ because from Eq. 12-1-13,

$$
t_{pq} = \min \{V[S]; \quad S \in \{S(p;q)\}\} \tag{12-2-4}
$$

Since S_1 separates p and q, S_1 is a cut-set in Eq. 12-2-4. Furthermore, S_1 is a corresponding cut-set of t_{ij} and t_{ij} is the smallest terminal capacity in G by assumption; $V[S_1]$ is also the smallest among the values of all cut-sets which separate p and q. Hence $t_{pq} = t_{ij}$. Thus a principal partition of terminal capacity matrix T as in Eq. 12-2-2 is possible by rearranging rows and columns:

$$
T = \begin{bmatrix} T_a & T_1 \\ T_1{}^t & T_b \end{bmatrix} \tag{12-2-5}
$$

such that T_a consists of rows and columns corresponding to vertices in Ω_1 and T_b consists of rows and columns corresponding to vertices in $\bar{\Omega}_1$. Note that each (r, s) entry in T_1 represents terminal capacity t_{rs}. Hence each entry in T_1 is equal to t_{ij}.

Let $t_{i'j'}$ be the smallest entry in the resultant main submatrices T_a and T_b. Without the loss of generality, let $t_{i'j'}$ be in T_a. Let S_2 where

$$S_2 = \mathcal{E}(\Omega_2 \times \bar{\Omega}_2) \tag{12-2-6}$$

be the corresponding cut-set of $t_{i'j'}$. Then any terminal capacity t_{uv}, where $u \in (\Omega_2 \cap \Omega_1)$ and $v \in (\bar{\Omega}_2 \cap \Omega_1)$, must be equal to $t_{i'j'}$ because the class of cut-sets that separate u and v contains S_2, and the value of S_2 is the smallest among the values of all cut-sets in the class by assumption. Now a principal partition of T_a is possible as

$$T = \begin{bmatrix} T_{aa} & T_{a1} & \\ \hline T_{a1}{}^t & T_{ab} & T_1 \\ \hline & T_1{}^t & T_b \end{bmatrix} \tag{12-2-7}$$

where T_{aa} consists of rows and columns corresponding to the vertices in $\Omega_2 \cap \Omega_1$ and T_{ab} consists of rows and columns corresponding to the vertices in $\bar{\Omega}_2 \cap \Omega_1$. Thus the entries in T_{a1} are identical and are smallest among all entries in T_a.

Similarly, we can pick a terminal capacity which is smallest among all entries in the resultant main submatrices T_{aa}, T_{ab}, and T_b so that a principal partition can be applied, and so on, until there are no resultant main submatrices containing more than one entry. This property is also sufficient for a matrix to be a terminal capacity matrix.

Theorem 12-2-1. A symmetric matrix T of order n_v is a terminal capacity matrix of a nonoriented EWC net if and only if we can obtain a principal partition of the matrix and principal partitions of all resultant main submatrices which contain more than one entry.

Proof. From the previous discussion, we know that matrix T must have the property given in the theorem, so we only need to prove the sufficiency part of the theorem. In the other words, we assume that there is a process of obtaining principal partitions of matrix T and all resultant main submatrices containing more than one entry. Now we are going to show that there is a nonoriented EWC net whose terminal capacity matrix is a given matrix T. From the definition of a principal partition, we can see that whenever we perform a principal partition to a matrix T_r as

$$T_r = \begin{bmatrix} T_{ra} & T_{r1} \\ T_{r1}{}^t & T_{rb} \end{bmatrix} \tag{12-2-8}$$

we are dividing a set of vertices corresponding to the rows of T_r into two sets, one consisting of vertices corresponding to the rows of T_{ra} and the other consisting of vertices corresponding to the rows of T_{rb}. Hence for a sequence of principal partitions applied to a given matrix T until all resultant main submatrices consist of diagonal entries d, we have a sequence of classes of sets of vertices. Let this sequence of classes of sets be $\Gamma_0, \Gamma_1, \ldots, \Gamma_{n_v - 1}$. For example, for a given matrix T where

$$T = \begin{array}{c} \\ 1 \\ 2 \\ 3 \\ 4 \end{array} \begin{array}{c} \begin{array}{cccc} 1 & 2 & 3 & 4 \end{array} \\ \begin{bmatrix} d & 1 & 1 & 1 \\ 1 & d & 3 & 2 \\ 1 & 3 & d & 2 \\ 1 & 2 & 2 & d \end{bmatrix} \end{array}$$

a sequence of principal partitions that can be applied is as follows:

$$\begin{array}{c} \\ 1 \\ 2 \\ 3 \\ 4 \end{array} \begin{array}{c} \begin{array}{cccc} 1 & 2 & 3 & 4 \end{array} \\ \left[\begin{array}{c|ccc} d & 1 & 1 & 1 \\ \hline 1 & d & 3 & 2 \\ 1 & 3 & d & 2 \\ 1 & 2 & 2 & d \end{array}\right] \end{array} \qquad \begin{array}{c} \\ 1 \\ 2 \\ 3 \\ 4 \end{array} \begin{array}{c} \begin{array}{cccc} 1 & 2 & 3 & 4 \end{array} \\ \left[\begin{array}{c|ccc} d & 1 & 1 & 1 \\ \hline 1 & d & 3 & 2 \\ 1 & 3 & d & 2 \\ \hline 1 & 2 & 2 & d \end{array}\right] \end{array} \qquad \left[\begin{array}{c|c|c|c} d & 1 & 1 & 1 \\ \hline 1 & d & 3 & 2 \\ \hline 1 & 3 & d & 2 \\ \hline 1 & 2 & 2 & d \end{array}\right]$$

Thus the sequence of classes of sets of vertices is

$$\Gamma_0 = \{(1, 2, 3, 4)\}, \qquad \Gamma_1 = \{(1), (2, 3, 4)\},$$
$$\Gamma_2 = \{(1), (2, 3), (4)\}, \qquad \Gamma_3 = \{(1), (2), (3), (4)\}$$

It is clear that there exist exactly two sets in Γ_k which are not in Γ_{k-1} and there exists exactly one set in Γ_{k-1} which is not in Γ_k. Let these two sets in Γ_k be called *generated sets* and the one set in Γ_{k-1} which is not in Γ_k be the *generator*. Then we can state that there exists one generator and two generated sets in each class Γ_p except in Γ_0 which consists of one generator and in $\Gamma_{n_v - 1}$, which does not contain the generator. For each pair of generated sets, there exists a set of identical terminal capacities which are from a vertex in one of these two generated sets to a vertex in the other set. These identical terminal capacities are all entries of the two submatrices obtained by the principal partition. We call any one of these terminal capacities the *corresponding terminal capacity* of the two generated sets in Γ_p for $p = 1, 2, \ldots,$ $n_v - 1$. For example, in Γ_1 given above, sets (1) and (2, 3, 4) are generated sets whose corresponding terminal capacity is 1. In Γ_2, (2, 3) and (4) are generated sets whose corresponding terminal capacity is 2. In Γ_3, (2) and (3) are generated sets whose corresponding terminal capacity is 3.

Now we are ready to give a method of constructing an EWC net from a given matrix with $\Gamma_0, \Gamma_1, \ldots, \Gamma_{n_v-1}$ being a sequence of classes of sets of vertices. Note that n_v is the number of columns in a given matrix.

Let Ω_{11} and Ω_{12} be two generated sets in Γ_1. Then we treat Ω_{11} as one vertex and Ω_{12} as one vertex and connect one edge between them with its edge capacity equal to the corresponding terminal capacity of these two generated sets. Next, let Ω_{21} and Ω_{22} be the two generated sets in Γ_2 and Ω_{11} be the generator in Γ_1. Then we make these two generated sets Ω_{21} and Ω_{22} as two vertices by cutting vertex Ω_{11}. The edge which was connected to vertex Ω_{11} can be connected to either of new vertices, Ω_{21} and Ω_{22}. Between these two new vertices, we connect one edge whose edge capacity is equal to the corresponding terminal capacity of the two generated sets in Γ_2. Now the process becomes clear. That is, we cut a vertex v_p representing the generator in Γ_p to two vertices v_{p1} and v_{p2} to represent two generated sets in Γ_{p+1}. Then we change the connection of edges, which were incident at vertex v_p, to either v_{p1} or v_{p2} and connect a new edge between v_{p1} and v_{p2} whose edge capacity is equal to the corresponding terminal capacity of the two generated sets in Γ_{p+1} for all $p = 1, 2, \ldots, n_v - 1$. The final EWC net by this process consists of $n_v - 1$ edges and n_v vertices. It is clear that this EWC net is a connected nonoriented linear graph. Hence this EWC net is a *tree* as far as the structure is concerned. This means that every cut-set consists of one edge and it is easily seen that this EWC net has a terminal capacity matrix equal to a given matrix.   QED

Example 12-2-2.   From a given matrix

$$
T = \begin{array}{c c} & \begin{array}{c c c c} 1 & 2 & 3 & 4 \end{array} \\ \begin{array}{c} 1 \\ 2 \\ 3 \\ 4 \end{array} & \left[\begin{array}{c c c c} d & 1 & 1 & 1 \\ 1 & d & 3 & 2 \\ 1 & 3 & d & 2 \\ 1 & 2 & 2 & d \end{array} \right] \end{array}
$$

and the sequence of classes $\Gamma_0, \Gamma_1, \Gamma_2, \Gamma_3$ given previously, we can form an EWC net as follows:

1. From $\Gamma_1 = \{(1), (2, 3, 4)\}$, we obtain an EWC net consisting of one edge whose edge capacity is the corresponding terminal capacity of (1) and (2, 3, 4) as in Fig. 12-2-1a.

2. From $\Gamma_2 = \{(1), (2, 3), (4)\}$, we cut vertex (2, 3, 4) into two vertices (2, 3) and (4) and insert one edge whose edge capacity is equal to the corresponding terminal capacity of (2, 3) and (4) as shown in Fig. 12-2-1b.

3. From $\Gamma_3 = \{(1), (2), (3), (4)\}$, we cut vertex (2, 3) into two vertices (2)

and (3), and insert one edge between these two vertices with the edge capacity being equal to the corresponding terminal capacity of the generated sets in Γ_3. The resultant EWC net shown in Fig. 12-2-1c is one whose terminal capacity is equal to the given matrix.

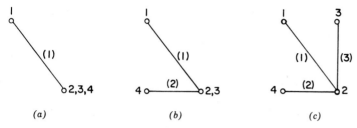

(a)   *(b)*   *(c)*

Fig. 12-2-1. A construction of an EWC net.

In the process of constructing an EWC net by using generator Ω_p in Γ_p and the generated sets Ω_a and Ω_b in Γ_{p+1}, we cut only vertex Ω_p into two vertices Ω_a and Ω_b. In addition to cutting a vertex, if we split edges connected to vertex Ω_p, we can obtain a different EWC net. Consider vertex Ω_p in Fig. 12-2-2 in which edges $e_1, e_2, \ldots, e_k$ are connected. Suppose we cut vertex Ω_p

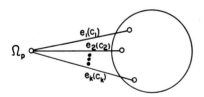

Fig. 12-2-2. An EWC net with vertex Ω_p.

into two Ω_a and Ω_b and split edge e_q into e_{q1} and e_{q2} with edge capacities of edges e_{q1} and e_{q2} being equal to $c_q/2$ where c_q is the edge capacity of edge e_q for all $q = 1, 2, \ldots, k$ as shown in Fig. 12-2-3. Next we insert edge e' between

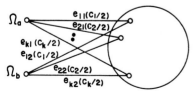

Fig. 12-2-3. An EWC net with vertices Ω_a and Ω_b.

these new vertices with the edge capacity c' of e' not equal to the corresponding terminal capacity t of generated sets Ω_a and Ω_b, but equal to

$$t - \sum_{q=1}^{k} \frac{c_q}{2}$$

Example 12-2-3. Suppose the given matrix is the one in Example 12-2-2. Then the modified process of obtaining a desired EWC net is as follows:

1. Between vertices (1) and (2, 3, 4), insert edge e_1 whose edge capacity is equal to the corresponding terminal capacity of generated sets (1) and (2, 3, 4) in Γ_1 as shown in Fig. 12-2-4a.

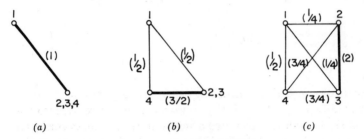

(a)   (b)   (c)

Fig. 12-2-4.   Process of constructing an EWC net.

2. Cut vertex (2, 3, 4) into two vertices (2, 3) and (4) because the generated sets in Γ_2 are (2, 3) and (4). Also split edge e_1 into e_{11} and e_{12} whose edge capacities are equal to one-half of the edge capacity of e_1. Then insert edge e_2 between vertices (2, 3) and (4) whose edge capacity is equal to the corresponding terminal capacity of (2, 3) and (4) minus one-half as shown in Fig. 12-2-4b.

3. Cut vertex (2, 3) into two vertices (2) and (3). Also split edges e_{12} and e_2 to e_{121}, e_{122}, e_{21}, and e_{22} with edge capacities of each of these edges being equal to one-half of the edge capacity of the corresponding edge. Then insert edge e_3 whose edge capacity is equal to the corresponding terminal capacity of generated sets (3) and (4) minus one-half of the sum of edge capacities of edges connected to vertex (2) as shown in Fig. 12-2-4c. The result is an EWC net whose terminal capacity matrix is equal to the given matrix.

We can see from the foregoing examples that there may be many EWC nets whose terminal capacity matrices are a given matrix. Hence it is reasonable to define indexes which show some differences among these EWC nets. One of such indexes is given by

$$I = \sum_{e_p \in G} \mathscr{W}_p c_p \qquad\qquad (12\text{-}2\text{-}9)$$

where c_p is the edge capacity of edge e_p and the summation is for all edge capacities in a communication net G. Here $\mathscr{W}_p$ is called a *cost* per unit of edge capacity of edge e_p. When $\mathscr{W}_p = 1$ for all p, the index I becomes

$$I_u = \sum_{e_p \in G} c_p \qquad (12\text{-}2\text{-}10)$$

Suppose we calculate index I_u for all EWC nets whose terminal capacity matrices are the same. Then there will be an EWC net whose index I_u is the smallest. This means that under the *unit cost*, this EWC net is the cheapest to construct. One way to find a cheapest EWC net is to obtain the lowest bound I_0 of indices I_u and to construct an EWC net whose index I_u is equal to I_0. To do this, we define I_0 as

$$I_0 = \min \{I_u \text{ of all EWC nets satisfying a given terminal capacity matrix}\} \qquad (12\text{-}2\text{-}11)$$

To obtain I_0, we take a vertex r in an EWC net G. Let t_{rj} be the largest among all terminal capacities t_{rq} for all vertices q ($\neq r$). Since we can assign a flow ψ_{rj} which is equal to t_{rj} from vertex r in G,

$$\psi_{rj} \leq V[S(r)] \qquad (12\text{-}2\text{-}12)$$

where $S(r)$ is an incidence set with respect to vertex r (Section 2-1). In other words, the sum of edge capacities of edges which are connected to vertex r must be at least equal to ψ_{rj} in order that an EWC net G can handle flow ψ_{rj}.

Let $r = 1, 2, \ldots, n$ be all vertices in an EWC net G. Then we can see that $\frac{1}{2} \sum_{r=1}^{n} V[S(r)]$ is the sum of all edge capacities in G, or

$$I_u = \frac{1}{2} \sum_{r=1}^{n} V[S(r)] \qquad (12\text{-}2\text{-}13)$$

Consider a terminal capacity matrix T. Let t_{ri_r} be the largest terminal capacity in the rth row of T. Then from Eq. 12-2-12,

$$t_{ri_r} \leq V[S(r)] \qquad (12\text{-}2\text{-}14)$$

for any EWC net satisfying T. Let I_a be defined as

$$I_a = \frac{1}{2} \sum_{r=1}^{n} t_{ri_r} \qquad (12\text{-}2\text{-}15)$$

By Eqs. 12-2-13 and 12-2-14, we can see that index I_u of any EWC net (satisfying a given terminal capacity matrix T) cannot be less than I_a, that is,

$$I_a \leq I_0 \leq I_u \qquad (12\text{-}2\text{-}16)$$

Suppose there exists an EWC net G_0 such that the largest terminal capacity from vertex r (i.e., the largest terminal capacity among t_{rq} for all vertices q in G_0) is equal to the value of incidence set $S(r)$ with respect to vertex r for all vertices r in G_0. Then Eq. 12-2-14 becomes

$$t_{ri_r} = V[S(r)] \qquad (12\text{-}2\text{-}17)$$

for all vertices r. Thus I_u of G_0 will be equal to I_0. A question is whether there exists such an EWC net. The answer is yes. In fact, an EWC net G (which is a complete graph) obtained by the modified process discussed previously satisfies Eq. 12-2-17 for all vertices. This can be shown by first noting that for any vertex r, there exists class Γ_q in sequence $\Gamma_0, \Gamma_1, \ldots, \Gamma_{n-1}$ such that one of two generated sets consists only of vertex r. Let Ω_q be the other generated set. Let t_q be the corresponding terminal capacity of generated set (r) and Ω_q.

Consider the steps for obtaining an EWC net by the modified process. That is, for any vertex v in a net, there exists a Step q at which vertex v is separated from a set of vertices. Hence we can assume that the EWC net in Fig. 12-2-5 is the resultant net by Step $q - 1$ where G_2 corresponds to generator Ω_{q-1}.

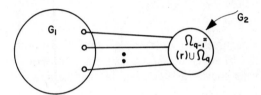

Fig. 12-2-5. EWC net at Step $q - 1$.

Note that $\Omega_{q-1} = (r) \cup \Omega_q$ so that vertex r is separated from Ω_{q-1} at Step q as shown in Fig. 12-2-6. To complete Step q, we give the capacity to an edge connected between vertex r and set Ω_q so that the value of incidence set $S(r)$ with respect to vertex r is equal to t_q. All remaining steps are for dividing either set Ω_q or vertices in G_1, which will not change the value of incidence set $S(r)$.

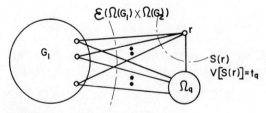

Fig. 12-2-6. EWC net at Step q.

Terminal capacity t_q must be the largest among all terminal capacities from vertex r, because it is impossible to assign a flow which is larger than t_q by Eq. 12-2-12. Thus we can see that

$$t_q = V[S(r)] \qquad (12\text{-}2\text{-}18)$$

It can be seen that the preceding result holds for any vertex in G. Thus I_u of G obtained by the modified process is equal to I_a, which also shows that $I_0 = I_a$ given by Eq. 12-2-16.

Example 12-2-4. A terminal capacity matrix T given in Example 12-2-2 is

$$T = \begin{matrix} & \begin{matrix} 1 & 2 & 3 & 4 \end{matrix} \\ \begin{matrix} t \\ 2 \\ 3 \\ 4 \end{matrix} & \begin{bmatrix} d & 1 & 1 & 1 \\ 1 & d & 3 & 2 \\ 1 & 3 & d & 2 \\ 1 & 2 & 2 & d \end{bmatrix} \end{matrix}$$

The largest terminal capacity in each row is shown in the following table:

Row	The largest terminal capacity
1	1
2	3
3	3
4	2

The sum of all these largest terminal capacities is 9. Thus

$$I_0 = \frac{1}{2} \sum_{r=1}^{4} t_{r_i r} = \frac{9}{2}$$

In Example 12-2-3, we obtained the EWC net in Fig. 12-2-4c by the modified process. This EWC net satisfies the above terminal capacity matrix. Index I_u of this net is

$$I_u = \frac{1}{2} + \frac{1}{4} + \frac{1}{4} + \frac{3}{4} + \frac{3}{4} + 2 = \frac{9}{2}$$

which is equal to I_0.

12-3   RELATIONSHIP AMONG TERMINAL CAPACITIES

There is a very interesting property of the corresponding cut-sets of terminal capacities. Let i, j, k, and m be vertices in a nonoriented EWC net. Let

$S_1 = \mathscr{E}(\Omega_1 \times \overline{\Omega}_1)$ be a corresponding cut-set of terminal capacity t_{ij} and $S_2 = \mathscr{E}(\Omega_2 \times \overline{\Omega}_2)$ be a corresponding cut-set of terminal capacity t_{km}. Also let

$$\Omega_1 = (\Omega_i, \Omega_m) \tag{12-3-1}$$

$$\overline{\Omega}_1 = (\Omega_k, \Omega_j) \tag{12-3-2}$$

$$\Omega_2 = (\Omega_i, \Omega_k) \tag{12-3-3}$$

and

$$\overline{\Omega}_2 = (\Omega_j, \Omega_m) \tag{12-3-4}$$

where $i \in \Omega_i$, $j \in \Omega_j$, $k \in \Omega_k$, and $m \in \Omega_m$ as shown in Fig. 12-3-1. Hence

$$S_1 = \mathscr{E}(\Omega_i \times \Omega_k) \cup \mathscr{E}(\Omega_i \times \Omega_j) \cup \mathscr{E}(\Omega_m \times \Omega_k) \cup \mathscr{E}(\Omega_m \times \Omega_j) \tag{12-3-5}$$

and

$$S_2 = \mathscr{E}(\Omega_i \times \Omega_m) \cup \mathscr{E}(\Omega_i \times \Omega_j) \cup \mathscr{E}(\Omega_k \times \Omega_m) \cup \mathscr{E}(\Omega_k \times \Omega_j) \tag{12-3-6}$$

When two cut-sets are in this situation, as shown in Fig. 12-3-1, we say that

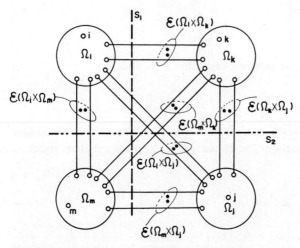

Fig. 12-3-1. An EWC net with Ω_i, Ω_j, Ω_k, and Ω_m.

two cut-sets are *crossing each other*. In other words, if the number of maximal connected subgraphs is increased more than two when we delete all edges in two cut-sets, then these two cut-sets are said to be crossing each other. For convenience, we define

$$V[\mathscr{E}(\Omega_p \times \Omega_q)] = \sum_{e_r \in \mathscr{E}(\Omega_p \times \Omega_q)} c_r \qquad (12\text{-}3\text{-}7)$$

that is, $V[\mathscr{E}(\Omega_p \times \Omega_q)]$ is the sum of edge capacities of all edges in set $\mathscr{E}(\Omega_p \times \Omega_q)$.

Since S_1 is a corresponding cut-set of terminal capacity t_{ij}, when we assign the maximum flow ψ_{ij} (equal to t_{ij}) from vertex i to vertex j, S_1 is a basic saturated cut-set. Thus whether or not S_2 is a basic saturated cut-set under the assignment of ψ_{ij}, we have

$$V[\mathscr{E}(\Omega_i \times \Omega_k)] + V[\mathscr{E}(\Omega_m \times \Omega_k)] \le V[\mathscr{E}(\Omega_k \times \Omega_j)] \qquad (12\text{-}3\text{-}8)$$

Hence

$$V[\mathscr{E}(\Omega_i \times \Omega_k)] \le V[\mathscr{E}(\Omega_k \times \Omega_j)] \qquad (12\text{-}3\text{-}9)$$

Similarly,

$$V[\mathscr{E}(\Omega_i \times \Omega_m)] \ge V[\mathscr{E}(\Omega_m \times \Omega_j)] \qquad (12\text{-}3\text{-}10)$$

On the other hand, S_2 is a basic saturated cut-set when we assign the maximum flow ψ_{km} (which is t_{km}) from vertex k to vertex m, we have

$$V[\mathscr{E}(\Omega_i \times \Omega_k)] \ge V[\mathscr{E}(\Omega_i \times \Omega_m)] \qquad (12\text{-}3\text{-}11)$$

and

$$V[\mathscr{E}(\Omega_k \times \Omega_j)] \le V[\mathscr{E}(\Omega_m \times \Omega_j)] \qquad (12\text{-}3\text{-}12)$$

Hence from Eqs. 12-3-9, 12-3-10, 12-3-11, and 12-3-12, we have

$$V[\mathscr{E}(\Omega_i \times \Omega_k)] \le V[\mathscr{E}(\Omega_k \times \Omega_j)] \le V[\mathscr{E}(\Omega_m \times \Omega_j)]$$
$$\le V[\mathscr{E}(\Omega_i \times \Omega_m)] \le V[\mathscr{E}(\Omega_i \times \Omega_k)] \qquad (12\text{-}3\text{-}13)$$

These must be equal and

$$V[\mathscr{E}(\Omega_i \times \Omega_j)] = V[\mathscr{E}(\Omega_m \times \Omega_k)] = 0 \qquad (12\text{-}3\text{-}14)$$

when S_1 and S_2 are crossing each other. Note that from Eqs. 12-3-5 and 12-3-6 that if Eq. 12-3-13 satisfies with equality and Eq. 12-3-14 satisfies, then $t_{ij} = t_{km}$. However, under this circumstance, we can see from Fig. 12-3-1 that instead of using S_2 as a corresponding cut-set of t_{km}, we can use

$$S_2' = \mathscr{E}(\Omega_i \times \Omega_k) \cup \mathscr{E}(\Omega_j \times \Omega_k) \qquad (12\text{-}3\text{-}15)$$

as a corresponding cut-set of t_{km}. Note that under this condition,

$$\mathscr{E}(\Omega_m \times \Omega_k) = \emptyset$$

Hence the following lemma can be obtained.

Lemma 12-3-1. Let S_1 be a corresponding cut-set of terminal capacity t_{ij}. Then for any terminal capacity t_{km}, there exists a corresponding cut-set of t_{km} such that this cut-set and S_1 do not cross each other.

With this result, we can discuss an interesting property of terminal capacities, the triangular relationship, which is given by the following theorem.

Theorem 12-3-1. Let i, j, and k be vertices in an EWC net. Then

$$t_{ij} \geq \min (t_{ik}, t_{kj}) \tag{12-3-16}$$

Proof. Let S_1 be a corresponding cut-set of t_{ij} where $S_1 = \mathscr{E}(\Omega_1 \times \overline{\Omega}_1)$, $i \in \Omega_1$, and $j \in \overline{\Omega}_1$. Assume $k \in \Omega_1$ as shown in Fig. 12-3-2a. Then by the

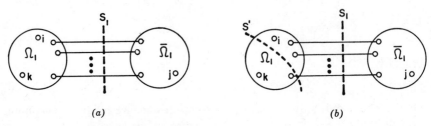

(a)   (b)

Fig. 12-3-2.   Location of vertices i, k, and j.

previous lemma, there exists a corresponding cut-set S' of t_{kj} such that S' and S_1 do not cross each other. Now S' must separate vertices i and k as shown in Fig. 12-3-2b except when $S_1 = S'$, in which case the proof becomes obvious. Thus in order that S' be a basic saturated cut-set when the maximum flow equal to t_{kj} is assigned,

$$V[S'] \leq V[S_1] \tag{12-3-17}$$

or

$$t_{kj} \leq t_{ij} \tag{12-3-18}$$

Thus Eq. 12-3-16 is correct.

When $k \in \overline{\Omega}_1$ as shown in Fig. 12-3-3a, then a corresponding cut-set S'' of t_{ik}, which does not cross with S_1 and which must separate vertices j and k because of the assumption that $S'' \neq S_1$, must be located as shown in Fig. 12-2-3b. Thus in order that S'' be a basic saturated cut-set when flow ψ_{ik} equal to t_{ik} is assigned,

$$V[S''] \leq V[S_1] \tag{12-3-19}$$

or

$$t_{ik} \leq t_{ij} \tag{12-3-20}$$

Hence Eq. 12-3-16 is correct.   QED

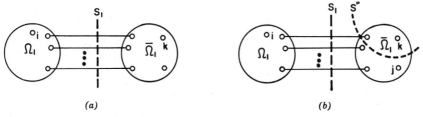

Fig. 12-3-3. Location of vertices i, k, and j.

We can easily see that Theorems 12-2-1 and 12-3-1 are equivalent.

12-4 CLASS W OF CUT-SETS

Consider a connected EWC net G consisting of n_v vertices. Let $S_1, S_2, \ldots,$ S_{n_v-1} be linearly independent cut-sets which separate i and j. Suppose assigning flow ψ_{ij} to G causes all of these cut-sets S_p for $p = 1, 2, \ldots, n_v - 1$ to be basic saturated cut-sets. Then it is clear that there exists a set of flows $\psi_{r_{ij}}$ which gives ψ_{ij} and every path $P_{r_{ij}}$ whose flow $\psi_{r_{ij}}$ is nonzero passes through each cut-set S_p exactly once.* In general, there is always a class of linearly independent cut-sets in an EWC net such that by assigning nonzero flows $\psi_{r_{ij}}$ only to those paths $P_{r_{ij}}$ which pass through every cut-set in the class exactly once, we can obtain a maximum flow from i to j. One such class which is not a subclass of another such class is a class W of cut-sets. We can obtain a class W of cut-sets by the following process. The class of cut-sets $S_{11}, S_{12}, \ldots, S_{1k_1}$ are those whose values $\sum\limits_{e_p \in S} c_p$ are the smallest among those of all cut-sets that separate i and j. Note that we are considering only cut-sets that separate vertices i and j. Suppose cut-sets $S_{11}, S_{12}, \ldots, S_{1m_1}$ are linearly independent among $S_{11}, S_{12}, \ldots, S_{1m_1}, \ldots, S_{1k_1}$. It is clear that when a maximum flow from i to j is given to G, these k_1 cut-sets become basic saturated cut-sets.

We modify EWC net G by multiplying all edge capacities by α^1 if the edge is in any of the cut-sets S_{1p} ($p = 1, 2, \ldots, m_1$). Let the resultant EWC net be $G(\alpha^1)$. Next we choose the smallest α^1 that satisfies $1 \leq \alpha^1 < \infty$ such that we obtain new cut-sets $S_{21}, S_{22}, \ldots, S_{2k_2}$ whose values are minimum of all of those that separate i and j in $G(\alpha^1)$. Let that value be α_0^1 and the EWC net with $\alpha^1 = \alpha_0^1$ be $G(\alpha_0^1)$. Note that values of cut-sets $S_{11}, S_{12}, \ldots, S_{1k_1}$ are also minimum in $G(\alpha_0^1)$. Let $S_{21}, S_{22}, \ldots, S_{2m_2}$ and cut-sets $S_{11}, S_{12}, \ldots, S_{1m_1}$ be linearly independent among $S_{11}, \ldots, S_{1k_1}, S_{21}, \ldots, S_{2m_2}, \ldots, S_{2k_2}$.

---

* "A path P passes through a cut-set S exactly once" means that $P \cap S$ has exactly one edge.

This procedure is repeated by multiplying edge capacities by α^2, to all edges in the cut-sets being considered, and then selecting the smallest value α_0^2 which produces new cut-sets whose values are the smallest among all those that separate i and j in the resulting net $G(\alpha_0^1\alpha_0^2)$.

The general pattern is now apparent. Let

$$\{S_0\} = \{S_{11}, \ldots, S_{1m_1}, S_{21}, \ldots, S_{2m_2}, \ldots, S_{p-11}, \ldots, S_{p-1m_{p-1}}\} \quad (12\text{-}4\text{-}1)$$

Suppose $S_{p1}, S_{p2}, \ldots, S_{pk_p}$ in $G(\alpha_0^1\alpha_0^2\cdots\alpha_0^{p-1})$ be the cut-sets whose values are the smallest among all cut-sets which separate i and j in the EWC net when $\alpha^{p-1} = \alpha_0^{p-1}$ and none of these are the smallest when $\alpha^{p-1} < \alpha_0^{p-1}$. Let $S_{p1}, \ldots, S_{pm_p} (m_p \leq k_p)$ and those in $\{S_0\}$ are linearly independent among all cut-sets which being considered. Then we include $S_{p1}, \ldots, S_{pm_p}$ in class $\{S_0\}$.

We continue the process just described until one of the following two cases occurs:

CASE 1. There is no cut-set in $G(\alpha_0^1\cdots\alpha_0^{p-1})$ which separates i and j and one cannot be obtained by a linear combination of the cut-sets in $\{S_0\}$. Then a class W of cut-sets is $\{S_0\}$ in Eq. 12-4-1.

CASE 2. There exists at least one cut-set in $G(\alpha_0^1\cdots\alpha_0^{p-1})$ which separates i and j but which cannot be obtained by a linear combination of the cut-sets in $\{S_0\}$ in Eq. 12-4-1. In order to be Case 2, we require that there be no $\alpha^{p-1}, 1 \leq \alpha^{p-1} < \infty$, which will produce at least one new linearly independent cut-set S_{pr} with respect to $\{S_0\}$ in Eq. 12-4-1 in $G(\alpha_0^1\cdots\alpha_0^{p-1})$ whose value is the smallest of those of cut-sets which separate i and j. Suppose the values of cut-sets $S_{p1}, S_{p2}, \ldots, S_{pk_p}$ in $G(\alpha_0^1\cdots\alpha_0^{p-1})$ cannot be minimum for any value of α_0^{p-1} $(1 \leq \alpha_0^{p-1} < \infty)$. Also suppose that $S_{p1}, S_{p2}, \ldots, S_{pm_p}$ and those in $\{S_0\}$ are linearly independent among $S_{11}, S_{12}, \ldots, S_{1k_1}, S_{21}, \ldots, S_{p-11}, \ldots, S_{p-1k_{p-1}}, S_{p1}, \ldots, S_{pk_p}$. Then a class W of cut-sets with respect to vertices i and j consists of all the cut-sets in $\{S_0\}$ and cut-sets $S_{p1}, S_{p2}, \ldots, S_{pm_p}$.

To illustrate the determination of the class W of cut-sets, consider the EWC net in Fig. 12-4-1a. We can see that $S_{11} = (a, b)$ is the only cut-set whose value is minimum among all those cut-sets that separate i and j. By multiplying edge capacities of edges a and b by $\alpha_0^1 = 2$, we produce a new cut-set $S_{21} = (c, e)$ whose value is minimum in $G(\alpha_0^1 = 2)$ as shown in Fig. 12-4-1b. Multiplying edge capacities of all edges in S_{11} and S_{21} in $G(\alpha_0^1)$ by α^2 and setting $\alpha_0^2 = 3$, we have $G(\alpha_0^1 = 2, \alpha_0^2 = 3)$ as shown in Fig. 12-4-1c. Now there exists a new cut-set $S_{31} = (a, d, e)$ whose value is now the minimum of all applicable cut-sets. Recall that we are dealing only with cut-sets which separate vertices i and j. Thus class W of cut-sets with respect to i and j consists of (a, b), (a, d, e), and (c, e).

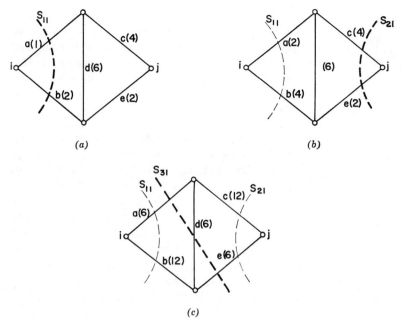

Fig. 12-4-1. EWC nets (*a*) G, (*b*) $G(\alpha_0^1)$, and (*c*) $G(\alpha_0^1 \alpha_0^2)$.

The class W of cut-sets and flow ψ_{ij} have a clear relationship which is given by the following theorem.

Theorem 12-4-1. For a given EWC net G, there exists a class of linearly independent cut-sets which separate i and j and a class $\{P_r\}$ of paths from i to j of which every path passes each cut-set in the class once, so that by assigning nonzero flows to paths in $\{P_r\}$ gives any permissible flow ψ_{ij} not exceeding the terminal capacity t_{ij}.

Proof. We consider the class W of cut-sets described previously and the modified net $G(\alpha_0^1 \cdots \alpha_0^{p-1})$ derived from the given EWC net G. There are two possible cases.

CASE 1.   For this case, the cut-sets in class W are those in S_0 in Eq. 12-4-1. In $G(\alpha_0^1 \cdots \alpha_0^{p-1})$, all these cut-sets in class W are basic saturated cut-sets when a maximum flow ψ_{ij} is assigned. Hence there exists a set of flows ψ'_{ij} that gives maximum flow ψ_{ij} from i to j in $G(\alpha_0^1 \cdots \alpha_0^{p-1})$ in such a way that each path with a nonzero flow passes each of these cut-sets once. Let $\psi'_{ij}(e)$ be the set of edge flows corresponding to the set of flows $\psi'_{r_{ij}}$. Dividing each of these path flows by

$$A = \alpha_0^1 \alpha_0^2 \cdots \alpha_0^{p-1} \tag{12-4-2}$$

is equivalent to dividing the flow assigned to every edge by A. Thus the flow in edge e becomes $\psi'_{ij}(e)/A$. On the other hand, the edge capacity of any edge in G is not smaller than $1/A$ times the edge capacity of the edge in the modified net $G(\alpha_0{}^1\cdots\alpha_0^{p-1})$. Thus the set of path flows $\psi_{r_{ij}}/A$ can be assigned to G. This clearly gives a maximum flow ψ_{ij} from i to j in G. Any other flow from i to j can be obtained by assigning K times every flow ψ'_{ij}/A where $0 \le K \le 1$, and so the theorem is true for this case.

CASE 2. When class W of cut-sets contains cut-sets $S_{p1}, S_{p2}, \ldots, S_{pm_p}$ which are not in $\{S_0\}$ in Eq. 12-4-1, there exists at least one edge in $G(\alpha_0{}^1\cdots\alpha_0^{p-2})$ whose edge capacity is the same as that in the original net G. Let these edges be $e_1, e_2, \ldots, e_q$. Also let $\bar{G}(\alpha_0{}^1\cdots\alpha_0^{p-2})$ be an EWC net obtained by removing these edges from $G(\alpha_0{}^1\cdots\alpha_0^{p-2})$. Then in $\bar{G}(\alpha_0{}^1\cdots\alpha_0^{p-2})$, the cut-sets corresponding to $S_{p1}, S_{p2}, \ldots, S_{pm_p}$ become basic saturated cut-sets under the same conditions that $S_{p-11}, S_{p-12}, \ldots, S_{p-1m_{p-1}}$ do because every edge in $S_{p1}, \ldots, S_{pm_p}$ which is not multiplied by α_0^{p-2} is removed to obtain $\bar{G}(\alpha_0{}^1\cdots\alpha_0^{p-2})$ from G. Thus as far as $\bar{G}(\alpha_0{}^1\cdots\alpha_0^{p-2})$ is concerned, Case 1 does apply. Since a class of flows assigned to paths in $\bar{G}(\alpha_0{}^1\cdots\alpha_0^{p-2})$ can be assigned to the same paths in G, the theorem is true for this case.

QED

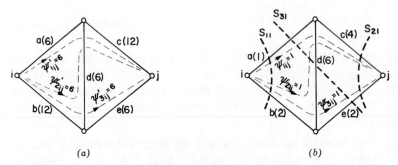

(a)   (b)

Fig. 12-4-2. Assigning flow ψ_{ij} by using $G(\alpha_0^1\alpha_0^2)$. (a) $G(\alpha_0^1\alpha_0^2)$; (b) G.

Consider EWC net $G(\alpha_0{}^1\alpha_0{}^2)$ in Fig. 12-4-1c. The only way we can assign the maximum flow ψ_{ij} to this communication net is by assigning flows $\psi'_{1_{ij}} = 6$, $\psi'_{2_{ij}} = 6$, and $\psi'_{3_{ij}} = 6$ as shown in Fig. 12-4-2a. Since $\alpha_0{}^1\alpha_0{}^2 = 6$, assigning flows $\psi_{1_{ij}} = 1$, $\psi_{2_{ij}} = 1$, and $\psi_{3_{ij}} = 1$ as shown in Fig. 12-4-2b gives the maximum flow ψ_{ij} and obviously cut-sets (a, b), (a, d, e), and (c, e) are those linearly independent cut-sets such that each of these paths to which nonzero flows are assigned passes once.

12-5   ORIENTED EWC NET

When each edge in an EWC net has an orientation indicating that the flow through the edge is in the specified direction only, it is called an *oriented EWC net*. Assigning flow $\psi_{r_{ij}}$ to a path $P_{r_{ij}}$ from vertex i to vertex j is the same as that of nonoriented EWC nets except that the path must be a *directed path* from i to j. Similarly, assigning flow ψ_{ij} to an oriented EWC net is the same as that of a nonoriented EWC net and the terminal capacity from i to j is the maximum flow that can be assigned from i to j. We should note that in the case of oriented EWC nets, terminal capacity t_{ij} may not be equal to terminal capacity t_{ji}.

In order to discuss the properties of oriented EWC nets, we define a semicut as follows.

Definition 12-5-1.   Let $S_{ij} = \mathscr{E}(\Omega_i \times \overline{\Omega}_i) \cup \mathscr{E}(\overline{\Omega}_i \times \Omega_i)$ be a cut-set in an oriented linear graph where $i \in \Omega_i$ and $j \in \overline{\Omega}_i$. Then *semicuts* s_{ij} and s_{ji} of S_{ij} are

$$s_{ij} = \mathscr{E}(\Omega_i \times \overline{\Omega}_i) \qquad (12\text{-}5\text{-}1)$$

and

$$s_{ji} = \mathscr{E}(\overline{\Omega}_i \times \Omega_i) \qquad (12\text{-}5\text{-}2)$$

Note that s_{ij} and s_{ji} are edge disjoint sets such that

$$s_{ij} \cup s_{ji} = S_{ij} \qquad (12\text{-}5\text{-}3)$$

As in the case of nonoriented EWC nets, an edge in an oriented EWC net is said to be saturated if the edge flow of the edge is equal to the edge capacity of the edge. However, the definition of a basic saturated cut-set with respect to an assigned flow ψ_{ij} in an oriented EWC net is somewhat different from that of a nonoriented EWC net.

Definition 12-5-2.   A cut-set S_{ij} in an oriented EWC net is a basic saturated cut-set under an assigned flow ψ_{ij} (from i to j) if every edge in semicut s_{ij} of S_{ij} is saturated and no edge in semicut s_{ji} of S_{ij} has nonzero edge flow.

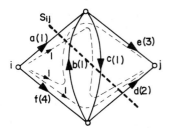

Fig. 12-5-1.   An oriented EWC net and flow ψ_{ij}.

For example, if we assign flow ψ_{ij} as shown in Fig. 12-5-1, then every edge in cut-set $S_{ij} = (a, b, c, d)$ is saturated. However, S_{ij} is not a basic saturated cut-set. This is because semicuts s_{ij} and s_{ji} of S_{ij} are

$$s_{ij} = (a, b, d) \qquad \text{and} \qquad s_{ji} = (c)$$

and there is nonzero edge flow $\psi_{ij}(c)$ assigned to edge c.

When we assign flow ψ_{ij} as shown in Fig. 12-5-2, cut-set $S_{ij} = (a, b, c, d)$ becomes a basic saturated cut-set. With these definitions, we can prove Theorem 12-5-1.

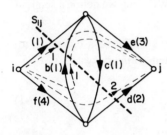

Fig. 12-5-2. Oriented EWC net in Fig. 12-5-1 with a different way of assigning flow ψ_{ij}.

Theorem 12-5-1. If and only if there exists a basic saturated cut-set which separate i and j, the flow ψ_{ij} assigned to an oriented EWC net is maximum.

The proof is exactly the same as that of Theorem 12-1-2 except we are using basic saturated cut-sets of oriented EWC nets. We can see readily that there exists at least one cut-set which is a basic saturated cut-set separating i and j for every possible way of assigning the maximum flow ψ_{ij}. This cut-set is called the *corresponding cut-set* of terminal capacity t_{ij}.

From Theorem 12-5-1, we obtain another theorem.

Theorem 12-5-2. Terminal capacity t_{ij} is equal to

$$t_{ij} = \min \{V[s_{p_{ij}}]; \quad \text{semicut } s_{p_{ij}} \text{ of } S_{p_{ij}} \in \{S(i;j)\} \qquad (12\text{-}5\text{-}4)$$

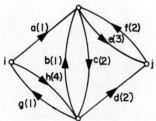

Fig. 12-5-3. An oriented EWC net G.

Example 12-5-1. From oriented EWC net G in Fig. 12-5-3, $\{S(i;j)\}$ consists of the following cut-sets which separate i and j:

$$S_{1_{ij}} = (a, g, h), \qquad S_{2_{ij}} = (a, b, c, d), \qquad S_{3_{ij}} = (d, e, f),$$

and

$$S_{4_{ij}} = (b, c, e, f, g, h)$$

Then semicuts $s_{p_{ij}}$ of these cut-sets are

$$s_{1_{ij}} = (a, h), \qquad s_{2_{ij}} = (a, b, d), \qquad s_{3_{ij}} = (d, e), \qquad \text{and} \quad s_{4_{ij}} = (c, e, h)$$

Hence

$$t_{ij} = \min \{V(s_{p_{ij}})\} = \min (5, 4, 5, 9) = 4$$

On the other hand,

$$s_{1_{ji}} = (g), \qquad s_{2_{ji}} = (c), \qquad s_{3_{ji}} = (f), \qquad \text{and} \quad s_{4_{ji}} = (b, f, g)$$

Hence

$$t_{ji} = \min \{V(s_{p_{ji}})\} = \min (1, 2, 2, 4) = 1$$

It is clear that a terminal capacity matrix of an oriented EWC net is not generally symmetric. However, using the same procedure as that employed to prove the necessity part of Theorem 12-2-1, we can obtain principal partitions of a terminal capacity matrix and all resultant main submatrices which contain more than one entry.

Theorem 12-5-3. A terminal capacity matrix of an oriented EWC net can be principally partitioned. Furthermore, all resultant main submatrices containing more than one entry can be principally partitioned.

Example 12-5-2. The terminal capacity matrix T of the oriented EWC net

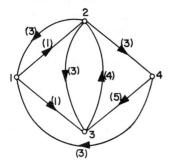

Fig. 12-5-4. Oriented EWC net.

in Fig. 12-5-4 and all resultant main submatrices can be principally partitioned as shown below:

$$
T = \begin{array}{c} \\ 1 \\ 2 \\ 3 \\ 4 \end{array}
\begin{array}{cccc}
1 & 2 & 3 & 4
\end{array}
\left[\begin{array}{c|c|c|c}
d & 2 & 2 & 2 \\ \hline
6 & d & 3 & 3 \\ \hline
4 & 4 & d & 3 \\ \hline
3 & 7 & 5 & d
\end{array}\right]
$$

In the case of nonoriented EWC nets, a terminal capacity matrix is symmetric. Hence the maximum number of different terminal capacities that a net can have is $n_v - 1$ where n_v is the number of vertices in the net. On the other hand, it is not difficult to see that the maximum number of different terminal capacities in an oriented EWC net will be larger than $n_v - 1$.

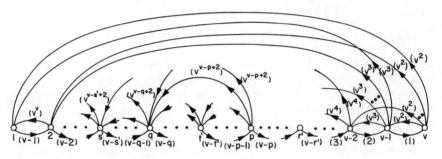

Fig. 12-5-5. An oriented EWC net.

Consider the oriented EWC net in Fig. 12-5-5 where p and q are vertices. Consider flow $\psi_{pq} = t_{pq}$ consists of $\psi_{1_{pq}}$, $\psi_{2_{pq}}$, $\psi_{3_{pq}}$, and $\psi_{4_{pq}}$ (i.e., $\psi_{pq} = \psi_{1_{pq}} + \psi_{2_{pq}} + \psi_{3_{pq}} + \psi_{4_{pq}}$) where $\psi_{1_{pq}}$ is the flow that passes through the edge connected from p to q, $\psi_{2_{pq}}$ is the flow that passes through the edges connected from vertex p to vertices $r' > p$, $\psi_{3_{pq}}$ is the flow that passes through the edges connected from vertex p to vertices $s' < q$, and $\psi_{4_{pq}}$ is the flow that passes through the edges connected from vertex p to vertices t' where $q < t' < p$. Since the edge capacity of the edge connected from vertex p to vertex q is v^{v-p+2}, $\psi_{1_{pq}} = v^{v-p+2}$. Flow $\psi_{2_{pq}}$ must pass through the edge connected from vertex p to vertex $p + 1$. The edge capacity of the edge connected from vertex $p + 1$ to vertex p is larger than that of the edge from vertex p to vertex $p + 1$, which tells us that $\psi_{2_{pq}} = v - p$. For $\psi_{3_{pq}}$, we consider all possible directed paths from p to s' ($< q$). Since all of these paths must contain the edge connected from vertex $q - 1$ to vertex q whose edge capacity is $v - q - 1$, and the edge capacity of the edge connected from p to $q - 1$ is larger than $v - q - 1$, $\psi_{3_{pq}} = v - q - 1$. Finally, every directed path from vertex p to vertex q passing through vertex t' but not vertices r' and s' contains the edge from vertex p to vertex t' whose edge capacity is v^{v-p+2} and the edge capacity of the edge from vertex t' to vertex q is larger than v^{v-p+2} for all t',

$$\psi_{4_{pq}} = (p - q - 1)v^{v-p+2}$$

Thus $t_{pq} = \psi_{pq}$ is equal to

$$t_{pq} = (p - q)v^{v-p+2} + 2v - p - q - 1 \qquad \text{for } p \neq v \quad \text{and} \quad q \neq 1$$
$$(12\text{-}5\text{-}5)$$

When $p = v$ or $q = 1$, $\psi_{2_{pq}}$ or $\psi_{3_{pq}}$ is zero, so that

$$t_{pq} = \begin{cases} (p - q)v^{v-p+2} + v - q - 1 & \text{for } p = v, \quad q \neq 1 \\ (p - q)v^{v-p+2} + v - p & \text{for } p \neq v, \quad q = 1 \quad (12\text{-}5\text{-}6) \\ (p - q)v^{v-p+2} & \text{for } p = v, \quad q = 1 \end{cases}$$

From these, we can see that only if $p_1 = p_2$ and $q_1 = q_2$,

$$t_{p_1 q_1} = t_{p_2 q_2} \qquad\qquad\qquad (12\text{-}5\text{-}7)$$

We can see also that all t_{pq} $(p > q)$ are larger than v. For t_{qp}, we have

$$t_{qp} = t_{p-1 p} = v - p - 1 \qquad\qquad (12\text{-}5\text{-}8)$$

which is less than v. Thus we have the terminal capacity matrix of the net in Fig. 12-5-5 in which upper diagonal entries consist of $v - 1$ different values and lower diagonal entries are all different. Now we can state that there is an oriented EWC net of n_v vertices (and connected) which has exactly $(n_v - 1)(n_v + 2)/2$ distinct terminal capacities.

Particular submatrices called S-submatrices of a terminal capacity matrix are closely related to sets in $\{S(i; j)\}$ and are very important for studying the realizability of a matrix to be a terminal capacity matrix of an oriented EWC net.

Definition 12-5-3. Let Ω_1 and $\bar{\Omega}_1$ be a pair of disjoint sets of vertices such that $\Omega_1 \cup \bar{\Omega}_1 = \Omega$ where Ω is the set of all vertices in the net. Then an S-submatrix is a submatrix of T obtained by deleting rows corresponding to all vertices in $\bar{\Omega}_1$ and columns corresponding to all vertices in Ω_1.

It is important to note that an S-submatrix does not contain d as an entry. For example, consider the terminal capacity matrix T given in Example

12-5-2. Hence $\Omega = (1, 2, 3, 4)$. Let $\Omega_1 = (1, 2)$. Then $\bar{\Omega}_1 = (3, 4)$ and an S-submatrix with these sets is

$$
\begin{array}{cc}
 & \begin{array}{cc} 3 & 4 \end{array} \\
\begin{array}{c} 1 \\ 2 \end{array} & \begin{bmatrix} 2 & 2 \\ 3 & 3 \end{bmatrix}
\end{array}
$$

If we take $\Omega_1 = (3, 4)$ and $\bar{\Omega}_1 = (1, 2)$, then we have

$$
\begin{array}{cc}
 & \begin{array}{cc} 1 & 2 \end{array} \\
\begin{array}{c} 3 \\ 4 \end{array} & \begin{bmatrix} 4 & 4 \\ 3 & 7 \end{bmatrix}
\end{array}
$$

On the other hand, if we choose $\Omega_1 = (2, 3, 4)$, then $\bar{\Omega}_1 = (1)$ and an S-submatrix is

$$
\begin{array}{cc}
 & \begin{array}{c} 1 \end{array} \\
\begin{array}{c} 2 \\ 3 \\ 4 \end{array} & \begin{bmatrix} 6 \\ 4 \\ 3 \end{bmatrix}
\end{array}
$$

Let S be a cut-set in an oriented EWC net. Then by using Ω_1 and $\bar{\Omega}_1$ of $S = \mathscr{E}(\Omega_1 \times \bar{\Omega}_1) \cup \mathscr{E}(\bar{\Omega}_1 \times \Omega_1)$, we can obtain an S-submatrix of T. Thus there is a definite relationship between terminal capacities and S-submatrices of T, as given by the following theorem.

Theorem 12-5-4. For every entry in T except d, there exists at least one S-submatrix such that the entry is the largest entry in the S-submatrix.

Proof. Let t_{pq} be a terminal capacity which is obviously an entry in terminal capacity matrix T. Let $S_{pq} = \mathscr{E}(\Omega_p \times \bar{\Omega}_p) \cup \mathscr{E}(\bar{\Omega}_p \times \Omega_p)$ be the corresponding cut-set of t_{pq} where $p \in \Omega_p$ and $q \in \bar{\Omega}_p$. It is clear that the S-submatrix obtained from T by deleting the rows corresponding to all vertices in $\bar{\Omega}_p$ and the columns corresponding to all vertices in Ω_p contains t_{pq}. Suppose there exists an entry t_{ru} in the S-submatrix which is larger than t_{pq}. In order for t_{ru} to be in the S-submatrix, r must be in Ω_p and u must be in $\bar{\Omega}_p$. Thus cut-set S_{pq} separates r and u. This means that the semicut s_{pq} is one of the semicuts in Eq. 12-5-4 which determines t_{ru}. Hence it is impossible to have $t_{ru} > t_{pq}$.   QED

Theorem 12-5-5. Let t_{pq} be a terminal capacity. If there exist k S-submatrices which contain t_{pq} and t_{pq} is the largest entry in each of these k S-submatrices for $k \geq 1$, then at least one of these k S-submatrices determines a corresponding cut-set of t_{pq}. [We say that an S-submatrix determines a cut-set if $\mathscr{E}(\Omega_1 \times \bar{\Omega}_1) \cup \mathscr{E}(\bar{\Omega}_1 \times \Omega_1)$ is a cut-set where Ω_1 is the set of all vertices corresponding to the rows and $\bar{\Omega}_1$ is the set of all vertices corresponding to the columns of the S-submatrix.]

The proof of this theorem is obvious because we know that there exists an S-submatrix obtained from Ω_p and $\bar{\Omega}_p$ of a corresponding cut-set

$$S_{pq} = \mathscr{E}(\Omega_p \times \bar{\Omega}_p) \cup \mathscr{E}(\bar{\Omega}_p \times \Omega_p)$$

of t_{pq}.

With the properties of S-submatrices given by Theorems 12-5-4 and 12-5-5, we can show that a necessary and sufficient condition for a terminal capacity matrix of the nonoriented EWC net given by Theorem 12-3-1 is not true for oriented EWC nets as follows. First, we note that Theorem 12-3-1, which is

$$t_{ij} \geq \min(t_{ik}, t_{kj}) \tag{12-5-9}$$

for any three vertices, i, j, and k, holds for oriented EWC nets. This can be shown in almost same way as the proof of the theorem. However, the relationship given by Eq. 12-5-9 is not sufficient for realizability of terminal capacity matrix of an oriented EWC net, which can be shown by proving that the matrix

$$R = \begin{array}{c} \\ a \\ b \\ c \\ d \end{array} \begin{array}{c} \begin{array}{cccc} a & b & c & d \end{array} \\ \begin{bmatrix} d & 3 & 1 & 1 \\ t_2 & d & 1 & 1 \\ t_3 & t_1 & d & 2 \\ t_4 & t_0 + \epsilon & t_0 & d \end{bmatrix} \end{array} \tag{12-5-10}$$

where $t_4 > t_3 > t_2 > t_1 > t_0 + \epsilon > t_0 > 3$ and $0 < \epsilon < 1$ is not realizable as a terminal capacity matrix. It can be seen that R satisfies Theorem 12-5-4 and the relationship given by Eq. 12-5-9. There is only one S-submatrix R_1 in which 1 is the largest entry:

$$R_1 = \begin{array}{c} \\ a \\ b \end{array} \begin{array}{c} \begin{array}{cc} c & d \end{array} \\ \begin{bmatrix} 1 & 1 \\ 1 & 1 \end{bmatrix} \end{array} \tag{12-5-11}$$

The vertex sets Ω_1 and $\bar{\Omega}_1$ corresponding to R_1 are

$$\Omega_1 = (a, b) \tag{12-5-12}$$

and

$$\bar{\Omega}_1 = (c, d) \tag{12-5-13}$$

which give the corresponding cut-sets $S_{ac} = \mathscr{E}(\Omega_1 \times \bar{\Omega}_1) \cup \mathscr{E}(\bar{\Omega}_1 \times \Omega_1)$ as shown in Fig. 12-5-6. Note that $s_{ac} = \mathscr{E}(\Omega_1 \times \bar{\Omega}_1)$ is the semi-cut whose value is equal to t_{ac}, t_{bc}, t_{ad}, and t_{bd}.

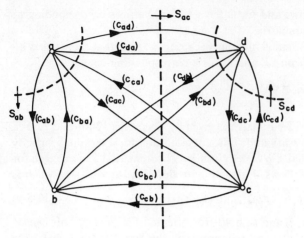

Fig. 12-5-6. An oriented EWC net for R.

There exists only one S-submatrix R_2 in which 2 is the largest entry and similarly there exists only one S-submatrix R_3 in which 3 is the largest as

$$R_2 = \begin{matrix} & d \\ a \\ b \\ c \end{matrix} \begin{bmatrix} 1 \\ 1 \\ 2 \end{bmatrix} \qquad (12\text{-}5\text{-}14)$$

and

$$R_3 = a \begin{matrix} b & c & d \\ [3 & 1 & 1] \end{matrix} \qquad (12\text{-}5\text{-}15)$$

These S-submatrices give cut-sets S_{cd} and S_{ab} as shown in Fig. 12-5-6 where $V[s_{cd}] = 2$ and $V[s_{ab}] = 3$. Thus

$$t_{ac} = 1 = c_{ad} + c_{ac} + c_{bc} + c_{bd} \qquad (12\text{-}5\text{-}16)$$

and

$$t_{ab} = 3 = c_{ab} + c_{ad} + c_{ac} \qquad (12\text{-}5\text{-}17)$$

Since the following S-submatrix is the only S-submatrix in R in which t_0 is the largest, the corresponding cut-set of t_0 must be $S_{dc} = \mathscr{E}(\Omega_d \times \overline{\Omega}_d)$ where $\Omega_d = (a, b, d)$ and $\overline{\Omega}_d = (c)$ as shown in Fig. 12-5-7:

$$R_4 = \begin{matrix} & c \\ a \\ b \\ d \end{matrix} \begin{bmatrix} 1 \\ 1 \\ t_0 \end{bmatrix} \qquad (12\text{-}5\text{-}18)$$

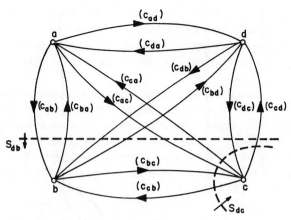

Fig. 12-5-7. EWC net with S_{dc} and S_{db}.

Thus

$$t_0 = c_{ac} + c_{bc} + c_{dc} \qquad (12\text{-}5\text{-}19)$$

Since there is only one S-submatrix R_5 in which $t_0 + \epsilon$ is the largest where

$$R_5 = \begin{array}{c} \\ a \\ d \end{array}\begin{array}{c} b \quad\; c \\ \left[\begin{array}{cc} 3 & 1 \\ t_0 + \epsilon & t_0 \end{array} \right] \end{array} \qquad (12\text{-}5\text{-}20)$$

S_{db} in Fig. 12-5-7 is the corresponding cut-set of $t_0 + \epsilon$. Hence

$$t_0 + \epsilon = c_{ab} + c_{ac} + c_{db} + c_{dc} \qquad (12\text{-}5\text{-}21)$$

From Eqs. 12-5-19 and 12-5-21, we have

$$\epsilon = c_{ab} + c_{db} - c_{bc} \qquad (12\text{-}5\text{-}22)$$

Since every edge capacity is nonnegative, Eq. 12-5-16 gives

$$c_{ad} + a_{ac} \le 1 \qquad (12\text{-}5\text{-}23)$$

Thus with Eq. 12-5-17, Eq. 12-5-23 gives

$$3 \ge c_{ab} \ge 2 \qquad (12\text{-}5\text{-}24)$$

Equation 12-5-16 gives

$$c_{bc} \le 1 \qquad (12\text{-}5\text{-}25)$$

Hence from Eqs. 12-5-24, 12-5-25, and 12-5-22, we have

$$\epsilon \ge c_{db} + 1 \qquad (12\text{-}5\text{-}26)$$

Since c_{db} is nonnegative, $\epsilon \geq 1$, which contradicts the assumption that $0 < \epsilon < 1$. Hence R is not realizable as a terminal capacity matrix of an oriented EWC net. Thus the relationship among terminal capacities given by Eq. 12-5-9 is not sufficient for the realizability of terminal capacity matrix of an oriented EWC net.

To obtain a realizability condition for a terminal capacity matrix of an oriented EWC net, we introduce a particular way of modifying EWC nets called the *shifting algorithm*. To explain the algorithm, it is convenient if we define the symbol $\Omega(g)$ to indicate the set of all vertices which are the endpoints of edges in a linear graph g. For example, consider the oriented EWC net in Fig. 12-5-7. Let g be a subgraph consisting of edges whose edge capacities are c_{cb}, c_{bd}, and c_{dc}. Then $\Omega(g) = (b, c, d)$.

In order to see how we modify an EWC net by the shift algorithm, let $C = (e_1, e_2, \ldots, e_p)$ be a directed circuit in an oriented EWC net and let $C' = (e'_1, e'_2, \ldots, e'_p)$ be another directed circuit such that edges e_r and e'_r are in parallel and the orientations of e'_r and e_r are opposite each other for $r = 1, 2, \ldots, p$ as shown in Fig. 12-5-8. We can see that the directed circuits C and C' have

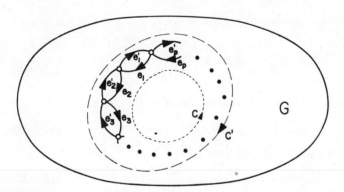

Fig. 12-5-8. Oriented EWC net G.

the properties that (1) $\Omega(C) = \Omega(C')$ and (2) the orientations of C and C' are opposite each other. Now by subtracting capacity δ from the edge capacity of every edge in directed circuit C and adding δ to the edge capacity of every edge in directed circuit C', we have the new EWC net G' shown in Fig. 12-5-9 where c_r and c'_r are the edge capacities of edges e_r and e'_r, respectively, for $r = 1, 2, \ldots, p$. This process of changing edge capacities is called the *shifting algorithm*. Using this, next show that G and G' have the same terminal capacities between any two vertices.

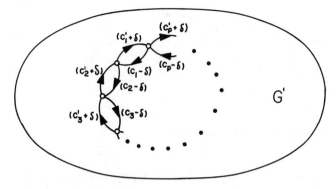

Fig. 12-5-9. EWC net G'.

By Eq. 12-5-4, if we can show that the value of every semicut in G is equal to that of the corresponding semicut in G', we can guarantee that the terminal capacities between any two vertices in G and in G' are the same. Take any semicut s. Suppose none of edges in C is in s. Then the values of semicuts $V(s)$ in G and in G' are obviously the same. Let S be the cut-set such that s is one of two semicuts of S. Since C is a circuit, there is an even number $2k$ of edges in C which are in S. Moreover, because C is a directed circuit, one half of these edges is in semicut s, that is, k edges in C are in s. Similarly, $2k$ edges of C' are in S and k edges of these are in s. Thus the number of edges in s whose capacities are increased by δ and the number whose capacities are decreased by δ are the same. This means that $V(s)$ of semicut s in G and in G' are the same.

Theorem 12-5-6. The shifting algorithm keeps the terminal capacities invariant.

It can be seen that we can change edge capacities of an EWC net to negative numbers by the use of the shifting algorithm. By the definition, however, any net having negative numbers as edge capacities cannot be called an EWC net. Hence we introduce the pseudo-EWC net.

Definition 12-14-1. A linear graph G is a *pseudo-EWC net* if (1) the weight of each edge called an edge capacity in G is a real number, (2) the absence of an edge is equivalent to having zero as the weight of the edge, and (3) the terminal capacity t_{ij} from vertex i to vertex j is nonnegative for every i and j in G where t_{ij} is defined by

$$t_{ij} = \min \{V(s_{p_{ij}}); \quad \text{semicut } s_{p_{ij}} \text{ of all cut-sets } S_{p_{ij}} \text{ which separate } i \text{ and } j\}$$

$$(12\text{-}5\text{-}27)$$

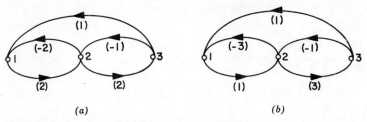

Fig. 12-5-10. (*a*) Pseudo-EWC net; (*b*) nonpseudo-EWC net.

For example, the linear graph in Fig. 12-5-10*a* is a pseudo-EWC net but that in Fig. 12-5-10*b* is not because there exists a semicut s_{21} whose value is negative.

Before studying the properties of pseudo-EWC nets that lead to a realizability condition for terminal capacity matrices, we will see whether any pseudo-EWC net can become a communication net by the successive application of the shifting algorithm. In other words, is it possible to make all negative edge capacities in a pseudo-EWC net nonnegative by the use of the shifting algorithm? The answer is no, because there exists a pseudo-EWC net as shown in Fig. 12-5-11 which cannot become an EWC net by the use of the

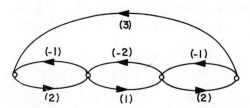

Fig. 12-5-11. A pseudo-EWC net.

shifting algorithm. However, any pseudo-EWC net satisfying a simple restriction can become a communication net by satisfying a simple restriction on the shifting algorithm which is given by the following theorem.

Theorem 12-5-7. If and only if a pseudo-EWC net G satisfies

$$c_{pq} + c_{qp} \geq 0 \tag{12-5-28}$$

for all vertices p and q, then G becomes an EWC net by the shifting algorithm.

Proof. The necessity is obvious from the definition of the shifting algorithm, that is, from an EWC net, any pseudo-EWC net obtained by the shifting algorithm will satisfy the above condition.

To prove the sufficiency, we define the *pseudo-index I* of a pseudo-EWC net as

$$I = -\left(\sum_{(r)} c_r\right) \tag{12-5-29}$$

where c_r is a negative edge capacity and $\sum_{(r)}$ means the "sum of all possible." In other words, $-I$ is the sum of all negative edge capacities. With this definition, we can prove the theorem by showing that the shifting algorithm produces a pseudo-EWC net G' from any pseudo-EWC net G having nonzero pseudo-index I such that the pseudo-index I' of G' is less than I.

Let e_{pq} be an edge whose edge capacity is negative. Then there exists a directed path P from vertex p to vertex q such that every edge in P has positive edge capacity. If this were not the case, there would be a semicut s_{pq} whose value is negative. Let $P = (e_{p1}, e_{12}, \ldots, e_{kq})$ as shown in Fig. 12-5-12.

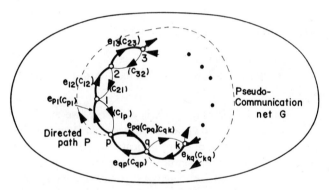

Fig. 12-5-12.   A pseudo-EWC net and path P.

Also let $c_{p1}, c_{12}, \ldots, c_{kq}$ be the edge capacities of edges $e_{p1}, e_{12}, \ldots, e_{kq}$, respectively. Note that by assumption c_{qp} is positive. Since $c_{p1}, c_{12}, \ldots, c_{kq}$ and c_{qp} are all positive, we have a positive δ such that

$$\delta = \min(c_{p1}, c_{12}, \ldots, c_{kq}, c_{qp}) > 0 \tag{12-5-30}$$

Now we can reduce all edge capacities of edges $e_{qp}, e_{p1}, e_{12}, \ldots, e_{kq}$ (which form a directed circuit) by δ and increase all edge capacities of edges $e_{pq}, e_{qk}, \ldots, e_{21}, e_{1p}$ (which form the corresponding directed circuit) by δ as shown in Fig. 12-5-13. This process is the shifting algorithm and the pseudo-index of the resultant net is clearly less than that of the original net. Note that we are considering pseudo-EWC nets having a finite number of edges.

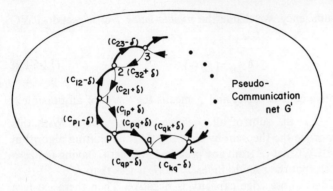

Fig. 12-5-13. The resultant net by the shifting algorithm.

Hence the number of directed circuits is also finite. Thus we can reduce the pseudo-index to zero by a finite number of successive application of the shifting algorithm.   QED

With Theorem 12-5-7, we can state Theorem 12-5-8.

Theorem 12-5-8. If there exists a pseudo-EWC net satisfying Eq. 12-5-28 whose terminal capacity matrix is T, T is realizable as a terminal capacity matrix of an oriented EWC net.

To show that the converse of Theorem 12-5-8 is also true, we employ a particular kind of pseudo-EWC net known as a fundamental structure. To define such a structure, we use the symbol e_{pq} to indicate the edge connected from vertex p to vertex q. (Note that the orientation of the edge e_{pq} is from p to q). Similarly, the symbol c_{pq} is used for the edge capacity of edge e_{pq}. We say that a pseudo-EWC net G consisting of vertices $1, 2, \ldots, n$ is of a fundamental structure (or a fundamental pseudo-EWC net) if G satisfies the following conditions:

1. For $p < q$, there are no edges connected from p to q for $q > p + 1$.
2. All edge capacities are positive except those of edges $e_{p+1\ p}$ for $p = 1, 2, \ldots, n - 1$.
3. $c_{p+1\ p} + c_{p\ p+1} \geq 0$ for $p = 1, 2, \ldots, n - 1$.

For example, the net in Fig. 12-5-14a is of a fundamental structure but the nets in Fig. 12-5-14b and c do not have a fundamental structure. This is because for the net in b, $c_{21} + c_{12} < 0$ which violates condition 3, and for the net in c, there is e_{13} which violates condition 1 and e_{31} whose edge capacity is negative, which violates condition 2. Note that an oriented EWC net is also a pseudo-EWC net.

From the preceding definition, a fundamental pseudo-EWC net consisting

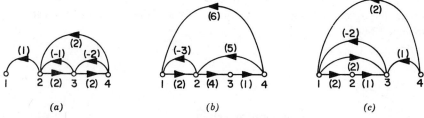

Fig. 12-5-14. Examples of pseudo-communication nets.

of n vertices must have the structure shown in Fig. 12-5-15. Note that this structure is exactly the same as that of the oriented EWC net in Fig. 12-5-5.

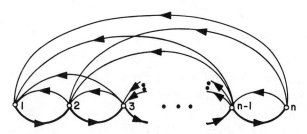

Fig. 12-5-15. Fundamental structure.

To show that any oriented EWC net can be changed to a fundamental pseudo-EWC net by the use of the shifting algorithm, first recall that we can apply the principal partitions to a terminal capacity matrix T and all resultant main submatrices. Suppose a terminal capacity matrix T of an oriented EWC net G has been principally partitioned as

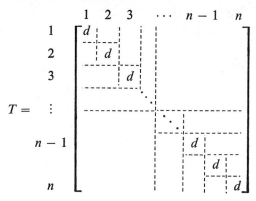

We draw EWC net G so that all vertices are placed on a straight line in the order that agrees with the order of the columns of the above matrix as shown in Fig. 12-5-16.

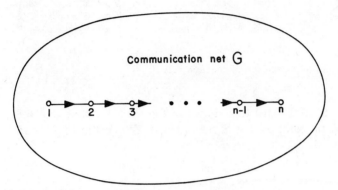

Fig. 12-5-16. A particular way of drawing G.

Example 12-5-3. Consider oriented EWC net G in Fig. 12-5-17. The terminal capacity matrix of G can be principally partitioned as

$$
T = \begin{array}{c} \\ i \\ j \\ k \\ p \end{array}
\begin{array}{cc}
\begin{array}{cccc} i & j & k & p \end{array} \\
\left[\begin{array}{c|c|c|c}
d & 5 & 4 & 2 \\
\hline
7 & d & 4 & 2 \\
\hline
6 & 6 & d & 2 \\
\hline
5 & 5 & 5 & d
\end{array} \right]
\end{array}
$$

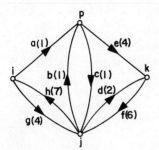

Fig. 12-5-17. An oriented EWC net.

By redrawing the net according to the order of columns of the terminal capacity matrix, the EWC net becomes as shown in Fig. 12-5-18.

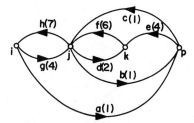

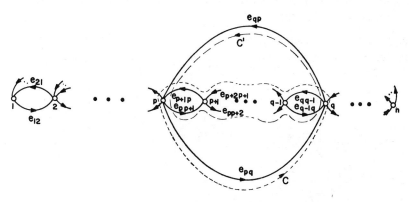

Fig. 12-5-18. Drawing of G according to the order of vertices given by the order of columns of T.

In EWC net G in Fig. 12-5-16, we take any edge e_{pq} where $q > p + 1$, whose edge capacity is larger than zero. For the shifting algorithm, we choose directed circuit C to be $(e_{pq}, e_{qq-1}, e_{q-1q-2}, \ldots, e_{p+1p})$ as shown in Fig. 12-5-19.

Fig. 12-5-19. Directed circuit C and the corresponding directed circuit C'.

The corresponding directed circuit C' is also shown in the figure. Now we apply the shifting algorithm so that the edge capacity c_{pq} of edge e_{pq} becomes zero by subtracting c_{pq} from the edge capacity of every edge in C as shown in Fig. 12-5-19. Note that we add c_{pq} to the edge capacity of every edge in C'. By this process, we can delete from the net any edge e_{pq} where $q > p + 1$. By repeating this process, we can delete all edges e_{pq} for $q > p + 1$ to obtain a pseudo-EWC net. Note that by this process, we may produce negative edge capacities in the net. However, the edges whose capacities are negative will only be those connected from vertex p to vertex $p - 1$ for $p = 2, 3, \ldots, n$. Furthermore, the sum $S = c_{pp-1} + c_{p-1p}$ of edge capacities of edges e_{pp-1} and e_{p-1p} is nonnegative in an oriented EWC net, and by the above process, whenever the edge capacity of edge e_{pp-1} is reduced by δ (> 0), the edge

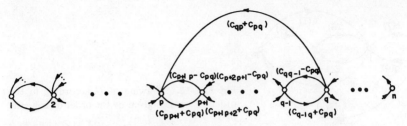

Fig. 12-5-20.   Result of the shifting algorithm.

capacity of edge e_{p-1p} is increased by δ. Thus the sum S will not be changed, and the resultant pseudo-EWC net is of a fundamental structure.

Theorem 12-5-9.   Any oriented EWC net G can be changed to a fundamental pseudo-EWC net G'. Furthermore, the sum of edge capacities of edges e_{pq} and e_{qp} in G and G' are the same.

Example 12-5-4.   Consider the EWC net in Fig. 12-5-18 (Example 12-5-3). Taking directed circuits C and C' as shown in Fig. 12-5-21, we can delete

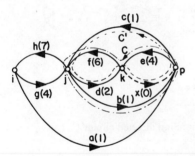

Fig. 12-5-21.   Directed circuits C and C'.

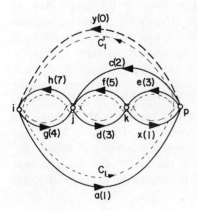

Fig. 12-5-22.   Directed circuits C_1 and C_1'.

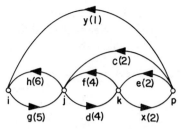

Fig. 12-5-23. A fundamental pseudo-EWC net.

edge b by the shifting algorithm as shown in Fig. 12-5-22. (Note the change of edge capacities in the figure.) Now by using directed circuits C_1 and C_1' shown in the figure, we can delete edge a by the shifting algorithm to obtain a fundamental pseudo-EWC net as shown in Fig. 12-5-23.

We know that any oriented EWC net can be changed to a fundamental pseudo-EWC net. Hence Theorem 12-5-8 can be modified.

Theorem 12-5-10. If and only if there exists a pseudo-EWC net satisfying Eq. 12-5-28 whose terminal capacity matrix is a given matrix T, then T is realizable as a terminal capacity matrix of an oriented EWC net.

12-6 LOSSY EWC NET

There are several physical systems in which not all flow reaches its destination. One reason may be the existence of defects in the systems such as leaks in a pipeline. To indicate such systems, we define a lossy oriented EWC net in which an edge efficiency α_p with $0 \le \alpha_p \le 1$ is given to every edge. To understand an edge efficiency α_p, suppose edge e_p is connected between vertices k and m whose orientation is from k to m. Then flow $\alpha_p \psi$ will be reached at vertex m through edge e_p when flow ψ is entered to edge e_p from vertex k. Hence flow $(1 - \alpha_p)\psi$ can be considered the flow which is lossed by edge e_p.

In order to discuss the properties of lossy EWC nets, we define the following: $V[s(c)]$ and $V[s(\alpha c)]$ indicate the values of semicut s defined by

$$V[s(c)] = \sum_{e_p \in s} c_p \qquad (12\text{-}6\text{-}1)$$

and

$$V[s(\alpha c)] = \sum_{e_p \in s} \alpha_p c_p \qquad (12\text{-}6\text{-}2)$$

where α_p and c_p are the edge efficiency and the edge capacity of edge e_p.

Since loss of flow occurs in edges, we assign ψ_{ij} to a lossy EWC net G as follows. Let $P_{r_{ij}}$ be a directed path from vertex i to vertex j in G whose edge train is $(e_1 e_2 \cdots e_n)$. Then assigning a flow $\psi_{r_{ij}}$ to path $P_{r_{ij}}$ means that adding flow $\psi'(e_p)$ to edge flow of edge e_p for $p = 1, 2, \ldots, n$ which are

$$\psi'(e_1) = \psi_{r_{ij}}, \qquad \psi'(e_2) = \alpha_1 \psi'(e_1), \qquad \ldots, \qquad \psi'(e_n) = \alpha_{n-1} \psi'(e_{n-1})$$
$$(12\text{-}6\text{-}3)$$

under the condition that $\psi'(e_p)$ must satisfy

$$\psi'(e_p) \leq c_p - \psi_o(e_p) \qquad (12\text{-}6\text{-}4)$$

for $p = 1, 2, \ldots, n$ where $\psi_o(e_p)$ is the flow that had been assigned to edge e_p previously. Thus the total flow assigned to edge e_p becomes $\psi'(e_p) + \psi_o(e_p)$.

Let $P_{1_{ij}}, P_{2_{ij}}, \ldots, P_{m_{ij}}$ be all the possible directed paths from vertex i to vertex j in G. Then we can assign flows $\psi_{1_{ij}}, \psi_{2_{ij}}, \ldots, \psi_{m_{ij}}$ to $P_{1_{ij}}, P_{2_{ij}}, \ldots, P_{m_{ij}}$, respectively, in order to assign this total flow $\psi_{ij} = \sum_{p=1}^{m} \psi_{p_{ij}}$. It is clear that any flow ψ_{ij} assigned to G can be broken into flows $\psi_{1_{ij}}, \psi_{2_{ij}}, \ldots, \psi_{m_{ij}}$, which are assigned to directed paths $P_{1_{ij}}, P_{2_{ij}}, \ldots, P_{m_{ij}}$, respectively.

Since a lossy oriented EWC net becomes a (lossless) oriented EWC net by letting $\alpha_p = 1$ for all edge efficiencies, any property of lossy EWC net which is dependent on α is closely related to that of a lossless net. We will show that there exists a cut-set which determines the maximum flow in a lossy EWC net which was the case for lossless. The definitions of saturated edges, saturated cut-sets, and basic saturated cut-sets in lossy EWC nets are the same as those of lossless EWC nets. That is, if the total edge flow $\psi(e)$ entering into edge e is equal to the edge capacity of the edge, then the edge is said to be saturated. If every edge in semicut s_{ij} of cut-set S_{ij} which separates i and j is saturated, then S_{ij} is said to be saturated. Furthermore, if semicut s_{ji} of saturated cut-set S_{ij} consists of edges which have zero edge flows, then S_{ij} is said to be a basic saturated cut-set. With these definitions, we can now discuss lossy EWC nets.

Theorem 12-6-1. Suppose flow ψ_{ij} from vertex i to vertex j has been assigned to a lossy EWC net G. If there exists no saturated cut-set which separates i and j, then there exists a directed path from i to j such that an additional flow $\psi > 0$ can be assigned to the path.

The proof of this theorem is exactly the same as the proof of Lemma 12-1-1. It is clear from the proof that if there exists a saturated cut-set which separates i and j when a flow ψ_{ij} is assigned, then we cannot assign any additional non-zero flow from i to j.

Theorem 12-6-2. There exists a flow ψ_{ij} from i to j such that assigning it to a lossy oriented EWC net produces at least one basic saturated cut-set.

Proof. Suppose by assigning a flow ψ_{ij} to a lossy oriented EWC net G, we have saturated cut-sets S_{ij} but none of these are basic saturated cut-sets. Then using the same procedure as one in the proof of Theorem 12-1-1, we can modify the way of assigning flow ψ_{ij} so that we can give additional flow from i to j. Hence we can show that we can increase flow from i to j until there is a basic saturated cut-set which separates vertices i and j.   QED

In the case of lossless EWC nets, whenever we have a basic saturated cut-set S_{ij}, flow ψ_{ij} is maximum. On the other hand, even if there is a basic saturated cut-set produced by assigning flow ψ_{ij}, the flow may not be maximum if the EWC net is lossy. This can be seen easily by the following example. Consider the lossy EWC net shown in Fig. 12-6-1 in which the first number in the

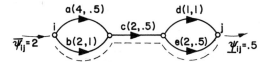

Fig. 12-6-1. A lossy EWC net.

parenthesis of each edge is the edge capacity and the second number is the edge efficiency of the edge. By assigning flow as shown in the figure, cut-set $S_{ij} = (c)$ becomes a basic saturated cut-set. However, neither the flow entering to vertex i nor the flow leaving vertex j is maximum because we can assign larger flow from i to j as shown in Fig. 12-6-2.

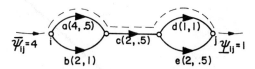

Fig. 12-6-2. Flow ψ_{ij} to a lossy EWC net.

This example indicates that we may have many flows ψ_{ij}; by assigning them we will have basic saturated cut-sets S_{ij}. Moreover, the flow ψ_{ij} entering vertex i is different from the flow leaving vertex j when ψ_{ij} is assigned. Hence we define terminal capacities of lossy EWC nets, which are somewhat different from those of lossless EWC nets, as follows.

Definition 12-6-1. A maximum flow ψ_{ij} from i to j which can be assigned to a net G in order to receive the maximum flow at j is called a *source terminal capacity* symbolized by $\bar{t}_{ij}$. The maximum flow which will be received at vertex j when the maximum flow ψ_{ij} is assigned is called a *sink terminal capacity* symbolized by $\underline{t}_{ij}$.

It must be noted that there exists a cut-set S_{ij} such that assigning any flow ψ_{ij}, which produces at least one basic saturated cut-set, will make S_{ij} a basic saturated cut-set. This cut-set is called a *corresponding cut-set* of such a flow. Since S_{ij} is clearly a basic saturated cut-set when the maximum flow ψ_{ij} is assigned, we say that S_{ij} is a corresponding cut-set of source and sink terminal capacities $\bar{t}_{ij}$ and $\underline{t}_{ij}$. With these definitions, we have another theorem.

Theorem 12-6-3. For any vertices i, j, and k in a lossy oriented EWC net, either

$$\underline{t}_{ij} \geq \underline{t}_{kj} \tag{12-6-5}$$

or

$$\bar{t}_{ij} \geq \bar{t}_{ik} \tag{12-6-6}$$

Proof. Let $S_{ij} = \mathscr{E}(\Omega_1 \times \overline{\Omega}_1) \cup \mathscr{E}(\overline{\Omega}_1 \times \Omega_1)$ be a corresponding cut-set of $\bar{t}_{ij}$ and $\underline{t}_{ij}$ where $i \in \Omega_1$ and $j \in \overline{\Omega}_1$. Let $S_{ik} = \mathscr{E}(\Omega_2 \times \overline{\Omega}_2) \cup \mathscr{E}(\overline{\Omega}_2 \times \Omega_2)$ be a corresponding cut-set of $\bar{t}_{ik}$ where $i \in \Omega_2$ and $k \in \overline{\Omega}_2$. Also let $S_{kj} = \mathscr{E}(\Omega_3 \times \overline{\Omega}_3) \cup \mathscr{E}(\overline{\Omega}_3 \times \Omega_3)$ be a corresponding cut-set of $\underline{t}_{kj}$ where $k \in \Omega_3$ and $j \in \overline{\Omega}_3$.

CASE 1.   Vertex k is in Ω_1. There are two situations, depending on whether $j \in \overline{\Omega}_2$ or $j \in \Omega_2$.
   a. When $j \in \overline{\Omega}_2$, either $i \in \Omega_3$ or $i \in \overline{\Omega}_3$ as shown in Fig. 12-6-3. For $i \in \Omega_3$, S_{kj} and S_{ij} must be basic saturated cut-sets for maximum flow ψ_{ij}. Thus

$$\underline{t}_{ij} = \underline{t}_{kj} \tag{12-6-7}$$

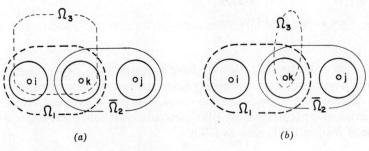

(a)   (b)

Fig. 12-6-3. Location of saturated cut-sets.

For $i \in \overline{\Omega}_3$, S_{ij} may or may not be a basic saturated cut-set for maximum flow ψ_{kj}. Thus

$$t_{ij} \geq t_{kj} \qquad (12\text{-}6\text{-}8)$$

b. When $j \in \Omega_2$, the situation depends on whether $i \in \Omega_3$ or $i \in \overline{\Omega}_3$ as shown in Fig. 12-6-4. For $i \in \Omega_3$, S_{ij} must be a basic saturated cut-set under maximum flow ψ_{kj}, which gives

$$t_{ij} = t_{kj} \qquad (12\text{-}6\text{-}9)$$

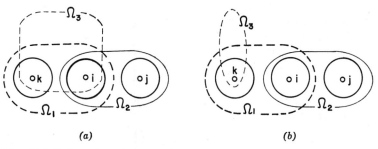

| (a) | (b) |

Fig. 12-6-4. Location of saturated cut-sets.

For $i \in \overline{\Omega}_3$, S_{ij} need not be a basic saturated cut-set under ψ_{kj}. Hence

$$t_{ij} \geq t_{kj} \qquad (12\text{-}6\text{-}10)$$

CASE 2. Vertex k is in $\overline{\Omega}_1$. We need to consider only the case when $\overline{\Omega}_2 \subset \overline{\Omega}_1$. Thus either $j \in \Omega_2$ or $j \in \overline{\Omega}_2$. Suppose $j \in \Omega_2$ as shown in Fig. 12-6-5. Then S_{ij} need not be a basic saturated cut-set under ψ_{ik}. Thus

$$\bar{t}_{ij} \geq \bar{t}_{ik} \qquad (12\text{-}6\text{-}11)$$

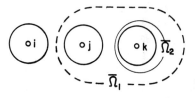

Fig. 12-6-5. Location of basic saturated cut-sets.

If $j \in \overline{\Omega}_2$, S_{ij} must be a basic saturated cut-set under ψ_{ik}. Thus

$$\bar{t}_{ij} = \bar{t}_{ik} \qquad (12\text{-}6\text{-}12)$$

$$\text{QED}$$

Next we define matrices that are similar to the terminal capacity matrices of lossless EWC nets.

Definition 12-6-2. A source terminal capacity matrix $\overline{T}$ is defined as

$$\overline{T} = [\bar{t}_{ij}] \qquad (12\text{-}6\text{-}13)$$

and a sink terminal capacity matrix $\underline{T}$ is

$$\underline{T} = [\underline{t}_{ij}] \qquad (12\text{-}6\text{-}14)$$

where $\bar{t}_{ii} = \underline{t}_{ii} = d$. Note that even if lossy EWC nets are nonoriented, $\overline{T}$ and $\underline{T}$ are generally nonsymmetric.

As we define in the case of lossless EWC nets, we define an S-submatrix $\overline{M}$ of $\overline{T}$ by deleting either row i or column i of $\overline{T}$ for all i. We also define an S-submatrix $\underline{M}$ of $\underline{T}$ in a similar way. We call S-submatrices $\overline{M}$ and $\underline{M}$ a "pair of S-submatrices" of $\overline{T}$ and $\underline{T}$ if (1) row i to $\overline{T}$ is in $\overline{M}$, row i of $\underline{T}$ is in $\underline{M}$, and (2) if column j of $\overline{T}$ is in $\overline{M}$, column j of $\underline{T}$ is in $\underline{M}$. With these definitions, we can obtain a property of a source and sink terminal capacity matrices as follows.

Theorem 12-6-4. For any entries $\underline{t}_{rs}$ and $\bar{t}_{rs}$ in $\underline{T}$ and $\overline{T}$, there exists a pair of S-submatrices $\underline{M}$ and $\overline{M}$ of $\underline{T}$ and $\overline{T}$ such that $\underline{t}_{rp}$ is the largest in column p for all p in $\underline{M}$ and $\bar{t}_{qs}$ is the largest in row q for all q in $\overline{M}$.

Proof. Let $S_{rs} = \mathscr{E}(\Omega_r \times \overline{\Omega}_r) \cup \mathscr{E}(\overline{\Omega}_r \times \Omega_r)$ be a corresponding cut-set of $\bar{t}_{rs}$ where $r \in \Omega_r$ and $s \in \overline{\Omega}_r$. Let $\underline{M}$ and $\overline{M}$ be the pair of S-submatrices of $\underline{T}$ and $\overline{T}$ such that the rows corresponding to the vertices in Ω_r and the columns corresponding to the vertices in $\overline{\Omega}_r$ are in $\underline{M}$ and $\overline{M}$. Then by the proof of Theorem 12-6-3, we have

$$\underline{t}_{rs} \geq \underline{t}_{ks} \qquad \text{for all} \quad k \in \Omega_r \qquad (12\text{-}6\text{-}15)$$

and

$$\bar{t}_{rs} \geq \bar{t}_{rk} \qquad \text{for all} \quad k \in \overline{\Omega}_r \qquad (12\text{-}6\text{-}16)$$

Thus the theorem is true for column s in $\underline{M}$ and row r in $\overline{M}$.

Consider $\underline{t}_{rp}$ where $p \in \overline{\Omega}_r$. Let $S_{rp} = \mathscr{E}(\Omega_{r'} \times \overline{\Omega}_{r'}) \cup \mathscr{E}(\overline{\Omega}_{r'} \times \Omega_{r'})$ be a corresponding cut-set of $\underline{t}_{rp}$ where $r \in \Omega_{r'}$, and $p \in \overline{\Omega}_{r'}$. If S_{rs} need not be a basic saturated cut-set under $\underline{t}_{rp}$, then $\overline{\Omega}_{r'}$ is a subset of $\overline{\Omega}_r$. Thus $\underline{t}_{rp} \geq \underline{t}_{kp}$ for all $k \in \Omega_r \subset \Omega_{r'}$. On the other hand, if S_{rs} is a basic saturated cut-set under $\underline{t}_{rp}$, we consider S_{rs} rather than S_{rp} as a corresponding cut-set of $\underline{t}_{rp}$. Hence

$$\underline{t}_{rp} \geq \underline{t}_{kp} \qquad \text{for all} \quad k \in \Omega_r \qquad (12\text{-}6\text{-}17)$$

Thus $\underline{t}_{rp}$ is the largest in column p for all p in $\underline{M}$. The proof for $\bar{t}_{qs} \geq \bar{t}_{qk}$ for all $k \in \overline{\Omega}_r$ is almost the same as this case.   QED

Note that for the lossless case, Theorem 12-6-4 becomes the S-submatrix condition of terminal capacity matrices given by Theorem 12-5-4.

The fundamental properties of lossy EWC nets now have been introduced. There remain many unsolved problems, one of the most important being to determine the realizability conditions for a matrix to be T or $\bar{T}$ of a lossy EWC net. No simple method has been given for evaluating t_{ij} and $\bar{t}_{ij}$. It is possible in some lossy EWC nets that a flow from i which is less than $\bar{t}_{ij}$ can result in a flow which is equal to t_{ij} at j. Hence we can define $\bar{\bar{t}}_{ij}$, which indicates the minimum required flow at i in order to receive the flow equal to t_{ij} at j. Then it is easily seen that Theorem 12-6-3 will hold with $\bar{\bar{t}}_{ij}$ and t_{ij}. Furthermore, if we define $\bar{\bar{T}} = [\bar{\bar{t}}_{ij}]$, then Theorem 12-6-4 will hold with T and $\bar{\bar{T}}$.

Clearly, every theorem given in this section is applicable to nonoriented lossy EWC nets. However, some modifications must be made in the definitions, for example, in that given for a basic saturated cut-set.

12-7 FLOW RELIABILITY OF EWC NET

When edges in an EWC net have a possibility of failure to transmit, all of a given flow cannot be expected to reach a destination all the time. However, some of the flow may be able to reach a destination most of the time. A percentage of a given flow which can be expected to reach a destination under possible failures in a net is known as a *flow reliability*; it depends on the structure (topology) of an EWC net, the reliability of edges, the amount, and an entry and a destination of a given flow. To obtain an expression for flow reliability, consider an EWC net G representing a practical system such as a data communication system. Flow ψ_{ij} in such a system indicates an amount of data transmitted from i to j *per unit of time*. Suppose flow ψ_{ij} is transmitted by G in T period of time. Then the total flow Φ_{ij} handled by G during T will be

$$\Phi_{ij} = T\psi_{ij} \tag{12-7-1}$$

If ψ_{ij} is equal to the maximum flow from i to j in G (which is the terminal capacity t_{ij}), then the total flow is

$$\Phi_{ij} = Tt_{ij} \tag{12-7-2}$$

which is clearly the maximum amount of flow from i to j during T.

Suppose edge e failed to transmit data for ΔT during T. However, all other edges are operating properly during the same period T. Then the total flow $\Phi_{ij}(\bar{e})$ under the continuous transmission of the maximum flow will be

$$\Phi_{ij}(\bar{e}) = Tt_{ij} - \Delta T[t_{ij} - t_{ij}(\bar{e})] \tag{12-7-3}$$

where $t_{ij}(\bar{e})$ is the maximum flow from i to j in G when edge e is absent. The ratio $R(t_{ij})$ of the total flow $\Phi_{ij}(\bar{e})$ with failure of edge e during ΔT and that Φ_{ij} without any failure is

$$R(t_{ij}) = 1 - \frac{\Delta T}{T}\left[1 - \frac{t_{ij}(\bar{e})}{t_{ij}}\right] \qquad (12\text{-}7\text{-}4)$$

Example 12-7-1. Consider the EWC net in Fig. 12-7-1. Suppose $\psi_{ij} = t_{ij} = 3$ is transmitted during $T = 10$. Then the total flow $\Phi_{ij} = 30$. Suppose edge a is down for $\Delta T = 3$ during the period T. Then the total flow $\Phi_{ij}(\bar{a}) = 27$ and $R(t_{ij})$ for this case will be 0.9.

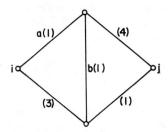

Fig. 12-7-1.   EWC net.

By replacing $\Delta T/T$ by probability $f(e)$ of failure of edge e, Eq. 12-7-4 becomes

$$R(t_{ij}) = 1 - f(e)[1 - T(\bar{e})] \qquad (12\text{-}7\text{-}5)$$

where

$$T(\bar{e}) = \frac{t_{ij}(\bar{e})}{t_{ij}} \qquad (12\text{-}7\text{-}6)$$

where $T(\bar{e})$ is called the *threshold level* under failure of edge e. In order to obtain $R(t_{ij})$ under the possibility of failure of several edges, we define the symbol $T(\overline{e_{p_1}e_{p_2}\cdots e_{p_k}})$ as follows.

Definition 12-7-1. The symbol $T(\overline{e_{p_1}e_{p_2}\cdots e_{p_k}})$ indicates the threshold level under failure of edges $e_{p_1}, e_{p_2}, \ldots, e_{p_k}$, which is equal to

$$T(\overline{e_{p_1}e_{p_2}\cdots e_{p_k}}) = \frac{t_{ij}(\overline{e_{p_1}e_{p_2}\cdots e_{p_k}})}{t_{ij}} \qquad (12\text{-}7\text{-}7)$$

where $t_{ij}(\overline{e_{p_1}e_{p_2}\cdots e_{p_k}})$ is the maximum flow from i to j when edges e_{p_1}, $e_{p_2}, \ldots, e_{p_k}$ are deleted.

As an example, the threshold level $T(\bar{a})$ under failure of edge a in Fig. 12-7-1 is $T(\bar{a}) = \frac{2}{3}$. In the same figure, the threshold level under the failure of edges a and b is $T(\overline{ab}) = \frac{1}{3}$.

When each edge e_p in net G has a probability $f(e_p)$ of failure (or a reliability $r(e_p) = 1 - f(e_p)$), then $R(t_{ij})$ under the maximum flow is

$$R(t_{ij}) = 1 - \sum F(e_{p_1}e_{p_2}\cdots e_{p_k})[1 - T(\overline{e_{p_1}e_{p_2}\cdots e_{p_k}})] \qquad (12\text{-}7\text{-}8)$$

where $F(e_{p_1}e_{p_2}\cdots e_{p_k})$ is the probability that exactly edges $e_{p_1}, e_{p_2}, \ldots, e_{p_k}$ fail and $\sum$ means the sum of all possible cases of failure of edges. Equation 12-7-8 is called a "flow reliability under the maximum flow."

Example 12-7-2. Consider the EWC net in Fig. 12-7-2. For input i and output j, the threshold levels are

$$T(\bar{a}) = \tfrac{1}{5}, \qquad T(\bar{b}) = \tfrac{4}{5}, \qquad T(\bar{c}) = 0, \qquad T(\overline{ab}) = 0,$$
$$T(\overline{ac}) = 0, \qquad T(\overline{bc}) = 0, \qquad \text{and} \quad T(\overline{abc}) = 0$$

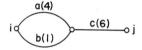

Fig. 12-7-2. EWC net.

Let the probability of failure of each edge be

$$f(a) = .1, \qquad f(b) = .2, \qquad \text{and} \quad f(c) = .3$$

Then the probabilities of failure appearing in Eq. 12-7-8 are

$$F(a) = .056, \qquad F(b) = .126, \qquad F(c) = .216$$
$$F(ab) = .014, \qquad F(ac) = .024, \qquad F(bc) = .054$$

and

$$F(abc) = .006$$

Hence the flow reliability under the maximum flow $t_{ij} = 5$ is

$$R(t_{ij}) = .616$$

Instead of transmitting the maximum flow, suppose we transmit flow $\psi_{ij} \leq t_{ij}$. Also suppose only one edge e fails for ΔT. Then Eq. 12-7-3 becomes

$$\Phi_{ij}(\bar{e}) = T\psi_{ij} - \Delta T[\psi_{ij} - \psi_{ij}(\bar{e})] \qquad (12\text{-}7\text{-}9)$$

where

$$\psi_{ij}(\bar{e}) = \min \{\psi_{ij}, t_{ij}(\bar{e})\} \qquad (12\text{-}7\text{-}10)$$

This is true because if $\psi_{ij} \leq t_{ij}(\bar{e})$, then the failure of edge e will not influence the transmission of ψ_{ij} by reassigning a given flow. On the other hand, if $\psi_{ij} > t_{ij}(\bar{e})$, the maximum amount of flow which the remaining net can handle is clearly $t_{ij}(\bar{e})$ but not ψ_{ij}. This result is under the assumption that there is no storage place for flow. In other words, if there is storage such that a part of flow, which cannot be transmitted because of failure of edge e, can be stored until the edge is fixed, then it will be possible to transmit these stored flow after edge e is fixed as long as $\psi_{ij} < t_{ij}$. Hence the total flow $\Phi_{ij}(\bar{e})$ with storage places will be larger than that given by Eq. 12-7-9. However, in practical cases, the amount of flow $\Delta T[\psi_{ij} - \psi_{ij}(\bar{e})]$ to be stored would be too large to handle. If $t_{ij} - \psi_{ij}$ is small, to transmit $\Delta T[\psi_{ij} - \psi_{ij}(\bar{e})]$ may take a long time after edge e is fixed. Thus we assume that there are no storage places in this section. By this assumption, $R(\psi_{ij})$ under a given flow ψ_{ij} in Eq. 12-7-4 becomes

$$R(\psi_{ij}) = 1 - \frac{\Delta T}{T}\left[1 - \min\left\{1, \frac{t_{ij}(\bar{e})}{\psi_{ij}}\right\}\right] \qquad (12\text{-}7\text{-}11)$$

By replacing $\Delta T/T$ by the probabilities of failure of edges, the foregoing equation becomes

$$R(\psi_{ij}) = 1 - \sum F(e_{p_1}e_{p_2}\cdots e_{p_k})\left[1 - \min\left\{1, \frac{t_{ij}}{\psi_{ij}}T(\overline{e_{p_1}e_{p_2}\cdots e_{p_k}})\right\}\right] \quad (12\text{-}7\text{-}12)$$

which is the flow reliability under flow ψ_{ij}. For example, if $\psi_{ij} = 4$, the flow reliability $R(\psi_{ij} = 4)$ of the EWC net in Fig. 12-7-2 is

$$R(\psi_{ij} = 4) = .644$$

which is higher than that under the maximum flow.

From Eq. 12-7-12, it can be seen that there are three distinct techniques of improving the flow reliability $R(\psi_{ij})$. The first technique is to use more reliable media in a net so that the probability $f(e)$ of each edge will be smaller. However, in general, the cost of a medium increases exponentially as the reliability $[1 - f(e)]$ of a medium increases, when the reliability of a medium is already high enough. Hence it would be a rather expensive way to increase the flow reliability by this technique. The second way is to increase the term $\min\{1, (t_{ij}/\psi_{ij})T(\overline{e_{p_1}e_{p_2}\cdots e_{p_k}})\}$ in Eq. 12-7-12 by reducing ψ_{ij}, that is, to reduce the amount of flow transmitted by a net. However, this is also not an economical way because decreasing a flow to be transmitted means to increase the cost of transmission per unit of flow.

The third technique is to change the configuration of a net. There are two ways, one of which is to change the structure of a given net, in other words, to redesign a net so that the flow reliability becomes higher under a given specification. This is probably very useful when a net is at the designing stage. However, actual reconstruction of an existing net may not be practical. The other method is to maintain its structure but interchange probabilities $f(e)$ of failure assigned to edges. This is equivalent to interchange the locations of media of a similar kind. By doing so,

$$\sum F(e_{p_1} e_{p_2} \cdots e_{p_k})[1 - \min \{1, (t_{ij}/\psi_{ij})T(\overline{e_{p_1} e_{p_2} \cdots e_{p_k}})\}]$$

may change. Hence it is possible to increase the flow reliability by this method.

Example 12-7-3.   Consider the EWC net in Fig. 12-7-3. The threshold levels are

$$
\begin{array}{llll}
T(\bar{a}) = \tfrac{1}{4}, & T(\bar{b}) = \tfrac{3}{4}, & T(\bar{c}) = \tfrac{3}{4}, & T(\bar{d}) = \tfrac{2}{4} \\
T(\overline{ab}) = 0, & T(\overline{ac}) = \tfrac{1}{4}, & T(\overline{ad}) = \tfrac{1}{4}, & T(\overline{bc}) = \tfrac{3}{4} \\
T(\overline{bd}) = \tfrac{2}{4}, & T(\overline{cd}) = 0, & T(\overline{abc}) = 0, & T(\overline{abd}) = 0 \\
T(\overline{acd}) = 0, & T(\overline{bcd}) = 0, & \text{and} \quad T(\overline{abcd}) = 0
\end{array}
$$

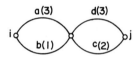

Fig. 12-7-3.   EWC net.

With the probabilities of failure of edges being

$$f(a) = .1, \qquad f(b) = .2, \qquad f(c) = .3, \qquad \text{and} \quad f(d) = .4$$

the flow reliability $R(t_{ij})$ under the maximum flow is

$$R(t_{ij}) = .621$$

If we reduce the amount of flow ψ_{ij} to 3, the flow reliability will be increased to

$$R(\psi_{ij} = 3) = .7273$$

Further reduction of ψ_{ij} to 2 and to 1, respectively, gives

$$R(\psi_{ij} = 2) = .823$$

and

$$R(\psi_{ij} = 1) = .8582$$

Instead of reducing ψ_{ij}, if we replace edge a to a new location as shown in

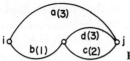

Fig. 12-7-4.   Relocation of edge a.

Fig. 12-7-4, the flow reliability under the maximum flow (with the same probabilities of failure of edges) is changed to

$$R(t_{ij}) \text{ of a new net} = .8648$$

which is a large improvement.

If instead of changing its structure we interchange only the probabilities $f(a)$ and $f(d)$ of edges a and d as $f(a) = .4$ and $f(d) = .1$, the flow reliability $R(t_{ij})$ under the maximum flow will be changed to

$$R(t_{ij}) = .704$$

which is quite an improvement from .621.

Suppose interchange of any probabilities of failure of edges is permissible. Then how can we obtain the largest flow reliability? An answer to this question may not be simple. However, if we assume that the probabilities of failures of edges are very small, an answer, as given by the following theorem, becomes very simple.

Theorem 12-7-1.   Let e_r be an edge for $r = 1, 2, \ldots, n$ in an EWC net. Suppose the probabilities $f(e_r)$ of failure of edges are very small. Then the flow reliability $R(\psi_{ij})$ is the maximum if

$$f(e_1) \le f(e_2) \le \cdots \le f(e_n)$$

implies

$$T(\bar{e}_1) \le T(\bar{e}_2) \le \cdots \le T(\bar{e}_n)$$

As an example, consider the EWC net in Fig. 12-7-3. Let $f(a) = .1$, $f(b) = .2$, $f(c) = .25$, and $f(d) = .15$. Then $f(a) \le f(d) \le f(b) \le f(c)$. The threshold levels are $T(\bar{a}) = \frac{1}{4}$, $T(\bar{b}) = \frac{3}{4}$, $T(\bar{c}) = \frac{3}{4}$, and $T(\bar{d}) = \frac{2}{4}$. Thus by the above theorem, interchange of probabilities $f(a)$, $f(b)$, $f(c)$, and $f(d)$ does not increase the flow reliability $R(\psi_{ij})$. To illustrate, if $\psi_{ij} = 3$, $R(\psi_{ij} = 3) = .939$. If we interchange probabilities of failure of a and d as $f(a) = .15$ and $f(d) = .1$, the flow reliability $R(\psi_{ij} = 3)$ becomes .929.

Proof of Theorem 12-7-1.   By the assumption that $f(e_r)$ is very small for all r, Eq. 12-7-12 can be simplified to

$$R(\psi_{ij}) \approx 1 - \sum F(e_p)[1 - Q(\bar{e}_p)] \qquad (12\text{-}7\text{-}13)$$

where

$$F(e_p) = f(e_p) \sum_{\substack{r=1 \\ r \neq p}}^{n} [1 - f(e_r)] \qquad (12\text{-}7\text{-}14)$$

and

$$Q(\bar{e}_p) = \min\left\{1, \frac{t_{ij}}{\psi_{ij}} T(\bar{e}_p)\right\} \qquad (12\text{-}7\text{-}15)$$

It is clear from Eq. 12-7-14 that the relationship

$$f(e_1) \leq f(e_2) \leq \cdots \leq f(e_n)$$

gives

$$F(e_1) \leq F(e_2) \leq \cdots \leq F(e_n)$$

It is also obvious from Eq. 12-7-13 that smaller $\sum F(e_p)[1 - Q(\bar{e}_p)]$ gives larger $R(\psi_{ij})$. Hence the theorem will be proven if we can show that $\sum F(e_p)[1 - Q(\bar{e}_p)]$ is smallest when $T(\bar{e}_1) \leq T(\bar{e}_2) \leq \cdots \leq T(\bar{e}_n)$ is satisfied.

Consider two sets of positive numbers $\{a_p; \ p = 1, 2, \ldots, n\}$ and $\{b_p; \ p = 1, 2, \ldots, n\}$. Let $a_1 \leq a_2 \leq \cdots \leq a_n$ and $b_n \leq b_{n-1} \leq \cdots \leq b_2 \leq b_1$. Let sum S_j be defined as follows:

$$S_j = \sum_{r=1}^{n} a_{j_r} b_r \qquad (12\text{-}7\text{-}16)$$

where $(j_1 j_2 \cdots j_n)$ is a permutation of $(1\ 2 \cdots n)$. Let $\{S_j\}$ be a set of sums S_j produced by employing all possible permutations of $(1\ 2 \cdots n)$. Let sum S_0 be

$$S_0 = \sum_{r=1}^{n} a_r b_r \qquad (12\text{-}7\text{-}17)$$

It can be seen that any sum S_p in $\{S_j\}$ other than S_0 can be expressed as

$$S_p = \sum_{r=1}^{m} a_{j_r} b_r + \sum_{r=m+1}^{n} a_r b_r \qquad (12\text{-}7\text{-}18)$$

where $(j_1 j_2 \cdots j_m)$ is a permutation of $(1\ 2 \cdots m)$ and $j_m \neq m$. When $m = n$, the second expression in the right-hand side of Eq. 12-7-18 will be absent. Let $j_t \ (1 \leq t \leq m)$ be m. Then S_p can be rewritten as

$$S_p = \sum_{\substack{r=1 \\ r \neq t}}^{m-1} a_{j_r} b_r + a_m b_t + a_{j_m} b_m + \sum_{r=m+1}^{n} a_r b_r \qquad (12\text{-}7\text{-}19)$$

By interchanging a_{j_m} and a_m, we have a new sum S_p':

$$S_p' = \sum_{\substack{r=1 \\ r \neq t}}^{m-1} a_{j_r} b_r + a_{j_m} b_t + a_m b_m + \sum_{r=m+1}^{n} a_r b_r \qquad (12\text{-}7\text{-}20)$$

Since $a_{j_m} \leq a_m$ and $b_m \leq b_t$, it is clear that $S'_p \leq S_p$. Similarly, if S'_p is not S_o, we can obtain another sum S''_p such that $S''_p \leq S'_p$. Hence we can state that any sum S_j in $\{S_j\}$ satisfies $S_j \geq S_o$. In other words, sum S_o is the smallest in $\{S_j\}$.

By substituting $a_r = F(e_r)$ and $b_r = [1 - Q(\bar{e}_r)]$, we can state that $\sum F(e_r)[1 - Q(\bar{e}_r)]$ is the minimum when

$$[1 - Q(\bar{e}_1)] \geq [1 - Q(\bar{e}_2)] \geq \cdots \geq [1 - Q(\bar{e}_n)]$$

satisfies. Since $[1 - Q(\bar{e}_r)]$ is nonnegative, the foregoing result gives

$$Q(\bar{e}_1) \leq Q(\bar{e}_2) \leq \cdots \leq Q(\bar{e}_n)$$

From the definition of $Q(\bar{e}_r)$ in Eq. 12-7-15, this relationship leads to

$$T(\bar{e}_1) \leq T(\bar{e}_2) \leq \cdots \leq T(\bar{e}_n)$$

which proves the theorem.   QED

By this theorem, we can assign probabilities of failure of edges so that the flow reliability becomes the maximum. In practical systems, not all probabilities of failure of edges can be interchanged. However, this theorem will hold if we consider only interchangeable edges. For example, if edges e_1, $e_2, \ldots, e_k$ can interchange their probabilities and edges $e'_1, e'_2, \ldots, e'_m$ have interchangeable probabilities among themselves, then we can interchange probabilities of $e_1, e_2, \ldots, e_k$ to obtain the highest available flow reliability first, then we interchange probabilities of $e'_1, e'_2, \ldots, e'_m$ to obtain the maximum flow reliability.

The insertion of additional media to a net for increasing the maximum flow is often employed. Here we investigate the effect of such an insertion on the flow reliability, first considering an example. We have an EWC net in Fig. 12-7-2 (Example 12-7-2). We have calculated the flow reliability $R(t_{ij})$ of

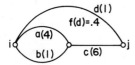

Fig. 12-7-5.   Insertion of edge d.

this net which was as .616 where $t_{ij} = 5$. Suppose we insert edge d between i and j whose edge capacity is 1 and probability $f(d)$ of failure is .4 as shown in Fig. 12-7-5. Then the flow reliability $R(t_{ij})$ becomes .6091, which shows that the flow reliability is reduced. On the other hand, the flow to be transmitted is increased to 6. If we maintain the amount of flow to be transmitted, the flow reliability $R(\psi_{ij} = 5)$ is

$$R(\psi_{ij} = 5) = .6713$$

which shows that the flow reliability is increased by the insertion of edge d. It is obvious that the flow reliability will not be decreased by an insertion of edges as long as the flow transmitted is the same.

Can we increase both the flow reliability and the amount of flow by inserting edges? To answer this, we investigate the flow reliability when two EWC nets are connected in parallel.

Theorem 12-7-2. In the parallel connection shown in Fig. 12-7-6, let R_1 and R_2 be the flow reliabilities of EWC nets G_1 and G_2 under the maximum flow

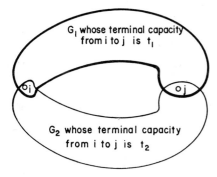

Fig. 12-7-6. Two nets in parallel.

from i to j, respectively. Also let t_1 and t_2 be the terminal capacities of G_1 and G_2, respectively. Then the flow reliability R_0 under the maximum flow from i to j of the entire net G_0 is

$$R_0 = \frac{[t_1 R_1 + t_2 R_2]}{t_1 + t_2} \qquad (12\text{-}7\text{-}21)$$

Proof. The flow reliability under a flow ψ_{ij} of an EWC net G given by Eq. 12-7-12 can be rewritten as

$$R(\psi_{ij}) = P + \sum F(e_{p_1} e_{p_2} \cdots e_{p_k}) \min \left\{ 1, \frac{t_{ij}(\overline{e_{p_1} e_{p_2} \cdots e_{p_k}})}{\psi_{ij}} \right\} \qquad (12\text{-}7\text{-}22)$$

where

$$P = \prod_{e_p \in G} [1 - f(e_p)] \qquad (12\text{-}7\text{-}23)$$

Using this equation, we can express the flow reliabilities R_0, R_1, and R_2 under the maximum flow from i to j of G_0, G_1, and G_2, respectively, as

$$R_0 = P_0 + \sum \frac{F_{0u} t_{0u}}{t_0} \tag{12-7-24}$$

$$R_1 = P_1 + \sum \frac{F_{1u} t_{1u}}{t_1} \tag{12-7-25}$$

and

$$R_2 = P_2 + \sum \frac{F_{2u} t_{2u}}{t_2} \tag{12-7-26}$$

where t_0 is the terminal capacity from i to j of G_0 and F_{ru} and t_{ru} for $r = 0$, 1, 2 are the shorthand notations of

$$F_{ru} = F(e_{p_1} e_{p_2} \cdots e_{p_k}) \tag{12-7-27}$$

and

$$t_{ru} = t_{ij} \overline{(e_{p_1} e_{p_2} \cdots e_{p_k})} \tag{12-7-28}$$

Note that

$$P_r = \prod_{e_p \in G_r} [1 - f(e_p)] \tag{12-7-29}$$

for $r = 0$, 1, 2.

Since

$$P_0 = P_1 P_2 \tag{12-7-30}$$

$$\sum F_{0u} = P_2 \sum F_{1u} + P_1 \sum F_{2u'} + \sum F_{1u} \sum F_{2u'} \tag{12-7-31}$$

and

$$t_{0u} = t_{1u} + t_{2u'} \tag{12-7-32}$$

the flow reliability R_0 can be expressed as

$$R_0 = P_1 P_2 + P_2 \sum F_{1u} \left(\frac{t_{1u} + t_2}{t_0} \right) + P_1 \sum F_{2u} \left(\frac{t_{2u} + t_1}{t_0} \right)$$
$$+ \sum F_{1u} \sum F_{2u'} \left(\frac{t_{1u} + t_{2u'}}{t_0} \right) \tag{12-7-33}$$

We can see that the terminal capacity t_0 of the entire net G_0 can be expressed as

$$t_0 = t_1 + t_2 \tag{12-7-34}$$

Hence Eq. 12-7-33 becomes

$$R_0 = \frac{(t_1 + t_2)}{t_0} P_1 P_2 + P_2 \sum F_{1u} \frac{t_2}{t_0} + P_2 \sum F_{1u} \frac{t_{1u}}{t_0}$$

$$+ P_1 \sum F_{2u} \frac{t_1}{t_0} + P_1 \sum F_{2u} \frac{t_{2u}}{t_0} + \sum F_{1u} \sum F_{2u'} \frac{t_{1u}}{t_0}$$

$$+ \sum F_{1u'} \sum F_{2u} \frac{t_{2u'}}{t_0}$$

$$= \frac{t_1 P_1 P_2}{t_0} + \frac{t_2 P_1 P_2}{t_0} + \frac{t_1 P_1 \sum F_{2u}}{t_0} + \frac{t_2 P_2 \sum F_{1u}}{t_0}$$

$$+ \frac{t_1}{t_0} \sum F_{1u} \frac{t_{1u}}{t_1} + \frac{t_2}{t_0} \sum F_{2u} \frac{t_{2u}}{t_2}$$

$$= \frac{t_1}{t_0} \left(P_1 + \sum F_{1u} \frac{t_{1u}}{t_1} \right) + \frac{t_2}{t_0} \left(P_2 + \sum F_{2u} \frac{t_{2u}}{t_2} \right)$$

$$= \frac{t_1 R_1 + t_2 R_2}{t_1 + t_2} \tag{12-7-35}$$

QED

If we consider G_1 as a given net and G_2 as a net to be inserted, then it is possible to have a higher flow reliability with increasing flow by proper choice of R_2 and t_2 by Eq. 12-7-35, which answers our previous question.

Example 12-7-4. EWC net G_0 in Fig. 12-7-7a can be considered as two nets

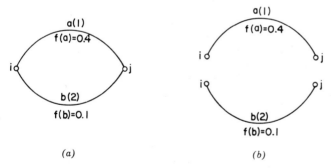

(a)   (b)

Fig. 12-7-7. A net consisting of two nets in parallel. (a) An EWC net; (b) two nets G_1 and G_2.

G_1 and G_2 as shown in Fig. 12-7-7b in parallel. The flow reliabilities R_1 and R_2 under the maximum flow from i to j of G_1 and G_2 are

$$R_1 = .6$$

and

$$R_2 = .9$$

The terminal capacities of G_1 and G_2 are

$$t_1 = 1$$

and

$$t_2 = 2$$

Hence by Theorem 12-7-2, the flow reliability R_0 under the maximum flow $t_0 = 3$ from i to j of a given net is

$$R_0 = \frac{[.6(1) + .9(2)]}{3} = .8$$

When two EWC nets G_1 and G_2 are in series, the relationship between the flow reliability of the entire net and those of G_1 and G_2 is given by the following theorems.

Theorem 12-7-3. In the series connection shown in Fig. 12-7-8, let R_1 be the flow reliability under a flow ψ_{ik} of G_1 and R_2 be the flow reliability under a

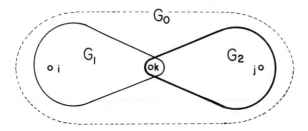

Fig. 12-7-8. Two nets in series.

flow ψ_{kj} of G_2 with $\psi_{ik} = \psi_{kj}$. Then the flow reliability R_0 under a flow $\psi_{ij} = \psi_{ik}$ of the entire net G_0 is bounded by

$$R_1 R_2 \leq R_0 \leq \min \{R_1 + (R_2 - 1)P_1, \; R_2 + (R_1 - 1)P_2\} \quad (12\text{-}7\text{-}36)$$

where

$$P_r = \prod_{e_p \in G_r} [1 - f(e_p)] \qquad \text{for} \quad r = 1, 2 \qquad (12\text{-}7\text{-}37)$$

Proof.   In the series connection, the flow reliability R_0 under a flow ψ_{ij} of the entire net can be expressed as

$$R_0 = P_1 P_2 + P_1 \sum F_{2u} \min\left(1, \frac{t_{2u}}{\psi_{kj}}\right)$$

$$+ P_2 \sum F_{1u} \min\left(1, \frac{t_{1u}}{\psi_{ik}}\right) + \sum F_{1u} \cdot \sum F_{2u'} \min\left(1, \frac{t_{1u}}{\psi_{ik}}, \frac{t_{2u'}}{\psi_{kj}}\right) \quad (12\text{-}7\text{-}38)$$

Since

$$\min\left(1, \frac{t_{2u}}{\psi_{kj}}\right) \cdot \min\left(1, \frac{t_{1u}}{\psi_{ik}}\right) \leq \min\left(1, \frac{t_{1u}}{\psi_{ik}}, \frac{t_{2u}}{\psi_{kj}}\right) \quad (12\text{-}7\text{-}39)$$

where $\psi_{ik} = \psi_{kj} = \psi_{ij}$ by assumption, Eq. 12-7-38 can be changed to

$$R_0 \geq P_1 P_2 + P_1 \sum F_{2u} \min\left(1, \frac{t_{2u}}{\psi_{kj}}\right) + P_2 \sum F_{1u} \min\left(1, \frac{t_{1u}}{\psi_{ik}}\right)$$

$$+ \left[\sum F_{1u} \min\left(1, \frac{t_{1u}}{\psi_{ik}}\right)\right] \cdot \left[\sum F_{2u} \min\left(1, \frac{t_{2u}}{\psi_{kj}}\right)\right]$$

$$= R_1 R_2 \quad (12\text{-}7\text{-}40)$$

which proves the left-hand side of Eq. 12-7-36.

With

$$\min\left(1, \frac{t_{1u}}{\psi_{ik}}\right) \geq \min\left(1, \frac{t_{1u}}{\psi_{ik}}, \frac{t_{2u}}{\psi_{kj}}\right) \quad (12\text{-}7\text{-}41)$$

Eq. 12-7-38 becomes

$$R_0 \leq P_1 P_2 + P_1 \sum F_{2u} \min\left(1, \frac{t_{2u}}{\psi_{kj}}\right) + P_2 \sum F_{1u} \min\left(1, \frac{t_{1u}}{\psi_{ik}}\right)$$

$$+ \sum F_{2u'} \sum F_{1u} \min\left(1, \frac{t_{1u}}{\psi_{ik}}\right)$$

$$= P_1 P_2 + P_1 \sum F_{2u} \min\left(1, \frac{t_{2u}}{\psi_{kj}}\right) + \sum F_{1u} \min\left(1, \frac{t_{1u}}{\psi_{ik}}\right)$$

$$= P_1(R_2 - 1) + R_1 \quad (12\text{-}7\text{-}42)$$

By interchanging the subscripts, Equation 12-7-42 gives

$$R_0 \leq P_2(R_1 - 1) + R_2 \quad (12\text{-}7\text{-}43)$$

Since both equations must hold, the right-hand side of Eq. 12-7-36 is true, which proves the theorem.   QED

Example 12-7-5.   EWC net G_0 in Fig. 12-7-9 consists of two nets G_1 and G_2 in series. By computing the flow reliability R_1 under $\psi_{ik} = 4$ of G_1, we have

$$R_1 = .875$$

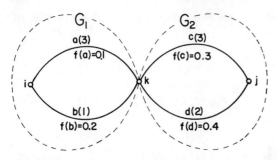

Fig. 12-7-9. Two nets G_1 and G_2 in series.

The flow reliability R_2 under $\psi_{kj} = 4$ is

$$R_2 = .72$$

Hence the product of R_1 and R_2 gives

$$R_1 R_2 = .63$$

Since

$$P_1 = .72$$

and

$$P_2 = .42$$

we have

$$P_1(R_2 - 1) + R_1 = .6734$$

and

$$P_2(R_1 - 1) + R_2 = .6675$$

Thus the flow reliability R_0 under $\psi_{ij} = 4$ of a given net G_0 is bounded by (Theorem 12-7-3)

$$.63 \leq R_0 \leq .6675$$

We can see from Eqs. 12-7-12 and 12-7-22 that it is a very time-consuming task to calculate the flow reliability of an EWC net. Theorem 12-7-2 gives a method of obtaining a flow reliability of a net when it can be considered as two nets in parallel. However, this does not give the flow reliability under a flow which is less than the maximum flow. Theorem 12-7-3, on the other hand, gives the upper and the lower bounds of the flow reliability of a net when it can be considered as two nets in series. Again, this theorem may not help to obtain an accurate enough flow reliability. Furthermore, a net may not be decomposable into series and parallel nets. Hence these two theorems may not help much in obtaining an approximate quantity of a flow reliability of a net. There is no simple method to obtain a flow reliability of a net at the present time; however, under some restrictions, we can obtain a reasonable approximation for a flow reliability as given by the next theorem.

Theorem 12-7-4. Let R be the flow reliability under a flow ψ_{ij} from i to j of an EWC net G. Let $S = (b_1, b_2, \ldots, b_n)$ be the corresponding cut-set of a terminal capacity t_{ij} of G. Let $(e_1, e_2, \ldots, e_m)$ be a set of all edges in $G - S$ (i.e., it is a set of all edges in G except those in S). Suppose the following conditions are satisfied:

   1. Probability $f(e_r)$ of failure of edge e_r is very small for $r = 1, 2, \ldots, m$.
   2. The deletion of any edge e_r does not reduce the terminal capacity from i to j.

Then the flow reliability under ψ_{ij} is very close to that under ψ_{ij} of G' where G' is obtained from G by setting probability $f(e_r)$ of failure of edge e_r to zero for $r = 1, 2, \ldots, m$.

In addition to Conditions 1 and 2, if probability $f(b_s)$ of failure of edge b_s is small for $s = 1, 2, \ldots, n$, then flow reliability R under the maximum flow can be approximated by

$$R \approx \sum_{s=1}^{n} \frac{r(b_s)c_s}{\sum\limits_{u=1}^{n} c_u} \qquad (12\text{-}7\text{-}44)$$

where $r(b_s)$ is the reliability of edge b_s which is equal to $1 - f(b_s)$.

Example 12-7-6. Consider the EWC net in Fig. 12-7-10. We can see that Conditions 1 and 2 satisfy for this net. Note that the corresponding cut-set S of terminal capacity t_{ij} consists of edges a, b, and c. Since probabilities of

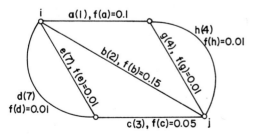

Fig. 12-7-10.   An EWC net.

failure of edges a, b, and c are small, we can use Eq. 12-7-44 to approximate the flow reliability R_0 under the maximum flow from i to j:

$$R_0 \approx \frac{.9(1) + .85(2) + .95(3)}{1 + 2 + 3} = .909$$

By using Eq. 12-7-12, the actual quantity of R_0 is .9047. Hence the approximation given by Eq. 12-7-44 gives a reasonable flow reliability.

Proof of Theorem 12-7-4. By Condition 1 we have

$$\sum F(e_{p_1} e_{p_2} \cdots e_{p_k}) \left[1 - \min \left\{ 1, \frac{t_{ij}}{\psi_{ij}} T(\overline{e_{p_1} e_{p_2} \cdots e_{p_k}}) \right\} \right]$$

$$\approx \sum_{b_p \in S} F(b_{p_1} b_{p_2} \cdots b_{p_w}) \left[1 - \min \left\{ 1, \frac{t_{ij}}{\psi_{ij}} T(\overline{b_{p_1} b_{p_2} \cdots b_{p_w}}) \right\} \right]$$

$$+ \sum_{e_u \in (G-S)} f(e_u) \left[1 - \min \left\{ 1, \frac{t_{ij}}{\psi_{ij}} T(\bar{e}_u) \right\} \right] \qquad (12\text{-}7\text{-}45)$$

Furthermore, by Condition 2, $t_{ij}(\bar{e}_u) = t_{ij}$. Hence

$$f(e_u) \left[1 - \min \left\{ 1, \frac{t_{ij}}{\psi_{ij}} T(\bar{e}_u) \right\} \right] = 0 \qquad (12\text{-}7\text{-}46)$$

Note that

$$T(\bar{e}_u) = \frac{t_{ij}(\bar{e}_u)}{t_{ij}} \qquad (12\text{-}7\text{-}47)$$

Thus the flow reliability R given by Eq. 12-7-12 becomes

$$R \approx 1 - \sum_{b_p \in S} F_s(b_{p_1} b_{p_2} \cdots b_{p_w}) \left[1 - \min \left\{ 1, \frac{t_{ij}}{\psi_{ij}} T(\overline{b_{p_1} b_{p_2} \cdots b_{p_w}}) \right\} \right] \qquad (12\text{-}7\text{-}48)$$

which is the flow reliability of G' obtained from G by setting all $f(e_r)$ of edge e_r to zero. This proves the first part of the theorem.

When $f(b_s)$ of all edges b_s in S are also small, Eq. 12-7-48 can be simplified further:

$$R \approx 1 - \sum F_s(b_s) \left[1 - \min \left\{ 1, \frac{t_{ij}}{\psi_{ij}} T(\bar{b}_s) \right\} \right]$$

$$\approx 1 - \sum f(b_s) \left[1 - \min \left\{ 1, \frac{t_{ij}}{\psi_{ij}} T(\bar{b}_s) \right\} \right] \qquad (12\text{-}7\text{-}50)$$

When $\psi_{ij} = t_{ij}$, which is the maximum flow,

$$T(\bar{b}_u) = 1 - \frac{c_u}{\sum\limits_{r=1}^{n} c_r} \qquad (12\text{-}7\text{-}51)$$

where c_u and c_r are the edge capacity of edges b_u and b_r, respectively. Note that

$$t_{ij} = \sum_{r=1}^{n} c_r \tag{12-7-52}$$

Hence the flow reliability $R(t_{ij})$ under the maximum flow is

$$R(t_{ij}) \approx 1 - \sum f(b_u)\left[1 - \left(1 - \frac{c_u}{\sum\limits_{r=1}^{n} c_r}\right)\right]$$

$$= 1 - \frac{\sum\limits_{u=1}^{n} f(b_u)c_u}{\sum\limits_{r=1}^{n} c_r} = \frac{\sum\limits_{u=1}^{n} [1 - f(b_u)]c_u}{\sum\limits_{r=1}^{n} c_r} = \frac{\sum\limits_{u=1}^{n} r(b_u)c_u}{\sum\limits_{r=1}^{n} c_r} \tag{12-7-53}$$

which proves the theorem.   QED

PROBLEMS

1. Find an assignment of flow $\psi_{ij} = 6$ from i to j for the EWC nets in Fig. P-12-1.

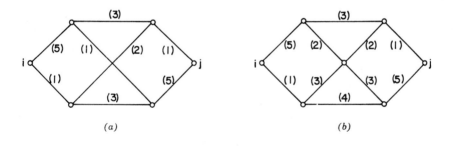

(a)   (b)

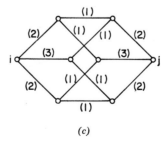

(c)

Fig. P-12-1. EWC nets (a) G_1, (b) G_2, and (c) G_3.

2. What is the maximum flow ψ_{ij} that can be transmitted from i to j in the EWC nets in Fig. P-12-2?

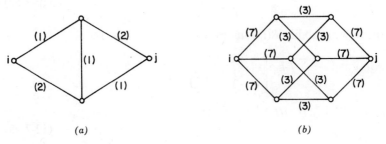

(a)   (b)

Fig. P-12-2. EWC nets (a) G_1 and (b) G_2.

3. Find the terminal capacity matrix of each of the EWC nets in Fig. P-12-3.

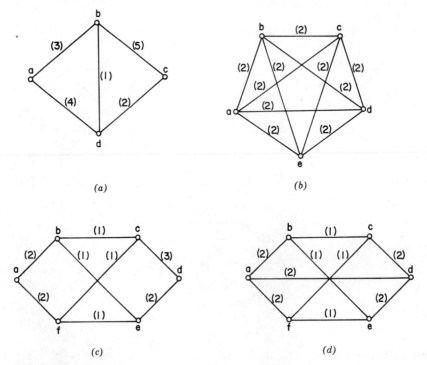

(a)   (b)

(c)   (d)

Fig. P-12-3. EWC nets (a) G_1, (b) G_2, (c) G_3, and (d) G_4.

4. Synthesize nonoriented EWC nets whose terminal capacity matrices are

(a)

$$
T = \begin{array}{c} \\ 1 \\ 2 \\ 3 \\ 4 \end{array}
\begin{array}{cccc}
1 & 2 & 3 & 4 \\
\begin{bmatrix} d & 2 & 1 & 2 \\ 2 & d & 1 & 3 \\ 1 & 1 & d & 1 \\ 2 & 3 & 1 & d \end{bmatrix}
\end{array}
$$

(b)

$$
T = \begin{array}{c} \\ 1 \\ 2 \\ 3 \\ 4 \end{array}
\begin{array}{cccc}
1 & 2 & 3 & 4 \\
\begin{bmatrix} d & 2 & 3 & 2 \\ 2 & d & 2 & 4 \\ 3 & 2 & d & 2 \\ 2 & 4 & 2 & d \end{bmatrix}
\end{array}
$$

5. Synthesize the cheapest nonoriented EWC net under the unit cost for each of the following terminal capacity matrices:

(a)

$$
T = \begin{array}{c} \\ 1 \\ 2 \\ 3 \\ 4 \end{array}
\begin{array}{cccc}
1 & 2 & 3 & 4 \\
\begin{bmatrix} d & 1 & 1 & 5 \\ 1 & d & 3 & 1 \\ 1 & 3 & d & 1 \\ 5 & 1 & 1 & 1 \end{bmatrix}
\end{array}
$$

(b)

$$
T = \begin{array}{c} \\ 1 \\ 2 \\ 3 \\ 4 \\ 5 \end{array}
\begin{array}{ccccc}
1 & 2 & 3 & 4 & 5 \\
\begin{bmatrix} d & 3 & 2 & 2 & 3 \\ 3 & d & 2 & 2 & 5 \\ 2 & 2 & d & 3 & 2 \\ 2 & 2 & 3 & d & 2 \\ 3 & 5 & 2 & 2 & d \end{bmatrix}
\end{array}
$$

6. Obtain the terminal capacity matrix of each of the oriented EWC nets in Fig. P-12-6.

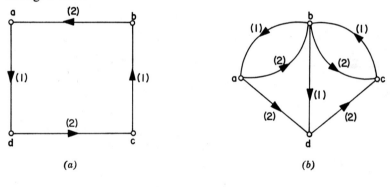

(a)   (b)

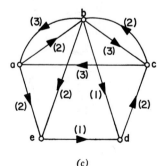

(c)

Fig. P-12-6. Oriented EWC nets (a) G_1, (b) G_2, and (c) G_3.

7. Which nets in Fig. P-12-7 are pseudo-EWC nets?

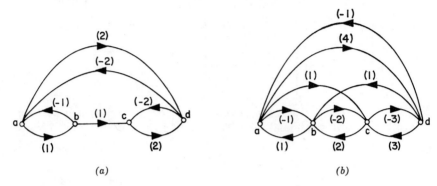

(a)   (b)

Fig. P-12-7. Nets (a) G_1 and (b) G_2.

8. Obtain a fundamental pseudo-EWC net of each of the oriented EWC nets in Fig. P-12-8.

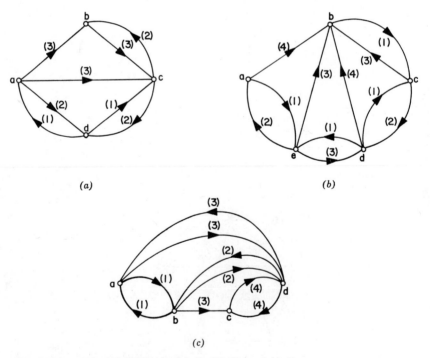

(a)   (b)

(c)

Fig. P-12-8. Oriented EWC nets (a) G_1, (b) G_2, and (c) G_3.

9. Obtain an oriented EWC net from each of the pseudo-EWC nets in Fig. P-12-9 by the shifting algorithm.

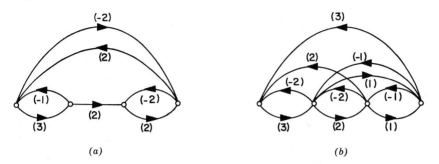

Fig. P-12-9. Pseudo-EWC nets (*a*) G_1 and (*b*) G_2.

10. Find terminal capacity matrices $\overline{T}$ and $\underline{T}$ of the lossy EWC nets in Fig. P-12-10.

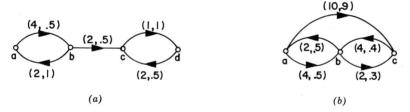

Fig. P-12-10. Lossy EWC nets (*a*) G_1 and (*b*) G_2.

11. Let R be the flow reliability under ψ_{ij} of an EWC net G. Let G' be an EWC net obtained from G by inserting edge e. Prove that the flow reliability R' under ψ_{ij} of G' is not smaller than R when edge e has nonzero capacity and probability $f(e)$ of failure of edge e is less than one.

12. Calculate the flow reliability R under the maximum flow from i to j of the EWC nets in Fig. P-12-12 where the first number inside of the bracket is the edge capacity and the second number is the probability of failure of the edge.

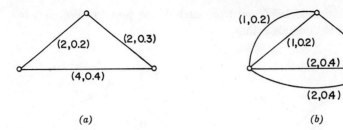

(a)   (b)

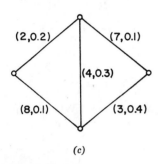

(c)   (d)

Fig. P-12-12.

13. Calculate the flow reliability R under $\psi_{ij} = 4$ and $\psi_{ij} = 2$ of EWC nets in Fig. P-12-12.

14. Find the upper and the lower bounds of the flow reliability R under the maximum flow from i to j of the EWC nets in Fig. P-12-14 by using Theorem 12-7-3.

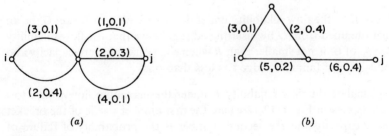

(a)   (b)

Fig. P-12-14.

15. Calculate the flow reliability R under the maximum flow from i to j of the EWC nets in Fig. P-12-15 using Theorem 12-7-2.

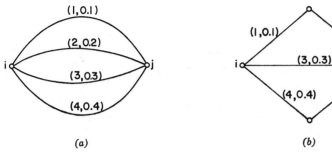

(a)   (b)

Fig. P-12-15.

CHAPTER
13
COMMUNICATION NET THEORY—VERTEX WEIGHTED CASE

13-1 TERMINAL CAPACITY FOR NONORIENTED CASE

Consider a station V in a typical data communication system where each station has many input and output terminals linking stations in the system. When several data are coming at input terminals of station V, these are collected properly. If data are coded, these may be decoded at station V, then data may be modified. For example, if station V is an intermediate station which can supply additional information and if a particular incoming datum required such additional information before it could be sent out to another station, then requested information is inserted to the datum at station V. When data have different destinations, they are grouped so that data in one group can be transmitted to the same output terminal of station V. If necessary, these data are coded and stored in station V until the transmission to a next station becomes possible.

Each station usually consists of equipment for handling data, and there is an upper bound on the number of data that can be handled by a station per unit of time. However, the links (transmission lines) between stations may have enough capacity for transmission of data so that the data flow will not be limited by links. That is, in such a data communication system, a flow of data is limited only by stations. To represent such a system, it is natural to use a linear graph where each vertex rather than each edge has a capacity called a *vertex capacity* indicating the upper bound of a flow through the vertex. Such a linear graph is called a *VWC (Vertex Weighted Communication) net*. Each edge in a VWC net indicates a link between two stations and if edges are nonoriented, flow through these edges can be in either direction. When all edges are nonoriented, it is called a *nonoriented VWC net*. Figure

494

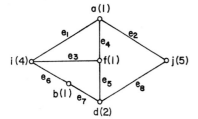

Fig. 13-1-1.   A nonoriented VWC net.

13-1-1 is an example of a nonoriented VWC net where vertex capacity is indicated by the number inside of the parenthesis at each vertex.

As in EWC nets discussed in Chapter 12, a flow is assigned by the use of a path. That is, we can assign a flow ψ_{ij} from i to j to a path P_{ij} if

$$\psi_{ij} \leq c_v - \psi_0(v) \qquad \text{for all vertices } v \text{ in } P_{ij} \qquad (13\text{-}1\text{-}1)$$

where $\psi_0(v)$ is the flow that has been assigned to vertex v previously. For example, we can assign flow ψ_{ij} to the VWC net in Fig. 13-1-1 as follows.

First, we assign $\psi_{1_{ij}} = 1$ to path $P_{1_{ij}} = (e_1, e_4, e_5, e_8)$, which is valid because

$$1 \leq \min \{c_i, c_a, c_d, c_f, c_j\}$$

where c_i is the vertex capacity of vertex i, c_a is the vertex capacity of vertex a, and so on. When we assign $\psi_{1_{ij}}$, we have

$$\psi_0(i) = \psi_0(a) = \psi_0(d) = \psi_0(f) = \psi_0(j) = 1 \qquad \text{and} \quad \psi_0(b) = 0$$

Next we assign $\psi_{2_{ij}} = 1$ to path $P_{2_{ij}} = (e_6, e_7, e_8)$, which is valid because

$$1 \leq \min \{c_i - \psi_0(i), c_b - \psi_0(b), c_d - \psi_0(d), c_j - \psi_0(j)\}$$

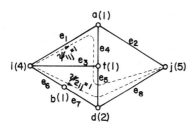

Fig. 13-1-2.   Assignment of $\psi_{1_{ij}} + \psi_{2_{ij}}$.

The result is shown in Fig. 13-1-2. We can see that no more flow from i to j can be assigned. Hence the total flow under this assignment is 2. If we assign differently, the total flow can be increased as shown in Fig. 13-1-3.

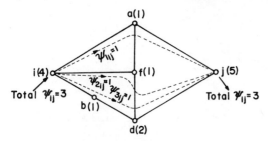

Fig. 13-1-3.   Assignment of $\psi_{ij} = 3$.

In the case of EWC nets, the maximum flow can be calculated if we know cut-sets. Similarly, the maximum flow in a VWC net can be determined if we know so-called vertex-cuts. In order to define vertex-cuts, we must explain the meaning of deleting vertices from a linear graph.

Definition 13-1-1.   Let Ω be a set of vertices in a linear graph G. Then *deleting all vertices in* Ω means to deleting all vertices in Ω and all edges incident at these vertices from G.

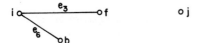

Fig. 13-1-4.   Deletion of vertices a and d.

For example, if we delete vertices a and d from the linear graph in Fig. 13-1-1, the resulting graph does not contain edges e_1, e_2, e_4, e_5, e_7, and e_8 and vertices a and d (Fig. 13-1-4).

A definition of vertex-cut is also needed.

Definition 13-1-2.   Suppose there exists at least one path from vertex i to vertex j in a linear graph. Then a vertex-cut separating i and j is a minimal set of vertices such that the deletion of all vertices in the set destroys all paths from i to j. When the resulting graph after deletion of vertices does not contain either i or j, then all paths between i and j are considered to be destroyed.

Consider the linear graph in Fig. 13-1-1. Set (a, d, f) is not a vertex-cut separating i and j because deleting only vertices a and d will destroy all paths between i and j. However, deleting either a or d (but not both a and d) will not destroy all paths between i and j. Hence (a, d) is a vertex-cut separating i and j (see Fig. 13-1-4). All vertex-cuts separating i and j in the linear graph in Fig. 13-1-1 are (i), (j), (a, b, f), and (a, d).

In order to study the maximum flow in a VWC net, we next define a saturated vertex-cut.

Definition 13-1-3. A vertex is said to be saturated when a flow assigned to the vertex is equal to the capacity of the vertex. A vertex-cut is said to be saturated (or called a saturated vertex-cut) if all vertices in the set are saturated.

As an example of a saturated vertex, consider a flow assigned as in Fig. 13-1-2. Vertex a is saturated because a flow of 1 is assigned to vertex a and the capacity of vertex a is 1. Note that if a vertex is saturated, we cannot assign any more flow to pass through the vertex. In the same figure, vertex-cuts (a, d) and (a, b, f) are saturated. However, we can see that the existence of saturated vertex-cuts under an assignment does not indicate that we have the maximum flow. On the other hand, we can say that no additional flow from i to j can be assigned if there is a saturated vertex-cut separating i and j.

Theorem 13-1-1. If there is a saturated vertex-cut separating vertices i and j when a flow ψ_{ij} from i to j is assigned, no more additional (nonzero) flow from i to j can be assigned.

Proof. Since any path from i to j will pass through a vertex in a vertex-cut separating i and j, and since all vertices in the vertex-cut are saturated, it is obvious that no additional flow from i to j can be assigned.   QED

Similar to basic saturated cut-sets for EWC nets, we define basic saturated vertex-cuts as follows.

Definition 13-1-4. A saturated vertex-cut separating i and j is basic (or called a basic saturated vertex-cut) under an assignment by which the vertex-cut becomes saturated, if every path from i to j, to which a nonzero flow is assigned, contains exactly one vertex in the vertex-cut.

Consider the flow assignment in Fig. 13-1-2. Set (a, d) is not a basic saturated vertex-cut because a path (e_1, e_4, e_5, e_8), to which a nonzero flow $\psi_{1_{ij}}$ is assigned, passes both vertices a and d. In the flow assignment in Fig. 13-1-3, a set (a, d) is a basic saturated vertex-cut. Also set (a, b, f) is a basic saturated vertex-cut. With this definition, we have the following important theorem.

Theorem 13-1-2. For a nonoriented VWC net, a flow from i to j is the maximum if and only if there exists a basic saturated vertex-cut separating i and j.

Before proving this theorem, we study a related lemma.

Lemma 13-1-1: Let W_1 and W_2 be vertex-cuts separating vertices i and j. Let G_2 be a maximal connected subgraph containing vertex j and G_1 be the remaining subgraph in the resultant graph obtained by deleting all vertices in W_1. Note that G_1 and G_2 may consist of vertices only. If W_2 contains vertices in both G_1 and G_2, then there exists a vertex-cut W separating i and j such that $W \subset W_1 \cup W_2$ and W does not contain any vertex in G_1.

Proof.   Consider the linear graph in Fig. 13-1-5 where all vertices in the linear graph are in the disjoint sets Ω_{11}, Ω_{12}, Ω_{21}, Ω_{22}, Ω_A, Ω_I, Ω_B, U_A, and U_B, vertex-cut $W_1 = \Omega_A \cup \Omega_I \cup \Omega_B$, vertex-cut $W_2 = U_A \cup \Omega_I \cup U_B$, and $i \in \Omega_{11}$. Note that G_1 consists of all vertices in Ω_{11}, Ω_{12}, and U_A and all edges connected between these vertices, and G_2 consists of all vertices in Ω_{21}, Ω_{22}, and U_B and all edges connected between these vertices. Also note that the linear graph in the figure is general enough for use in proving the lemma.

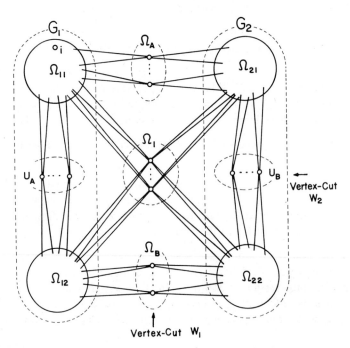

Fig. 13-1-5.   Vertex-cuts W_1 and W_2.

Since $j \in G_2$, j can be in either Ω_{21} or Ω_{22}. When $j \in \Omega_{21}$, it is clear that set $\Omega_A \cup U_B \cup \Omega_I$ contains a set which is a desired vertex-cut separating i and j. Similarly, when $j \in \Omega_{22}$, set $U_B \cup \Omega_I \cup \Omega_B$ contains a set which is a desired vertex-cut separating i and j.   QED

Definition 13-1-5. Let W_0 be a saturated vertex-cut separating i and j, G_2 be a maximal connected subgraph containing vertex j, and G_1 be the remaining subgraph in the resultant graph obtained by deleting all vertices in W_0. Then W_0 is said to be the *closest saturated vertex-cut* to vertex j if there are no other saturated vertex-cuts separating i and j which contain vertices in G_2.

Vertex-cut (a, d) in Fig. 13-1-2 is the closest saturated vertex-cut to j because there are no other saturated vertex-cuts separating i and j which contains vertex j. Note that G_2 in this case consists of one vertex, which is j. Also note that the existence of such a vertex-cut is guaranteed by Lemma 13-1-1.

With this definition, we can prove Theorem 13-1-2 as follows. It is obvious that we can assign additional flow from i to j if there are no saturated vertex-cuts separating i and j. Thus if the maximum flow from i to j has been assigned to a nonoriented VWC net G, we can assume that there is at least one saturated vertex-cut separating i and j.

Suppose there are k saturated vertex-cuts separating i and j but none of these are basic saturated vertex-cuts. Let W_0 be the closest saturated vertex-cut to j as shown in Fig. 13-1-6. Since W_0 is not a basic saturated vertex-cut,

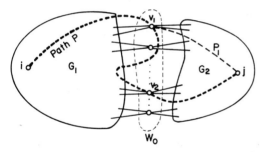

Fig. 13-1-6. Saturated vertex-cut W_0 and paths P and P_1'.

there exists a path from i to j which passes more than one vertex in W_0 to which a nonzero flow $\psi_{1_{ij}}$ has been assigned. Let this path be P and the first vertex in W_0 which passes be v_1 as shown in the figure. Then there exists a path P_1 from v_1 to j which passes only vertices in G_2 to which a nonzero flow from v_1 to j can be assigned. Let P' be a path from i to j formed by path P_1 and a portion of path P from i to v_1. By changing the flow which has been assigned to P to $\psi_{1_{ij}} - \delta$ and assigning δ to path P', we can maintain the same amount of flow from i to j for some nonzero δ. However, because there is at least one vertex in W_0 whose flow assignment has been reduced by δ, W_0 is no longer saturated. Thus we are reducing the number of saturated vertex-cuts and producing no basic saturated vertex-cuts by reassigning a flow from i to j. This is possible as long as there is at least one saturated vertex-cut but no basic saturated vertex-cuts separating i and j. Thus we can reduce the number k of saturated vertex-cuts to zero. Since we can assign an additional flow when there are no saturated vertex-cuts separating i and j, the assigned flow cannot be the maximum, which is a contradiction. Thus there must be at least one basic saturated vertex-cut separating i and j. The proof of the converse is left to the reader.

To obtain the maximum flow from vertex-cuts, we define the value of a vertex-cut as follows.

Definition 13-1-6. The symbol $V[W_i]$ indicates the sum of the vertex capacities of all vertices in set W_i, which is called the *value* of set W_i.

As an example, the value of a vertex-cut (a, d) of the linear graph in Fig. 13-1-1 is $V[(a, d)] = 1 + 2 = 3$.

When the maximum flow has been assigned from i to j, there is a basic saturated vertex-cut separating i and j by Theorem 13-1-2. Each path from i to j to which a nonzero flow has been assigned to obtain the maximum flow passes exactly one vertex in a basic saturated vertex-cut by definition. Hence if W_0 is a basic saturated vertex-cut separating i and j, then the value of W_0 must be equal to the maximum flow from i to j. As in the case of EWC nets, if we use a terminal capacity t_{ij} to indicate the maximum flow from i to j, we have

$$t_{ij} = V[W_0] \tag{13-1-2}$$

Let $\{W\}$ be the all possible vertex-cuts separating i and j. Then we can see that

$$V[W_0] \leq V[W] \quad \text{for every } W \text{ in } \{W\} \tag{13-1-3}$$

Theorem 13-1-3. For a nonoriented VWC net, terminal capacity t_{ij} from i to j is equal to

$$t_{ij} = \min \{V[W]; \quad W \in \{W\}\} \tag{13-1-4}$$

Note that the similarity between this and Theorem 12-1-2.

Example 13-1-1. Consider nonoriented VWC net G in Fig. 13-1-1. All vertex-cuts separating i and j are (i), (j), (a, b, f), and (a, d). The values of these vertex-cuts are

$$V[(i)] = 4$$
$$V[(j)] = 5$$
$$V[(a, b, f)] = 3$$

and

$$V[(a, d)] = 3$$

Thus the terminal capacity t_{ij} from i to j is

$$t_{ij} = \min \{4, 5, 3, 3\} = 3$$

When the maximum flow from i to d is under consideration, we use all vertex-cuts separating i and d, which are (i), (d), (a, b, f), and (b, f, j). Then the terminal capacity t_{id} from i to d is

$$t_{id} = \min \{4, 2, 3, 7\} = 2$$

To discuss the relationship among terminal capacities, it is important to have at least one vertex-cut that is basic saturated under any assignment of the maximum flow from a vertex to another vertex. Such a vertex-cut is called the corresponding vertex-cut of a terminal capacity.

Definition 13-1-7. If W_0 is a basic saturated vertex-cut separating i and j for every assignment of the maximum flow from i to j, then W_0 is called the *corresponding vertex-cut* of terminal capacity t_{ij}.

To show the existence of the corresponding vertex-cut for t_{ij}, we consider otherwise and obtain a contradiction as follows.

Let $\{W_{11} W_{12} \cdots W_{1k_1}\}$ be a set of all basic saturated vertex-cuts separating i and j produced by an assignment of the maximum flow. Let this assignment be accomplished by assigning $\psi_{r_{ij}}$ to path P_{r_1} from i to j for $r = 1, 2, \ldots, m_1$. Let $\{W_{21} W_{22} \cdots W_{2k_2}\}$ be a set of all basic saturated vertex-cuts separating i and j, which are produced by another assignment of the maximum flow. Suppose this assignment is accomplished by assigning $\tilde{\psi}_{s_{ij}}$ to path P_{s2} from i to j for $s = 1, 2, \ldots, m_2$. Now we assume that $\{W_{11} W_{12} \cdots W_{1k_1}\} \cap \{W_{21} W_{22} \cdots W_{2k_2}\} = \emptyset$. By assigning $\alpha \psi_{r_{ij}}$ to P_{r1} and $(1 - \alpha)\tilde{\psi}_{s_{ij}}$ to P_{s2} for $r = 1, 2, \ldots, m_1$ and $s = 1, 2, \ldots, m_2$ where $0 < \alpha < 1$, we will have a new assignment of the maximum flow from i to j. Since any W_{1p} $(1 \leq p \leq k_1)$ is not in $\{W_{21} W_{22} \cdots W_{2k_2}\}$, the new assignment will not make W_{1p} a basic saturated vertex-cut. Similarly, any W_{2q} $(1 \leq q \leq k_2)$ is not in $\{W_{11} W_{12} \cdots W_{1k_1}\}$, thus the new assignment will not make W_{2q} a basic saturated vertex-cut. Furthermore, this new assignment will not make any other vertex-cut a basic saturated vertex-cut. We see that there are no basic saturated vertex-cuts by this new assignment. This means that the flow is not the maximum, which is a contradiction. Thus there is at least one vertex-cut which is in both $\{W_{11} W_{12} \cdots W_{1k_1}\}$ and $\{W_{21} W_{22} \cdots W_{2k_2}\}$.

Suppose $\{W_{p1} W_{p2} \cdots W_{pk_p}\}$ is a set of basic saturated vertex-cuts separating i and j which are produced by assignment $\psi_{ij}^{(p)}$ of the maximum flow for $p = 1, 2, \ldots, n$. By the similar argument, we can show that there is at least one basic saturated vertex-cut W_0 in all of these sets. That is,

$$W_0 \subset \bigcap_{p=1}^{n} \{W_{p1} W_{p2} \cdots W_{pk_p}\} \qquad (13\text{-}1\text{-}5)$$

and we can conclude that there is the corresponding vertex-cut of a terminal capacity. With this corresponding vertex-cut, we can prove the following important theorem.

Theorem 13-1-4.   For any three vertices, i, j, and k,

$$t_{ij} \geq \min \{t_{ik}, t_{kj}\} \qquad (13\text{-}1\text{-}6)$$

in a nonoriented VWC net.

Proof.   Consider the linear graph in Fig. 13-1-7 where W_0 is the corresponding vertex-cut of terminal capacity t_{ij}.

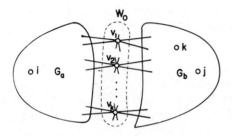

Fig. 13-1-7.   Vertices i, j, and k and vertex-cut W_0.

CASE 1.   Suppose vertex k is in G_b. Then if the corresponding vertex-cut W_1 of t_{ik} is also a vertex-cut separating i and j, $V[W_1]$ must be equal to $V[W_0]$. Thus $t_{ij} = t_{ik}$. If W_1 is not a vertex-cut separating i and j as shown in Fig. 13-1-8, then t_{ik} cannot be larger than t_{ij} because W_0 is also a vertex-cut separating i and k. Thus the theorem is true for this case.

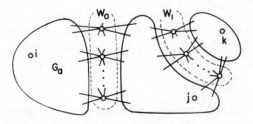

Fig. 13-1-8.   Vertex-cuts W_0 and W_1.

CASE 2.   Suppose vertex k is in W_0. Then it is clear that $t_{ij} \geq t_{ik}$. Thus Eq. 13-1-6 is true for this case.

CASE 3. Suppose vertex k is in G_a. Then instead of considering t_{ik}, if we consider t_{kj}, we will have the same result as in Case 1. That is, $t_{ij} \geq t_{kj}$. Thus the theorem is true for this case.

When either i or j forms a vertex-cut W_0, we have the same situation as Cases 1, 2, or 3 depending on the location of vertex k. Hence the theorem is true.   QED

This triangular relationship is exactly the same as that for EWC nets. As an example, consider the VWC net in Fig. 13-1-1. In Example 13-1-1, we found terminal capacities $t_{ij} = 3$ and $t_{id} = 2$. It is clear that for this case $t_{ij} \geq \min\{t_{id}, t_{dj}\}$. Let us see whether $t_{id} \geq \min\{t_{ij}, t_{jd}\}$ holds. Since $t_{ij} > t_{id}$, if Theorem 13-1-4 is true, we must have $t_{id} \geq t_{jd}$. Calculating t_{jd} by using Theorem 13-1-3 as

$$t_{jd} = \min\{2, 5\} = 2$$

we can see that it is true.

13-2 TERMINAL CAPACITY MATRIX OF NONORIENTED VWC NET

A terminal capacity matrix of a VWC net is defined exactly as is that of an EWC net: Each entry of a terminal capacity matrix T is

$$t_{ij} = \text{terminal capacity from } i \text{ to } j \qquad \text{for } i \neq j$$

and

$$t_{ii} = d$$

(Often t_{ii} is defined as the vertex capacity of vertex i. However, the symbol d is used here so that a terminal capacity matrix has the same form for both EWC and VWC nets.)

As an example, consider the VWC net in Fig. 13-2-1. The terminal capacity matrix T is

$$
T = \begin{array}{c c} & \begin{array}{c c c c c} a & b & c & d & e \end{array} \\ \begin{array}{c} a \\ b \\ c \\ d \\ e \end{array} & \left[\begin{array}{c c c c c} d & 2 & 1 & 4 & 6 \\ 2 & d & 1 & 2 & 2 \\ 1 & 1 & d & 1 & 1 \\ 4 & 2 & 1 & d & 4 \\ 6 & 2 & 1 & 4 & d \end{array}\right] \end{array}
$$

Since we are dealing with nonoriented VWC nets, the terminal capacity matrix is symmetric. Furthermore, because of the existence of the corresponding vertex-cut of a terminal capacity, the matrix has the principal

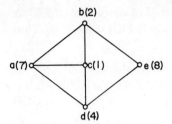

Fig. 13-2-1.   VWC net.

partionality property which is identical to that for the terminal capacity matrices of EWC nets.

Theorem 13-2-1.   Let T be a terminal capacity matrix of a nonoriented VWC net. Then a principal partition can be applied to T and to every resultant main submatrix of order more than one obtained by the principal partitions.

The proof is very similar to that used for proving the necessary part of Theorem 12-2-1 of EWC nets.

As an example, by rearranging rows and columns of the terminal capacity matrix of the VWC net in Fig. 13-2-1 to

$$T = \begin{array}{c} \\ c \\ b \\ d \\ a \\ e \end{array} \begin{array}{c} \begin{array}{ccccc} c & b & d & a & e \end{array} \\ \left[\begin{array}{c:c:ccc} d & 1 & 1 & 1 & 1 \\ \hdashline 1 & d & 2 & 2 & 2 \\ \hdashline 1 & 2 & d & 4 & 4 \\ \hdashline 1 & 2 & 4 & d & 6 \\ \hdashline 1 & 2 & 4 & 6 & d \end{array} \right] \end{array}$$

we can see that Theorem 13-2-1 holds.

Consider the following matrix:

$$M = \left[\begin{array}{c:c:cc} d & 2 & 1 & 1 \\ \hdashline 2 & d & 1 & 1 \\ \hdashline 1 & 1 & d & 3 \\ \hdashline 1 & 1 & 3 & d \end{array} \right]$$

We can show that it is not a terminal capacity matrix of a nonoriented VWC net consisting of four vertices. However, principal partitions can be applied to the matrix and all resultant main submatrices. This indicates that the condition in Theorem 13-2-1 is not sufficient for a matrix to be a terminal capacity matrix of VWC nets which have the same number of vertices as the order of the matrix. (Recall that the same condition given by Theorem 12-2-1 is also sufficient for EWC nets.)

One reason is that the terminal capacities of VWC nets have peculiar properties which those of EWC do not have. To find these properties, let W_0 be the corresponding vertex-cut of a terminal capacity t_{ij} from i to j as shown in Fig. 13-2-2 where W_0 is assumed to be $(v_1, v_2, \ldots, v_k)$. Also let $c_1, c_2, \ldots, c_k$ be the vertex capacities of vertices $v_1, v_2, \ldots, v_k$, respectively. Note that $c_1 + c_2 + \cdots + c_k$ is equal to t_{ij}.

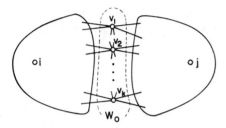

Fig. 13-2-2. Vertex-cut W_0.

To find a terminal capacity t_{iv_1} from i to v_1, we recall how to assign a maximum flow. That is, when we assign the maximum flow $\psi_{ij} = \sum_{r=1}^{n} \psi_{r_{ij}}$ from i to j, each $\psi_{r_{ij}}$ is assigned to a path P_r from i to j. Suppose that flows $\psi_{1_{ij}}, \psi_{2_{ij}}, \ldots, \psi_{m_{ij}}$ ($m \leq n$) have been assigned to paths $P_1, P_2, \ldots, P_m$ from i to j all of which pass through vertex v_1. Also

$$\sum_{r=1}^{m} \psi_{r_{ij}} = c_1 \tag{13-2-1}$$

which is true because vertex v_1 is saturated by the assignment. Since the maximum flow from i to v_1 can be assigned by using the part of the same flow $\psi_{1_{ij}}, \psi_{2_{ij}}, \ldots, \psi_{m_{ij}}$ to assigning to the part of paths $P_1, P_2, \ldots, P_m$, the maximum flow from i to v_1 cannot be less than c_1. Moreover, t_{iv_1} cannot be larger than c_1. Hence $t_{iv_1} = c_1$. This is also true for every terminal capacity t_{iv_p} ($p = 1, 2, \ldots, k$).

Theorem 13-2-2. Let $W_0 = (v_1, v_2, \ldots, v_k)$ be the corresponding vertex-cut of a terminal capacity t_{ij}. Then $t_{iv_p} = c_p$ for $p = 1, 2, \ldots, k$ and

$$t_{ij} = \sum_{p=1}^{k} t_{iv_p} \tag{13-2-2}$$

where c_p is the vertex capacity of vertex v_p.

Theorem 13-2-3. Suppose terminal capacity t_{ij} is not equal to c_i or c_j, which are the vertex capacities of i and j, respectively. Then there are terminal capacities t_{iv_p} for $p = 1, 2, \ldots, m$ such that

$$t_{ij} = \sum_{p=1}^{m} t_{iv_p} \qquad (13\text{-}2\text{-}3)$$

Another interesting property is given by the following lemma.

Lemma 13-2-1. Let t_{ij} be the smallest terminal capacity in a VWC net. Then

$$t_{ij} = \min \{c_i, c_j\} \qquad (13\text{-}2\text{-}4)$$

where c_i and c_j are vertex capacities of vertices i and j, respectively.

With this lemma, we should always be able to partition a terminal capacity matrix t of a nonoriented VWC net as

$$T = \begin{bmatrix} d & \vdots & T_{11} \\ \cdots & \cdots & \cdots \\ T_{11}{}^t & \vdots & T_{22} \end{bmatrix} \qquad (13\text{-}2\text{-}5)$$

where T_{11} is a row matrix consisting of identical entries that are the smallest in T. For convenience, we call this partition a *simple partition*. Note that a simple partition has at most one resultant main submatrix T_{22} consisting of more than one entry. A sufficient condition of a terminal capacity matrix can be stated by the use of simple partitions as follows.

Theorem 13-2-4. If simple partition can be applied to a matrix and all resultant main submatrices containing more than one entry, then the matrix is a terminal capacity matrix of a nonoriented VWC net.

Proof. By the condition, a matrix M can be expressed as

$$
M =
\begin{array}{c}

\end{array}
\begin{array}{c}
1 \\ 2 \\ \\ \vdots \\ \\ n+1
\end{array}
\begin{bmatrix}
d & & & & M_1 & & \\
\hline
 & d & & & M_2 & & \\
\hline
M_1{}^t & & & & & & \\
 & M_2{}^t & & \ddots & & & \\
 & & & & d & | & M_n \\
 & & & & \hline & & \\
 & & & & M_n{}^t & | & d
\end{bmatrix}
\qquad (13\text{-}2\text{-}6)
$$

where $M_r = [m_{rr+1} m_{rr+2} \cdots m_{rn}]$ is a row matrix for $r = 1, 2, \ldots, n$ and

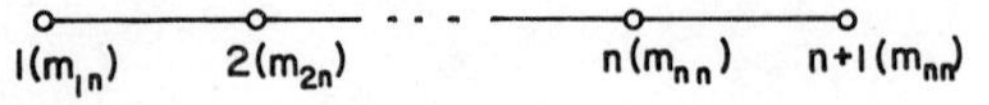

$1(m_{1n})$ $2(m_{2n})$ $\cdots$ $n(m_{nn})$ $n+1(m_{nn})$

Fig. 13-2-3. A VWC net.

$m_{1n} \le m_{2n} \le \cdots \le m_{nm}$. Note that M_n consists of one entry m_{nn}. A VWC net whose terminal capacity matrix is M is shown in Fig. 13-2-3.   QED

If a VWC net can have more vertices than the order of a given matrix, we can design a net rather easily. For example, a matrix

$$
M = \begin{array}{c} \\ a \\ b \\ c \\ d \end{array}
\begin{array}{c} \begin{array}{cccc} a & b & c & d \end{array} \\
\left[\begin{array}{cc:c:c} d & 2 & 1 & 1 \\ \hdashline 2 & d & 1 & 1 \\ \hdashline 1 & 1 & d & 3 \\ \hdashline 1 & 1 & 3 & d \end{array} \right] \end{array}
$$

which cannot be a terminal capacity matrix of a VWC net of four vertices can be a part of a terminal capacity matrix of the VWC net in Fig. 13-2-4.

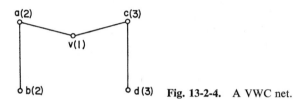

Fig. 13-2-4.   A VWC net.

A condition for a matrix to be a submatrix of a terminal capacity matrix is given by Theorem 13-2-5.

Theorem 13-2-5.   If principal partitions can be applied to a matrix M and to all resultant main submatrices of order more than 2, then M is a submatrix of a terminal capacity matrix of a VWC net.

Proof.   Consider an EWC net whose terminal capacity matrix is M. Note that the condition in this theorem is sufficient for EWC nets. We will change this EWC net to a VWC net whose terminal capacity matrix contains M as a submatrix by the following steps:

STEP 1.   Assign each vertex a vertex capacity of ∞.
STEP 2.   Change each edge e by a series of two edges e' and e'' as shown in Fig. 13-2-5 with the capacity of the common vertex v_e of these edges equal to

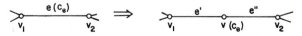

Fig. 13-2-5.   Transformation of edge e to series of edges.

the edge capacity c_e of edge e. We can see that the resultant is a VWC net whose terminal capacity matrix contains M as its submatrix.   QED

As an example, consider matrix M given previously. An EWC net whose terminal capacity is M is shown in Fig. 13-2-6. First we give ∞ to every vertex

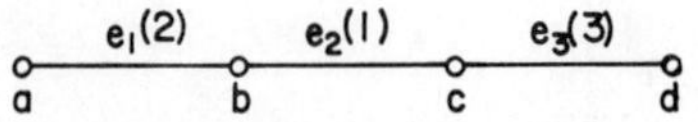

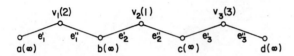

Fig. 13-2-6.   EWC net.

by Step 1. Then by Step 2, each edge e_i is replaced by series edges e_i' and e_i'' for $i = 1, 2, 3$. Then give edge capacity of e_i to vertex v_i as a vertex capacity

Fig. 13-2-7.   VWC net.

for $i = 1, 2, 3$ to obtain the VWC net in Fig. 13-2-7. The terminal capacity matrix of the resultant VWC net is

$$
T =
\begin{array}{c}
\\ a \\ b \\ c \\ d \\ v_1 \\ v_2 \\ v_3
\end{array}
\begin{array}{c}
\begin{array}{ccccccc} a & b & c & d & v_1 & v_2 & v_3 \end{array} \\
\left[
\begin{array}{cccc|ccc}
d & 2 & 1 & 1 & 2 & 1 & 1 \\
2 & d & 1 & 1 & 2 & 1 & 1 \\
1 & 1 & d & 3 & 1 & 1 & 3 \\
1 & 1 & 3 & d & 1 & 1 & 3 \\
\hline
2 & 2 & 1 & 1 & d & 1 & 1 \\
1 & 1 & 1 & 1 & 1 & d & 1 \\
1 & 1 & 3 & 3 & 1 & 1 & d
\end{array}
\right]
\end{array}
$$

where a given matrix M is a submatrix.

13-3   ORIENTED VWC NET

When every edge in a VWC net is oriented, the net is called an *oriented VWC net*. For example, the net in Fig. 13-3-1 is an oriented VWC net. The

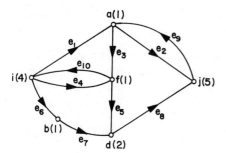

Fig. 13-3-1. Oriented VWC net.

orientation of each edge indicates the direction of flow which this edge can transmit. Hence, in order to assign a flow from vertex i to vertex j, we must assign a flow to a directed path from i to j. For example, we can assign flow $\psi_{ij} = 3$ to the net in Fig. 13-3-1 as follows. To a directed path $P_1 = (e_1, e_2)$, we assign flow $\psi_{1_{ij}} = 1$, which is permissible because

$$1 \leq \min \{c_i, c_a, c_j\}$$

Then we assign flow $\psi_{2_{ij}} = 1$ to a directed path $P_2 = (e_4, e_5, e_8)$, which is valid because

$$1 \leq \min \{c_i - \psi_0(i), c_f - \psi_0(f), c_d - \psi_0(d), c_j - \psi_0(j)\}$$

where $\psi_0(i) = 1$, $\psi_0(f) = 0$, $\psi_0(d) = 0$, $\psi_0(j) = 1$, $c_i = 4$, $c_f = 1$, $c_d = 2$, and $c_j = 5$. Finally, we assign flow $\psi_{3_{ij}} = 1$ to a directed path $P_3 = (e_6, e_7, e_8)$ as shown in Fig. 13-3-2.

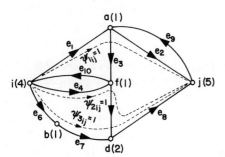

Fig. 13-3-2. Assignment of $\psi_{ij} = 3$.

For oriented VWC nets, the following vertex-semicuts rather than vertex-cuts are employed for obtaining the maximum flow topologically.

Definition 13-3-1. A *vertex-semicut* from i to j is a minimal set of vertices such that the deletion of all vertices in the set destroys all directed paths from

i to j. When the resultant graph after deleting vertices does not contain either i or j, all directed paths from i to j are considered to be destroyed by definition.

As an example, a set (a, d) in the oriented VWC net in Fig. 13-3-1 is a vertex-semicut from i to j as well as a vertex-cut separating i and j. (Recall the semicuts in Section 12-5.) Let $S = \mathscr{E}(\Omega_1 \times \overline{\Omega}_1) \cup \mathscr{E}(\overline{\Omega}_1 \times \Omega_1)$ be a cut-set separating i and j where $i \in \Omega_1$. Then S can be decomposed into two semicuts s_{ij} and s_{ji}:

$$s_{ij} = \mathscr{E}(\Omega_1 \times \overline{\Omega}_1) \tag{13-3-1}$$

and

$$s_{ji} = \mathscr{E}(\overline{\Omega}_1 \times \Omega_1) \tag{13-3-2}$$

which have a property that

$$s_{ij} \cap s_{ji} = \emptyset \tag{13-3-3}$$

However, vertex-semicuts may not have such a property. For example, set (a, d) in the oriented VWC net in Fig. 13-3-1 is a vertex-cut separating i and j and subset (a) is a vertex-semicut from j to i. However, subset (d) is not a vertex-semicut from i to j. A vertex-semicut from i to j which is a subset of (a, d) is (a, d) itself. Thus these two vertex-semicuts have a vertex in common.

With vertex-semicuts, we have a theorem similar to Theorem 12-5-2.

Theorem 13-3-1.   A terminal capacity t_{ij} from i to j of an oriented VWC net is equal to

$$t_{ij} = \min \{V(w_{ij})\} \tag{13-3-4}$$

where $\{V(w_{ij})\}$ is a set of the values of all vertex-semicuts from i to j. Note that t_{ij} is equal to the maximum flow from i to j and the value of a vertex-semicut is the sum of the vertex capacities of all vertices in the vertex-semicut.

The proof of this theorem is the same as that of Theorem 13-1-3 except that vertex-semicuts rather than vertex-cuts are used. Also defining saturated and basic saturated vertex-semicuts need some cautions.

Example 13-3-1.   Consider the oriented VWC net in Fig. 13-3-3. To obtain terminal capacity t_{ij} from i to j, first we collect all vertex-semicuts from i to j, (i), (a, c), (b, c), (b, d), and (j). The values of these vertex-semicuts are

$$V[(i)] = 4, \qquad V[(a, c)] = 4, \qquad V[(b, c)] = 5, \qquad V[(b, d)] = 3,$$

and

$$V[(j)] = 5$$

Hence terminal capacity t_{ij} is

$$t_{ij} = \min \{4, 4, 5, 3, 5\} = 3$$

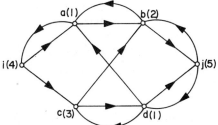

Fig. 13-3-3. Oriented VWC net.

by Theorem 13-3-1. To obtain terminal capacity t_{ji} from j to i, we collect all vertex-semicuts from j to i as $\{(j), (b, d), (a), (i)\}$. Then t_{ji} is

$$t_{ji} = \min V[(j)], V[(b, d)], V[(a)], V[(j)]$$
$$= \min \{5, 3, 1, 4\} = 1$$

Theorem 13-1-4 also holds for oriented VWC nets.

Theorem 13-3-2.   For any vertices i, j, and k in an oriented VWC net,

$$t_{ij} \geq \min (t_{ik}, t_{kj}) \tag{13-3-5}$$

The proof of this theorem is the same as that of Theorem 13-1-4 except that vertex-semicuts rather than vertex-cuts are used. As an example, consider the oriented VWC net in Fig. 13-3-3. Terminal capacity t_{ib} is

$$t_{ib} = \min \{V[(i)], V[(a, c)], V[(b)]\}$$
$$= \min \{4, 4, 2\} = 2$$

Since $t_{ij} = 3$, we have

$$t_{ij} \geq \min \{t_{ib}, t_{bj}\} = \min \{2, t_{bj}\}$$

The definition of a terminal capacity matrix T of an oriented VWC net is the same as that of a nonoriented VWC net and the property of a terminal capacity matrix given by Theorem 13-2-1 is also true for an oriented VWC net. As an example, a terminal capacity matrix of the oriented VWC net shown in Fig. 13-3-1 is

$$
T = \begin{array}{c}
 \\ a \\ f \\ b \\ j \\ d \\ i
\end{array}
\begin{array}{c}
\begin{array}{cccccc} a & f & b & j & d & i \end{array} \\
\left[
\begin{array}{c|c|c|c|cc}
d & 1 & 1 & 1 & 1 & 1 \\ \hline
1 & d & 1 & 1 & 1 & 1 \\ \hline
1 & 1 & d & 1 & 1 & 1 \\ \hline
1 & 1 & 1 & d & 1 & 1 \\ \hline
1 & 1 & 1 & 2 & d & 1 \\ \hline
1 & 1 & 1 & 3 & 2 & d
\end{array}
\right]
\end{array}
$$

which shows one way of applying principal partitions. Note that a terminal capacity matrix of an oriented VWC net generally is not symmetric.

13-4   GENERATION OF VERTEX-CUTS AND VERTEX-SEMICUTS

As we saw in previous sections, vertex-cuts separating i and j and vertex-semicuts from i to j are important sets of vertices in VWC nets. Especially when we need to find a terminal capacity from i to j, it would be convenient to know all vertex-cuts separating i and j for the nonoriented case and all vertex-semicuts from i to j for the oriented case. Hence, if we can generate them as cut-sets separating i and j given in Section 5-1, we will have an easier task of obtaining terminal capacities of VWC nets. Let us review how we generate all cut-sets separating i and j.

We learned in Section 2-3 that all cut-sets, edge disjoint unions of cut-sets, and the empty set form a group under the ring sum. Hence we can generate all cut-sets by knowing generators. Furthermore, if we classify cut-sets and edge disjoint unions of cut-sets into two groups $\{S(i;j)\}$ and $\{S(ij;.)\}$ where $\{S(i;j)\}$ consists of all cut-sets separating i and j and edge disjoint unions of cut-sets which contain an odd number of cut-sets separating i and j, and $\{S(ij;.)\}$ consists of all cut-sets and edge disjoint unions of cut-sets other than those belonging to $\{S(i;j)\}$, we have

1. If S_1 and S_2 are in $\{S(i;j)\}$, then

$$S_1 \oplus S_2 \text{ is in } \{S(ij;.)\}$$

2. If S_1' and S_2' are in $\{S(ij;.)\}$, then

$$S_1' \oplus S_2' \text{ is in } \{S(ij;.)\}$$

3. If S_1 is in $\{S(i;j)\}$ and S_2' is in $\{S(ij;.)\}$, then

$$S_1 \oplus S_2' \text{ is in } \{S(i;j)\}$$

These properties allow us to generate those cut-sets separating i and j from a set of generators only if we know which of these generators separate i and j. Especially if we choose a set of generators $S_0, S_1, \ldots, S_m$ $(m = n_e - n_v + \rho)$ where only S_0 separates i and j, then we know that all cut-sets separating i and j are in the collection

$$\{S(i;j)\} = \{S_0 \oplus S_k; \quad S_k \in \{S'\}\} \tag{13-4-1}$$

where $\{S'\}$ consists of the empty set, cut-sets $S_1, S_2, \ldots, S_m$ and the sets obtained by the ring sum of all possible combinations of $S_1, S_2, \ldots, S_m$. Furthermore, if S_a in collection $\{S_0 \oplus S_k; \quad S_k \in \{S'\}\}$ is not a cut-set, then there exists another set S_b in the same collection such that $S_a \supset S_b$. Thus in the case of EWC nets, terminal capacity is equal to

$$t_{ij} = \min \{V[S_0 \oplus S_k]; \quad S_k \in \{S'\}\} \qquad (13\text{-}4\text{-}2)$$

Unfortunately, a collection of vertex-cuts separating i and j does not have such nice properties. Hence generation of all vertex-cuts separating i and j is not as simple as that of all cut-sets separating i and j. However, because of properties given in the next few theorems, we can use the generation of cut-sets separating i and j to obtain all vertex-cuts separating i and j.

Theorem 13-4-1. For a VWC net G, let W be a vertex-cut separating i and j. Let G_1 and G_2 be the maximal connected subgraphs obtained from G by deleting all vertices in W. Let vertex j be in G_2 except when $W = (j)$; then G_2 is the null graph (which contains no vertices). Then if $\Omega(G_2) \neq \emptyset$, $\mathscr{E}(W \times \Omega(G_2))$ is a cut-set separating i and j where $\Omega(G_2)$ is the set of all vertices in G_2.

As an example, consider the VWC net in Fig. 13-4-1a. For a vertex-cut (a, b) separating i and j, the maximal connected subgraphs G_1 and G_2 are shown in Fig. 13-4-1b. Hence $\Omega(G_2) = (c, d, j)$ and $\mathscr{E}(W \times \Omega(G_2)) =$

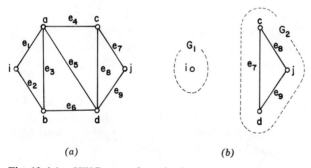

(a)   (b)

Fig. 13-4-1. VWC net and maximal connected subgraphs G_1 and G_2.

(e_4, e_5, e_6) is a cut-set separating i and j. For a vertex-cut (a, d), $\Omega(G_2) = (c, j)$ and $\mathscr{E}(W \times \Omega(G_2)) = (e_4, e_8, e_9)$ which is a cut-set separating i and j. For vertex-cut (i), $\Omega(G_2) = (a, b, c, d, j)$ and $\mathscr{E}(W \times \Omega(G_2)) = (e_1, e_2)$ which is also a cut-set separating i and j.

Proof of Theorem 13-4-1. From the definition of a vertex-cut, every path from i to j must pass at least one vertex in vertex-cut W. Since every edge in $\mathscr{E}(W \times \Omega(G_2))$ is incident at a vertex in W, every path from i to j must contain at least one edge in $\mathscr{E}(W \times \Omega(G_2))$. Hence the deletion of all edges in $\mathscr{E}(W \times \Omega(G_2))$ from a given net will destroy all paths from i to j. Thus $\mathscr{E}(W \times \Omega(G_2))$ contains a cut-set separating i and j. Let G_1' and G_2' be the subgraphs in the graph obtained by deleting all edges in $\mathscr{E}(W \times \Omega(G_2))$ such that G_2' is a maximal connected subgraph and contains vertex j. It is easily seen that G_2' and G_2 are the same. Moreover, $\Omega(G_1') = \Omega(G_1) \cup W$. Hence G_1' consists of edges connected between two vertices in $\Omega(G_1) \cup W$. Now G_1' must be connected because if it is not, there exists at least one vertex in W such that no path from i to j passes the vertex, which contradicts the assumption that W is a vertex-cut separating i and j. Since $\mathscr{E}(W \times \Omega(G_2))$ is a set of edges which connect G_1' and G_2', $\mathscr{E}(W \times \Omega(G_2))$ is a minimal set of edges which makes a connected graph separated by deletion of all edges in the set. Thus $\mathscr{E}(W \times \Omega(G_2))$ is a cut-set separating i and j.   QED

By Theorem 13-4-1, we can obtain a cut-set $\mathscr{E}(W \times \Omega(G_2))$ from each vertex-cut W except $W = (j)$. For convenience, we call cut-set $\mathscr{E}(W \times \Omega(G_2))$ the corresponding cut-set of vertex-cut W.

Definition 13-4-1. Set $\mathscr{E}(W \times \Omega(G_2))$ is called the *corresponding cut-set* of vertex-cut W where G_2 is a maximal connected subgraph containing vertex j which is obtained by deleting all vertices in W.

We want to know whether for any cut-set S separating i and j, there is a vertex-cut separating i and j such that S is the corresponding cut-set of the vertex-cut. The answer is no. As an example, set $S = (e_2, e_3, e_4, e_5)$ in Fig. 13-4-1a is a cut-set separating i and j. However, there is no vertex-cut separating i and j whose corresponding cut-set is S. On the other hand, we have Theorem 13-4-2.

Theorem 13-4-2. Let S be a cut-set separating i and j of a connected linear graph G. Let G_1 and G_2 be the corresponding subgraphs of S (i.e., G_1 and G_2 are the maximal connected subgraphs obtained from G by deleting all edges in S). Suppose $i \in G_1$. Then set W of vertices given by

$$W = \Omega(S) \cap \Omega(G_1) \tag{13-4-3}$$

contains a vertex-cut separating i and j where $\Omega(S)$ is a set of the endpoints of all edges in S.

As an example, $S = (e_5, e_6, e_7, e_8)$ in Fig. 13-4-1a is a cut-set separating i and j. Now $\Omega(S)$ is (a, b, c, d, j) and $\Omega(G_1)$ is (i, a, b, c). Hence set W given by Eq. 13-4-3 is (a, b, c). Clearly this set contains vertex-cut (a, b) separating i and j.

Proof of Theorem 13-4-2. It is only necessary to show that the deletion of all vertices in W will destroy all paths from i to j. Since each edge in S is connected between a vertex in G_1 and a vertex in G_2, set W given by Eq. 13-4-3 has one of two endpoints of each edge in S. Hence deleting all vertices in W will delete all edges in S. We know that the deletion of all edges in S destroys all paths from i to j. Thus the deletion of all vertices in W destroys all paths from i to j. QED

For convenience, set W in Theorem 13-4-2 is called the corresponding vertex-set of a cut-set S. For example, consider cut-sets S_1, S_2, S_3, and S_4 of the VWC net in Fig. 13-4-1a where $S_1 = (e_1, e_2)$, $S_2 = (e_1, e_3, e_6)$, $S_3 = (e_2, e_3, e_4, e_5)$, and $S_4 = (e_4, e_5, e_6)$. The corresponding vertex-sets are

$$W(S_1) = (i)$$
$$W(S_2) = (i, b)$$
$$W(S_3) = (i, a)$$

and

$$W(S_4) = (a, b)$$

Suppose S is not a cut-set but an edge disjoint union of cut-sets in $\{S(i; j)\}$ in Eq. 13-4-1. Then we can form a vertex-set W corresponding to S as follows. Let G_1 and G_2 be disjoint subgraphs obtained by deleting all edges in S such that G_1 is a maximal connected subgraph containing vertex i. Then set W is the collection of the endpoints of all edges in S which are also in G_1.

Note that the deletion of all edges in S with produce ρ (> 2) maximal connected subgraphs, and one of these maximal connected subgraphs which contains vertex i must be chosen as G_1. As an example, an edge disjoint union of cut-sets S in $\{S(i; j)\}$ in Fig. 13-4-1a is $S = (e_1, e_3, e_4, e_5, e_7, e_9)$. Then subgraph G_1 is as shown in Fig. 13-4-2. Hence, vertex-set W will be (i, b, c, d).

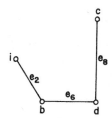

Fig. 13-4-2. Subgraph G_1.

For any set S in $\{S(i; j)\}$, a vertex-set $W(S)$ can be defined as follows.

Definition 13-4-2. Let S be a set in $\{S(i; j)\}$. Let G_1 be a maximal connected subgraph containing vertex i when all edges in S are deleted. Then the corresponding vertex-set of a set S, symbolized by $W(S)$, is given by

$$W(S) = \Omega(S) \cap \Omega(G_1) \qquad (13\text{-}4\text{-}4)$$

where $\Omega(S)$ is a collection of the endpoints of all edges in S and $\Omega(G_1)$ is a collection of all vertices in G_1.

With this definition, we can extend Theorem 13-4-2.

Theorem 13-4-3. Let S be a set in $\{S(i;j)\}$. Then the corresponding vertex-set $W(S)$ of set S contains a vertex-cut separating i and j.

The proof of this theorem is left to the reader.

Since $\{S(i;j)\}$ can be generated easily, if we can relate $\{S(i;j)\}$ and a terminal capacity t_{ij} of a VWC net, we will have an easy time of obtaining t_{ij}. The next theorem gives such a relationship.

Theorem 13-4-4. A terminal capacity t_{ij} of a VWC net is equal to

$$t_{ij} = \min\{V(j), V[W(S)]; \quad S \in \{S(i;j)\}\} \qquad (13\text{-}4\text{-}5)$$

where $V[W(S)]$ is the sum of the vertex capacities of all vertices in set $W(S)$.

Proof. For any vertex-cut W separating i and j except $W = (j)$, there exists S in $\{S(i;j)\}$ which is the corresponding cut-set by Theorem 13-4-1. We can see from Definition 13-4-2 that the vertex-cut $W(S)$ given by Eq. 13-4-4 is a vertex-cut W whose corresponding cut-set is S. Thus the collection $\{V(j), V[W(S)]; \quad S \in \{S(i;j)\}\}$ contains collection $\{V(W); \quad$ all vertex-cut W separating i and $j\}$. Furthermore, for any set S' in $\{S(i;j)\}$, set, $W(S')$ contains a vertex-cut, say W, by Theorem 13-4-3. Hence

$$\min\{V(j), V[W(S)]; \quad S \in \{S(i;j)\}\}$$
$$= \min\{V(W); \quad \text{all vertex-cut } W \text{ separating } i \text{ and } j\} \qquad (13\text{-}4\text{-}6)$$

Since the right-hand side of Eq. 13-4-6 is equal to the terminal capacity t_{ij} from i to j by Theorem 13-1-3, this theorem is true.   QED

By this theorem, we can obtain a terminal capacity of a VWC net by generating collection $\{S(i;j)\}$ as shown in the following example.

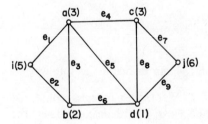

Fig. 13-4-3.   VWC net.

Example 13-4-1. From the VWC net in Fig. 13-4-3, all incidence sets except that corresponding to vertex j are

$$S(i) = (e_1, e_2), \quad S(a) = (e_1, e_3, e_4, e_5), \quad S(b) = (e_2, e_3, e_6),$$
$$S(c) = (e_4, e_7, e_8), \quad \text{and} \quad S(d) = (e_5, e_6, e_8, e_9)$$

Since only $S(i)$ is the cut-set separating i and j, collection $\{S(i; j)\}$ can be expressed by Eq. 13-4-1. Hence all sets in $\{S(i; j)\}$, $W(S)$, and $V[W(S)]$ are those shown in Table 13-4-1. Thus

Table 13-4-1 $\{S(i; j)\}$, $W(S)$, and $V[W(S)]$

$\{S(i; j)\}$	$W(S)$	$V(j)$ and $V[W(S)]$
	(j)	6
$S(i) = (e_1, e_2)$	(i)	5
$S(i) \oplus S(a) = (e_2, e_3, e_4, e_5)$	(i, a)	8
$S(i) \oplus S(b) = (e_1, e_3, e_6)$	(i, b)	7
$S(i) \oplus S(c) = (e_1, e_2, e_4, e_7, e_8)$	(i)	5
$S(i) \oplus S(d) = (e_1, e_2, e_5, e_6, e_8, e_9)$	(i)	5
$S(i) \oplus S(a) \oplus S(b) = (e_4, e_5, e_6)$	(a, b)	5
$S(i) \oplus S(a) \oplus S(c) = (e_2, e_3, e_5, e_7, e_8)$	(i, a, c)	11
$S(i) \oplus S(a) \oplus S(d) = (e_2, e_3, e_4, e_6, e_8, e_9)$	(i, a, d)	9
$S(i) \oplus S(b) \oplus S(c) = (e_1, e_3, e_4, e_6, e_7, e_8)$	(i, b)	7
$S(i) \oplus S(b) \oplus S(d) = (e_1, e_3, e_5, e_8, e_9)$	(i, b, d)	8
$S(i) \oplus S(c) \oplus S(d) = (e_1, e_2, e_4, e_5, e_6, e_7, e_9)$	(i)	5
$S(i) \oplus S(a) \oplus S(b) \oplus S(c) = (e_5, e_6, e_7, e_8)$	(a, b, c)	8
$S(i) \oplus S(a) \oplus S(b) \oplus S(d) = (e_4, e_8, e_9)$	(a, d)	4
$S(i) \oplus S(a) \oplus S(c) \oplus S(d) = (e_2, e_3, e_6, e_7, e_9)$	(i, a, c, d)	12
$S(i) \oplus S(b) \oplus S(c) \oplus S(d) =$		
$\quad (e_1, e_3, e_4, e_5, e_7, e_9)$	(i, b, c, d)	11
$S(i) \oplus S(a) \oplus S(b) \oplus S(c) \oplus S(d) = (e_7, e_9)$	(c, d)	4

$$t_{ij} = \min \{V(j), V[W(S)]; \quad S \in \{S(i; j)\}\} = 4$$

When a VWC net is oriented, we should use vertex-semicuts as discussed in the previous section. To obtain vertex-semicuts, we define the corresponding vertex-semicut as follows.

Definition 13-4-3. Let g be a linear graph and $\Omega(g)$ be a set of all vertices in g. Then $\Omega^+(g)$ is a subset of $\Omega(g)$ such that every edge in g is connected from a vertex in $\Omega^+(g)$. In other words, $\Omega^+(g)$ is a set of all vertices in g whose outgoing degree $d^+(v)$ is nonzero.

With this definition, the corresponding vertex-semiset of a set S in $\{S(i; j)\}$ can be defined.

Definition 13-4-4.   Let S be a set in $\{S(i;j)\}$ of an oriented VWC net G. Let G_1 be a maximal connected subgraph containing vertex i of a linear graph obtained from G by deleting all edges in S. Then the corresponding vertex-semiset, symbolized by $w(S)$, of set S is a set of vertices given by

$$w(S) = \Omega^+(S) \cap \Omega(G_1) \tag{13-4-7}$$

We next show that for any vertex-semicut from i to j, there is a cut-set S separating i and j such that $w(S)$ is the vertex-semicut from i to j.

Theorem 13-4-5.   For any vertex-semicut w from i to j except $w = (j)$ in an oriented VWC net, there exists a cut-set S separating i and j such that w is as given in Eq. 13-4-7.

Proof.   We form a set of edges s_{ij} by taking all edges satisfying the following two conditions:

1. Each edge is connected from a vertex in w (its orientation is from a vertex in w).

2. Each edge is in at least one directed path from i to j which passes exactly one vertex in w.

It is clear that s_{ij} is a semicut from i to j. We know that there is another semicut s_{ji} from j to i such that $s_{ij} \cap s_{ji} = \emptyset$ and $s_{ij} \cup s_{ji} = S$, which is a cut-set separating i and j. With this cut-set S, we can obtain a maximal connected subgraph G_1 containing vertex i. By Theorem 13-4-2, vertex-set $\Omega(S) \cap \Omega(G_1)$ contains a vertex-cut W separating i and j. Hence deleting all vertices in $\Omega(S) \cap \Omega(G_1)$ destroys all directed paths from i to j and all directed paths from j to i. In order to destroy all directed paths from i to j but not necessarily to destroy directed paths from j to i, we only need to replace $\Omega(S)$ by $\Omega^+(S)$. In other words, deleting all vertices in $\Omega^+(S) \cap \Omega(G_1)$ will destroy all directed paths from i to j. Since $S = s_{ij} \cup s_{ji}$,

$$\Omega^+(S) \cap \Omega(G_1) = \Omega^+(s_{ij}) \cap \Omega(G_1) \tag{13-4-8}$$

Furthermore, we know that

$$\Omega^+(s_{ij}) \cap \Omega(G_1) = w \tag{13-4-9}$$

Hence the theorem is true.   QED

As an example, we have the oriented VWC net in Fig. 13-4-4. Consider a vertex-semicut (a, b). Semicut s_{ij} is $s_{ij} = (e_4, e_5, e_6)$ and semicut s_{ji} is $s_{ji} = (e_5', e_{10}', e_{11}')$. Hence $S = (e_4, e_5, e_6, e_5', e_{10}', e_{11}')$. Subgraph G_1 consists of vertices i, a, and b and edges e_1, e_1', e_2, e_2', and e_3. Vertex-set $\Omega(S)$ is $\Omega(S) = (a, b, c, d, i, j)$ and

$$\Omega(S) \cap \Omega(G_1) = (i, a, b)$$

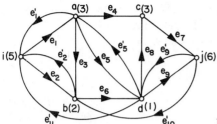

Fig. 13-4-4.   Oriented VWC net.

Set $\Omega^+(S)$ is $\Omega^+(S) = (a, b, d, j)$. Hence

$$\Omega^+(S) \cap \Omega(G_1) = (a, b)$$

which gives the given vertex-semicut. Note that it is also equal to $\Omega^+(s_{ij}) \cap \Omega(G_1)$.

Similar to Theorem 13-4-2, we have Theorem 13-4-6.

Theorem 13-4-6.   Let S be a set in $\{S(i; j)\}$ of an oriented VWC net. Let G_1 be a maximal connected subgraph containing vertex i of a linear graph obtained by deleting all edges in S. Then $\Omega^+(S) \cap \Omega(G_1)$ contains a vertex-semicut from i to j.

The proof of this theorem is left to the reader.

With Theorems 13-4-5 and 13-4-6, we can obtain another important theorem.

Theorem 13-4-7.   A terminal capacity t_{ij} from i to j of an oriented VWC net is equal to

$$t_{ij} = \min \{V(j), V[w(S)]; \quad S \in \{S(i; j)\}\} \qquad (13\text{-}4\text{-}10)$$

where $w(S)$ is given by Eq. 13-4-7.

The proof of this theorem is almost the same as that of Theorem 13-4-4. By this theorem, we can find a terminal capacity of an oriented VWC net by obtaining $\{S(i; j)\}$ as shown in the next example.

Example 13-4-2.   Consider the oriented VWC net in Fig. 13-4-4. All incidence sets except that corresponding to vertex j are

$$S(i) = (e_1, e_1', e_2, e_2', e_{11}')$$
$$S(a) = (e_1, e_1', e_3, e_4, e_5, e_5')$$
$$S(b) = (e_2, e_2', e_3, e_6, e_{10}')$$
$$S(c) = (e_4, e_7, e_8)$$

and

$$S(d) = (e_5, e_5', e_6, e_8, e_9, e_9', e_{11}')$$

Since only $S(i)$ separates i and j, collection $\{S(i;j)\}$ can be expressed by Eq. 13-4-1. Hence all sets in $\{S(i;j)\}$, $w(S)$, and $V[w(S)]$ are those shown in Table 13-4-2. From this table, we can see that $t_{ij} = 4$.

Table 13-4-2 $\{S(i;j)\}$, $w(S)$, $V(j)$, and $V[w(S)]$

$\{S(i;j)\}$	$w(S)$	$V(j)$ and $V[w(S)]$
	(j)	6
$S(i)$	(i)	5
$S(i) \oplus S(a) = (e_2, e'_2, e_3, e_4, e_5, e'_5, e'_{11})$	(i, a)	8
$S(i) \oplus S(b) = (e_1, e'_1, e_3, e_6, e'_{10}, e'_{11})$	(i, b)	7
$S(i) \oplus S(c) = (e_1, e'_1, e_2, e'_2, e_4, e_7, e_8, e'_{11})$	(i)	5
$S(i) \oplus S(d) = (e_1, e'_1, e_2, e'_2, e_5, e'_5, e_6, e_8, e_9, e'_9)$	(i, d)	6
$S(i) \oplus S(a) \oplus S(b) = (e_4, e_5, e'_5, e_6, e'_{10}, e'_{11})$	(a, b)	5
$S(i) \oplus S(a) \oplus S(c) = (e_2, e'_2, e_3, e_5, e'_5, e_7, e_8, e'_{11})$	(i, a, c)	11
$S(i) \oplus S(a) \oplus S(d) = (e_2, e'_2, e_3, e_4, e_6, e_8, e_9, e'_9)$	(i, a, d)	9
$S(i) \oplus S(b) \oplus S(c) = (e_1, e'_1, e_3, e_4, e_6, e_7, e_8, e'_{10}, e'_{11})$	(i, b)	7
$S(i) \oplus S(b) \oplus S(d) = (e_1, e'_1, e_3, e_5, e'_5, e_8, e_9, e'_9, e'_{10})$	(i, d)	6
$S(i) \oplus S(c) \oplus S(d) = (e_1, e'_1, e_2, e'_2, e_4, e_5, e'_5, e_6, e_7,$ $e_9, e'_9)$	(i, c, d)	9
$S(i) \oplus S(a) \oplus S(b) \oplus S(c) = (e_5, e'_5, e_6, e_7, e_8, e'_{10}, e'_{11})$	(a, b, c)	8
$S(i) \oplus S(a) \oplus S(b) \oplus S(d) = (e_4, e_8, e_9, e'_9, e'_{10})$	(a, d)	4
$S(i) \oplus S(a) \oplus S(c) \oplus S(d) = (e_2, e'_2, e_3, e_6, e_7, e_9, e'_9)$	(i, a, c, d)	12
$S(i) \oplus S(b) \oplus S(c) \oplus S(d) = (e_1, e'_1, e_3, e_4, e_5, e'_5, e_7$ $e_9, e'_9, e'_{10})$	(i, c, d)	9
$S(i) \oplus S(a) \oplus S(b) \oplus S(c) \oplus S(d) = (e_7, e_9, e'_9, e'_{10})$	(c, d)	4

PROBLEMS

1. Prove the sufficient part of Theorem 13-1-2.
2. Obtain a terminal capacity matrix of the VWC nets in Fig. P-13-2.
3. Synthesize a VWC net having four vertices whose terminal capacity matrices are the following matrices:

(a)
$$\begin{array}{c} \\ a \\ b \\ c \\ d \end{array} \begin{array}{cccc} a & b & c & d \\ \left[\begin{array}{cccc} d & 1 & 1 & 1 \\ 1 & d & 2 & 2 \\ 1 & 2 & d & 3 \\ 1 & 2 & 3 & d \end{array}\right] \end{array}$$

(b)
$$\begin{array}{c} \\ a \\ b \\ c \\ d \end{array} \begin{array}{cccc} a & b & c & d \\ \left[\begin{array}{cccc} d & 3 & 1 & 2 \\ 3 & d & 1 & 2 \\ 1 & 1 & d & 1 \\ 2 & 2 & 1 & d \end{array}\right] \end{array}$$

(c)
$$\begin{array}{c} \\ a \\ b \\ c \\ d \end{array} \begin{array}{cccc} a & b & c & d \\ \left[\begin{array}{cccc} d & 1 & 1 & 1 \\ 1 & d & 3 & 2 \\ 1 & 3 & d & 2 \\ 1 & 2 & 2 & d \end{array}\right] \end{array}$$

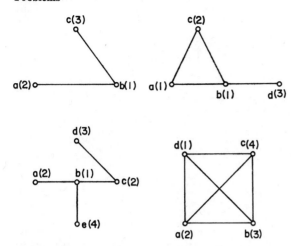

Fig. P-13-2.

4.   Synthesize a VWC net whose terminal capacity matrix contains matrix M as a submatrix where

$$M = \begin{array}{c} \\ a \\ b \\ c \\ d \\ e \end{array} \begin{array}{ccccc} a & b & c & d & e \\ \left[\begin{array}{ccccc} d & 5 & 4 & 1 & 1 \\ 5 & d & 4 & 1 & 1 \\ 4 & 4 & d & 1 & 1 \\ 1 & 1 & 1 & d & 2 \\ 1 & 1 & 1 & 2 & d \end{array}\right] \end{array}$$

5.   Prove Theorem 13-3-1.

6.   Prove Theorem 13-3-2.

7.   Obtain terminal capacity t_{ij} from i to j of the VWC nets in Fig. P-13-7 by generating $\{S(i; j)\}$ first.

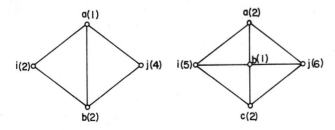

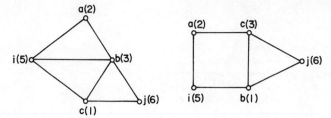

Fig. P-13-7.

8. Prove Theorem 13-4-3.

9. Prove Theorem 13-4-6.

10. Prove Theorem 13-4-7.

11. Obtain terminal capacity t_{ij} from i to j of the oriented VWC nets in Fig. P-13-11 by generating $\{S(i;j)\}$ first.

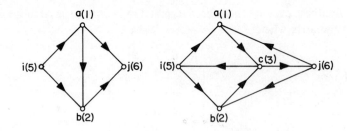

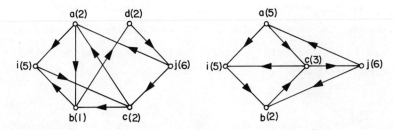

Fig. P-13-11.

CHAPTER
14
SYSTEM DIAGNOSIS

14-1 DISTINGUISHABILITY

The systems we are considering here are those that can be represented by blocks and links where a block performs a certain (known) function and a link transmits information between two blocks. Computer structures, computer programs, and control systems are typical examples. By using vertices for blocks and edges for links, we can obtain linear graphs corresponding to such systems. For example, a microprogrammed control unit of a computer shown in Fig. 14-1-1a is represented by an oriented linear graph as shown in Fig. 14-1-1b.

The terminals from which information can be fed into a system are called *entry terminals* and the terminals at which information can be taken out of a system are called *exit terminals*. By assuming that each entry (and exit) terminal belongs to one block, we can define an entry vertex as a vertex whose corresponding block has entry terminals. Similarly, an exit vertex is one whose corresponding block has exit terminals.

Definition 14-1-1. In a linear graph, a vertex at which an external signal can enter is called an *entry vertex* and a vertex from which a signal can leave is called an *exit vertex*.

We make the following assumption for studying a situation when blocks in a system become defective.

ASSUMPTION. When either an undistorted or a distorted signal is passed through a fault vertex, the signal will be distorted. On the other hand, when an undistorted (distorted) signal is passed through a faultless vertex, the signal will be undistorted (distorted), where a fault vertex is defined as follows.

Definition 14-1-2. A *fault vertex* is a vertex corresponding to a defected block and a *faultless vertex* is a vertex corresponding to a normally functioning block.

523

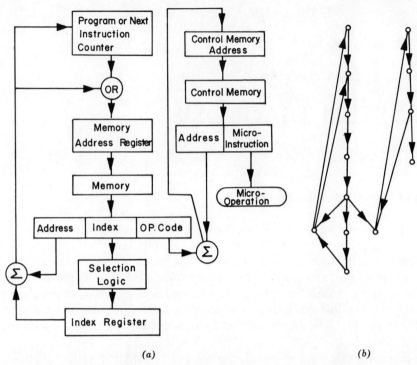

Fig. 14-1-1. A system and corresponding oriented graph. (*a*) A microprogrammed control unit; (*b*) the corresponding oriented graph.

With this assumption, we can see that once a signal is distorted, the signal cannot be undistorted. Hence by monitoring an output signal, we can find out whether the signal has passed a fault vertex or not. In practice, this assumption does not hold for all cases. On the other hand, by this assumption many theoretical results on diagnosis can be obtained easily. Furthermore, these results may be modified to cover particular cases where the assumption does not hold. In order to indicate which vertices influence a signal that is reached at a particular terminal, we use the symbol $\Omega(i \times j)$ defined as follows.

Definition 14-1-3. Let i and j be vertices in an oriented graph G. The symbol $\Omega(i \times j)$ indicates a set of vertices such that vertex p is in $\Omega(i \times j)$ if and only if there exists a connected directed M-graph of type $M(i \times j)$ which contains vertex p. A set $\Omega(i \times j)$ is called a *measurable set*.

Example 14-1-1. Consider the oriented graph in Fig. 14-1-2. A measurable set $\Omega(1 \times 3)$ is $(1, 2, 3, 4)$ because directed path $P_1 = (a, b)$ from 1 to 3

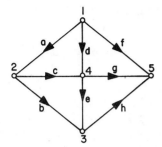

Fig. 14-1-2. Oriented graph.

passes vertices 1, 2, and 3 and directed path $P_2 = (d, e)$ from 1 to 3 passes vertices 1, 3, and 4, but there are no connected directed M-graphs of type $M(1 \times 3)$ which contain vertex 5. Note that a directed path from i to j is a connected directed M-graph of type $M(i \times j)$ (see Definition 6-6-6). A measurable set $\Omega(1 \times 4)$ is $(1, 2, 4)$.

Definition 14-1-4. Let Ω be a set of all vertices in an oriented graph. Then

$$\overline{\Omega(i \times j)} = \Omega - \Omega(i \times j) \tag{14-1-1}$$

For example, $\overline{\Omega(1 \times 4)}$ of the oriented graph in Fig. 14-1-2 is $(3, 5)$.

Consider the case when a measurable set $\Omega(i \times j)$ does not contain all vertices in an oriented graph G. Suppose we inject a signal from vertex i. If the signal which is monitored at vertex j indicates the existence of fault vertices, then the fault vertices must be in $\Omega(i \times j)$. On the other hand, if the signal which is monitored at vertex j indicates the absence of fault vertices but the existence of fault vertices has been known, then it is clear that $\overline{\Omega(i \times j)}$ contains fault vertices.

Suppose monitoring at vertex j indicates the absence of fault vertices but monitoring at vertex k indicates the existence of fault vertices when a signal is injected at vertex i. Then we know that a set

$$\overline{\Omega(i \times j)} \cap \Omega(i \times k)$$

contains fault vertices.

If monitoring signals at vertices j and k under injection of a signal at vertex i are the only tests we can perform and if a fault vertex exists, then we can determine which one of the following sets contains the fault vertex by the preceding test:

$$\Omega(i \times j) \cap \Omega(i \times k)$$
$$\Omega(i \times j) \cap \overline{\Omega(i \times k)}$$
$$\overline{\Omega(i \times j)} \cap \Omega(i \times k)$$
$$\Omega(i \times j) \cap \overline{\Omega(i \times k)}$$

These sets form a collection called a *D*-partition of measurable sets $\Omega(i \times j)$ and $\Omega(i \times k)$.

Definition 14-1-5. For given k measurable sets $\Omega(i_p \times j_p), p = 1, 2, \ldots, k$, a collection of sets

$$\left\{ \bigcap_{p=1}^{k} \overbrace{\Omega(i_p \times j_p)}; \quad \begin{array}{c} \text{all possible combination of } \overbrace{\Omega(i_p \times j_p)}, \\ p = 1, 2, \ldots, k \text{ where } \overbrace{\Omega(i_p \times j_p)} \text{ is either} \\ \Omega(i_p \times j_p) \text{ or } \overline{\Omega(i_p \times j_p)} \end{array} \right\}$$

is called a *D-partition* of those measurable sets.

Example 14-1-2. Consider the oriented graph in Fig. 14-1-2. For the measurable sets

$$\Omega(1 \times 3) = (1, 2, 3, 4)$$
$$\Omega(1 \times 4) = (1, 2, 4)$$

and

$$\Omega(2 \times 5) = (2, 3, 4, 5)$$

a *D*-partition consists of

$$\Omega(1 \times 3) \cap \Omega(1 \times 4) \cap \Omega(2 \times 5) = (2, 4)$$
$$\Omega(1 \times 3) \cap \Omega(1 \times 4) \cap \overline{\Omega(2 \times 5)} = (1)$$
$$\Omega(1 \times 3) \cap \overline{\Omega(1 \times 4)} \cap \Omega(2 \times 5) = (3)$$
$$\Omega(1 \times 3) \cap \overline{\Omega(1 \times 4)} \cap \overline{\Omega(2 \times 5)} = \emptyset$$
$$\overline{\Omega(1 \times 3)} \cap \Omega(1 \times 4) \cap \Omega(2 \times 5) = \emptyset$$
$$\overline{\Omega(1 \times 3)} \cap \Omega(1 \times 4) \cap \overline{\Omega(2 \times 5)} = \emptyset$$
$$\overline{\Omega(1 \times 3)} \cap \overline{\Omega(1 \times 4)} \cap \Omega(2 \times 5) = (5)$$
$$\overline{\Omega(1 \times 3)} \cap \overline{\Omega(1 \times 4)} \cap \Omega(2 \times 5) = \emptyset$$

A *D*-partition has the following property.

Theorem 14-1-1. Any two nonempty sets in a *D*-partition of a collection of measurable sets have no vertices in common.

Proof. We note that $\Omega(p \times q)$ and $\overline{\Omega(p \times q)}$ have no vertices in common. By Definition 14-1-5, any two sets R_1 and R_2 in a *D*-partition have at least one measurable set $\Omega(p \times q)$ such that $\Omega(p \times q)$ is in the expression for R_1 and $\overline{\Omega(p \times q)}$ is in the expression for R_2. Thus R_1 and R_2 have no vertices in common.   QED

Recall that a measurable set $\Omega(p \times q)$ corresponds to a measurement performed by injecting a signal at vertex p and monitoring the signal at vertex

q. Hence if a set in a D-partition $\{D\}$ consists of one vertex v and if only vertex v is false, then we can predict that v is the false vertex by the measurements corresponding to a given collection of measurable sets by which $\{D\}$ is obtained. On the other hand, if two vertices v_1 and v_2 are together in a set in a D-partition under a collection M of measurable sets, and if one of these two vertices is fault, then it is impossible to determine which of these vertices is fault just by the measurement corresponding to measurable sets in M. Thus a given collection of measurable sets determine whether or not fault vertices can be located. Hence we define k-distinguishability as follows.

Definition 14-1-6. An oriented graph is *k-distinguishable* under a collection M of measurable sets if (1) there exists a set R in a D-partition of M such that R contains k vertices, and (2) there are no sets in a D-partition of M which contain more than k vertices.

For example, the oriented graph in Fig. 14-1-2 is 2-distinguishable under a collection M of measurable sets $\Omega(1 \times 3)$, $\Omega(1 \times 4)$, and $\Omega(2 \times 5)$ because there is $(2, 4)$ in a D-partition $\{D\}$ of M and there are no sets in $\{D\}$ consisting of more than two vertices.

Note that an oriented graph G being k-distinguishable under a collection M of measurable sets means that there exists a set of k vertices such that if one of these vertices is fault, the measurement corresponding to these measurable sets cannot determine which one of these k vertices is the fault vertex but these measurements will indicate k vertices which contain a fault vertex.

14-2 TEST POINT

We have defined entry and exit terminals of a system. Any other terminal in a system is called an *inner terminal*, that is, an inner terminal is a part of a block where a link is located which is neither an entry terminal nor an exit terminal. Hence each vertex in a linear graph corresponding to a system indicates either an entry, exit, or inner terminal.

If any inner terminal in a system can be either an entry terminal or an exit terminal, for measurement, then any measurable set $\Omega(p \times q)$ where p and q are vertices can be used as a member of a collection of measurable sets for diagnosis. However, even if all vertices can be entry and exit vertices, 1-distinguishability may not be possible. For example, oriented graph G in Fig. 14-2-1 is not 1-distinguishable under a collection of measurable sets $\Omega(p \times q)$ of all vertices p and q in G. In order to remove such difficulties, we use so-called test gates by which we can delete edges whenever necessary. The properties of test gates and diagnosis of systems with the use of test gates will be studied later.

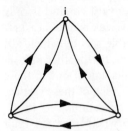

Fig. 14-2-1.   An oriented graph G.

Another obvious case where k-distinguishability for given k becomes impossible is when there are not enough entry and exit vertices. In this case, we use so-called test points to obtain additional measurable sets so that k-distinguishability becomes possible. In other words, we would like to consider a test point as an additional vertex which is placed on an edge so that we can inject as well as monitor a signal for diagnosis. However, we do not like to consider an inserted vertex as a member of a set of vertices in a given linear graph G when we consider the distinguishability of vertices of G. Therefore, we define a test point as follows.

Definition 14-2-1.   A test point being assigned to an edge means that a signal can be injected through the edge and a signal which passes through the edge can be monitored.

Let edge e be connected from vertex v_1 to vertex v_2. Then if a test point is assigned to edge e, we consider vertex $v(e^-)$ as an entry vertex and vertex $v(e^+)$ as an exit vertex.

Definition 14-2-2.   The symbols $v(e^+)$ and $v(e^-)$ indicate the vertices such that edge e is connected from $v(e^+)$ to $v(e^-)$.

For example, in the oriented graph in Fig. 14-1-2, $v(a^+) = 1$ and $v(a^-) = 2$. Similarly, $v(b^+) = 2$ and $v(b^-) = 3$.

For convenience, we have the following definition.

Definition 14-2-3.   The symbol $\Omega(a^{\pm} \times j)$ is a shorthand notation of $\Omega(v(a^{\pm}) \times j)$. Similarly, $\Omega(i \times a^{\pm})$ is a shorthand notation of $\Omega(i \times v(a^{\pm}))$, and $\Omega(a^{\pm} \times b^{\pm})$ is a shorthand notation of $\Omega(v(a^{\pm}) \times v(b^{\pm}))$. This shorthand notation is also used for directed M-graphs such as $M(a^{\pm} \times b^{\pm})$, which indicates $M(v(a^{\pm}) \times v(b^{\pm}))$.

Example 14-2-1.   Consider oriented graph G in Fig. 14-2-2 which has only one entry vertex 1 and only one exit vertex 2. Hence we have only one measurable set $\Omega(1 \times 2) = (1, 2, 3, 4)$. If we assign a test point to edge c, then we have sets $\Omega(1 \times c^+)$ and $\Omega(c^- \times 2)$ as additional measurable sets.

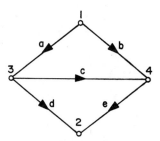

Fig. 14-2-2. An oriented graph.

To study the properties of test points, we consider a connected oriented graph which has only one entry vertex and one exit vertex. In other words, we will define an SEC graph as follows.

Definition 14-2-4. An *SEC (Single Entry Single Exit Connected) graph* is a connected oriented graph in which there is exactly one entry vertex and exactly one exit vertex.

Let i and j be the entry and the exit vertices in an SEC graph, respectively. Since a signal which is injected from vertex i can be obtained only at vertex j, any block that is represented by a vertex not in measurable set $\Omega(i \times j)$ will not contribute to the performance of a system corresponding to the SEC graph. Hence *we consider only an SEC graph whose measurable set $\Omega(i \times j)$ consists of all vertices in the SEC graph.*

Definition 14-2-5. An *SEC graph from i to j* means an SEC graph whose entry vertex is i and whose exit vertex is j.

First we study the relationship between test points and directed cut-sets.

Theorem 14-2-1. Let G be an SEC graph from i to j. Suppose $S(e)$ is a directed cut-set which separates vertices i and j and contains edge e. Further suppose that there is no other directed cut-set separating i and j that contains edge e. Let G_1 and G_2 be the maximal connected subgraphs obtained from G by deleting all edges in $S(e)$. Then $\Omega(i \times e^+) = \Omega(G_1)$ and $\Omega(e^- \times j) = \Omega(G_2)$.

Proof. Let p be a vertex in G_1. Suppose there exists a connected directed M-graph of type $M(e^- \times j)$ which contains vertex p. Then $S(e)$ is not a directed cut-set because at least one edge in $S(e)$ has the orientation from a vertex in $\Omega(G_2)$ to a vertex in $\Omega(G_1)$. Hence there is no directed M-graph of type $M(e^- \times j)$ containing a vertex in $\Omega(G_1)$.

Suppose p is not in any connected directed M-graph of type $M(i \times e^+)$. Let Ω_p be a set of all vertices such that for any vertex v' in Ω_p, there exists a directed path from p to v' in G_1. Since $\Omega(i \times j)$ contains vertex p, there is no directed M-graph of type $M(i \times e^+)$ which passes v' in Ω_p. Note that vertex $v(e^+)$ is not in Ω_p. Also we can see that vertex i is not in Ω_p. Consider a cut-set $S'(e)$ where

$$S'(e) = \mathscr{E}\{[\Omega(G_1) - \Omega_p] \times \overline{[\Omega(G_1) - \Omega_p]}\}$$

$$\cup \; \mathscr{E}\{[\overline{\Omega(G_1) - \Omega_p}] \times [\Omega(G_1) - \Omega_p]\} \qquad (14\text{-}2\text{-}1)$$

Since $S(e)$ is a directed cut-set, $\mathscr{E}[\overline{\Omega(G_1)} \times \Omega(G_1)]$ is empty. Hence $\mathscr{E}\{\overline{\Omega(G_1)} \times [\Omega(G_1) - \Omega_p]\}$ is empty. An edge connected between a vertex v' in Ω_p and a vertex v'' in $[\Omega(G_1) - \Omega_p]$ cannot have its orientation from v' to v'' because if it does v'' must be in Ω_p. Hence $\mathscr{E}\{\Omega_p \times [\Omega(G_1) - \Omega_p]\}$ is empty, and so $\mathscr{E}\{[\overline{\Omega(G_1) - \Omega_p}] \times [\Omega(G_1) - \Omega_p]\}$ is empty. This means that $S'(e)$ is a directed cut-set separating i and j, which is a contradiction because $S(e)$ is the only directed cut-set containing edge e by assumption. Thus p must be in at least one connected directed M-graph of type $M(i \times e^+)$. Similarly, we can take a vertex in G_2 to prove that the theorem is true.   QED

When there is more than one directed cut-set containing edge e and separating i and j, another theorem applies.

Theorem 14-2-2. Let G be an SEC graph from i to j. Suppose k directed cut-sets $S_1(e)$, $S_2(e)$, ..., $S_k(e)$ are those which contain edge e among all directed cut-sets separating i and j. Then $\Omega(i \times e^+) = \Omega(G_1')$ and $\Omega(e^- \times j) = \Omega(G_2')$ where G_1' and G_2' are two maximal connected subgraph obtained from G by deleting all edges in $S_1(e)$, $S_2(e)$, ..., $S_k(e)$.

Proof. A proof for any vertex in G_1' and G_2' would be exactly the same as the proof of Theorem 14-2-1. Therefore we only need to prove that any vertex which is neither in G_1' nor in G_2' cannot be in either $\Omega(i \times e^+)$ or $\Omega(e^- \times j)$. Consider a subgraph G_3' which is obtained by deleting all vertices in G_1' and G_2' and all edges which are connected to these vertices. We can see easily that there exist exactly two cut-sets in $S_1(e)$, $S_2(e)$, ..., $S_k(e)$ such that by deleting all edges in these two cut-sets, we have three subgraphs G_1', G_2', and G_3'. Let these cut-sets be $S_1(e)$ and $S_k(e)$. Also let v' be a vertex in G_3'. If v' is in a connected directed M-graph of type $M(i \times e^+)$, then either $S_1(e)$ or $S_k(e)$ is not a directed cut-set. Thus v' cannot be in any connected directed M-graph of type $M(i \times e^+)$. Similarly, we can show that v' cannot be in any connected directed M-graph of type $M(e^- \times j)$. Thus the theorem is true.
                                                                                            QED

In the proof of Theorem 14-2-2, we use two directed cut-sets $S_1(e)$ and $S_k(e)$ so that the deletion of all edges in these two cut-sets produces G_1', G_2', and G_3'. We say that two directed cut-sets $S_1(e)$ and $S_k(e)$ are covered by edge e, or edge e covers $S_1(e)$ and $S_k(e)$. When there exists only one cut-set $S(e)$ as in Theorem 14-2-1, we then say that edge e covers directed cut-set $S(e)$.

Definition 14-2-6. Edge e is said to cover a directed cut-set S if S is the only directed cut-set which separates i and j and contains edge e. Edge e covers directed cut-sets $S_1 = \mathscr{E}(\Omega_1 \times \overline{\Omega_1})$ and $S_2 = \mathscr{E}(\Omega_2 \times \overline{\Omega_2})$ if and only if (1) S_1 and S_2 separate i and j, (2) S_1 and S_2 contain edge e, and (3) any other cut-set $S' = \mathscr{E}(\Omega_3 \times \overline{\Omega_3})$ separating i and j and containing edge e satisfies the condition

$$(\Omega_1 \cup \Omega_2) \supset \Omega_3 \supset (\Omega_1 \cap \Omega_2) \qquad (14\text{-}2\text{-}2)$$

Example 14-2-2. Consider the SEC graph from 1 to 6 shown in Fig. 14-2-3a. $S(b)$ is the only directed cut-set which contains edge b and separates 1 and 6. Hence edge b covers directed cut-set $S(b)$. By deleting all edges in $S(b)$, we have G_1 and G_2 as shown in Fig. 14-2-3. Thus $\Omega(1 \times b^+) = \Omega(G_1)$ and $\Omega(b^- \times 6) = \Omega(G_2)$.

$S_1(h)$, $S_2(h)$, and $S_3(h)$ are the all directed cut-sets which contain edge h and separate vertices 1 and 6. From the figure, we can see that

$$S_1(h) = \mathscr{E}(\Omega_1 \times \overline{\Omega_1}) = \mathscr{E}((1, 2, 4) \times \overline{(124)})$$
$$S_2(h) = \mathscr{E}(\Omega_2 \times \overline{\Omega_2}) = \mathscr{E}((1, 2, 3, 4, 5) \times \overline{(12345)})$$

and

$$S_3(h) = \mathscr{E}(\Omega_3 \times \overline{\Omega_3}) = \mathscr{E}((1, 2, 3, 4) \times \overline{(1234)})$$

Hence

$$[\Omega_1 \cup \Omega_2 = (1, 2, 3, 4, 5)] \supset [\Omega_3 = (1, 2, 3, 4)] \supset [\Omega_1 \cap \Omega_2) = (124)]$$

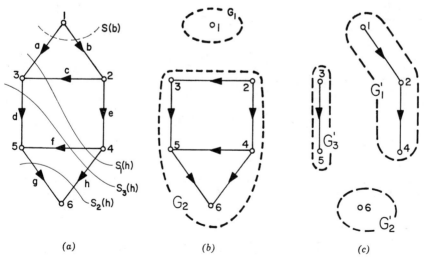

Fig. 14-2-3. Directed cut-sets covered by edges. (*a*) An SEC graph; (*b*) G_1 and G_2; (*c*) G_1', G_2', and G_3'.

Thus edge h covers $S_1(h)$ and $S_2(h)$. Subgraphs G'_1, G'_2, and G'_3 are shown in Fig. 14-2-3c. Note that $\Omega(1 \times h^+) = \Omega(G'_1)$ and $\Omega(h^- \times 6) = \Omega(G'_2)$.

The converse of Theorems 14-2-1 and 14-2-2 is also true.

Theorem 14-2-3. Let G be an SEC graph from i to j. Let $\Omega(i \times e^+)$ and $\Omega(e^- \times j)$ be measurable sets. If $\Omega(i \times e^+) \cap \Omega(e^- \times j) = \emptyset$, then there exists a directed cut-set $S(e)$ and $S'(e)$ containing edge e such that $\Omega(G_e) = \Omega(i \times e^+)$ and $\overline{\Omega(G'_e)} = \Omega(e^- \times j)$ where

$$S(e) = \mathscr{E}(\Omega(G_e) \times \overline{\Omega(G_e)}) \quad \text{and} \quad S'_e(e) = \mathscr{E}(\Omega(G'_e) \times \overline{\Omega(G'_e)})$$

Proof. Consider a set of edges $S = \mathscr{E}(\Omega(i \times e^+) \times \overline{\Omega(i \times e^+)})$ as shown in Fig. 14-2-4. If edge e is not in S, then $\Omega(i \times e^+)$ contains vertex $v(e^-)$.

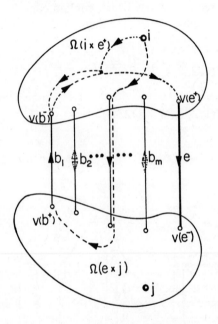

Fig. 14-2-4. Set S.

Since $\Omega(e^- \times j)$ contains vertex $v(e^-)$ by assumption, we have $\Omega(i \times e^+) \cap \Omega(e^- \times j) \neq \emptyset$. Hence edge e must be in S. Let $S = (b_1, b_2, \cdots, b_m, e)$. Suppose S is not a directed cut-set. Then at least one edge, say b_1, has its orientation from a vertex $v(b_1^+)$ in $\Omega(e^- \times j)$ to a vertex $v(b_1^-)$ in $\Omega(i \times e^+)$. However, this is impossible for the following reasons:

1. There exists at least one connected directed M-graph of type $M(i \times e^+)$ which contains vertex $v(b_1^-)$ since $v(b_1^-)$ is in $\Omega(i \times e^+)$.

2. There exists at least one directed path from i to $v(b_1^+)$ because G is an SEC graph from i to j.

3. Hence there exists a directed path from i to $v(e^+)$ which contains edge b_1. This means that $v(b_1^+)$ must be in $\Omega(i \times e^+)$, which is the contradiction.

In other words, S is a directed cut-set containing edge e. Furthermore, the deletion of all edges in S will give a maximal connected subgraph G_e which consists of edges in $\mathscr{E}(\Omega(i \times e^+) \times \Omega(i \times e^+))$. Similarly, we can prove the existence of directed cut-set $S'(e)$ by using $\Omega(e^- \times j)$.   QED

If the condition $\Omega(i \times e^+) \cap \Omega(e^- \times j) = \emptyset$ in Theorem 14-2-3 does not satisfy, then the next theorem shows that there are no directed cut-sets separating i and j which contain edge e.

Theorem 14-2-4.   Let G be an SEC graph from i to j. Suppose measurable sets $\Omega(i \times e^+)$ and $\Omega(e^- \times j)$ have the property that

$$\Omega(i \times e^+) \cap \Omega(e^- \times j) \neq \emptyset$$

Then there exist no directed cut-set separating i and j which contains edge e.

Proof.   Let v be a vertex in both $\Omega(i \times e^+)$ and $\Omega(e^- \times j)$. Then there exists a connected directed M-graph of type $M(i \times e^+)$ which contains vertex v. There also exists a connected directed M-graph of type $M(e^- \times j)$ which contains vertex v as shown in Fig. 14-2-5. Hence there exists a directed

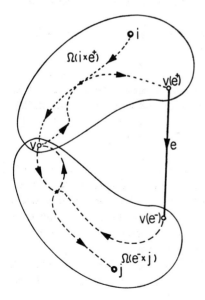

Fig. 14-2-5.   Existence of directed circuit containing e.

circuit containing edge e. Thus there are no directed cut-sets containing edge e.

QED

An important problem is how edges should be chosen so that when test points are assigned to these edges, 1-distinguishability can be achieved. The following definitions help with this problem.

Definition 14-2-7.   For a set E of edges, *measurable sets under E* means the all possible measurable sets when we assign test points to all edges in E.

Definition 14-2-8.   An SEC graph from i to j being 1-distinguishable under set E of edges means that when test points are assigned to every edge in set E, measurable sets under E will be 1-distinguishable.

As an example, consider the SEC graph from 1 to 6 in Fig. 14-2-3. Suppose set E consists of edges a and h. Then the measurable sets under E are

$$\Omega(1 \times a^+) = (1)$$
$$\Omega(a^- \times 6) = (3, 5, 6)$$
$$\Omega(1 \times h^+) = (1, 2, 4)$$
$$\Omega(h^- \times 6) = (6)$$
$$\Omega(a^- \times h^+) = \emptyset$$

and

$$\Omega(h^- \times a^+) = \emptyset$$

Since vertices 3 and 5 are together in a set in the D-partition for these measurable sets, G is not 1-distinguishable.

The next theorem gives a condition on a set E of edges which will not achieve 1-distinguishability.

Theorem 14-2-5.   Let G be an SEC graph from i to j. Also let E be a set of edges and $\{S\}$ be a collection of all directed cut-sets covered by edges in E. Then if there exists an edge in G which is not in any directed cut-set in $\{S\}$, G is not 1-distinguishable under E.

Proof.

CASE 1.   Assume a pair of measurable sets such that $\Omega(i \times b^+) \cap \Omega(b^- \times j) \neq \emptyset$. Then there is a directed circuit C containing edge b by Theorem 14-2-4. We can see that if any edge in C is in a connected directed M-graph of type $M(i \times e^+)$ or type $M(e^- \times j)$, then there is another connected directed M-graph of the same type which contains all edges in C where e is an edge in E. Hence either all vertices belonging to C are together in a measurable set or none of the vertices belonging to C is in a measurable set. Thus a set in the D-partition for the measurable sets under E contains all vertices belonging to C and G is not 1-distinguishable under E. Since edges in C cannot be in any directed cut-set, the theorem is true for this case.

CASE 2. Suppose $\Omega(i \times e^+) \cap \Omega(e^- \times j) = \emptyset$ for every edge e in E. Let $\{S\}$ be a collection of directed cut-sets $S_1, S_2, \ldots, S_m$. Also let $S_p = \mathscr{E}(\Omega_p \times \overline{\Omega}_p)$ for $p = 1, 2, \ldots, m$. By Theorem 14-2-3, for any measurable sets $\Omega(i \times e^+)$ and $\Omega(e^- \times j)$, there is a directed cut-set S_r in $\{S\}$ such that $\Omega(i \times e^+) = \Omega_r$, and there is a directed cut-set S_u in $\{S\}$ such that $\Omega(e^- \times j) = \overline{\Omega}_u$. Since at least one edge, say edge b, is not in any directed cut-set in $\{S\}$, both $v(b^+)$ and $v(b^-)$ must be together in either Ω_p or $\overline{\Omega}_p$ for $p = 1, 2, \ldots, m$. Hence there is no measurable set which contains one of $v(b^+)$ and $v(b^-)$ but not both. Thus a set in a D-partition under a collection of measurable sets with E will contain $v(b^+)$ if it contains $v(b^-)$. Hence by the definition of k-distinguishability, G is not 1-distinguishable under E.   QED

Example 14-2-3.   Consider the SEC graph G from 1 to 5 shown in Fig. 14-2-6. Suppose $E = (a, b)$. Then S consists of $S_1 = \mathscr{E}((1) \times \overline{(1)})$, $S_2 = \mathscr{E}((1, 2) \times \overline{(1, 2)})$, and $S_3 = \mathscr{E}((1, 2, 3, 4) \times \overline{(1, 2, 3, 4)})$. Note that cut-set S_1 is

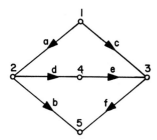

Fig. 14-2-6.   SEC graph G.

covered by edge a and cut-sets S_2 and S_3 are covered by edge b. Since edge e is not in any directed cut-set in $\{S\}$, G is not 1-distinguishable under E. We can check this by obtaining, first, all possible nonempty measurable sets:

$$\Omega(1 \times 5) = (1, 2, 3, 4, 5)$$
$$\Omega(1 \times a^+) = (1)$$
$$\Omega(a^- \times 5) = (2, 3, 4, 5)$$
$$\Omega(1 \times b^+) = (1, 2)$$
$$\Omega(b^- \times 5) = (5)$$

and

$$\Omega(a^- \times b^+) = (2)$$

From these measurable sets, we can obtain a D-partition in which we have

$$\Omega(1 \times 5) \cap \overline{\Omega(1 \times a^+)} \cap \Omega(a^- \times 5) \cap \overline{\Omega(1 \times b^+)} \cap \overline{\Omega(b^- \times 5)} \cap \overline{\Omega(a^- \times b^+)}$$
$$= (1, 2, 3, 4, 5) \cap (2, 3, 4, 5) \cap (2, 3, 4, 5) \cap (3, 4, 5) \cap (1, 2, 3, 4) \cap (1, 3, 4, 5)$$
$$= (3, 4)$$

Thus G is not 1-distinguishable under E.

Theorem 14-2-5 indicates that an SEC graph G is not 1-distinguishable if there is an edge that is not in a cut-set in $\{S\}$. Is G 1-distinguishable if every edge is in a cut-set in $\{S\}$? The answer is no, in general. An example is the SEC graph G in Fig. 14-2-7 where $E = (a)$. That is, $\{S\}$ consists of S_1 and S_2 which contains all edges in G. However, measurable sets are

$$\Omega(1 \times 4) = (1, 2, 3, 4)$$
$$\Omega(1 \times a^+) = (1)$$

and

$$\Omega(a^- \times 4) = (4)$$

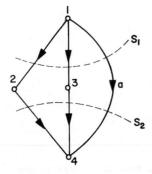

Fig. 14-2-7.   SEC graph G.

Thus vertices 2 and 3 are together in a set in a D-partition under these measurable sets. This means that G is not 1-distinguishable under E.

Further study of measurable sets may enlighten us about 1-distinguishability.

Theorem 14-2-6.   Let G be an SEC graph from i to j and E be a set of edges. Let $\{\Omega\}$ be a collection of all measurable sets produced by assigning test points to all edges in E. Then if and only if for every pair of vertices v_r and v_s in G there exists a measurable set $\Omega \in \{\Omega\}$ such that Ω contains either v_r or v_s but not both, G is 1-distinguishable under E.

Proof. Let $v_0, v_1, \ldots, v_n$ be all vertices in G. Consider a vertex v_0. Then for another vertex v_p in G, there exists a measurable set $\Omega_p \in \{\Omega\}$ such that Ω_p contains either v_0 or v_p but not both. Since this is true for all vertices in G, there exists a set D in a D-partition under E such that

$$D = \tilde{\Omega}_1 \cap \tilde{\Omega}_2 \cap \cdots \cap \tilde{\Omega}_m = (v_0)$$

Furthermore, this is true for every vertex in G. Thus every nonempty set in a D-partition consists of one vertex which is, by definition, 1-distinguishability.

Conversely, if G is 1-distinguishable under E, then every nonempty set in a D-partition consists of one vertex. Hence for a pair of vertices v_r and v_s, there is a set D in a D-partition under E which consists of v_r where

$$D = \tilde{\Omega}_1 \cap \tilde{\Omega}_2 \cap \cdots \cap \tilde{\Omega}_m \qquad (14\text{-}2\text{-}3)$$

in which there must be Ω_r $(1 \le r \le m)$ which contain either v_r or v_s but not both. Thus there exists a measurable set Ω_r in $\{\Omega\}$ such that Ω_r contains either v_r or v_s but not both. QED

Since it is easier to obtain a collection $\{S\}$ of all possible directed cut-sets separating i and j and covered by edges in E than to obtain a collection $\{\Omega\}$ of all possible measurable sets, we will restate Theorem 14-2-6 as follows.

Theorem 14-2-7. Let G be an SEC graph from i to j and E be a set of edges. Let $\{S\}$ be a collection of all possible directed cut-sets separating i and j and covered by edges in E. Then G is 1-distinguishable under E if and only if for any pair of vertices v_r and v_s, there exists a directed cut-set in $\{S\}$ which separates v_r and v_s.

Proof. From Theorem 14-2-6, we found that G is 1-distinguishable if and only if for any pair of vertices v_r and v_s, there exists a measurable set Ω such that Ω contains either v_r or v_s but not both. From Theorems 14-21-1 and 14-2-2, a measurable set Ω can be expressed as $\Omega(G')$ where G' is one of the two maximal connected subgraphs obtained by deleting all edges in a directed cut-set in $\{S\}$ if $\Omega(i \times e^+) \cap \Omega(e^- \times j) = \emptyset$, and clearly one of v_r and v_s is in $\Omega(G')$ and the other is in $\overline{\Omega(G')}$. Hence we need only show that $\Omega(i \times e^+) \cap \Omega(e^- \times j) = \emptyset$ satisfies for all edges in E.

By Theorem 14-2-4, if $\Omega(i \times e^+) \cap \Omega(e^- \times j) \ne \emptyset$, then there are no directed cut-sets in $\{S\}$ containing edge e. On the other hand, by the assumption in this theorem, there exists a set S in $\{S\}$ which separates endpoints $v(e^+)$ and $v(e^-)$ of edge e. Hence S is a directed cut-set containing edge e. Thus the condition $\Omega(i \times e^+) \cap (e^- \times j) = \emptyset$ must satisfy for every edge in E. This proves the necessary part of the theorem. The sufficient part is obvious from Theorem 14-2-6. QED

To find whether every pair of vertices can be separated by a directed cut-set in $\{S\}$ is not an easy task. Thus we begin by looking for a simpler necessary and sufficient condition for an SEC graph G to be 1-distinguishable under E. First we study the ordering of vertices.

Definition 14-2-9.   Two vertices v_r and v_s are ordered as $v_r > v_s$ if there exist no directed paths from v_s to v_r. Also $v_r > v_r$ for convenience.

As an example, consider the SEC graph in Fig. 14-2-6. We can order all vertices in the graph as

$$1 > 2 > 4 > 3 > 5$$

With this definition of ordering of vertices, we can state the following theorem.

Theorem 14-2-8.   Let G be an SEC graph from v_1 to v_n which consists of vertices $v_1, v_2, \ldots, v_n$. Suppose G contains no directed circuits. Then vertices in G can be ordered.

Before proving this theorem, we will state and prove the following theorem.

Theorem 14-2-9.   Let G be an SEC graph from v_1 to v_n. If there are no directed circuits in G, then all edges connected at v_1 have orientation away from v_1 and all edges connected at v_n have orientation toward v_n.

Proof.   Since we are considering only an SEC graph from v_1 to v_n which has the property that measurable set $\Omega(v_1 \times v_n)$ contains all vertices in the SEC graph, there is a directed path from v_1 to any vertex in G. Thus all edges connected at v_1 must have orientation away from v_1 in order that no directed circuits exist in G. Similarly, all edges connected at v_n must have orientation toward v_n so that G has no directed circuits.   QED

Now we are ready to prove Theorem 14-2-8.

Proof of Theorem 14-2-8.   The theorem is clearly true for $n = 2$. Suppose the theorem is true for $n = k$. Then for $n = k + 1$, we form an oriented graph G' from G by deleting vertex v_{k+1} and all edges connected at v_{k+1}. Let $v_k, v_{k-1}, \ldots, v_{k-p}$ in G' be the vertices such that all edges connected at these vertices have orientation toward these vertices. For $p = 0$, G' is an SEC graph from v_1 to v_k. Thus by assumption all vertices in G' can be ordered as $v_1 > v_2 > \cdots > v_k$. Then $v_1 > v_2 > \cdots > v_k > v_{k+1}$ is a desired order.
For $p \geq 1$, we insert edges $y_1, y_2, \ldots, y_p$ into G' where y_r is connected from v_{k-r} to v_k as shown in Fig. 14-2-8. This modified oriented graph is an SEC graph from v_1 to v_k. Thus this becomes exactly the same as the previous graph. Hence, in G with $y_1, y_2, \ldots, y_p$, all vertices can be ordered. The same

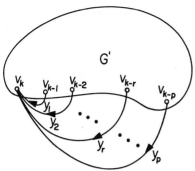

Fig. 14-2-8. Insertion of edges $y_1, y_2, \ldots, y_p$.

ordering of vertices is clearly satisfied when edges $y_1, y_2, \ldots, y_p$ are deleted. Thus the theorem is true. QED

With the ordering of vertices, we can obtain the following interesting property of an SEC graph with no directed circuits.

Theorem 14-2-10. Let G be an SEC graph from v_1 to v_n which consists of n vertices and contains no directed circuits. Then there exist $n - 1$ linearly independent directed cut-sets which separate v_1 and v_n.

Proof. Let $v_1 > v_2 > \cdots > v_n$ be the order by Theorem 14-2-8. Then cut-set S_r where

$$S_r = \mathscr{E}((v_1 v_2 \cdots v_r) \times (v_{r+1} \cdots v_n))$$

is a directed cut-set separating v_1 and v_n for $r = 1, 2, \ldots, n - 1$. These cut-sets are clearly linearly independent. QED

Example 14-2-4. Consider SEC graph from 1 to 6 in Fig. 14-2-9. The

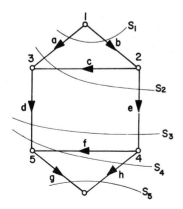

Fig. 14-2-9. SEC graph having no directed circuits.

order of vertices is $1 > 2 > 3 > 4 > 5 > 6$ and the five linearly independent directed cut-sets are

$$S_1 = \mathscr{E}((1) \times (2, 3, 4, 5, 6)) = (a, b)$$
$$S_2 = \mathscr{E}((1, 2) \times (3, 4, 5, 6)) = (a, c, e)$$
$$S_3 = \mathscr{E}((1, 2, 3) \times (4, 5, 6)) = (d, e)$$
$$S_4 = \mathscr{E}((1, 2, 3, 4) \times (5, 6)) = (d, f, h)$$

and

$$S_5 = \mathscr{E}((1, 2, 3, 4, 5) \times (6)) = (g, h)$$

Note that when directed circuits exist, all vertices cannot be ordered. If there are no directed paths from v_r to v_s and no directed paths from v_s to v_r, the ordering of vertices is not unique. As an example, another way of ordering vertices of the SEC graph in Fig. 14-2-9 is $1 > 2 > 4 > 3 > 5$.

Relating the ordering of vertices and 1-distinguishability is possible by the use of the following theorem.

Theorem 14-2-10. Let G be an SEC graph from i to j which contains n vertices. Let $S_p = \mathscr{E}(\Omega_p \times \overline{\Omega_p})$ be a directed cut-set separating i and j for $p = 1, 2, \ldots, n - 1$. Suppose $S_1, S_2, \ldots, S_{n-1}$ are linearly independent. Then for any pair of vertices v_r and v_s in G, there exists S_r $(1 \leq r \leq n - 1)$ which separates v_r and v_s.

Proof. Suppose there exists a pair of vertices v_r and v_s which cannot be separated by any of $S_1, S_2, \ldots, S_{n-1}$. Hence both v_r and v_s are together in either Ω_p or $\overline{\Omega_p}$ for all $p = 1, 2, \ldots, n - 1$. By coinciding v_r and v_s, we have a new oriented graph G'. It is clear that $S_1, S_2, \ldots, S_{n-1}$ of G' are the same as those of G. Thus $S_1, S_2, \ldots, S_{n-1}$ are linearly independent. However, G' consists of $n - 1$ vertices. Hence there are no more than $n - 2$ linearly independent cut-sets. Thus such a pair of vertices cannot exist in G.   QED

With these results, we can return to the problem of 1-distinguishability.

Theorem 14-2-12. Let G be an SEC graph from i to j which contains no directed circuits. Suppose the ordering of vertices is unique. Then G is 1-distinguishable under a set E of edges if and only if edges in E cover all linearly independent directed cut-sets which separate i and j.

Proof. Since the ordering of vertices is unique, linearly independent directed cut-sets separating i and j are uniquely determined. If any one of these directed cut-sets is not covered by edges in E, there are vertices which cannot be separated by any of these cut-sets covered by edges in E. Thus by Theorem 14-2-7, G is not 1-distinguishable under E, which contradicts the assumption. The converse can easily be proven by contradiction.   QED

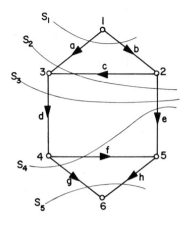

Fig. 14-2-10. An SEC graph.

Example 14-2-5. Consider the SEC graph from 1 to 6 shown in Fig. 14-2-10. Since the ordering of vertices of this SEC graph is unique, linearly independent directed cut-sets separating 1 and 6 are S_1, S_2, S_3, S_4, and S_5 as shown in the figure. By Theorem 14-2-12, in order that this SEC graph be 1-distinguishable under set E, we must choose edges in E such that these edges cover all of these directed cut-sets. For example, set E can be (a, d, g).

We have found now that $n - 1$ linearly independent directed cut-sets separating i and j are important for 1-distinguishability. Are there $n - 1$ linearly independent directed cut-sets separating i and j in an SEC graph G from i to j if G is 1-distinguishable under some set E of edges? The answer is yes and a reason is given by the following theorem. Note that from Theorem 14-2-8 if G has no directed circuits, there are $n - 1$ linearly independent directed cut-sets separating i and j.

Theorem 14-2-13. In an SEC graph from i to j, the vertices in a directed circuit are indistinguishable.

Proof. Since edges in a directed circuit cannot be in a directed cut-set, vertices in a directed circuit cannot be separated by a directed cut-set. Hence by Theorem 14-2-7 this theorem is true. QED

By this theorem, an SEC graph cannot be 1-distinguishable under a set E if it contains a directed circuit. Thus we can say that a 1-distinguishable SEC graph from i to j will have $n - 1$ linearly independent directed cut-sets separating i and j.

When the ordering of vertices in an SEC graph is unique, then Theorem 14-2-12 gives a simple necessary and sufficient condition for 1-distinguishability. Suppose the ordering of vertices in an SEC graph is not unique.

Then $n - 1$ linearly independent directed cut-sets being covered by edges in set E becomes a sufficient but not a necessary condition.

Theorem 14-2-14. Let G be an SEC graph from i to j consisting of n vertices. If $n - 1$ linearly independent directed cut-sets separating i and j are covered by edges in set E, then G is 1-distinguishable under set E.

The proof of this theorem follows directly from Theorem 14-2-11. To show that the above condition is not necessary, consider the SEC graph from

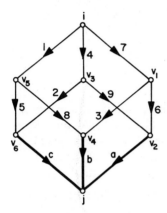

Fig. 14-2-11.   SEC graph G.

i to j shown in Fig. 14-2-11. A set E of edges is chosen to be $E = (a, b, c)$. A collection $\{S\}$ of directed cut-sets separating i and j which are covered by edges in E consists of S_1, S_2, S_3, and S_4 where

$$S_1 = (1, 2, 3, a) = \mathscr{E}((i, v_1, v_2, v_3) \times (v_4, v_5, v_6, j))$$
$$S_2 = (4, 5, 6, b) = \mathscr{E}((i, v_1, v_4, v_5) \times (v_2, v_3, v_6, j))$$
$$S_3 = (7, 8, 9, c) = \mathscr{E}((i, v_3, v_5, v_6) \times (v_1, v_2, v_4, j))$$

and

$$S_4 = (a, b, c) = \mathscr{E}((i, v_1, v_2, v_3, v_4, v_5, v_6) \times (j))$$

Note that S_1 and S_4 are covered by edge a, S_2 and S_4 are covered by edge b, and S_3 and S_4 are covered by edge c. Since any two vertices are separated by one of these four directed cut-sets, by Theorem 14-2-6 this SEC graph is 1-distinguishable under E. However, collection $\{S\}$ does not contain all linearly independent directed cut-sets separating i and j. Moreover, it is clear that edges in E do not cover all linearly independent directed cut-sets separating i and j.

When a set E of edges is a directed cut-set, a condition for 1-distinguishability becomes very simple as given in the following theorem.

Theorem 14-2-15. Let G be an SEC graph from i to j. Suppose a set E of edges is a directed cut-set separating i and j in G. Then G is 1-distinguishable under E if and only if every edge in G is in a cut-set in $\{S\}$ where $\{S\}$ is a collection of all directed cut-sets separating i and j covered by edges in E.

Note that in the preceding example, set E is a directed cut-set separating i and j. Also every edge is in a cut-set in collection $\{S\}$. Thus by Theorem 14-2-15, SEC graph G is 1-distinguishable under E.

Proof of Theorem 14-2-15. Suppose G is 1-distinguishable under E. Then every edge must be in at least one cut-set in Theorem 14-2-6. Hence we only need to prove that G is 1-distinguishable under E when every edge is in a cut-set in $\{S\}$.

Suppose G is not 1-distinguishable under E. Then there must be at least one set Ω_I of more than one vertex such that vertices in Ω_I cannot be separated by any cut-set in $\{S\}$. On the other hand, every edge is in at least one directed cut-set in $\{S\}$. Hence no edges will be connected between vertices in Ω_I. In order that there be such a set Ω_I, there must be two directed cut-sets $S_1 = \mathscr{E}(\Omega_1 \times \overline{\Omega_1})$ and $S_2 = \mathscr{E}(\Omega_2 \times \overline{\Omega_2})$ in $\{S\}$ such that

$$i \in \Omega_1 \quad \text{and} \quad j \in \overline{\Omega_1}$$
$$i \in \Omega_2 \quad \text{and} \quad j \in \overline{\Omega_2}$$

and

$$\Omega_I = \overline{\Omega_1} \cap \Omega_2$$

as shown in Fig. 14-2-12. Note that Ω_1, Ω_I, and $\overline{\Omega_2}$ are vertex disjoint sets which contain all vertices in G.

Note that there is an edge e in S_2 that is in E. If any edge e' in $\mathscr{E}(\Omega_I \times \overline{\Omega_2})$ is in E, it is easily seen that there is a directed cut-set covered by edge e' which separates Ω_I. Hence no edges in $\mathscr{E}(\Omega_I \times \overline{\Omega_2})$ belong to set E. Similarly, no edges in $\mathscr{E}(\Omega_1 \times \Omega_I)$ belong to set E. However, by assumption, E is a directed cut-set separating i and j. We know that edge e in S, which is in directed cut-set E, is located between a vertex in Ω_1 and a vertex in $\overline{\Omega_2}$, and we know that E cannot separate vertices in Ω_I. Furthermore, we can see that E cannot separate any vertices in $\overline{\Omega_2}$ and that E cannot separate any vertices in Ω_1 because S_1 and S_2 are directed cut-sets covered by edges in E. Thus E must be either S_1 or S_2. This means that E contains either $\mathscr{E}(\Omega_1 \times \overline{\Omega_2})$ or $\mathscr{E}(\Omega_1 \times \Omega_I)$, which is a contradiction. Hence no such set Ω_I exists in G, which proves the theorem. QED

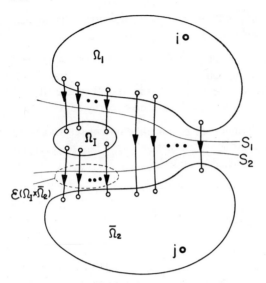

Fig. 14-2-12. SEC graph G which has Ω_I.

Let us study more about measurable sets that are directly related to actual measurement necessary to locate fault vertices. Let $\{\Omega\}$ be a collection of all possible measurable sets produced by assigning test points to all edges in a set E. We have 1-distinguishability of an SEC graph defined by sets in a D-partition obtained by $\{\Omega\}$. When we investigate measurable sets in $\{\Omega\}$, we will see that there are proper subcollections $\{\Omega'\}$ of $\{\Omega\}$ such that a D-partition obtained by $\{\Omega'\}$ is the same as that formed from $\{\Omega\}$. For example, consider SEC graph G in Fig. 14-2-13. Let $E = (a, b)$. Then $\{\Omega\}$ consists of $\Omega(i \times a^+) = (i)$, $\Omega(a^- \times j) = (1, 2, j)$, $\Omega(i \times b^+) = (i, 2)$, $\Omega(b^- \times j) = (j)$, and $\Omega(a^- \times b^+) = (2)$. Nonzero sets in a D-partition by the above measurable sets are

$$\Omega(i \times a^+) \cap \overline{\Omega(a^- \times j)} \cap \Omega(i \times b^+) \cap \overline{\Omega(b^- \times j)} \cap \overline{\Omega(a^- \times b^+)} = (i)$$

$$\overline{\Omega(i \times a^+)} \cap \Omega(a^- \times j) \cap \Omega(i \times b^+) \cap \overline{\Omega(b^- \times j)} \cap \Omega(a^- \times b^+) = (2)$$

$$\overline{\Omega(i \times a^+)} \cap \Omega(a^- \times j) \cap \overline{\Omega(i \times b^+)} \cap \Omega(b^- \times j) \cap \overline{\Omega(a^- \times b^+)} = (j)$$

$$\overline{\Omega(i \times a^+)} \cap \Omega(a^- \times j) \cap \overline{\Omega(i \times b^+)} \cap \overline{\Omega(b^- \times j)} \cap \overline{\Omega(a^- \times b^+)} = (1)$$

The same D-partition can be obtained by using only $\Omega(i \times a^+)$, $\Omega(i \times b^+)$, and $\Omega(b^- \times j)$. This means that we can obtain all necessary information for diagnosis by three measurements corresponding to $\Omega(i \times a^+)$, $\Omega(i \times b^+)$, and $\Omega(b^- \times j)$ rather than five measurements corresponding to all nonempty

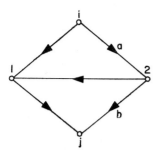

Fig. 14-2-13. SEC graph G.

measurable sets in $\{\Omega\}$. To study the number of measurements necessary for diagnosis, we define generators of $\{\Omega\}$ as follows.

Definition 14-2-8. Let $\{\Omega\}$ be a collection of all measurable sets produced by a set E. Also let $\{\Omega'\}$ be a subcollection of $\{\Omega\}$. Then $\{\Omega\}$ is called a *collection of generators* of $\{\Omega\}$ if (1) a D-partition obtained by $\{\Omega'\}$ is the same as that formed by $\{\Omega\}$ and (2) a D-partition obtained by any proper subcollection of $\{\Omega'\}$ is not the same as that formed by $\{\Omega\}$.

In the previous example, a collection $\{\Omega\}$ of nonzero measurable sets produced by set $E = (a, b)$ of SEC graph G in Fig. 14-2-13 is

$$\{\Omega\} = \{\Omega(i \times a^+), \Omega(a^- \times j), \Omega(i \times b^+), \Omega(b^- \times j), \Omega(a^- \times b^+)\}$$

A D-partition $\{D\}$ formed by $\{\Omega\}$ is

$$\{D\} = \{(i), (2), (j), (1)\}$$

A subcollection $\{\Omega(i \times a^+), \Omega(i \times b^+), \Omega(b^- \times j)\}$ of $\{\Omega\}$ also gives the same D-partition. However, any proper subcollection of

$$\{\Omega(i \times a^+), \Omega(i \times b^+), \Omega(b^- \times j)\}$$

does not give the same D-partition. Hence $\{\Omega(i \times a^+), \Omega(i \times b^+), \Omega(b^- \times j)\}$ is a collection of generators of $\{\Omega\}$.

For a given $\{\Omega\}$, there may be many collections of generators. Furthermore, the number of sets in each collection of generators of $\{\Omega\}$ may be different. Since each measurable set corresponds to a measurement for diagnosis, and a collection of generators of $\{\Omega\}$ gives a desired D-partition, the number of measurements necessary for diagnosis is the same as the number of sets in a collection of generators of $\{\Omega\}$.

With this definition of generators, we have a new theorem.

Theorem 14-2-16. For a set $E = (e_1, e_2, \ldots, e_k)$ of an SEC graph from i to j, a collection consisting of measurable sets $\Omega(i \times e_p^+)$ and $\Omega(e_p^- \times j)$ for $p = 1, 2, \ldots, k$ contains a collection of generators of $\{\Omega\}$.

Proof. Since a measurable set in $\{\Omega\}$ is one of $\Omega(i \times e_p^+)$, $\Omega(e_p^- \times j)$, and $\Omega(e_p^- \times e_q^+)$, it is necessary to show that $\Omega(e_p^- \times e_p^+)$ is not needed to obtain a collection of generators of $\{\Omega\}$. By the definition of measurable sets, $\Omega(e_p^- \times e_q^+)$ is a set of vertices which are in at least one connected directed M-graph of type $M(e_p^- \times e_q^+)$. On the other hand, $\Omega(i \times e_q^+)$ consists of vertices which are in at least one connected directed M-graph of type $M(i \times e_q^+)$. Furthermore, we know that there is a directed path from i to $v(e_p^-)$ in an SEC graph from i to j. Thus for any vertex v which is in a connected directed M-graph of type $M(e_p^- \times e_q^+)$, there is a connected directed M-graph of type $M(i \times e_q^+)$ which contains v. Hence

$$\Omega(e_p^- \times e_q^+) \subset \Omega(i \times e_q^+) \tag{14-2-4}$$

Similarly,

$$\Omega(e_p^- \times e_q^+) \subset \Omega(e_p^- \times j) \tag{14-2-5}$$

Thus

$$\Omega(e_p^- \times e_q^+) \subset \Omega(i \times e_q^+) \cap \Omega(e_p^- \times j) \tag{14-2-6}$$

Let v' be a vertex in $\Omega(i \times e_q^+) \cap \Omega(e_p^- \times j)$. Then there is a connected directed M-graph of type $M(e_p^- \times v')$. There is also a connected directed M-graph of type $M(v' \times e_q^+)$. Thus there is a connected directed M-graph of type $M(e_p^- \times e_q^+)$ which contains v'. Hence with Eq. 14-2-6,

$$\Omega(e_p^- \times e_q^+) = \Omega(i \times e_q^+) \cap \Omega(e_p^- \times j) \tag{14-2-7}$$

Thus if a collection contains $\Omega(i \times e_q^+)$ and $\Omega(e_p^- \times j)$, it is not necessary to have $\Omega(e_p^- \times e_q^+)$ in the collection to contain a collection of generators of $\{\Omega\}$. Similarly, we can show that $\Omega(e_p^- \times e_q^+)$, $\Omega(e_p^+ \times e_q^-)$, and $\Omega(e_p^+ \times e_q^+)$ are not necessary in the collection of generators of $\{\Omega\}$.   QED

Theorem 14-2-17. If an edge e in set E covers one directed cut-set, then only one of $\Omega(i \times e^+)$ and $\Omega(e^- \times j)$ can be in a collection of generators of $\{\Omega\}$.

Proof. Since edge e covers only one directed cut-set, $\Omega(i \times e^+) \cup \Omega(e^- \times j)$ is a set of all vertices. Thus

$$\Omega(i \times e^+) = \overline{\Omega(e^- \times j)} \tag{14-2-8}$$

which indicates that only one of these two measurable sets is needed to form a D-partition.   QED

For example, consider the SEC graph in Fig. 14-2-3 of Example 14-2-2. If set $E = (b, h)$, then edge b covers only one directed cut-set $S = (a, b)$. It can be seen that

$$\Omega(1 \times b^+) = (1) \quad \text{and} \quad \Omega(b^- \times 6) = (2, 3, 4, 5, 6)$$

This shows that $\Omega(1 \times b^+) = \overline{\Omega(b^- \times 6)}$. On the other hand, edge h covers two directed cut-sets $S_1(h)$ and $S_3(h)$. The measurable sets associated with h are

$$\Omega(1 \times h^+) = (1, 2, 4) \quad \text{and} \quad \Omega(h^- \times b) = (6)$$

We can see clearly that a collection of generators of $\{\Omega\}$ can be

$$\{\Omega(1 \times b^+), \Omega(1 \times h^+), \Omega(h^- \times 6)\}$$

From the preceding theorem, we can state Theorem 14-2-18.

Theorem 14-2-18. For a set $E = (e_1, e_2, \ldots, e_k)$, the number of sets in a collection of generators of $\{\Omega\}$ is at most $2k$.

14-3 TEST GATES

In the previous section, we observed that when a directed circuit is in an SEC graph, 1-distinguishability becomes impossible. Hence, in order to obtain 1-distinguishability for such an SEC graph, it is necessary to use some device other than test points. Such a device is a test gate defined as follows.

Definition 14-3-1. A *test gate being assigned to an edge e* means that edge e can be deleted whenever required.

For example, if we assign a test gate to edge e in the SEC graph in Fig. 14-3-1, then the measurable set $\Omega(i \times j) = (i, j, 1, 2)$ can be changed to $(i, 2, j)$ by the use of the test gate. That is, by the test gate, we can change G into the graph shown in Fig. 14-3-2.

In order to show the measurable set under the influence of test gates, we use a subscript E, $\Omega(i \times j)_E$, to indicate that test gates assigned to edges in E are in active.

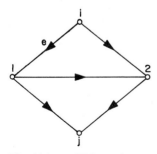

Fig. 14-3-1. SEC graph G.

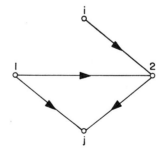

Fig. 14-3-2. Result of using test gate on edge e.

Definition 14-3-2. The symbol $\Omega(i \times j)_E$ is a measurable set of a subgraph obtained by deleting all edges in set E.

When test gates are assigned to all edges in a set E, we can obtain several measurable sets generated by the different choice of edges whose test gates are in active. In other words, deleting some edges in E may produce different measurable sets.

Example 14-3-1. Consider the SEC graph in Fig. 14-3-3. Suppose $E = (e_1, e_2, e_3, e_7)$. Then we have the following measurable sets:

$$\Omega(i \times j)_{e_1 e_2 e_3} = \Omega(i \times j)_{e_1 e_2} = \Omega(i \times j)_{e_3} = (i, v_2, v_3, j)$$
$$\Omega(i \times j)_{e_7} = (i, v_1, v_2, j)$$

and

$$\Omega(i \times j)_{e_3 e_7} = \Omega(i \times j)_{e_1 e_2 e_7} = \Omega(i \times j)_{e_1 e_2 e_3 e_7} = (i, v_2, j)$$

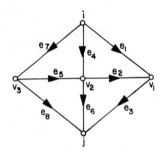

Fig. 14-3-3. SEC graph G.

In order to study properties of test gates, we consider an SEC graph employing test gates only. For convenience, we use the following definition.

Definition 14-3-3. The measurable sets by gate-set E means that all possible measurable sets are produced by assigning test gates to all edges in set E.

In Example 14-3-1, measurable sets $\Omega(i \times j)_{e_3}$, $\Omega(i \times j)_{e_7}$, and $\Omega(i \times j)_{e_3 e_7}$ are the measurable sets by gate-set (e_1, e_2, e_3, e_7).

The following theorem should be obvious from the all possible choices of edges whose test gates are inactive.

Theorem 14-3-1. The number of distinct measurable sets by gate-set E is at most $2^k - 1$ where k is the number of edges in set E.

Consider measurable set $\Omega(i \times j)_E$ of an SEC graph G from i to j. Suppose $\Omega(i \times j)_E$ is not the same as $\Omega(i \times j)_{E'}$ for every proper subset E' of E. Then we say that E is the minimum set for $\Omega(i \times j)_E$. For example, by considering $E = (e_1, e_2, e_3, e_7)$ in Example 14-3-1, we have $\Omega(i \times j)_E = (i, v_2, j)$. However, with $E' = (e_3, e_7)$, which is a proper subset of E, we have $\Omega(i \times j)_{E'}$, which is equal to $\Omega(i \times j)_E$. Hence E is not the minimum set for $\Omega(i \times j)_E$. It can be seen that E' is the minimum set for $\Omega(i \times j)_{E'}$.

In Example 14-3-1,

$$\Omega(i \times j)_{e_3e_7} = \Omega(i \times j)_{e_1e_2e_7} = (i, v_2, j)$$

where both (e_3, e_7) and (e_1, e_2, e_7) are the minimum sets for (i, v_2, j). This example indicates that the minimum set for a measurable set may not be unique. However, the minimum set has an interesting property which is given by the following theorem.

Theorem 14-3-2. Let E be the minimum set for $\Omega(i \times j)_E$ ($\neq \emptyset$) of an SEC graph G from i to j. Then by deleting all edges in E from G, the resultant graph is connected.

Proof. Let G' be the graph obtained by deleting all edges in E from G. Suppose G' is separated. Let G_1' and G_2' be the subgraphs of G' such that $\Omega(G_1') = \Omega(i \times j)_E$ and G_1' and G_2' are not connected. Note that $\Omega(i \times j)_E$ of G is equal to $\Omega(i \times j)$ of G'. Let v be a vertex in G_2'. Since G is an SEC graph from i to j, there exists at least one connected directed M-graph of type $M(i \times j)$ that contains vertex v in G_2'. Hence there exist at least two edges e_1 and e_2 in the M-graph which are neither in G_1' nor in G_2', as shown in Fig. 14-3-4. We can easily see that by the insertion of e_1 into G', $\Omega(i \times j)$ of the resultant graph is the same as $\Omega(i \times j)$ of G'. This means that $\Omega(i \times j)_E = \Omega(i \times j)_{E'}$ in G where $E' = E - (e_1)$. Hence E is not the minimum for

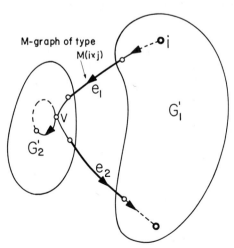

Fig. 14-3-4. G_1', G_2', and path P.

$\Omega(i \times j)_E$, which contradicts the assumption that E is the minimum for $\Omega(i \times j)_E$. Thus the theorem is true.   QED

From a collection of measurable sets by a gate-set E, we can obtain a D-partition as in the previous sections. For example, from a collection of measurable sets $\Omega(i \times j)_{e_3}$, $\Omega(i \times j)_{e_7}$, and $\Omega(i \times j)_{e_3 e_7}$ by a gate-set $E = (e_1, e_2, e_3, e_7)$, we can obtain a D-partition as

$$\Omega(i \times j)_{e_3} \cap \Omega(i \times j)_{e_7} \cap \Omega(i \times j)_{e_3 e_7} = (i, v_2, j)$$

$$\Omega(i \times j)_{e_3} \cap \Omega(i \times j)_{e_7} \cap \overline{\Omega(i \times j)_{e_3 e_7}} = \emptyset$$

$$\Omega(i \times j)_{e_3} \cap \overline{\Omega(i \times j)_{e_7}} \cap \Omega(i \times j)_{e_3 e_7} = \emptyset$$

$$\Omega(i \times j)_{e_3} \cap \overline{\Omega(i \times j)_{e_7}} \cap \overline{\Omega(i \times j)_{e_3 e_7}} = (v_3)$$

$$\overline{\Omega(i \times j)_{e_3}} \cap \Omega(i \times j)_{e_7} \cap \Omega(i \times j)_{e_3 e_7} = \emptyset$$

$$\overline{\Omega(i \times j)_{e_3}} \cap \Omega(i \times j)_{e_7} \cap \overline{\Omega(i \times j)_{e_3 e_7}} = (v_1)$$

$$\overline{\Omega(i \times j)_{e_3}} \cap \overline{\Omega(i \times j)_{e_7}} \cap \Omega(i \times j)_{e_3 e_7} = \emptyset$$

$$\overline{\Omega(i \times j)_{e_3}} \cap \overline{\Omega(i \times j)_{e_7}} \cap \overline{\Omega(i \times j)_{e_3 e_7}} = \emptyset$$

Note that a measurable set $\Omega(i \times j)_E$ has the same properties as that in the previous sections. That is, when a signal is injected at vertex i under acting all test gates assigned to edges in E, the signal that is monitored at vertex j will pass only those vertices in $\Omega(i \times j)_E$. Thus all properties about measurable sets in the previous sections are applicable to the measurable sets $\Omega(i \times j)_E$. For example, if there exists a pair of vertices v_r and v_s in an SEC graph G from i to j such that every measurable set (in a collection of all measurable sets of a gate-set E) which contains one of v_r and v_s will contain the other, then G is not 1-distinguishable (Theorem 14-2-6). Since every measurable set $\Omega(i \times j)_{E'}$ in a collection of measurable sets of a gate-set E contains both i and j, an SEC graph from i to j is not 1-distinguishable under any gate-set E. Obviously, there are SEC graphs that are 1-distinguishable under a gate-set, if we neglect one of vertices i and j. To study such SEC graphs and gate-sets, we need another definition.

Definition 14-3-4.   An SEC graph G from i to j is 1-(ij)-distinguishable under a gate-set E if by disregarding vertex j, G is 1-distinguishable under set E.

Example 14-3-2.   Take a gate-set $E = (b, d)$ for SEC graph G from 1 to 4 in Fig. 14-3-5. The measurable sets of E are

$$\Omega(1 \times 4)_b = (1, 2, 4)$$
$$\Omega(1 \times 4)_d = (1, 3, 4)$$

and

$$\Omega(1 \times 4)_{bd} = \emptyset$$

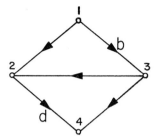

Fig. 14-3-5. SEC graph G.

A D-partition of these measurable sets consists of

$$\Omega(1 \times 4)_b \cap \Omega(1 \times 4)_d = (1, 4)$$
$$\Omega(1 \times 4)_b \cap \overline{\Omega(1 \times 4)_d} = (2)$$
$$\overline{\Omega(1 \times 4)_b} \cap \Omega(1 \times 4)_d = (3)$$
$$\overline{\Omega(1 \times 4)_b} \cap \overline{\Omega(1 \times 4)_d} = \emptyset$$

If we disregard vertex 4, each set in the above D-partition contains at most one vertex. Hence G is 1-distinguishable under E if we neglect vertex 4. Thus G is 1-(1, 4)-distinguishable under E.

Using the foregoing definition, we can modify Theorem 14-2-6 as follows, so that it will be applicable for the case of test gates.

Theorem 14-3-3. Let G be an SEC graph from i to j and E be a set of edges. Also let $\{\Omega\}$ be a collection of all measurable sets produced by assigning test gates to all edges in E. Then if and only if for every pair of vertices v_r and v_s in G without considering vertex j (i.e., $v_r \neq j$ and $v_s \neq j$) there exists a measurable set $\Omega \in \{\Omega\}$ such that Ω contains either v_r or v_s but not both, G is 1-(i, j)-distinguishable under E.

The proof is almost identical to that of Theorem 14-2-6.

A collection $\{\Omega\}$ of measurable sets produced by set $E = (b, d)$ of SEC graph G in Fig. 14-3-5 of Example 14-3-2 is

$$\{\Omega\} = \{\Omega(1 \times 4)_b, \Omega(1 \times 4)_d\}$$
$$= \{(1, 2, 4), (1, 3, 4)\}$$

Hence for any pair of vertices we choose except vertex 4, there exists a measurable set in $\{\Omega\}$ such that only one of these two vertices is in the set. Thus by Theorem 14-3-3, G is 1-(i, j)-distinguishable under (b, d).

Recall that a semicut separating i and j can be expressed as $\mathscr{E}(\Omega_i \times \overline{\Omega_i})$ where $i \in \Omega_i$, and $j \in \overline{\Omega_i}$. Now if $\mathscr{E}(\overline{\Omega_i} \times \Omega_i) = \emptyset$, then semicut $\mathscr{E}(\Omega_i \times \overline{\Omega_i})$ becomes a directed cut-set separating i and j. We will see by the next theorem that such a semicut is important for 1-(i, j)-distinguishability.

Theorem 14-3-4. Let G be an SEC graph from i to j which contains no edges connected between i and j. If a gate-set E does not contain a semicut separating i and j as a subset, G is not 1-(i, j)-distinguishable under gate-set E.

Proof. Since E does not contain a semicut separating i and j as a subset, there is a directed path P from i to j which contains no edges in E. By assumption, G has no edges connected between i and j. Hence path P must contain at least one vertex v other than i and j. Thus every measurable set by E contains vertex v together with i and j, and by Theorem 14-2-6, G is not 1-distinguishable even if vertex j is disregarded. Thus G is not 1-(i, j)-distinguishable under gate-set E.   QED

Example 14-3-3. Consider SEC graph G from i to j in Fig. 14-3-6. Let a

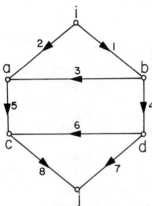

Fig. 14-3-6.   SEC graph G.

gate-set E be $(1, 4, 6)$. Since $(1, 4, 6)$ does not contain a semicut separating i and j, G is not 1-(i, j)-distinguishable under $(1, 4, 6)$ by Theorem 14-3-7. We now check this.

The measurable sets by gate-set $(1, 4, 6)$ are

$$\Omega(i \times j)_1 = \Omega(i \times j)_{14} = \Omega(i \times j)_{16} = \Omega(i \times j)_{146} = (i, a, c, j)$$
$$\Omega(i \times j)_6 = (i, a, b, c, d, j)$$

and

$$\Omega(i \times j)_4 = \Omega(i \times j)_{46} = (i, a, b, c, j)$$

A D-partition of these measurable sets contain a set D,

$$D = \Omega(i \times j)_1 \cap \Omega(i \times j)_4 \cap \Omega(i \times j)_{16}$$
$$= (i, a, c, j)$$

Thus G is not 1-(i, j)-distinguishable under gate-set (1, 4, 6).

By Theorem 14-3-4, a set E must contain a semicut from i to j in order that an SEC graph be 1-(i, j)-distinguishable under E. Hence, in order to study a set E of edges which gives a 1-(i, j)-distinguishable set, we assume that E is itself a directed cut-set $S = \mathscr{E}(\Omega_i \times \overline{\Omega_i})$ where $i \in \Omega_i$ and $j \in \overline{\Omega_i}$. Let $E = (e_1, e_2, \ldots, e_k)$. If we delete all edges except e_p $(1 \le p \le k)$ from SEC graph G, then a test signal which is injected at vertex i and monitored at vertex j must pass through edge e_p. This leads to the following theorem.

Theorem 14-3-5. Let G be an SEC graph from i to j. Let $E = (e_1, e_2, \ldots, e_k)$ be a directed cut-set $\mathscr{E}(\Omega_i \times \overline{\Omega_i})$ where $i \in \Omega_i$ and $j \in \overline{\Omega_i}$. Then

$$\Omega(i \times j)_{(\overline{e_p})} = \Omega(i \times e_p^+) \cup \Omega(e_p^- \times j) \qquad (14\text{-}3\text{-}1)$$

for $p = 1, 2, \ldots, k$ where $(\overline{e_p}) = E - (e_p)$. Let E_q be a subset of E. Then

$$\Omega(i \times j)_{\overline{E_q}} = \bigcup_{e_p \in E_q} \Omega(i \times j)_{(\overline{e_p})} \qquad (14\text{-}3\text{-}2)$$

where $\overline{E_q} = E - E_q$.

Proof. Recall that $\Omega(i \times e_p^+)$ is a set of vertices each of which is in at least one connected directed M-graph of type $M(i \times e_p^+)$ and $\Omega(e_p^- \times j)$ is a set of vertices which are in at least one connected directed M-graph of type $M(e_p^- \times j)$ (see Section 14-2). The first part of the theorem is obvious from the preceding discussion. For the second part, we note that a directed path from i to j which passes vertices in $\Omega(i \times j)_{\overline{E_q}}$ must contain exactly one edge in E_q. Thus every vertex in $\Omega(i \times j)_{\overline{E_q}}$ is in $\Omega(i \times j)_{(\overline{e_p})}$ for some $e_p \in E_q$. It is also clear that any vertex in $\Omega(i \times j)_{(\overline{e_p})}$ must be in $\Omega(i \times j)_{\overline{E_q}}$. QED

Theorem 14-3-5 not only gives the relationship between measurable sets of type $\Omega(i \times j)_{(\mathscr{E})}$ and measurable sets in Section 14-2 but also indicates that we only need to know measurable sets of type $\Omega(i \times j)_{(\mathscr{E})}$ to obtain a D-partition under a given set E.

Example 14-3-4. Consider a nonseparable SEC graph from i to j as shown in Fig. 14-3-7. Let $E = (e_1, e_2, e_3)$. We can see that

$$\Omega(i \times e_1^+) = (i, 1)$$
$$\Omega(e_1^- \times j) = (4, j)$$
$$\Omega(i \times e_2^+) = (i, 1, 2, 3)$$
$$\Omega(e_2^- \times j) = (4, 5, 6, j)$$
$$\Omega(i \times e_3^+) = (i, 3)$$

and

$$\Omega(e_3^- \times j) = (6, j)$$

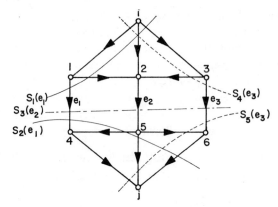

Fig. 14-3-7. A nonseparable SEC graph.

By Theorem 14-3-5, we can obtain

$$\Omega(i \times j)_{(e_2 e_3)} = \Omega(i \times j)_{\overline{(e_1)}} = \Omega(i \times e_1^+) \cup \Omega(e_1^- \times j) = (i, 1, 4, j)$$

$$\Omega(i \times j)_{(e_1 e_3)} = \Omega(i \times j)_{\overline{(e_2)}} = \Omega(i \times e_2^+) \cup \Omega(e_2^- \times j) = (i, 1, 2, 3, 4, 5, 6, j)$$

$$\Omega(i \times j)_{(e_1 e_2)} = \Omega(i \times j)_{\overline{(e_3)}} = \Omega(i \times e_3^+) \cup \Omega(e_3^- \times j) = (i, 3, 6, j)$$

$$\Omega(i \times j)_{(e_1)} = \Omega(i \times j)_{\overline{(e_2 e_3)}} = \Omega(i \times j)_{\overline{(e_2)}} \cup \Omega(i \times j)_{\overline{(e_3)}}$$
$$= (i, 1, 2, 3, 4, 5, 6, j)$$

$$\Omega(i \times j)_{(e_2)} = \Omega(i \times j)_{\overline{(e_1 e_3)}} = \Omega(i \times j)_{\overline{(e_1)}} \cup \Omega(i \times j)_{\overline{(e_3)}} = (i, 1, 3, 4, 6, j)$$

and

$$\Omega(i \times j)_{(e_3)} = \Omega(i \times j)_{\overline{(e_1 e_2)}} = \Omega(i \times j)_{\overline{(e_1)}} \cup \Omega(i \times j)_{\overline{(e_2)}}$$
$$= (i, 1, 2, 3, 4, 5, 6, j)$$

When we employ test points, we find by Theorem 14-2-15 that an SEC graph G from i to j is 1-distinguishable under a set E of edges if E is a directed cut-set separating i and j and every edge in G is in at least one directed cut-set in $\{S\}$ covered by E. This is not the case when we assign test gates on edges in E. For example, collection $\{S\}$ of directed cut-sets covered by $E = (e_1, e_2, e_3)$ of the SEC graph in Fig. 14-3-7 consists of cut-sets S_1, S_2, S_3, S_4, and S_5 as shown in the figure. Note that E is a directed cut-set separating i and j. It is easily seen that every edge in the SEC graph is in a cut-set in $\{S\}$. On the other hand, by considering set $(1, 4)$ of vertices 1 and 4, we can see that either $(1, 4) \subset \Omega(i \times j)_E$ or $(1, 4) \subset \overline{\Omega(i \times j)_{E'}}$ for every subset E' of E. Hence, by Theorem 14-3-3, G is not 1-(i, j)-distinguishable under (e_1, e_2, e_3). This example indicates that the condition requiring all edges be in $\{S\}$ is not sufficient to say that an SEC graph is 1-(i, j)-distinguishable when we use test gates. However, we have the following properties.

Theorem 14-3-6. Let G be an SEC graph from i to j. Suppose a set E is a directed cut-set $\mathscr{E}(\Omega \times \bar{\Omega})$ such that either Ω consists only of vertex i or $\bar{\Omega}$ consists only of vertex j. Then G is 1-(i, j)-distinguishable if and only if every edge in G is in a directed cut-set in $\{S\}$ where $\{S\}$ is a collection of directed cut-sets covered by E.

Proof. By Theorem 14-3-5,

$$\Omega(i \times j)_{\overline{(e_p)}} = \Omega(i \times e_p^+) \cup \Omega(e_p^- \times j) \qquad (14\text{-}3\text{-}3)$$

where $e_p \in E$. Suppose $\bar{\Omega}$ consists only of vertex j. Then $\Omega(e_p^- \times j) = (j)$. Thus $\Omega(i \times j)_{\overline{(e_p)}} = \Omega(i \times e_p^+) \cup (j)$ for all edges $e_p \in E$. Hence without vertex j, the measurable set for e_p with test gates is the same as that for e_p with test points. When Ω consists only of vertex i, we have $\Omega(i \times j)_{(e_p)} = (i) \cup \Omega(e_p^- \times j)$. This shows that the measurable set for e_p with test gates and that for e_p with test points are the same when we disregard vertex i. Thus a necessary and sufficient condition for an SEC graph from i to j be 1-distinguishable under E with test points can be applied for this case, which proves the theorem. QED

Example 14-3-5. Consider the nonseparable SEC graph in Fig. 14-3-8. Collection $\{S\}$ of directed cut-sets covered by $E = (e_1, e_2, e_3)$ consists of four cut-sets S_1, S_2, S_3, and S_4 where

$$S_1 = \mathscr{E}\{(i, 1, 2, 4) \times \overline{(i, 1, 2, 4)}\}$$
$$S_2 = \mathscr{E}((i, 1, 3, 5) \times \overline{(i, 1, 3, 5)})$$
$$S_3 = \mathscr{E}((i, 2, 3, 6) \times \overline{(i, 2, 3, 6)})$$

and

$$S_4 = \mathscr{E}((\overline{(j)}) \times (j))$$

as shown in the figure.

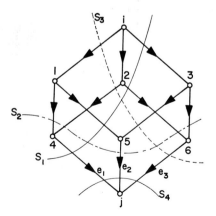

Fig. 14-3-8. A nonseparable SEC graph.

Since every edge in G is in a cut-set in $\{S\}$, G is 1-(i, j)-distinguishable under (e_1, e_2, e_3) by Theorem 14-3-6.

Theorem 14-3-7. Let G be an SEC graph from i to j. Suppose set $E = \mathcal{E}(\Omega \times \bar{\Omega})$ is a directed cut-set. If every edge in G is in a cut-set in $\{S\}$, then G is either 1-(i, j)-distinguishable or 2-(i, j)-distinguishable under E where $\{S\}$ is a collection of directed cut-sets covered by E.

Proof. Consider an SEC graph G' obtained from G by coinciding all vertices in Ω. Then it is clear that a subcollection of $\{S\}$ is a collection of directed cut-sets in G' covered by E and every edge in G' is in a cut-set in the subcollection. Since $E = \mathcal{E}(\Omega' \times \bar{\Omega'})$ is a directed cut-set in G' where $\Omega' = (i)$ and $\bar{\Omega'} = \bar{\Omega}$, G' is 1-(i, j)-distinguishable under E by Theorem 14-3-6. Similarly, by coinciding all vertices in $\bar{\Omega}$, we obtain a new SEC graph G'' which is 1-(i, j)-distinguishable under E. Thus any two vertices in Ω cannot be together in a set in a D-partition under E. Similarly, any two vertices in $\bar{\Omega}$ cannot be together in a set in a D-partition under E. However, there is no guarantee for two vertices to be in the different sets in a D-partition under E if one vertex is from Ω and the other is from $\bar{\Omega}$. Thus G is either 1-$(i\,j)$-distinguishable or 2-(i, j)-distinguishable under E.   QED

Consider the SEC graph G from i to j shown in Fig. 14-3-7. If $E = (e_1, e_2, e_3)$, then $\{S\} = \{S_1, S_2, S_3, S_4, S_5\}$, which is a collection of directed cut-sets covered by E. It is clear that every edge in G is in a cut-set in $\{S\}$. Thus G is either 1-(i, j)-distinguishable or 2-(i, j)-distinguishable under E by Theorem 14-3-7. Note that sets in a D-partition under E are (i, j), $(1, 4)$, $(2, 5)$, and $(3, 6)$.

PROBLEMS

1. Obtain the D-partition under the following measurable sets where $\Omega_1 \cup \bar{\Omega}_1 = (1, 2, 3, 4, 5)$:

$$\Omega_1 = (1, 2, 3, 4)$$
$$\Omega_2 = (1, 3, 5)$$
$$\Omega_3 = (2, 3, 4)$$

and

$$\Omega_4 = (1, 3, 4)$$

2. Find all measurable sets $\Omega(i \times a^+)$, $\Omega(a^- \times j)$, $\Omega(i \times b^+)$, $\Omega(b^- \times j)$, $\Omega(a^+ \times b^+)$, $\Omega(a^+ \times b^-)$, $\Omega(a^- \times b^+)$, $\Omega(a^- \times b^-)$, $\Omega(b^+ \times a^+)$, $\Omega(b^+ \times a^-)$, $\Omega(b^- \times a^+)$, and $\Omega(b^- \times a^-)$ under set $E = (a, b)$ of SEC graphs from i to j in Fig. P-14-2.

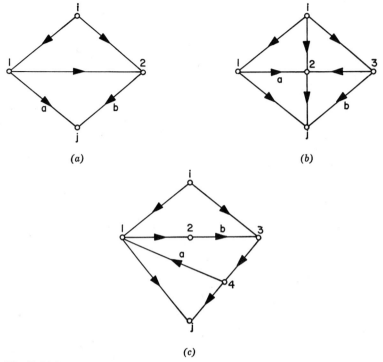

(a)

(b)

(c)

Fig. P-14-2.

3. Are those SEC graphs from i to j in Problem 2 1-distinguishable under $E = (a, b)$?

4. Find all directed cut-sets covered by $E = (a, b)$ of the SEC graphs in Problem 2.

5. Find all directed cut-sets covered by $E = (a, b)$ of SEC graph from i to j in Fig. P-14-5. Is this 1-distinguishable under $E = (a, b)$?

6. Is the SEC graph from i to j in Fig. P-14-6 1-distinguishable under $E = (a, b, c, d)$?

7. Obtain all measurable sets by gate-set $E = (a, b, c)$ of SEC graphs from i to j in Fig. P-14-7.

8. Are the SEC graphs in Problem 7 1-distinguishable under gate-set $E = (a, b, c)$?

9. Suppose we assign test points to edges a and b and test gates to edges c and d in a SEC graph from i to j in Fig. P-14-6.

(a) Obtain all measurable sets.

(b) Is this SEC graph 1-distinguishable under this assignment?

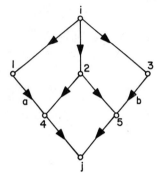

Fig. P-14-5.

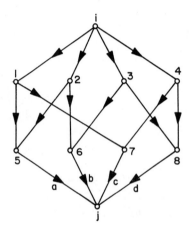

Fig. P-14-6.

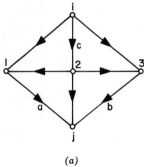

(a)

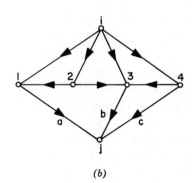

(b)

Fig. P-14-7.

BIBLIOGRAPHY

Chapters 1 and 2

Abramson, H. D., "An Algorithm for Finding Euler Circuits in Complete Graphs of Prime Order," *Theory of Graphs* (International Symposium, Rome, 1966), Gordon & Breach, Inc., New York, 1967, pp. 1–7.

Ball, W. W., *Mathematical Recreations and Essays*, The Macmillan Co., New York, 1956.

Beckenbach, E. F., *Applied Combinatorial Mathematics*, John Wiley & Sons, Inc., New York, 1964.

Beineke, L. W., and Plummer, M. D., "On the 1-Factors of a Non-Separable Graph," *J. Comb. Theory*, **2**, 285–289 (1967).

Berge, C., *The Theory of Graphs and Its Applications*, John Wiley & Sons, Inc., New York, 1962.

Branin, F. H., Jr., "The Inverse of the Incidence Matrix of a Tree and the Formulation of the Algebraic-First-Order Differential Equations of an RLC Network," *IEEE Trans. Circuit Theory*, **CT-10**, 543–544 (Dec. 1963).

Busacker, R. B., and Saaty, T. L., *Finite Graphs and Networks: An Introduction with Applications*, McGraw-Hill Book Co., Inc., New York, 1965.

Carrádi, K. A., and Hajnal, A., "On the Maximal Number of Independent Circuits in a Graph," *Acta Math. Acad. Sci. Hungary*, **14**, 423–439 (1963).

Dirac, G. A., "Note on the Colouring of Graphs," *Math. Z.*, **54**, 347–353 (1951).

Dirac, G. A., "Theorems Related to the Four Color Conjecture," *J. London Math. Soc.*, **29**, 143–149 (1954).

Dirac, G. A., "On the Maximal Number of Independent Triangles in Graphs," *Abh. Math. Sem. Univ. Hamburg*, **26**, 78–82 (1963).

Erdös, P., and Katona, G., *Theory of Graphs* (Proceedings of the Colloquium, Tihany, Hungary, Sept. 1966), Academic Press, Inc., New York, 1968.

Erdös, P., and Pósa, L., "On the Maximal Number of Disjoint Circuits of a Graph," *Publ. Math., Debrecen.*, **9**, 3–12 (1962).

Euler, L., "Solutio Problematis ad Geometriam Situs Pertinantis," *Academimae Petropolitanse*, **8**, 128–140 (1736).

Euler, L., "The Königsberg Bridges," *Sci. Amer.*, **189**(1), 66–70 (July 1953).

Fiedler, M., *Theory of Graphs and Its Applications* (Proceedings of the Symposium, Smolenice, Czechoslovakia, June 1963), Academic Press, Inc., New York, 1964.

Harary, F., Graph Theory and Theoretical Physics (Proceedings of NATO Summer School, Frascati, Italy, 1964), Academic Press, Inc., 1967.

Harary, F., *Graph Theory*, Addison-Wesley Publ. Co., Reading, Mass., 1969.

Harary, F., and Beineke, L., *A Seminar on Graph Theory*, Holt, Rinehart & Winston, Inc., New York, 1967.

Howlett, F., "Konigsberg Bridge Problem," *Pi Mu Epsilon J.*, 3, 218–223 (1961).

Kasai, T., "A New Group of Incomplete Graphs," *Trans. IECEJ (Japan)*, 51A(5), 204–205 (May 1968).

Kasai, T., "Generalization of a Fundamental Group in Incomplete Graphs," *Trans. IECEJ (Japan)*, 52A(11), 459–460 (Nov. 1969).

Konig, D., *Theorie der endlichen und unendlichen Graphen*, Leipzig, 1936.

Mayeda, W., "Application of Mathematical Logic to Network Theory," Circuit Theory Group, University of Illinois.

Mayeda, W., "Properties of Classes of Paths," Report No. R-212, C.S.L., University of Illinois, Urbana, May 1964.

Moon, J. W., "On Edge-Disjoint Cycles in a Graph," *Can. Math. Bull.*, 7, 519–523 (1964).

Newman, J. R., "Leonard Euler and the Konigsberg Bridges," *Sci. Amer.*, 189(1), 66–70 (1953).

Onodera, R., *Introduction to Graph Theory*, Morikita Co., Japan, 1968 (Japanese).

Ore, O., *Theory of Graphs*, American Mathematical Society, Providence, R.I., 1962.

Ore, O. *Graphs and Their Uses*, Random House, New York, 1963.

Quast, J., and Schuh, F., "A Number of Path Problems," *Simon Stevin*, 27, 201–211 (1950).

Reed, M. B., "The Seg: A New Class of Subgraphs," *IRE Trans. Circuit Theory*, CT-12, 162–168 (June 1955).

Rényi, A., "On Connected Graphs. I," *Magyar Tud, Akad. Mat. Kutató Int. Közl.*, 4(1), 73–85 (1959).

Ringel, G., *Färbungsprobleme auf Flächen und Graphen*, Veb Deutscher Verlag der Wissenachaften, Berlin, 1959.

Roberts, S. M., and Flores, B., "Systematic Generation of Hamiltonian Circuits," *Comm. ACM*, 9, 690–694 (Sept. 1966).

Rosenstiehl, P., *Theory of Graphs* (International Symposium, Rome, July 1966), Gordon & Breach, Inc., New York, 1967.

Sabidussi, G., "Sequence of Euler Graphs," *Can. Math. Bull.*, 9, 177–182 (1966).

Seshu, S., and Reed, M. B., *Linear Graphs and Electrical Networks*, Addison-Wesley Publ. Co., Inc., Reading, Mass., 1961.

Thorelli, L. E., "An Algorithm for Computing All Paths in a Graph," *BIT*, 6(4), 347–349 (1966).

Tutte, W. T., "On Hamilton Circuits," *J. London Math. Soc.*, 21, 98–101 (1946).

Watkins, M. E., and Mesner, D. M., "Cycles and Connectivity in Graphs," *Can. J. Math.*, 19, 1319–1328 (1967).

Whitney, H., "Congruent Graphs and Connectivity of Graphs," *Amer. J. Math.*, 54, 150–168 (1932).

Yau, S. S., "Generation of All Hamiltonian Circuits, Paths and Centers of a Graph and Related Problems," *IEEE Trans. Circuit Theory*, **CT-14**, 79–81 (Mar. 1967).

Chapters 3 and 4

Adam, A., "Isomorphism Problem for a Special Class of Graphs," *Res. Prob.* 2–10, *J. Comb. Theory*, **2**(3), 393 (May 1967).

Adkisson, V. W., and Maclane, S., "Planar Graphs Whose Homeomorphisms Can Be Extended for Any Mapping on the Sphere," *Amer. J. Math.* **59**, 823–832 (1937).

Adkisson, V. W., and Maclane, S., "Fixed Points and the Extensions of Homeomorphisms of a Planar Graph," *Amer. J. Math.*, **60**, 611–639 (1939).

Artzy, R., "Self-Dual Configurations and Their Levi Graphs," *Proc. Amer. Math. Soc.*, **7**(2), 299–303 (Apr. 1956).

Ash, R. B., and Kim, W. H., "On the Realizability of a Circuit Matrix," *IRE Trans. Circuit Theory*, **6**(2), 219–223 (June 1959).

Auslander, L., and Trent, H. M., "Incidence Matrices and Linear Graphs," *J. Math. Mech.*, **8**(5), 827–835 (1959).

Ball, W. W. R., *Mathematical Recreations and Essays*, The Macmillan Company, New York, 1960.

Bapesmara Rao, V. V., and Murti, F. K., "Comment on the Construction of a Pair of *M*-Submatrices of a Cut-Set Matrix," *IEEE Trans. Circuit Theory*, **CT-16**(1), 141–142 (Feb. 1969).

Battle, J., Harary, F., and Kodama, Y., "Every Planar Graph With Nine Points Has a Nonplanar Complements," *Bull. Amer. Math. Soc.*, **68**(6), 569–571 (Nov. 1962).

Behgad, M., "A Criterion for the Planarity of the Total Graph of a Graph," *Proc. Cambridge Philos. Soc.*, **63**, 679–681 (1967).

Beineke, L. W., "The Decomposition of Complete Graphs into Planar Subgraphs," *Graph Theory and Theoretical Physics* (Proceedings NATO Summer School, Frascati, Italy, 1964), Academic Press, Inc., New York, 1967, pp. 139–154.

Belevitch, V., "On the Realizability of Graphs With Prescribed Circuit Matrices," *Switching Theory in Space Technology*, Edited by H. Aiken and W. F. Main, Stanford University Press, Stanford, Calif., 1963, pp. 126–144.

Berconici, M., "Formulas for the Number of Trees in a Graph," *IEEE Trans. Circuit Theory*, **CT-16**(1), 101–102 (Feb. 1969).

Boesch, F. T., "Cut-Set Matrices and the Cederbaum Algorithm," *IEEE Internat. Convention Record*, Part I, 257–262 (1964).

Bower, R., "On Sums of Valencies in Planar Graphs," *Can. Math. Bull.*, **9**, 111–114 (1966).

Brooks, R. L., "On Colouring the Nodes of a Network," *Proc. Cambridge Philos. Soc.*, **37**, 194–197 (1941).

Brown, D. P., "On the Rank and Nullity of Subgraphs," *SIAM J. Appl. Math.*, **16**, 387–394 (1968).

Brown, D. P., and Budner, A., "A Note on Planar Graphs," *J. Franklin Inst.*, **280**, 222–230 (1965).

Cauer, W., *Theorie der Linearen Wechselstromschaltungen*, Akademic Verlag, Berlin, 1954; English translation, McGraw-Hill Book Co., Inc., New York, 1958.

Cederbaum, I., "Matrices All of Whose Elements and Subdeterminants Are 1, −1, or 0," *J. Math. Phys.*, **36**, 351–361 (1958).

Cederbaum, I., "On Duality and Equivalence," *IRE Trans. Circuit Theory*, **CT-8**(4), 487–488 (Dec. 1961).

Chan, S. P., and Dunn, W. R., Jr., "An Algorithm for Testing the Planarity of a Graph," *IEEE Trans. Circuit Theory*, **CT-15**, 166–168 (June 1968).

Dirac, G. A., and Schuster, S., "A Theorem of Kuratowski," *Nederl. Akad. Wetensch. Proc. Sec. A*, **57**, 343 (1954).

Dunn, W. R., Jr., and Chan, S. P., "Realizability of a Planar Graph from its Circuit Matrix," Tenth Midwest Symposium on Circuit Theory, Purdue University, Lafayette, Ind., 1967.

Fisher, G. J., and Wing, O., "A Correspondence Between A Class of Planar Graphs and Bipartite Graphs," *IEEE Trans. Circuit Theory*, **CT-12**, 266–267 (June 1965).

Fisher, G. J., and Wing, O., "Computer Recognition and Extraction of Planar Graphs from the Incidence Matrix," *IEEE Trans. Circuit Theory*, **CT-13**, 154–163 (June 1966).

Foster, R. M., "Topologic and Algebraic Considerations in Network Synthesis," *Proc. Polytech. Inst. Brooklyn Symp. Modern Network Synthesis I*, 8–18 (Apr. 1952).

Fu, Y., "Realization on Circuit Matrices," *IEEE Trans. Circuit Theory*, **CT-12**, 604–607 (Dec. 1965).

Gould, R., "Graphs and Vector Spaces," *J. Math. Phys.*, **37**(3), 193–214 (1958).

Gould, R., "The Application of Graph Theory to the Synthesis of Contact Networks," Proceedings, International Symposium on Theory of Switching, 1957; and *Annals of the Computation Laboratory of Harvard University*, **29**, 1959.

Guillemin, E. A., *Introductory Circuit Theory*, John Wiley & Sons, Inc., New York, 1953.

Guillemin, E. A., "How to Grow Your Own Trees from Given Cut-Set or Tie-Set Matrices," *IRE Trans. Circuit Theory*, **CT-6**, 110–126 (May 1959).

Halin, R., and Jung, H. A., "Note on Isomorphisms of Graphs, *J. London Math. Soc.*, **42**, 254–256 (1967).

Halkias, C. C., and Kim, W. H., "Realization on Fundamental Circuit and Cut-Set Matrices," *IRE Internat. Convention Record*, Part II, 8–15 (1962).

Halton, J. H., "A Combinatorial Proof of a Theorem of Tutte," *Proc. Cambridge Philos. Soc.*, **62**, 683–684 (1966).

Harary, F., and Tutte, W. T., "A Dual Form of Kuratowski Theorem," *Can. Math. Bull.*, **8**, 17–20 (1965).

Harary, F., and Tutte, W. T., "Correction and Addendum to A Dual Form of Kuratowski's Theorem," *Can. Math. Bull.*, **8**, 373 (1965).

Hiraoka, T., "Theorem on a Regular Graph," *Bull. Educ. Fuc. Ibaraki Univ., Japan*, **7**, 39–45 (1957).

Iri, M., "A Necessary and Sufficient Condition for a Matrix to be the Loop or Cut-Set Matrix of a Graph and Practical Method for the Topological Synthesis of Networks," *RAAG Res. Note 50*, Tokyo, Japan, 1962.

Kirchhoff, G., "Über die Auflösung der Gleidungen, auf welche man bei der Untersuchungen der Linearen Verteilung Galvanisher Storöme geführt wird," *Poggendorf Ann. Physik*, **72**, 497–508 (1847); English translation, *IRE Trans. Circuit Theory*, **CT-5**, 4–7 (Mar. 1958).

Kishi, G., and Kajitani, Y., "Subsets of Trees which Determine the Original Graph," *Trans. IECEJ, Japan*, **52-A**(2), 69–76 (Feb. 1969).

Kishi, G., and Uetake, Y., "Rank of Edge Incidence Matrix," *IEEE Trans. Circuit Theory*, **CT-16**, 230–232 (May 1969).

Kuratowski, C., "Sur le Probleme des Courbes Gauches en Topologie," *Fund. Math.*, **15**, 271–281 (1930).

Lefschetz, S., "Planar Graphs and Related Topics," *Proc. Nat. Acad. Sci., U.S.A.*, **54**, 1763–1765 (1965).

Lempel, A., Eren S., and Cederbaum, I., "An Algorithm for Planarity Testing of Graphs," *Theory of Graphs* (International Symposium, Rome, 1966), Gordon & Breach, Inc., New York, 1967, pp. 215–232.

Malik, N. R., "Relationship Between Nonzero Determinants Formed from Vertex and Circuit Matrices," *IEEE Trans. Circuit Theory*, **CT-13**, 196 (June 1966).

Mayeda, W., "Necessary and Sufficient Conditions for Realizability of Cut-Set Matrices," *IRE Trans. Circuit Theory*, **CT-7**, 79–81 (Mar. 1960).

Mayeda, W., "Properties of the Non-Singular Matrices Which Transform Cut Set Matrices into Incidence Matrices," *IRE Trans. Circuit Theory*, **CT-10**, 128–131 (Mar. 1963).

Miller, W. C., "Inversion of Nonsingular Submatrix of an Incidence Matrix," *IEEE Trans. Circuit Theory*, **CT-10**, 132 (1963).

Okada, S., and Young, K., "Ambit Realization of Cut-Set Matrices into Graphs," Tech. Report Contract AF-19(604)-6620, Poly. Inst. Brooklyn, New York, 1961.

Parker, S. R., and Lohre, H. J., "A Direct Procedure for the Synthesis of Network Graphs from a Given Fundamental Loop or Cutset Matrix," *IEEE Trans. Circuit Theory*, **CT-16**, 221–223 (May 1969).

Piekarski, M., "On the Construction of a Pair of M-Submatrices of a Cut-Set Matrix," *IEEE Trans. Circuit Theory*, **CT-13**, 114–117 (Mar. 1966).

Rao, V. V. B., and Murti, V. G. K., "Comment on the Construction of a Pair of M-Submatrices of a Cut-Set Matrix," *IEEE Trans. Circuit Theory*, **CT-16**, 141–142 (Feb. 1969).

Rao, V. V. B., Rao, K. S., Sankaran, P., and Murti, V. G. K., "Planar Graphs and Circuits," *Matrix Tensor Q.*, **18**, 81–91 (1968).

Reddy, P. S., Murti, V. G. K., and Thulasiraman, K., "Realization of Modified Cut-Set Matrix and Application," *IEEE Trans. Circuit Theory*, **CT-17**, 475–486 (Nov. 1970).

Resh, J. A., "The Inverse of a Nonsingular Sub-Matrix of an Incidence Matrix," *IEEE Trans. Circuit Theory*, **CT-10**, 131–132 (Mar. 1963).

Seshu, S., and Reed, M. B., "On Cut Sets of Electrical Networks," *Proc. Second Midwest Symp. Circuit Theory*, Michigan State Univ., 1956, pp. 1.1–1.13.

Shirakawa, I., Takadashi, H., and Ozaki, H., "Planar Decomposition of a Class of Complete Bipartete Graphs," *Trans. IECEJ, Japan*, **51-A**(10), 373–377 (Oct. 1968).

Tutte, W. T., "A Theorem on Planar Graphs," *Trans. Amer. Math. Soc.*, **82**, 99–116 (May 1956).

Tutte, W. T., "A Homotopy Theorem for Matroids, I, II," *Trans. Amer. Math Soc.*, **88**, 144–174 (May 1958).

Tutte, W. T., "Matroids and Graphs," *Trans. Amer. Math. Soc.*, **90**, 527–552 (Mar. 1959).

Tutte, W. T., "An Algorithm for Determining Whether a Given Binary Matroid is Graphic," *Proc. Amer. Math. Soc.*, **11**, 905–917 (Dec. 1960).

Ungar, P., "A Theorem on Planar Graphs," *J. London Math. Soc.*, **26**, 256–262 (1951).

Veblen, O., *Analysis Situs*, American Mathematical Society, Cambridge Colloquim Publications, 1931.

Vidyasagar, M., "An Algebraic Method for Finding a Dual Graph of a Given Graph," *IEEE Trans. Circuit Theory*, **CT-17**, 434–436 (Aug. 1970).

Whitney, H., "Non-Separable and Planar Graphs," *Trans. Amer. Math. Soc.*, **34**, 339–362 (Apr. 1932).

Whitney, H., "A Set of Topological Invariants for Graphs," *Amer. J. Math.*, **55**, 231–235 (1933).

Whitney, H., "2-Isomorphic Graphs," *Amer. J. Math.*, **55**, 245–254 (1933).

Whitney, H., "Planar Graphs," *Fund. Math.*, **21**, 73–84 (1933).

Whitney, H., "On the Abstract Properties of Linear Dependence," *Amer. J. Math.*, **57**, 509–533 (1935).

Wing, O., "On Drawing a Planar Graph," *IEEE Trans. Circuit Theory*, **CT-13**, 112–114 (Mar. 1966).

Wing, O., and Kim, W. H., "The Path Matrix and Its Realizability," *IRE Trans. Circuit Theory*, **CT-6**, 267–272 (1959).

Chapters 5 and 6

Ayoub, J. N., and Frisch, I. T., "On the Smallest-Branch Cuts in Directed Graphs," *IEEE Trans. Circuit Theory*, **CT-17**, 249–250 (May 1970).

Brownlee, A., "Directed Graph Realization of Degree Pairs," *Amer. Math. Monthly*, **75**, 36–38 (Jan. 1968).

de Bruijn, N. G., and Ehrenfest, T. van A., "Circuits and Trees in Oriented Graphs," *Simon Stevin*, **28**, 203–217 (1951).

Chen, Y. C., and Wing, O., "Realization of a Directed Graph Having a Prescribed Terminal Connection Matrix," *IEEE Trans. Circuit Theory*, **CT-13**, 197–198 (June 1966).

Gallai, T., "On Directed Paths and Circuits," *Theory of Graphs* (Proceedings of the Colloquium held at Tihany, Hungary, Sept. 1966), Academic Press, Inc., New York, 1968, pp. 115–118.

Hakimi, S. L., "On the Degrees of the Vertices of a Directed Graph," *J. Franklin Inst.*, **279**, 290–308 (Apr. 1965).

Harary, F., Norman, R. Z., and Cartwright, D., *Structural Models: An Introduction to the Theory of Directed Graphs*, John Wiley & Sons, Inc., New York, 1965.

Kamae, T., "A Systematic Method of Finding All Directed Circuits and Enumerating All Directed Paths," *IEEE Trans. Circuit Theory*, **CT-14**, 166–171 (June 1967).

Lal, M., "Directed Hamiltonian Circuits," *IEEE Trans. Circuit Theory*, **CT-14**, 356–357 (Sept. 1967).

Lempel, A., and Cederbaum, I., "Minimum Feedback Arc and Vertex Sets of a Directed Graph," *IEEE Trans. Circuit Theory*, **CT-13**, 399–403 (Dec. 1966).

Mayeda, W., "Realizability of Fundamental Cut Set Matrices of Oriented Graphs," *IEEE Trans. Circuit Theory*, **CT-10**, 133–134 (March 1963).

Mayeda, W., "Pseudo-Cuts and Their Applications," CSL Report R-233, University of Illinois, Urbana, July 1964.

Nash-Williams, C. St. J. A., "Abelian Groups and Generalized Knights," *Proc. Cambridge Philos. Soc.*, **55**, 181–184 (1959).

Norman, R. L., "A Matrix Method for Location of Cycles of a Directed Graph," *Amer. Inst. Chem. Eng. J.*, **11**, 450–452 (1965).

Okada, S., "On Node and Mesh Determinants," *Proc. IRE*, **43**, 1527 (Oct. 1955).

Ore, O., "Studies in Directed Graphs, I," *Ann. Math.*, **63**, 383–406 (May 1956).

Ore, O., "Studies in Directed Graphs, II," *Ann. Math.*, **64**, 142–153 (July 1956).

Ore, O., "Studies in Directed Graphs, III," *Ann. Math.*, **68**, 526–549 (1958).

Percus, J. K., "Matrix Analysis of Oriented Graphs with Irreducible Feedback Loops," *IRE Trans. Circuit Theory*, **CT-12**, 117–127 (June 1955).

Shinoda, S., "Finding All Possible Directed Trees of a Directed Graph," *Trans. IECEJ, Japan*, **51-A**(7), 290–291 (July 1968).

Shinoda, S., "On Hamilton Circuits of a Directed Graph," *Trans. IECEJ, Japan*, **51-A**(7), 291–292 (July 1968).

Van Aardenne-Ehrenfest, T., and de Bruijn, N. G., "Circuits and Trees in Linear Oriented Graphs," *Simon Stevin*, **28**, 203–217 (1951).

Younger, D. H., "Minimum Feedback Arc Sets for a Directed Graph," *IEEE Trans. Circuit Theory*, **CT-10**, 238–245 (June 1963).

Chapters 7 and 8

Ali, A. A., "On the Sign of a Tree Pair," *IEEE Trans. Circuit Theory*, **CT-11**, 294–296 (June 1964).

Barksdale, G. L., "An Efficient Computational Technique for Determination of the Cut-Set Equations of a Network," *IEEE Trans. Circuit Theory*, **CT-13**, 339–340 (Sept. 1966).

Barrows, I. T., Jr., "Extension of Feusener's Method to Active Networks," CSL Report, University of Illinois, Aug. 1965.

Bedrosian, S. D., "Formulas for the Number of Trees in a Network," *IRE Trans. Circuit Theory*, **CT-8**, 363–364 (Sept. 1961).

Bellert, S., "Topological Analysis and Synthesis of Linear Systems," *J. Franklin Inst.*, **274**, 425–443 (Dec. 1962).

Bercovici, M., "The Number of Trees of a Linear Graph," Symposium of Analysis and Synthesis of Electrical Networks, Bucharest, Rumania, 1967.

Bercovici, M., "Formulas for the Number of Trees in a Graph," *IEEE Trans. Circuit Theory*, **CT-16**, 101–102 (Feb. 1969).

Brayshaw, G. S., "Topological Analysis of Networks Containing Nullators and Norators," *IEEE Trans. Circuit Theory*, **CT-6**, 226–227 (May 1969).

Brown, D. P., "New Topological Formulas for Linear Networks," *IEEE Trans. Circuit Theory*, **CT-12**, 358–365 (Sept. 1965).

Bryand, P. R., "A Topological Investigation of Network Determinants," IEE (London) Monograph 312R (Sept. 1958); *Proc. IEE (London)*, **106-C**, 16–22 (Mar. 1959).

Bryant, P. R., "Order of Complexity of Electrical Networks," IEE (London) Monograph 335E, June 1959; *Proc. IEE (London)*, **106-C**, 174–188 (Mar. 1959).

Bryant, P. R., "The Algebra and Topology of Electrical Networks," *Proc. IEE (London)*, **108-C**, 215–229 (1961).

Bryant, P. R., "Graph Theory Applied to Electrical Networks," *Graph Theory and Theoretical Physics* (Procedings NATO Summer School, Frascati, Italy, 1964), Academic Press, Inc., New York, 1967, pp. 111–138.

Carpenter, R. M., "Topological Analysis of Active Networks," Proceedings of the Institute on Modern Solid State Circuit Design, University of Santa Clara, 1966, pp. 50–61.

Chan, S. P., *Introductory Topological Analysis of Electrical Networks*, Holt, Rinehart & Winston, Inc., New York, 1969.

Chan, S. P., and Chan, S. G., "Modification of Topological Formulas," *IEEE Trans. Circuit Theory*, **CT-15**, 84–86 (Mar. 1968).

Chen, W. K., "Topological Analysis for Active Networks," *IEEE Trans. Circuit Theory*, **CT-12**, 85–91 (Mar. 1965).

Chen, W. K., "Note on Topological Analysis of Active Networks," *IEEE Trans. Circuit Theory*, **CT-13**, 438–439 (Dec. 1966).

Coates, C. L., "General Topological Formulas for Linear Network Functions," *IRE Trans. Circuit Theory*, **CT-5**, 30–42 (Mar. 1958).

Dawson, D., "Computational Aspects of Topological Approach to Active Linear Network Analysis," Proceedings of First Hawaii International Conference System Science, 1968.

Declaris, N., and Saeks, R., "Graph-Theoretic Foundations of Linear, Lumped, Finite Networks," *Elec. Engr. Res. Lab. Report TR-4*, EERL-51, Cornell University, Ithaca, N.Y., May 1966.

Dodd, G. G., "On Unistor Graphs," *IEEE Trans. Circuit Theory*, **CT-14**, 154–159 (June 1967).

Farber, L., and Malik, R., "A New Modification of Topological Formulas," *IEEE Trans. Circuit Theory*, **CT-16**, 89–91 (Feb. 1961).

Fujisawa, T., "On a Problem of Network Topology," *IRE Trans. Circuit Theory*, **CT-6**, 261–266 (Sept. 1959).

Hakimi, S. L., and Mayeda, W., "Proofs of Some Network Theorems by Topological Formulas," ITR No. 11, EERL, University of Illinois, 1958.

Hakimi, S. L., and Mayeda, W., "On Coefficients of Polynomials in Network Functions," *IRE Trans. Circuit Theory*, **CT-7**, 40–44 (Mar. 1960).

Hirayama, H., and Ohtsuki, T., "Topological Network Analysis by Digital Computer," *Trans. IECEJ, Japan*, **48**, 424–432 (Mar. 1965).

Hobbs, E. W., and MacWilliams, F. J., "Topological Network Analysis as a Computer Program," *IRE Trans. Circuit Theory*, **CT-6**, 135–136 (Mar. 1959).

Hoffman, A. J., "Generalization of a Theorem of König," *J. Washington Acad. Sci.*, **46**, 211–212 (1956).

Jong, M. T., "Topological Formulas for Networks Containing Operational Amplifiers," *IEEE Trans. Circuit Theory*, **CT-17**, 160–162 (Feb. 1970).

Jong, M. T., and Zobrist, G. W., "Topological Formulas for General Linear Networks," *IEEE Trans. Circuit Theory*, **CT-15**, 251–259 (Sept. 1968).

Kim, W. H., "Application of Graph Theory to the Analysis of Active and Mutually Coupled Networks," *J. Franklin Inst.*, **271**, 200–221 (Mar. 1961).

Ku, Y. H., "Resume of Maxwell's and Kirchhoff's Rules," *J. Franklin Inst.*, **253**, 211–224 (1952).

Mason, S. J., "Topological Analysis of Linear Non-Reciprocal Networks," *Proc. IRE*, **45**, 829–838 (June 1957).

Maxwell, J. C., *Electricity and Magnetism*, Vol. 1, Clarendon Press, Oxford, 1892.

Mayeda, W., "Digital Determination of Topological Quantities and Network Functions," ITR, No. 6, EERL, University of Illinois, 1957.

Mayeda, W., "Topological Formulas for Active Networks," ITR No. 8, EERL, University of Illinois, Jan. 1958.

Mayeda, W., "Topological Formulas for Active Networks," *Proc. Nat. Elec. Conf.*, **15**, 1–13 (1958).

Mayeda, W., "Application of Linear Graphs to Electrical Networks, Switching Networks and Communication Nets," CSL Report No. R-203, University of Illinois, July 1964.

Mayeda, W., "General Transformation for Network Analysis," Proceedings, First Annual Princeton Conference on Information Science and Systems, Mar. 1967.

Mayeda, W., "Graph Theoretical Analysis Using Network Equivalent Transformation," *Trans. IECEJ, Japan*, **50**, 115–120 (July 1967).

Mayeda, W., and Seshu, S., "Topological Formulas for Network Functions," Engr. Experimental Station Bulletin No. 446, University of Illinois, Nov. 1957.

Mayeda, W., and Van Valkenburg, M. E., "Network Analysis and Synthesis by Digital Computers," *IRE Wescon Convention Record*, Part 2, 1957, pp. 137–144.

Mayeda, W., and Van Valkenburg, M. E., "Analysis of Non-Reciprocal Networks by Digital Computers," *IRE Wesson Convention Record*, Part 2, 1958, pp. 70–75.

Myers, B. R., "Efficient Generation of Tree Admittance Products in a Cascade of Two-Port Networks," *Proc. IEE (London)*, **114**, 1641–1646 (Nov. 1967).

Nakagawa, N., "On the Evaluation of Graph Trees and Driving Point Admittance," *IRE Trans. Circuit Theory*, CT-5, 122–127 (June 1958).

Nathan, A., "Topological Rules for Linear Networks," *IEEE Trans. Circuit Theory*, **CT-12**, 344–358 (Sept. 1965).

Newcomb, R. W., "Topological Analysis with Ideal Transformers," *IEEE Trans. Circuit Theory*, **CT-10**, 457–458 (Sept. 1963).

Numata, J., "Modified Unister Graphs and Signal Flow Graphs," CSL Report No. R-261, University of Illinois, July 1965.

Ohtsuki, T., Ishizaki, Y., and Watanabe, H., "Topological Degrees of Freedom and Mixed Analysis of Electrical Networks," *IEEE Trans. Circuit Theory*, **CT-17**, 499–505 (Nov. 1970).

Okada, S., and Onodera, R., "On Network Topology I," *Bull. Yamagata Univ.*, **2**, 89–117 (1952).

Okada, S., and Onodera, R., "On Network Topology II," *Bull. Yamagata Univ.*, **2**, 191–206 (1953).

Percival, W. S., "Solution of Passive Electrical Networks by Means of Mathematical Tress," *Proc. IEE (London)*, *Part III*, **100**, 143–150 (1953).

Percival, W. S., "Graphs of Active Networks," *Proc. IEE (London)*, **102**, 270–278 (Apr. 1955).

Rode, F., and Chan, S. P., "Evaluation of Topological Formulas Using Digital Computers," *Electronic Letters*, **4**, 257–258 (1968).

Rothfarb, W., "Topological Formulas for Networks Containing Ideal 3-Terminal Transformers," *IEEE Trans. Circuit Theory*, **CT-12**, 421–423 (Sept. 1965).

Seacat, R. H., and Merkl, E. D., "The Application of Topological Methods to Active Networks," Proceedings of the Eighth Midwest Symposium on Circuit Theory, Colorado State University, Boulder, June, 1965.

Seshu, S., "Network Applications of Graph Theory—A Survey," Proceedings of the Fifth Midwest Symposium on Circuit Theory, 1961.

Shinoda, S., and Horiuchi, K., "On Theorems of the Directed Tree Enumeration," *Trans. IECEJ, Japan*, **52-A**(1), 40–41 (Jan. 1969).

Sinha, V. P., "Topological Formulas for Passive Transformerless 3-Terminal Networks Constrained by One Operational Amplifier," *IEEE Trans. Circuit Theory*, **CT-10**, 125–126 (Mar. 1963).

Sinha, V. P., "Correction to Topological Formulas for Passive Transformerless 3-Terminal Networks Constrained by One Operational Amplifier," *IEEE Trans. Circuit Theory*, **CT-13**, 123 (March 1966).

Su, Y. H., "Topological Formulas and the Order of Complexity for Networks with a Non-Reciprocal Element," *J. Franklin Inst.*, **286**, 204–224 (Sept. 1968).

Talbot, A., "Topological Analysis of General Linear Networks," *IEEE Trans. Circuit Theory*, **CT-12**, 170–180 (June 1965).

Talbot, A., "Topological Analysis for Active Networks," *IEEE Trans. Circuit Theory*, **CT-13**, 111–112 (Mar. 1966).

Tsang, N. F., "On Electrical Network Determinants," *J. Math Phys.*, **33**, 185–193 (July 1954).

Wang, C. L., and Tokad, Y., "Polygon to Star Transformations," *IRE Trans. Circuit Theory*, **CT-8**, 489–491 (1961).

Wang, R. T. P., "On the Sign of Common Tree Product," *IEEE Trans. Circuit Theory*, **CT-13**, 103–105 (Mar. 1966).

Watanabe, H., "A Method of Tree Expansion in Network Topology," *IRE Trans. Circuit Theory*, **CT-8**, 4–11 (Mar. 1961).

Weinberg, L., "Kirchhoff's Third and Fourth Laws," *IRE Trans. Circuit Theory*, **CT-5**, 8–30 (Mar. 1958).

Chapter 9

Bedrosian, S. D., "Generating Formulas for the Number of Trees in a Graph," *J. Franklin Inst.*, **277**, 313–326 (Apr. 1964).

Berger, I., "The Enumeration of Trees Without Duplication," *IEEE Trans. Circuit Theory*, **CT-14**, 417–418 (Dec. 1967).

Berger, I., and Nathan, A., "The Algebra of Sets of Trees, k-Trees, and Other Configurations," *IEEE Trans. Circuit Theory*, **CT-15**, 221–228 (Sept. 1968).

Char, J. P., "Generation of Trees, Two-Trees, and Storage of Master Forests," *IEEE Trans. Circuit Theory*, **CT-15**, 228–238 (Sept. 1968).

Chen, W. K., "On the Directed Trees and Directed k-Trees of a Digraph and Their Generation," *SIAM J. Appl. Math.*, **14**, 550–560 (1966).

Chen, W. K., "Generation of Trees by Algebraic Methods," *Electronics Letters*, **4**, 456–457 (Oct. 1968).

Chen, W. K., "Computer Generation of Trees and Co-Trees in a Cascade of Multi-terminal Networks," *IEEE Trans. Circuit Theory*, **CT-16**, 518–526 (Nov. 1969).

Clarke, L. E., "On Otter's Formula for Enumerating Trees," *Q. J. Math., Oxford Ser. 2*, **10**, 43–45 (1963).

Glicksman, S., "On the Representation and Enumeration of Trees," *Proc. Cambridge Philos. Soc.*, **59**, 509–517 (1963).

Hakimi, S. L., "On Trees of a Graph and Their Generation," *J. Franklin Inst.*, **272**, 347–359 (Nov. 1961).

Hakimi, S. L., and Green, D. G., "Generation and Realization of Trees and k-Trees," *IEEE Trans. Circuit Theory*, **CT-11**, 247–255 (June 1964).

Jong, M. T., Lau, H. C., and Zobrist, G. W., "Tree Generation," *Electronics Letters*, **2**, 318–319 (Aug. 1966).

Knuth, D. E., "Another Enumeration of Trees," *Can. J. Math.*, **20**, 1077–1086 (1968).

Malik, N. R., "Compound Matrices Applied to the Tree-Generating Problem," *IEEE Trans. Circuit Theory*, **CT-17**, 149–151 (Feb. 1970).

Mark, S. K., and Chen, W. K., "On the Generation of Trees, Cotrees, k-Trees and k-cotrees," *Proceedings of the First Annual Asilomar Conference on Circuit and Systems*, Santa Clara University, 1967, pp. 659–668.

Mayeda, W., "Reducing Computation Time in the Analysis of Networks by Digital Computer," *IRE Trans. Circuit Theory*, **CT-6**, 136–137 (Mar. 1959).

Mayeda, W., "Correction to Reducing Computation Time in the Analysis of Networks by Digital Computers," *IRE Trans. Circuit Theory*, **CT-6** (Dec. 1959).

Mayeda, W., "Generation of Trees and Complete Trees," CSL R-284, University of Illinois, Apr. 1966.

Mayeda, W., Hakimi, S. L., Chen, W. K., and Deo, N., "Generation of Complete Trees," *IEEE Trans. Circuit Theory*, **CT-15**, 101–105 (June 1968).

Mayeda, W., and Seshu, S., "Generation of Trees Without Duplications," *IEEE Trans. Circuit Theory*, **CT-12**, 181–185 (June 1965).

Minty, G. J., "A Simple Algorithm for Listing All the Trees of a Graph," *IEEE Trans. Circuit Theory*, **CT-12**, 120 (Mar. 1965).

Myers, B. R., and Auth, L. V., Jr., "The Number and Listing of All Trees in an Arbitrary Graph," *Proceedings of the Third Allerton Conference on Circuit and System Theory*, Oct. 1965, pp. 906–912.

Myers, B. R., and Tapia, M. A., "Generation of the Topologically Distinct Types of Trees," Proceedings of the Eighth Midwest Symposium on Circuit Theory, Nov. 1965.

O'Neil, P. V., "An Application of Feussner's Method to Tree Counting," *IEEE Trans. Circuit Theory*, **CT-13**, 336–339 (Sept. 1966).

O'Neil, P. V., "Enumeration of Spanning Trees in Certain Graphs," *IEEE Trans. Circuit Theory*, **CT-17**, 250–252 (May 1970).

O'Neil, P. V., and Slepian, P., "The Number of Trees in a Network," *IEEE Trans. Circuit Theory*, **CT-13**, 271–281 (Sept. 1966).

Otter, R., "The Number of Trees," *Ann. Math.*, **49**, 583–599 (July 1948).

Paul, A. J., Jr., "Generation of Directed Trees and 2-Trees Without Duplication," *IEEE Trans. Circuit Theory*, **CT-14**, 354–356 (Sept. 1967).

Piekarski, M., "Listing of All Possible Trees of a Linear Graph," *IEEE Trans. Circuit Theory*, **CT-12**, 124–125 (Mar. 1965).

Pullen, K. A., "On the Number of Trees for a Network," *IEEE Trans. Circuit Theory*, **CT-7**, 175–176 (June 1960).

Taki, I., Kasai, T., Yoneda, S., and Kusaka, H., "Total Number of Trees in Some Incomplete Graphs," *Trans. IECEJ, Japan*, **47**, 1908–1909 (Dec. 1964).

Taki, I., Kasai, T., Yoneda, S., and Kusaka, H., "Mh-Series on the Total Number of Trees in Incomplete Graphs," *Trans. IECEJ, Japan*, **48**, 90–91 (Jan. 1965).

Trent, H. M., "Note on the Enumeration and Listing of All Possible Trees in a Connected Linear Graph," *Proc. Nat. Acad. Sci., U.S.A.*, **40**, 1004–1007 (Oct. 1954).

Wing, O., "Enumeration of Trees," *IEEE Trans. Circuit Theory*, **CT-10**, 127–128 (Mar. 1963).

Zobrist, G. W., and Lago, G. V., "Method for Obtaining the Trees of a v Vertex Complete Graph from the Trees of a $v - 1$ Vertex Complete Graph," *Matrix and Tensor Q.*, **15**, 94 (Mar. 1965).

Chapter 10

Adamian, R. G., "Comments on System State Analysis and Flow Graph Diagrams in Reliability," *IEEE Trans. Reliability*, **R-16**, 138–139 (Dec. 1967).

Bichart, T. A., "Flowgraphs for Representation of Nonlinear Systems," *IRE Trans. Circuit Theory*, **CT-8**, 49–58 (Mar. 1961).

Brozozowski, J. A., and McCluskey, E. J., Jr., "Signal Flow Graph Techniques for Sequential Circuit State Diagrams," *IEEE Trans. Electronic Computers*, **EC-12**, 67–76 (Apr. 1963).

Burroughs, J. L., "Signals Flow Graph Analysis of Physical Systems," *Proc. Montana Acad. Sci.*, **18**, 43–48 (Mar. 1958).

Chen, W. K., "The Inversion of Matrices by Flow Graphs," *SIAM J. Appl. Math.*, **12**, 676–685 (1964).

Chen, W. K., "Flow Graphs: Some Properties and Methods of Simplification," *IEEE Trans. Circuit Theory*, **CT-12**, 128–130 (1965).

Chen, W. K., "On the Modifications of Flow Graphs," *SIAM J. Appl. Math.*, **13**, 493–505 (1965).

Chen, W. K., "On Flow Graph Solutions of Linear Algebraic Equations," *SIAM J. Appl. Math.*, **15**, 136–142 (1967).

Chien, R. T., "A Simplification of the Coates-Desoer Formula for the Gain of a Flow-Graph," *Proc. IEEE*, **53**, 1240–1241 (Sept. 1965).

Chow, Y., "Node Duplication: A Transformation of Signal-Flow Graphs," *IRE Trans. Circuit Theory*, **CT-6**, 233–234 (June 1959).

Chow, Y., and Cassignol, E., *Linear Signal-Flow Graphs and Applications*, John Wiley & Sons, Inc., New York, 1962.

Coates, C. L., "Flow Graph Solutions of Linear Algebraic Equations," *IRE Trans. Circuit Theory*, **CT-6**, 170–187 (June 1959).

Davis, M. C., "Engineering Applications of the Signal Flow Graph," *J. Amer. Soc. Naval Engr.*, **73**, 147–156 (1961).

Desoer, C. A., "Optimum Formula for the Gain of a Flow Graph, or a Simple Derivation of Coates' Formula," *Proc. IRE*, **48** 883–889 (May 1960).

Divieti, L., "A Method for the Determination of the Paths and the Index of a Signal Flow Graph," *Proceedings of the Seventh International Convention on Automation and Instrumentation* (Milan, Italy, Nov. 1964), Pergamon Press, Inc., London, 1966, pp. 77–94.

Dolazza, E., "System States Analysis and Flow Graph Diagram in Reliability," *IEEE Trans. Reliability*, **R-15**, 85–94 (Dec. 1966).

Feeser, L. J., and Feng, C. C., "Flow Graphs and Boundary Value Problems," *J. Franklin Inst.*, **284**, 251–261 (Oct. 1967).

Hall, D. W., "Signal Flow Graphs," *Methodology for Systems Engineers Session 14*, 7–15, D. Van Nostrand Co., Inc., Princeton, N.J., 1962, pp. 352–378.

Happ, W. W., "Signal Flow Graphs," *Proc. IRE*, **45**, 1293 (1957).

Happ, W. W., "Flowgraphs as a Teaching Aid," *IEEE Trans. Education*, **E-9**, 69–80 (June 1966).

Happ, W. W., and Burroughs, J. L., "Systematic Formulation of the Signal Flow Graph of a Complex Servo System," AIEE Conv. Paper, CP60-1101, San Diego, 1960.

Henning, B., "Flow Graph Analysis for Variable-Parameter Networks," *IRE Trans. Circuit Theory*, **CT-9** (Sept. 1962).

Hoskins, R. F., "Signal Flow Graphs, Application to Linear Circuit Analysis," *Electron. Radio Engr.*, 298–304 (Aug. 1959).

Hoskins, R. F., "Signal Flow Graph Analysis and Feedback Theory," *Proc. IEE (London)*, **108-C**, 12–19 (1961).

Huggins, W. H., "Signal Flow Graphs and Random Signals," *Proc. IRE*, **45**, 78–86 (Jan. 1957).

Jacob, J. P., "The Number of Terms in the General Gain Formulas for Coates and Mason Signal Flow Graphs," *IEEE Trans. Circuit Theory*, **CT-12**, 601–603 (Dec. 1965).

Lal, M., "On Physical Realizability of Signal Flow Graphs and Realization Techniques," CSL Report R-157, University of Illinois, Dec. 1962.

Mayeda, W., "Elimination of Subgraphs in a Signal Flow Graph," CSL Report R-339, University of Illinois, Feb. 1967; Proceedings of the Fifth Annual Allerton Conference on Circuit and System Theory, Oct. 1967.

Milic, M. M., "Flow-Graph Evaluation of the Characteristic Polynomial of a Matrix," *IEEE Trans. Circuit Theory*, **CT-11**, 423–424 (1964).

Nathan, A., "Algebraic Approach to Signal Flow Graphs," *Proc. IRE*, **46**, 1955–1956 (1958).

Nathan, A., "A Two-Step Algorithm for the Reduction of Signal Flow Graphs," *Proc. IRE*, **49**, 1431 (Sept. 1961).

Robichaud, L. P. A., Boisvert, M., and Robert, J., *Signal Flow Graphs and Applications*, Prentice-Hall, Inc., Englewood Cliffs, N.J., 1962.

Wing, O., "Signal Flow Graph and Network Topology—How to Avoid Them," *IRE Nat. Conv. Rec.*, **6**, 48–52 (1958).

Woodley, G. V., "The Flow Graph: A Short Cut to Network Simulation," *Electron. Design* (Jan. 1959).

Younger, D. H., "A Simple Derivation of Mason's Formula," *Proc. IRE*, **51**, 1043–1044 (1963).

Zadeh, L. A., "Signal-Flow Graphs and Random Signals," *Proc. IRE*, **45**, 1413–1414 (Oct. 1957).

Chapter 11

Adam, A., "Some Open Problems of the Switching Circuit Theory," *Theory of Graphs and Its Applications* (Proceedings of the Symposium held in Smolenice, Czechoslovakia), Publishigh House, Czechoslavak Acad. Sci., Prague, 1964, pp. 109–112.

Adam, A., "On the Repetition-Free Realization of Truth Functions by Two-Terminal Graphs II," *Studia Sci. Math.*, *Hungary*, **1**, 323–326 (1966).

Chan, S. P., "Application of Graph Theory to the Synthesis of Single Contact Networks," *J. Franklin Inst.*, **280**, 425–442 (Nov. 1965).

Iri, M., "Algebraic and Topological Foundations of the Analysis and Synthesis of Oriented Switching Circuits," *RAAG Memorandum*, **2**, 469–518 (1958) (Gakujutsu Bunken Fukyukai, Tokyo).

Lavallee, P., *Switching Networks by Linear Graph Theory*, Microwave Research Institute, Polytechnic Institute of Brooklyn, May 1963.

Mayeda, W., "Synthesis of Switching Functions by Linear Graph Theory," *IBM J. Res. Develop.*, **4**, 320–328 (July 1960).

Prabhakar, A., "On a Topological Formula for the Switching Function of a Two-Terminal Single-Contact Switching Network," *IEEE Trans. Circuit Theory*, **CT-11**, 412–413 (Sept. 1964).

Roginsky, V. N., "A Graphical Method for the Synthesis of Multiterminal Contact Networks," *Proceedings of the International Symposium on Theory of Switch. (Apr. 1957)*, Part 2, Harvard University Press, Cambridge, 1959, pp. 302–315.

Saltzer, C., "Algebraic Topological Methods for Contact Network Analysis and Synthesis," *Q. Appl. Math.*, **17**, 173–183 (1959).

Chapters 12 and 13

Ali, A. A., "Realizability Conditions of Special Types of Oriented Communication Nets," *IEEE Trans. Circuit Theory*, **CT-12**, 417–419 (Sept. 1965).

Ali, A. A., "On the Note on Completely Partitionable Terminal Capacity Matrix," *IEEE Trans. Circuit Theory*, **CT-13**, 323 (Sept. 1966).

Ali, A. A., "On the Analysis of Weighted Communication Nets," *IEEE Trans. Circuit Theory*, **CT-16**, 223–225 (May 1969).

Ali, A. A., "Relationships Between Semiprincipal Partitioning, Triangle Inequality, and the Presence of Min-Cut Matrices," *IEEE Trans. Circuit Theory*, **CT-17**, 151–153 (Feb. 1970).

Baran, P., "On Distributed Communication Networks," *IEEE Trans. Communication Systems*, **CS-12**, 1–9 (1964).

Barnard, H. M., "Note on Completely Partitionable Terminal Capacity Matrices," *IEEE Trans. Circuit Theory*, **CT-12**, 122–124 (Mar. 1965).

Baum, L. E., and Eagon, J. A., "Pseudo-Symmetric Connected Graph," *Can. J. Math.*, **18**, 237–239 (1966).

Beckenbach, E. F., "Network Flow Problems," *Applied Combinatorial Mathematics*, John Wiley & Sons, Inc., New York, 1964, Chapter 12, pp. 348–365.

Bellman, R., "On a Routing Problem," *Q. Appl. Math.*, **16**, 87–90 (1958).

Boesch, F. T., and Thomas, R. E., "Optimal Damage Resistance Communication Nets," *1968 IEEE Internat. Comm. Conf. Rec.*, 688–693 (June 1968).

Boldyreff, A. W., "Determination of Maximal Steady State Flow of Traffic Through a Railroad Network," *Operations Res.*, **3**, 443–465 (Nov. 1955).

Bollobás, B., "A Problem of the Theory of Communication Networks," *Acta Math. Acad. Sci., Hungary*, **19**, 75–80 (1968).

Boyer, D. D., and Hunt, D. J., A Modified Simplex Algorithm for Solving the Multi-Commodity Maximum Flow Problem, Logistics Res. Proj. Report T-211, George Washington University, Washington, D.C., Sept. 1968.

Burkov, V. N., "Method of Cuts in Transport Networks," *Tech. Cybernetics*, 1–12 (Aug. 1966).

Burns, W. C., Computation of Maximum Flow in Networks, M.S. Thesis, Naval Postgraduate School, Monterey, Calif., June 1968.

Cederbaum, I., "On Optimal Operation of Communication Nets," *J. Franklin Inst.*, **274**, 130–141 (Aug. 1962).

Chartrand, G., "A Graph-Theoretic Approach to a Communications Problem," *SIAM J. Appl. Math.*, **14**, 778–781 (1966).

Chien, R. T., "A Method for Computing Maximum Flow Through a Communication Net," *Proc. Sixth Nat. Symp. Comm. Systems*, 282–285 (1960).

Chien, R. T., "Synthesis of a Communication Net," *IBM J. Res. Devel.*, **4**, 311–320 (1960).

Dantzig, G. B., and Fulkerson, D. R., "On the Max-Flow Min-Cut Theorem of Networks," *Linear Inequalities and Related Systems*, Annals of Math. Study 38, Princeton University Press, 1956, pp. 215–221.

Davila, E. A., "Analysis and Synthesis of Node Weighted Networks from the Simultaneous Flow Standpoint," Proceedings of the Fourth Annual Allerton Conference on Circuit and System Theory, University of Illinois, Oct. 1966, p. 426.

Deo, N., "On Servicability of Communication Systems," *IEEE Trans. Communication Technology*, **COM-12**, 227–228 (Dec. 1964).

Deo, N., and Hakimi, S. L., "Minimum Cost Increase of the Terminal Capacities of a Communication Network," *IEEE Trans. Communication Technology*, **COM-14**, 63–64 (1966).

Elias, P., Feinstein, A., and Shannon, C. E., "A Note on the Maximum Flow Through a Network," *IRE Trans. Information Theory*, **IT-2**, 117–119 (Dec. 1956).

Ford, L. R., Jr., and Fulkerson, D. R., "Maximum Flow Through a Network," *Can. J. Math.*, **8**, 399–404 (1956).

Ford, L. R., Jr., and Fulkerson, D. R., "Solving the Transportation Problem," *Management Sci.*, **3**, 24–32 (1956).

Ford, L. R., Jr., and Fulkerson, D. R., "A Simple Algorithm for Finding Maximal Network Flows and an Application to the Hitchcock Problem," *Can. J. Math.*, **9**, 210–218 (1957).

Ford, L. R., Jr., and Fulkerson, D. R., "A Suggested Computation for Maximal Multicommodity Flows," *Operations Res.*, **6**, 419–433 (1958).

Ford, L. R., Jr., and Fulkerson, D. R., "Constructing Maximal Dynamic Flows from Static Flows," *Operations Res.*, **6**, 419–433 (1958).

Ford, L. R., Jr., and Fulkerson, D. R., "Networks Flow and Systems of Representatives," *Can. J. Math.*, **10**, 78–85 (1958).

Ford, L. R., Jr., and Fulkerson, D. R., *Flows in Networks*, Princeton University Press, Princeton, N.J., 1962.

Frank, H., "Maximally Reliable Node Weighted Graphs," Third Annual Princeton Conference on Information Science and Systems, Mar. 1969.

Frank, H., and Hakimi, S. L., "Probabilistic Flows Through a Communication Network," *IEEE Trans. Circuit Theory*, **CT-12**, 413–414 (Sept. 1965).

Frank, H., and Hakimi, S. L., "On the Optimum Synthesis of Statistical Communication Nets—Pseudo-Parametric Techniques," *J. Franklin Inst.*, **284**, 407–416 (1967).

Frank, H., and Hakimi, S. L., "Parametric Analysis of Statistical Communication Nets," *Q. Appl. Math.*, **26**, 249–263 (1968).

Frisch, I. T., and Sen, D. K., "Algorithm to Realize Direct Communication Nets," *IEEE Trans. Circuit Theory*, **CT-14**, 370–379 (Dec. 1967).

Frisch, I. T., and Shein, N. P., "Necessary and Sufficient Conditions for Realizability of Vertex-Weighted Communication Nets," *IEEE Trans. Circuit Theory*, **CT-16**, 496–502 (Nov. 1969).

Fu, Y., "A Note on the Reliability of Communication Networks," *SIAM J. Appl. Math.*, **10**, 469–474 (Sept. 1962).

Fu, Y., "On the Expected Value of Terminal Capacity for Probabilistic Communication Networks," *IEEE Trans. Circuit Theory*, **CT-10**, 134–137 (Mar. 1963).

Fujisawa, T., "Maximal Flow in a Lossy Network," Proceedings of the First Allerton Conference on Circuit and System Theory, University of Illinois, 1963.

Fulkerson, D. R., "Increasing the Capacity of a Network: The Parametric Budget Problem," *Management Sci.*, **5**, 472–483 (1959).

Fulkerson, D. R., "An Out-of-Kilter Method for Minimal Cost Flow Problems," *SIAM J. Appl. Math.*, **9**, 18–27 (1961).

Fulkerson, D. R., and Dantzig, G. B., "Computation of Maximal Flow in Networks," *Naval Res. Logist Q.*, **2**, 277–285 (1955).

Gale, D. A., "A Theorem on Flows in Networks," *Pacific J. Math.*, **7**, 1073–1082 (1957).

Gilbert, E. N., "Minimum Cost Communication Networks," *BSTJ*, **XLVI**(9), 2209–2227 (Nov. 1967).

Gomory, R. E., and Hu, T. C., "Multi-Terminal Network Flows," *SIAM J. Appl. Math.*, **9**, 551–570 (Dec. 1961).

Gomory, R. E., and Hu, T. C., "Synthesis of a Communication Network," *SIAM J. Appl. Math.*, **12**(2), 348–369 (June 1964).

Gupta, R. P., "On Flows in Pseudosymmetric Networks," *SIAM J. Appl. Math.*, **14**(2), (Mar. 1966).

Gupta, R. P., "Two Theorems on Pseudosymmetric Graphs," *SIAM J. Appl. Math.*, **15**, 168–171 (1967).

Hakimi, S. L., "Simultaneous Flows Through a Communication Network," *IRE Trans. Circuit Theory*, **CT-9**, 169–175 (June 1962).

Hakimi, S. L., "Comments on Simultaneous Flows Through a Communication Network," *IEEE Trans. Circuit Theory*, **CT-10** (June 1963).

Hirayama, H., and Uyehara, T., "Expected Value of Flow in Networks," *Trans. IECEJ, Japan*, **52-A**, 471–477 (Dec. 1969).

Hoang, T., "Some Theorems on Network Flows," *Theory of Graphs* (Proceedings of the

Colloquium Tihany, Hungary, 1966), Academic Press, Inc., New York, 1968, pp. 173–184.

Hu, T. C., "Multi-Commodity Network Flows," *Operations Res.* **11**, 344–360 (1963).

Iri, M., "A New Method of Solving Transportation Problems," *J. Operations Res. Soc., Japan,* **3**, 27–87 (1960).

Jewell, W. S., "Optimal Flow Through Networks with Gain," *Operations Res.,* **10**, 476–499 (July 1962).

Jewell, W. S., "Multi-Commodity Network Solutions," *Theory of Graphs* (International Symposium, Rome, 1966), Gordon & Breach, Inc., New York, 1967, pp. 183–192.

Kim, W. H., and Chien, R. T., *Topological Analysis and Synthesis of Communication Networks,* Columbia University Press, New York, 1962.

Mayeda, W., "Maximum Flow Through a Communication Net," Interim Tech. Report No. 13, Contract DA-11-022-ORD-1983, University of Illinois, Feb. 1959.

Mayeda, W., "Synthesis of a Communication Net," Interim Tech. Report No. 14, Contract DA-11-022-ORD-1983, University of Illinois, Feb. 1959.

Mayeda, W., "Terminal and Branch Capacity Matrices of a Communication Net," *IRE Trans. Circuit Theory,* **CT-7**, 261–269 (Sept. 1960).

Mayeda, W., "On Oriented Communication Nets," *IRE Trans. Circuit Theory,* **CT-9**, 261–267 (Sept. 1962).

Mayeda, W., "Application of Linear Graph Theory to Communication Nets," *Progress in Radio Science* 196-63, Vol. VI, Elsevier Publ. Co., New York, 1966, pp. 67–69.

Mayeda, W., "Maximum Flow under Controlled Edge Flows," Proc. IEEE International Conference on Communication, June 1968.

Mayeda, W., "Computerized Data Communication System and Vertex Weighted Linear Graph," *Proceedings of the 3rd International Conference on System Science,* University of Hawaii, Jan. 1970, pp. 1032–1035.

Mayeda, W., "Flow Reliability of Communication Net," *Proceedings of the Kyoto International Conference on Circuit and System Theory,* Kyoto, Japan, Sept. 1970, pp. 39–40.

Mayeda, W., and Van Valkenburg, M. E., "Properties of Lossy Communication Nets," *IEEE Trans. Circuit Theory,* **CT-12**, 334–338 (Sept. 1965).

Mayeda, W., and Van Valkenburg, M. E., "Set of Cut Sets and Optimum Flow," Third Colloquium on Microwave Communication, Budapest, Apr. 1966.

Minty, G. J., "Statistical Estimation of Flows in Networks," *IEEE Trans. Circuit Theory,* **CT-10**, 310–311 (June 1963).

Murata, T., "Analysis of Lossy Communication Nets by Modified Incidence Matrices," *Proceedings of the Third Annual Allerton Conference on Circuit and System Theory,* University of Illinois, Oct. 1965, pp. 751–761.

Myers, B. R., and Tapia, M. A., "Analysis and Synthesis of Node-Weighted Networks," Proceedings of the Conference on Electrical Network Theory, University of Newcastle upon Tyne, England, 1966.

Myers, B. R., and Tapia, M. A., "Flow Through Node-Weighted Networks," Proceedings of the Third Colloquium on Microwave Communication, Budapest, 1966.

Myers, B. R., and Tapia, M. A., "Minimal Realization of a Class of Max-Flow Min-Outage Weighted Nets," *Proceedings of the Fourth Annual Allerton Conference on Circuit and System Theory,* University of Illinois, Oct. 1966, pp. 416–425.

Onaga, K., "Stochastic Flow and Efficient Transmission Through Communication Nets," *Proceedings of the Third Annual Allerton Conference on Circuit and System Theory*, University of Illinois, Oct. 1965, pp. 762–771.

Onaga, K., "Dynamic Programming of Optimum Flows in Lossy Communication Nets," *IEEE Trans. Circuit Theory*, **CT-13**, 282 (Sept. 1966).

Onaga, K., "Optimum Flow in General Communication Networks," *J. Franklin Inst.*, **283**, 308–327 (Apr. 1967).

Onaga, K., "Bounds on the Average Terminal Capacity of Probabilistic Nets," *IEEE Trans. Inf. Theory*, **IT-14**, 766–768 (Sept. 1968).

Onaga, K., "A Multi-Commodity Flow Theorem," *Trans. IECEJ, Japan*, **53-A**, 350–356 (July 1970).

Pollack, M., "The Maximum Capacity Through a Network," *Operations Res.*, **8**, 733–736 (1960).

Resh, J. A., "On the Synthesis of Oriented Communication Nets," *IEEE Trans. Circuit Theory*, **CT-12**, 540–546 (Dec. 1965).

Rothfarb, W., and Frisch, I. T., "On the 3-Commodity Flows," *SIAM J. Appl. Math.*, **17**, 46–58 (Jan. 1969).

Rothfarb, W., Shein, N. P., and Frisch, I. T., "Common Terminal Multicommodity Flow," *Operations Res.*, **16**, 202–205 (Jan. 1968).

Rothschild, B., and Whinston, A., "Maximal Two-Way Flows," *SIAM J. Appl. Math.*, **15**, 1228–1238 (1967).

Sen, D. K., and Frisch, I. T., "Synthesis of Oriented Communication Nets," *Proceedings of the IEEE Symposium on Signal Transmission and Processing*, Columbia University, May 1965, pp. 90–101.

Shein, N. P., and Frisch, I. T., "Some Sufficient Conditions for Realizability of Directed Vertex Weighted Graphs," Third Annual Princeton Conference on Information Science and Systems, Mar. 1969.

Shein, N. P., and Frisch, I. T., "Sufficient Conditions for Realizability of Vertex-Weighted Directed Graphs," *IEEE Trans. Circuit Theory*, **CT-17**, 499–505 (Nov. 1970).

Sunaga, T., and Iri, M., "Theory of Communication and Transportation Networks," *RAAG Memorandum* (Gakujutsu Bunken Fukyokai, Tokyo), **2**, 444–468 (1958).

Tang, D. T., "Bi-Path Networks and Multicommodity Flow," *IEEE Trans. Circuit Theory*, **CT-11**, 468–474 (Dec. 1964).

Tang, D. T., and Chien, R. T., "Analysis and Synthesis of Oriented Communication Nets," *IRE Trans. Circuit Theory*, **CT-8**, 39–43 (Mar. 1961).

Tapia, M. A., "Analysis and Synthesis of Weighted Communication Networks," Tech. Report No. EE-665, University of Notre Dame, 1966.

Tomlin, J. A., "Minimum Cost Multicommodity Network Flows," *Operations Res.*, **14**, 45–51 (Jan. 1966).

Wing, O., and Chien, R. T., "Optimal Synthesis of a Communication Net," *IRE Trans. Circuit Theory*, **CT-8**, 44–49 (Mar. 1961).

Wollmer, R. D., "Maximizing Flow Through a Network With Node and Arc Capacities," *Transport. Sci.*, **2**, 213–232 (1968).

Yagyu, Y., "A Necessary Condition for the Realizability of Oriented Communication Nets," *IEEE Trans. Circuit Theory*, **CT-15**, 157–158 (June 1968).

Yau, S. S., "On the Structures of a Communication Net," *IRE Trans. Circuit Theory*, **CT-8**, 365–366 (Sept. 1961).

Yau, S. S., "A Generalization of the Cut-Set," *J. Franklin Inst.*, **273**, 31–48 (Jan. 1962).

Yau, S. S., "Synthesis of Radio-Communication Nets," *IRE Trans. Circuit Theory*, **CT-9**, 62–68 (Mar. 1962).

Chapter 14

Manning, E., "On Computer Self-Diagnosis: Part I and II," *IEEE Trans. Electronic Computers*, **EC-15**, 873–890 (Dec. 1966).

Mayeda, W., and Ramamoorthy, C. V., "Distinguishability Criteria in Oriented Graphs and Its Application to Computer Diagnosis-I," *IEEE Trans. Circuit Theory*, **CT-16**, 448–454 (Nov. 1969).

Mayeda, W., and Ramamoorthy, C. V., "Distinguishability Criteria in Oriented Graphs and Its Application to Computer Diagnosis-II," Proceedings of IEEE International Symposium on Circuit Theory, Georgia, 1970.

Preparata, E. P., Metze, G., and Chien, R. T., "On the Connection Assignment Problem of Diagnosible Systems," *IEEE Trans. Electronic Computers*, **EC-16**, 848–854 (Dec. 1967).

Ramamoorthy, C. V., "A Structural Theory of Machine Diagnosis," Proceedings of the Spring Joint Computer Conference, 1967.

Ramamoorthy, C. V., "Computer Fault Diagnosis Using Graph Theory," *Honeywell Computer J.* (Fall 1967).

Seshu, S., and Freeman, D., "Diagnosis of Asynchronous Sequential Switching Systems," *IRE Trans. Electronic Computers*, **EC-11**, 459–465 (Aug. 1962).

SYMBOLS

NOMENCLATURE